A SYSTEMIC-STRUCTURAL THEORY OF ACTIVITY

APPLICATIONS TO HUMAN PERFORMANCE AND WORK DESIGN

A SYSTEMIC-STRUCTURAL THEORY OF ACTIVITY

APPLICATIONS TO HUMAN PERFORMANCE AND WORK DESIGN

GREGORY BEDNY
ESSEX COUNTY COLLEGE, NEWARK, NJ, USA

WALDEMAR KARWOWSKI
UNIVERSITY OF LOUISVILLE, KENTUCKY, USA

CRC Press
Taylor & Francis Group
Boca Raton London New York

CRC Press is an imprint of the
Taylor & Francis Group, an **informa** business
A TAYLOR & FRANCIS BOOK

CRC Press
Taylor & Francis Group
6000 Broken Sound Parkway NW, Suite 300
Boca Raton, FL 33487-2742

First issued in hardback 2019

ISBN-13: 978-0-8493-9764-6 (hbk)

Library of Congress Cataloging-in-Publication Data

Bednyi, G. Z. (Grigorii Zakharovich)
A systemic-structural theory of activity : applications to human performance and work design / authors, Gregory Bedny and Waldemar Karwowski.
p. ; cm.
Includes bibliographical references and index.
ISBN-13: 978-0-8493-9764-6 (Hardcover)
ISBN-10: 0-8493-9764-2 (Hardcover)
1. Work--Psychological aspects. 2. Performance. 3. Human engineering. I. Karwowski, Waldemar, 1953-. II. Title.
[DNLM: 1. Psychological Theory. 2. Work--psychology. 3. Cognitive Science. 4. Human Engineering. BF 481 B412s 2006]

BF481.B365 2006
158.7--dc22 2005036795

Visit the Taylor & Francis Web site at
http://www.taylorandfrancis.com

and the CRC Press Web site at
http://www.crcpress.com

Dedication

To

Inna, Marina and Jeffrey – GB

Bernardette, Mateusz and Jessica – WK

Preface

Currently, the prevalent view within many of the human sciences, including the discipline of ergonomics, bestows the status of "scientific data" mainly on those facts and propositions that stem directly from empirical or experimental work. In practice, experiment subjugates theory, leaving to theory the modest function of data interpretation. Nevertheless, experimental data, just like data gathered by any other method, are only isolated elements that must be interpreted and synthesized by holistic theory. In the absence of theory, the myriad arrays of experimental data turn into a heap of disparate material that is difficult to generalize, let alone correctly interpret.

Psychology and human factors (ergonomics) are critically important contributors to the study of human performance. Since 1960, modern research, with an increasing reliance on the mental aspects of human performance, has discovered significant shortcomings in the traditional behavioral analysis of human work. As a result, cognitive studies and task analysis have largely replaced the traditional behavioral approach.

The traditional behavioral approach to the study of human work decomposes human tasks into behavioral elements that can be directly observed. In contrast, the cognitive approach concentrates on the study of cognitive processes and skills that cannot be as directly observed as overt behaviors. Many different techniques fall within the behavioral and cognitive approaches. There are no integral and unified theories behind the application of these techniques. Moreover, there is sincere skepticism of even the possibility of developing such a theory. Many practitioners of human factors note the significant gap between research and application. Hence, there is no comprehensive unified psychological theory that can be utilized as a general approach to the study of human work. Behavioral and cognitive approaches artificially separate cognition and behavior during the study of human performance. Further, there is also a separation of cognition and motivation. The separation of these major components of human behavior is not productive. In fact, human factors specialists and psychologists study human work holistically, where these three components are interdependent and cannot exist independently.

The study of numerous psychological functions, processes, and phenomena is in need of a theoretical foundation that would allow for the integration of these disparate psychological phenomena and data into a holistic, general theory. We propose that the systemic-structural theory of activity, which is derived from general theory of activity, qualifies as this unifying theory. The general and systemic-structural theories of activity will contribute to the development of additional experimentation in psychology. Such a theory, which precedes the conducting of experiment, will make the experiment both goal-directed and more effective.

In describing the kind of theory currently needed in experimental psychology we refer not to a microtheory but to a highly general theory that has a broad range of applications. Such a general theory must have a clearly worked out analysis of behavior, which would allow scientists to use not only experimental but also analytical methods of study. Such a theory is indispensable if the human sciences are to more fully understand human work, learning, and training. Despite the progress that cognitive psychology has made in these areas, it currently lacks a behavioral unit of analysis and, consequently, is rigidly tied to the experimental approach. This not only has a negative impact on the interpretation and description of experimental data, but it also makes the applications of psychological methods to design practically impossible.

Currently, cognitive psychology faces the challenge of developing not just experimental but also analytical methods of study. The need for this development becomes apparent when we consider the application of cognitive psychology to the problems of design in human performance, the man–machine system, or human–computer interaction and training. Here, we propose that the systemic-structural theory of activity and general theory of activity can be integrated with cognitive psychology to create a powerful theoretical and experimental approach. This advance would enable psychology to more effectively tackle the difficult problems of human work activity and learning.

This book discusses general activity theory (AT) and introduces systemic-structural activity theory (SSAT) and its applications to the study of human work. Such knowledge can be very useful not only to ergonomists or human factor specialists but also to the general audience of psychologists, industrial engineers, system designers, and other interested readers. The shift from cognitive psychology to activity theory does not reject cognitive psychology. In fact, the cognitive approach is considered an integral stage of analysis when cognition is regarded as a process. However, it is regarded as a necessary but not sufficient stage. Cognition has a complicated structure which consists of different cognitive actions.

Recently, activity theory has gained increasing popularity with professionals who study human work. The scientific study of labor productivity is an important aspect of the work of many professionals, such as engineers, managers, economists, psychologists, and the like. Since people are often the most critical contributors to the forming of relationships between productivity and job performance, our book focuses on the psychological and human factors engineering issues of productivity and their associated work system design challenges. The last several decades have demonstrated the dramatic technological changes that influence work conditions in all applied domains, including manufacturing, transportation, and human–computer interactions. These changes require new approaches to the study of human performance.

From these perspectives, both general and systemic-structural activity theories are particularly useful. SSAT advocates a systemic approach, acknowledging that cognition, behavior, and motivation are different facets of unitary human activity. Human work activity is considered, in this book, to be an integrative goal-directed system that unifies cognition, behavior, and motivation. The concept of goal is a particularly important distinction to be made in this area. The goal in AT is the conscious, desired result of a person's own actions or activity and is a cognitive component of activity. The motive or motivation, in general, is understood as the inducing or energetic component of activity, directing the activity towards the achievement of the desired goal.

The goal, accordingly, performs an integrative function, and the vector motive-goal gives activity a goal-directed character.

We have derived the main principles of SSAT from general activity theory, which was developed over the course of approximately 70 years in the former Soviet Union. While in this book we focus our attention on SSAT, we also explore the relationships between general activity theory and SSAT. We consider SSAT a grand theory, or framework, from which one can derive unified and standardized methods for the study of human performance. General AT is a powerful descriptive tool rather than a strong predictive theory. In contrast, SSAT, which has carefully developed units of analysis and principles of development of predictive models of human performance, can be regarded as a strong predictive theory for the study of human work systems.

Human work activity is a complex system which can be studied from many different perspectives using a variety of methods. All such methods should be organized into a system with stages and levels of analysis. Depending on the specificity of study, some of these stages and levels of analysis can be abbreviated or omitted entirely. Throughout the book we compare SSAT with other theories and derive from this comparison appropriate methods of analysis. We pay particular attention to activity descriptions during task performance and consider new methods of task description and analysis. The issues of ergonomics design and developing predictive activity models are also presented.

Cognitive psychology suggests different models of mental architecture. However, these models are not task-specific but simply describe cognitive mechanisms responsible for the generation of task-specific behavior. Such a modeling approach contradicts basic design principles that focus on task-specific strategies of task performance. In contrast, SSAT suggests a more general approach to developing interdependent activity models for each task-specific situation. Each model describes activity structure from a different perspective utilizing a different language of description. The analytical comparison of the structure of activity and the physical configuration of the equipment or user interface becomes a central component of ergonomic design from the SSAT perspective. In the creation of task-specific activity models, particular attention has been given to morphological and functional analysis. In the first case, cognitive and behavioral actions become major units of analysis. During functional analysis, the function blocks are suggested as the major units of analysis.

From the functional analysis point of view, activity is considered to be a goal-directed self-regulative system. Each function block becomes an object of special analysis. This gives a specialist the opportunity to describe different strategies of task performance. Functional analysis is considered to be a theoretical foundation for self-regulation theory of learning and training methods derived from it. Macro- and micro-levels of activity analysis and the interrelationships between personality and human work are also of importance in such analyses. From the perspective of activity theory, an individual develops himself during work activity. Work activity produces a major impact on a developing person. Individual style of activity is an important concept in the study of personality and human performance from activity perspectives.

SSAT offers a new direction for human sciences in general, and psychology and human factor disciplines in particular. The proposed theoretical framework can be successfully applied to the study of human work, ergonomics design, systems learning, and training, as well as to the development of individualized methods of human performance improvement. This book, therefore, can be useful to a broad spectrum of professionals and students in human sciences, engineering and business, including, but not limited to, psychologists, human factors specialists, ergonomists, human systems integrators, work sociologists, designers, human–computer interaction and usability researchers, industrial engineers, management professionals, and many others.

Gregory Bedny
Wayne, New Jersey

Waldemar Karwowski
Louisville, Kentucky

Authors

Gregory Z. Bedny, Ph.D, Sc.D, CPE, presently resides in Wayne, NJ. He works in Essex County College. He is an American citizen who emigrated from the former Soviet Union, where he earned his Doctorate degree (Ph.D) in industrial organizational psychology from the Educational University of Moscow and the Post-Doctorate degree (Sc.D) in experimental psychology from the Research Institute of General and Educational Psychology, National Academy of Pedagogical Sciences of the USSR. He is the author of multiple scholarly articles and five original scholarly books, the latest of which was coauthored with Dr. Meister in the U.S. in 1997. He was a guest editor of the journal *Theoretical Issues in Ergonomics Science*; 2004. Dr. Bedny is also the author of several textbooks in psychology and ergonomics. He is a Certified Professional Ergonomist and has been awarded an Honorary Doctorate of Science by the South Ukrainian State University. He can be reached at gbedny@optonline.net.

Waldemar Karwowski, Sc.D. Ph.D., CPE, P.E. is professor of industrial engineering and director of the Center for Industrial Ergonomics at the University of Louisville, Louisville, Kentucky. He holds an M.S. (1978) in production engineering and management from the Technical University of Wroclaw, Poland, and a Ph.D. (1982) in industrial engineering from Texas Tech University. Recently, he was awarded the Sc.D. (dr. hab.) degree in management science by the Institute for Organization and Management in Industry (ORGMASZ), Warsaw, Poland (June 2004).

Dr. Karwowski is a board certified professional ergonomist (BCPE). He served as secretary-general (1997–2000) and president (2000–2003) of the International Ergonomics Association (IEA). Dr. Karwowski is the author or co-author of over 300 scientific publications (including over 100 peer-reviewed archival journal papers). His research focuses on human–system integration and safety aspects of advanced manufacturing enterprises, human–computer interaction, prevention of work-related musculoskeletal disorders, and theoretical aspects of ergonomics science. Dr. Karwowski served as the editor of *Human Factors and Ergonomics in Manufacturing*, an international journal published by John Wiley & Sons (New York), and the editor-in-chief of *Theoretical Issues in Ergonomics Science* (TIES) (Taylor & Francis, Ltd., London). He can be reached at karwowski@louisville.edu.

Contents

1 Activity-Oriented Approaches in Psychology

1.1 ACTIVITY THEORY

1.1.1 OVERVIEW AND INTRODUCTION

The concept of activity (*deyatel'nost'*) plays a key role in Russian psychology and ergonomics. It is roughly comparable to the English term "behavior," but it is not the same and the differences are instructive. The construct of behavior emphasizes the similarities between human and nonhuman activities, and as a consequence it overlooks some of what is considered to be human functioning. The behavior construct is associated with stimulus-response psychology. Consequently, it is virtually unused in the former Soviet Union. Instead, the concept of activity that emphasizes the specificity of human behavior and its connection with internal mental processes plays a central role in Soviet psychology.

According to the basic Russian textbook of general psychology (Petrovsky, 1986) activity is an internal (cognitive) and external (behavioral) process regulated by a conscious goal. From the systemic-structural perspective, activity can be defined as a goal-directed system, in which cognition, behavior, and motivation are integrated and organized by a mechanism of self-regulation toward achieving a conscious goal. We consider first the general theory and then the systemic-structural theory of activity.

Activity theory is a theoretical framework for studying different forms of human praxis. It is a theory with a high level of generality that can be applied in different domains and has come to receive international recognition. Activity theory was developed as a psychological theory, but now attracts a wide audience of specialists including philosophers, anthropologists, educators, and linguists.

Developments in activity theory are associated with the works of Rubinshtein (1959), Leont'ev (1978), Vygotsky (1956), and other former Soviet Union psychologists. These scientists were the founders of general activity theory, which has been used to examine a number of different practical problems in such domains as education and performance. However, this theory does not provide an exact method, principles, or methodology for the study of human work: General activity theory is only a philosophical framework for studying human performance, although sometimes it helps in the discovery of a nontraditional method of solving a practical problem.

In order to overcome this drawback of general activity theory, it will be presented as systemic-structural activity theory — an outgrowth — general activity theory, which is employed to develop practical approaches to and methods for the study of human work and learning. Based on this theory, activity is considered to be a functionally, dynamically organized system. The basis for this theory is the work of the scientists Anokhin (1962) and Bernshtein (1966) from whose works are derived the principles of the systemic study of activity. Following them, Bedny (1987), Gordeeva and Zinchenko (1982), Konopkin (1980), Ponomarenko and Zavalova (1981), Pushkin (1965), Zarakovsky (1966), and others collected important data for the systemic study of activity. In this book, we present the original systemic-structural activity theory that utilizes both general activity theory and systemic principles for the study of activity developed relatively recently by these authors and their followers.

Systemic-structural activity theory is an original theory with a high-level of generality theory that presents a new approach to the study of work activity. It suggests not only a general conceptual framework for the study of human activity but also a theoretical method and technique derived from practical principles for the study of human work. It provides experimental as well as formalized and theoretical procedures, methods, and techniques. Within the systemic-structural theory of activity we can isolate different, narrower theoretical approaches: the cognitive approach, in which the concept of process is central; the morphological approach, in which mental and motor actions are the most important concept; and the functional approach, in which the major concepts are self-regulation and functional blocks.

In this chapter, we discuss general activity theory. A critical feature of activity is its relation to consciousness. Activity is not merely an external behavior; it is also inextricably linked with internal mental activity and consciousness of abstractions from a concrete situation that allow an individual to anticipate the sequences of other situations and provides insight into mental processes that guide conscious and volitional behavior (Rubinshtein, 1957). The unity of consciousness and behavior is a major principle of activity theory. Consciousness is treated as a psychological feature derived from human activity and thus must be understood in terms of social–historical development.

Activity theory also attends to unconscious processes. According to this theory, motivation and methods of performance may be either conscious or unconscious. However, the goal of an activity or task, and the goals of separate actions, must be conscious. Activity incorporates unconscious components beginning with Uznadze's (1961) work on the phenomenon of "set." A set is a predisposition toward a particular activity, composed of tendencies to perceive, interpret, and formulate a goal and act in terms of that predisposition. The goal of activity, the plan for its performance, and goal-directed activity as a whole can be formulated consciously or unconsciously. This sometimes involves voluntary processes, and sometimes is triggered unconsciously by external stimuli or the inner state of an individual. In either case, goals are conscious and guide the integration of all processes involved in goal attainment. In the case of involuntary triggering of the goal-formation process, orientation reflexes, emotional evaluative and motivational components of activity, and set become very influential. The relationship between conscious and unconscious processes has important implications in clinical psychology (where it figures prominently in the West) and for the study of human labor. When we study work activity, it is important to understand the transition between conscious and unconscious components, and it is particularly important to elucidate the ways in which unconscious levels of regulation can be elevated to conscious ones for the purpose of interventions.

Another aspect of activity that underscores its specifically human character is the use of tools, including symbolic tools. An animal may create a tool for a specific situation, but it is not used beyond this limited situation, and after use, it is abandoned. Further, tools are used by infrahuman species for the immediate satisfaction of their own needs and in isolation, not in social settings. Human beings, on the other hand, can consciously plan for future events, and even make tools to create other tools.

Human psychological processes are unique in terms of their social aspects. The psychological processes of infrahuman species develop according to the laws of biological evolution. The psychological processes of humans are also subject to the laws of social–historical development. Work activity is seen as a matrix for the development of human psychology, and human consciousness is seen as being formed through labor processes. Tools made by people determine their actions during a particular activity. One generation is able to transfer its learned experiences to another through action–operations carried out using specific tools. Thus, activity theory broadly resonates with early Marxian theory on the relationship of consciousness to "praxis" and suggests that human psychological processes are constituted by work activity. In this way, the study of human labor dominates the development of activity theory. Over time, what was first called "psychotechnics," then "work psychology" — and later "engineering psychology" — in the former Soviet Union was applied psychology and an influential source of basic psychological theory and research. The study of human work activity is also central to ergonomics.

Activity theory is formulated in terms of goal-directed actions rather than psychic processes or reactions. Action emerges as a major unit of activity. Activity is treated as a logically ordered system of mental and behavioral actions. In contrast to notions of processing, which emphasize what happens strictly in the psychological domain, the concept of action connects theory to the practical domain in ergonomics and other applications. One of the major obstacles to the effective application of psychology is the lack of an adequate concept of action. Theories, especially the so-called microtheories that are so prevalent are frequently of questionable utility. It is understandable therefore, that many practitioners in ergonomics attempt to go beyond academic theorizing. There is a risk to reduce ergonomics and human factors to a technology unrelated to science. Kurt Lewin (1951) said — "There is nothing more practical than a good theory." His statement is particularly relevant to psychology.

1.1.2 Vygotsky's Cultural–Historical Theory of the Development of Higher Mental Functions

We can divide the study of psychology in the early years of the 20th century into a phenomenological approach and a behaviorist approach. The first undertakes the study of psychological processes such as consciousness, higher neural processes, and the like through introspective techniques. The second emerged as a reaction to what was considered the unscientific nature of introspective methods. In the United States, this became known as behaviorism. In the Soviet Union the objective method of study was developed by physiologists. In contrast to American behaviorists, the Soviets always tried to correlate external behavior with brain mechanisms. Moreover, they aspired to bypass psychology using physiological methods — as in the significant achievements of Pavlov on "higher nervous activity" (Pavlov, 1927). Pavlov's work is well known in the United States under the rubric of behaviorism. In the former Soviet Union, his theory was not treated as behavioral science, but rather as a physiology in which conditioned reflexes were always correlated with brain functioning.

In Western Europe, Gestalt psychology developed and simultaneously eschewed the atomistic-introspective methods inaugurated by Wundt and his school, and the atomistic S–R (stimulus–response) methods of behaviorism. Gestalt and other techniques unacceptable to the Communist regime because they were not sufficiently materialistic. As a consequence, Pavlov's theory occupied a favored place within Communist ideology. However, it attempted to reduce complex human behavior to a simple reflex, ignoring, from the materialistic point of view, the important philosophical issue of consciousness. The cultural–historical environment shapes consciousness in specific ways. Accordingly, human behavior is not amenable to reduction to the models used for the study of animal behavior. Basov (1931) suggested replacing the notion of behavior with the notion of *deyatel'nost*, which is best translated into English as "activity." The challenge to psychology in Soviet society was to explain the emergence and function of consciousness in terms acceptable to Marxist doctrine.

In this social environment emerged Vygotsky, an extraordinarily talented, individual who generated fundamental research and theory consistent with Soviet Marxism that enabled psychology to advance as a discipline independent of physiology. Vygotsky's publications around 1930 formed the foundation of the social–cultural theory of higher mental functions. He attempted to reconstruct the origins of human behavior and consciousness. Marx's theory of society, historical materialism, played a fundamental role in Vygotsky's thinking. Marx stated that historical changes in society and material life produce changes in human consciousness and behavior. Another important aspect of theory with psychological implications is the role of human labor and tools in the development of society. Labor and the use of tools alters the nature not only of the object of the labor, but also of the laborer, the agent of the change. Vygotsky introduced the notions of tools, and signs, that are central to his psychological theorizing. He used these notions to explain the origins of consciousness and cognition in general. The study of human labor became a major concern of his scientific work. He introduced the notion of lower and higher mental functions to distinguish the mental functions of animals and humans. Higher mental functions specific to humans include voluntary attention, memory, and rational goal-directed thought. Vygotsky organized a collection of scientists under his supervision, many of whom went on to become prominent figures in their own right in the former Soviet Union, including Leont'ev, Luria, Gal'perin, Zaporzets, El'khonin, and others.

One of Vygotsky's earliest scientific works was *The Psychology of Arts* (1971), the first work in Soviet psychology to address the system of culture and its signs while studying psychological mental processes. Feelings previously within the individual became public through the arts. According to Vygotsky, the arts become an instrument for shaping personality. He wrote about the aesthetic responses that are elicited by art as a system of special signs. Later, signs become a point of departure for developing his social–cultural theory. In place of the dyad consciousness–behavior around which the major ideas of psychology had developed, Vygotsky introduced the triad consciousness–culture–behavior. Among the major determinants of psychological development is human labor, characterized by the use of tools. Tools emerge as social forces insofar as they have social–cultural, but not individual, psychological features. Tools encode particular types of operation that implicitly impose constraints and prescriptions upon the user. In other words, tools socially determine practical actions and mental operations. Vygotsky introduced the notion of an "instrumental act" to describe mental operations involved in the use of tools. In this way, he connected consciousness with human labor and practical activity.

Human beings act on nature indirectly through the use of special tools that possess a mediation function. Indeed, according to Vygotsky, this mediation through the use of tools constitutes a defining existential characteristic. Vygotsky (1962) called tools that mediate mental activity "signs." When individuals perform mental activities they use signs as tools in the same way they use physical tools to perform external activities. With the help of physical tools, people change the external environment and surrounding objects. Signs fulfill the role of internal psychological tools. They are directed to internal plans and change human psychological composition. Language is a major system of signs that mediate human mental activity. Because speech is considered to be the most important system of signs, social interactions and communication assume critical importance in human consciousness and cognitive functions. At the same time, Vygotsky attended to nonverbal signs: gestures, mathematical symbols, and the like. A sign is a symbol with a definite referent that has evolved through cultural history. The idea of using the sign as a psychological tool in Vygotsky's theory is one of the most successful examples of the application of semiotic ideas in psychology. The notion that a sign is a psychological tool made it possible to describe how the human mind develops.

The acquisition of a sign as a cultural tool empowers the regulation of one's own behavior. Through signs, an artifact, human psychology is constructed around a system of "meanings." Vygotsky' view of higher mental functions as having social–historical, rather than biological origins establishes his theory as an enduring achievement in psychology.

Humans in the first instance use an external tool, which later is transformed into an internal sign. This progression provides an opportunity to track how external material activity is transformed into an internal mental activity. This inspires the idea of integrating external behavioral and internal mental activities, which brings us to the notion of "internalization." In the West, the notion of internalization originate with Janet (1928) and Piaget (1952). However, the concept was neither clearly defined nor very important. In the former Soviet Union, it was fundamental to Vygotsky's theory, and later to activity theory. Vygotsky assert an inherent relationship between external and internal activity. In Vygotsky's view, internalization involves social processes and semiotic mechanisms — particularly language. He elucidated this as follows:

> Any function in child's cultural development appears twice, or on two planes, first on the social plane and then on the psychological plane. First it appears between people as an interpsychological category, and then within an intrapsychological category.
>
> (Vygotsky, 1960, pp. 197–198)

According to Vygotsky, the process of internalization is not merely a transfer of external processes to the internal plane. Rather, internalization is a transformational process accompanied by changes in the structure of activity. This transforming process depends upon cooperative labor and social interactions. Psychological tools perform their functions on the basis of their distinct meaning. The meaning of things and actions is acquired through external and internal tools. According to Vygotsky, internalization in this case is the transformation onto an internal mental plane of external performances during social interactions. Thus, internalizations are derived from social interactions.

The zone of proximal development (ZPD) is another important concept in Vygotsky's theory. This idea was first applied by Vygotsky in the context of instruction and testing. He asserted that the ZPD is the gap between a child's "actual development as determined by independent problem solving … [and] potential development as determined through problem solving under adult guidance or in collaboration with more capable peers" (Vygotsky, 1978).

On the basis of this notion, he called for "dynamic assessment," a two-step process in which, first, a child performs at the limits of his solitary competencies and then again assesses the limits of his capacity for his problem solving with the aid of an adult or peer expert. The difference between the independent and assisted solutions enables the assessment of "learning potential" as real ability, and thus of the ZPD. Vygotsky called this the "genetic study of psychological functions." The essence of the genetic method is that psychological functions are studied dynamically during their development. It is noteworthy that this method is also used in the former Soviet Union in the area of engineering psychology. For example, when evaluating new equipment, we use low-level personnel (contingent, of course, on safety requirements) and observe how they accomplish different tasks. During this procedure, psychologists can use different aids to assist naïve subjects to manage a task, observing how the aids affect performance. Here it is important to identify which information is more or less helpful. As a result, we can obtain a more authentic assessment of "learning ability," as opposed to an assessment of past experience and instruction. Inasmuch as this enables us to examine the deployment of learning capacities over time, it also enables us to change a "snapshot" of performance into a "movie" with an unfolding plot. In this way we can uncover underlying processes and capacities. From the point of view of an engineering psychologist's concern for the "fitness for use" of equipment the length of time required for mastery is the index of the design quality of the equipment. In the study of the dynamic development of diverse psychological functions, following Vygotsky's theory, we use the genetic method in Soviet ergonomics and engineering psychology.

The study of ZPD allowed Vygotsky to articulate an important principle of psychology, "learning precedes development." Tasks designed for the trainee should be positioned based on difficulty in the ZPD. This is contrary to Piaget's (1952) principle according to which "Development precedes learning." It is also contrary to Skinner (1974), according to whom development is learning. Vygotsky (1978) also distinguished between "scientific" and "everyday" concepts. He stated that development is achieved when the everyday version of a concept is transformed into a scientific version. For Vygotsky, the historical perspective is the point of departure for comprehending the development of consciousness.

Using the genetic method, Vygotsky studied the relationship among social, egocentric, and inner speech. Social speech is used during social interactions. Egocentric speech refers to the speech of small children addressed to themselves. With each year of development, children's speech becomes more intelligible to adults, and the relative weight of egocentric speech is reduced until at school age it can rarely be observed at all. Piaget believed that this kind of speech disappeared altogether. Using the genetic method of study, Vygotsky demonstrated that egocentric speech transforms into inner speech. Studies of the relationship between social and internal speech are relevant to the study of work activity — particularly in the development of instructions and training. Inner speech enables humans to plan and regulate their actions. It can be understood in terms of changing their functions in the individual activity. Vygotsky argued that phonetic and grammatical abbreviation of inner speech and its nonvocalization emerges in conjunction with changes in the function of speech from mediation during social interaction to mediation during individual activity.

Sokolov (1963) studied inner speech using electromyographs (EMGs) of the muscles of the throat and lips as people solved different mental problems. His studies showed that increasing task complexity is indexed by the levels of EMG intensity. This method is used to study problem solving and decision making. Vygotsky attended principally to language and speech. However, he did not assert a one-to-one match between thought and speech. Rather, these two functions overlap in verbal thought. For example, it is well known in engineering psychology that thought exists in the absence of verbalization. In such cases, other kinds of sign systems are critically important — especially visual signs. What is fundamental here is Vygotsky's claim that humans master themselves through the symbols of a cultural system.

In his studies of the origins of consciousness and human cognition, Vygotsky emphasized social interactions and social context. Currently prevailing cognitive psychology treats consciousness and human cognition as a process of individual development; however, for Vygotsky, cognition is not merely a property of individuals, but also a function that is developed and exists within a sociocultural context. This is similar to the understanding of mind as "extending beyond the skin" (Bateson, 1972).

Vygotsky's work in some ways resembles Piaget's (1952). Genetic principles are fundamental to Piaget's stage theory of development and Vygotsky's social–historical theory. At the same time as these theories supplement each other, they possess significant differences. Piaget's theory emerges from his background in biology. He comprehended development as the outcome of interactions between biologically given regulative principles of assimilation and accommodation. Development is seen in terms of discrete stages with qualitatively different characteristics: the sensory-motor stage up to 2 years of age, the preoperational stage through 7 years, the concrete operational stage through age 12, and the formal-operational stage beginning at adolescence. Development is an interaction between an individual and the environment. One factor that distinguishes the thinking of the child from that of the adult is the child's egocentrism. The child cannot look at things from the perspective of others or critically evaluate himself. He cannot understand that people can see things in different ways. Two important aspects of development for Piaget are that of socialization and moral reasoning. Later, Piaget formulated development in terms of the acquisition of logical–mathematical operations, considered as an interdependent system of individual operations. Mental operations emerge from external operations that become mental with the help of the process of internalization (Piaget, 1952). However, they become mental only when coordinated into a reversible system of symbolic operations and actions. For each operation there exists an opposite operation that restores prior states of mind.

A child's thought is formed from these operations. In Piaget's theory the sociocultural aspects of development are ignored. Development is regarded as an isolated interaction between the child and the external environment. The major process of development is the dyadic interaction of individuals with objects. This approach affords psychologists no insight into the cognitive processes of cultural and historical evolution. According to Vygotsky, a triadic relationship must be examined that includes the subject as a social entity, the process of a subject's acquisition through tools and signs of products of historical development, and an objective world amenable to alteration through "praxis." In order to comprehend the development of higher cognitive functions and consciousness it is necessary to transcend their formulation as "closed" individual phenomena. With the help of goal-directed activity and systems of action individuals interact with the social world and transform it while shaping themselves.

Vygotsky's theory challenged the behaviorist and introspective methods. His object of study was neither the external behavior nor the internal world of the subject. Rather it was goal-directed activity constitutive of mental or cognitive processes. Sociohistorically developed human labor is a genetically derived form of activity through which human cognition evolved. Thus, psychotechnics (later, work psychology or engineering psychology) is not simply an applied field but an area of basic theory and research in Soviet psychology (Vygotsky, 1930).

1.1.3 Theory of Activity: Its Emergence and Development

In the 1930s Stalin's efforts to enforce more rigid controls on Soviet culture and society resulted in a reexamination of the existing schools of psychology from an ideological perspective. Many schools were subjected to brutal attacks and some were eliminated altogether. Different psychological schools engaged in struggles with each other. Vygotsky's theory was denounced as "bourgeois." During this period, some aspects of psychology progressed, but psychotechnics, the precursor of Soviet work psychology and engineering psychology, was part of the general decline in scientific psychology between 1935 and 1955. Those aspects of psychology that advanced Stalin's ideological and political interests and were simultaneously practical and useful grew. Activity theory emerged in this environment. The object-oriented practical aspects of activity theory were the major focus at this time. Psychologists attempted to discern the similarities between the process of labor and mental processes and between the tools of labor and psychological tools. These correspond to the Marxist interpretation of the development of the mind. The sign-mediation aspect of mental functions was suppressed during this period. However, psychologists did obtain new data from the study of mental functions and external, practical material activity. Rubinshtein (1935) and Leont'ev (1947) were leaders in this area. Rubinshtein demonstrated that thinking processes are controlled by direct practical-material manipulation of objects, as well as by speech acts. Practical actions are embedded in the ongoing evaluation of the results of manipulations of external objects, and by the same token are involved in the ongoing monitoring of mental actions.

From this follows Rubinshtein's idea about dual interaction. Every human act changes not only the object and situation but the subject as well. Through activity the subject not only changes the situation but also develops the self. In the process of dual interaction the instruments and the products of action are changed, which in turn changes the subject. Activity is never completely preplanned or predetermined but is formed as a flexible, dynamic process. The interaction between the external or behavioral and internal or cognitive occurs through a unitary and uninterrupted process. Rubinshtein (1935) formulated the principle of personality in activity theory. According to this principle, the psyche is always tied to the existence of a particular person. While studying the social origins of the human mind psychologists should take into consideration the real individual and his social existence. Therefore the individual and the social are interdependent. Rubinshtein analyzed the interrelationship of external behavior and internal cognition through the prism of individual activity, which is always unique. The social determination of consciousness occurs not from "without" but rather from "within," as a result of the sociocultural existence of the individual.

From this follows another important principle of activity: external influences impact the subject through his internal condition rather than directly. The origin of consciousness cannot be reduced to internalization of the social. An individual is an active subject who constantly changes the objective world and culture and, based on this, changes himself. This idea is closely related to the concept of feedback influences and self-regulation in general. The idea of self-regulation was introduced into activity theory later under the influence of the work of Anokhin (1955) and Bernshtein (1947). From the work of Rubinshtein (1957) it follows that the individual–psychological aspects of activity were not sufficiently considered in the theory of Vygotsky. Criticizing Vygotsky's cultural–historical theory, Rubinshtein argued that the psychological characteristics of the individual are not completely derived from the social environment. The social aspects of experience are integrated into individual activity. The social depends on the individual, just as the individual depends on the social. In the same social environment different individuals act differently and are impacted by the social environment in different ways. From this follows a different concept of mental development from that proposed by Vygotsky and Leont'ev. All of the above-described ideas are related to an important principle in activity theory referred to as the "personality principle."

In his book *Existence and Consciousness* (1957), Rubinshtein argued that objects cannot exist without a subject. According to this assertion, the objective world, including the various things in it, exists independently of the subject. Things become objects only through their interactions with subjects. Objects arise from the material world through the process of activity. In the 1950s, this was a dangerous assertion to make in the former Soviet Union because it went against the materialistic philosophical doctrine dominating Communistic ideology at the time. From the perspective of this materialistic doctrine, Rubinshtein could have been accused of idealism. Idealism went against Marxist ideology, which was enforced by government.

Rubinshtein, further, did not accept Vygotsky's or Leont'ev's concept of internalization which contradicted the Personality Principle. He insisted that the subject does not internalize readymade standards but rather utilized exploration and interaction with the objective world as the source of reflection. During this process of dynamic reflection human consciousness develops. Rubinshtein (1935) further stated that intellect cannot be reduced merely to theoretical operations of conceptual thought since practice and intellect are interconnected. Rubinshtein introduced the unity of consciousness and practical activity as a vital principle of activity theory. For Vygotsky the major principle of the medium of mental development was sign, speech, and language; for Rubinshtein practical activity assumed the same fundamental role. Children acquire speech through both social–verbal interaction and during practical manipulation of the external environment.

Regarding the interrelationship of Rubinsthien's work to that of Leont'ev and Vygotsky, several points are worth noting. The work of Rubinshtein has much in common with the work of Leont'ev; however, there are important differences. These differences include: conceptions of the relationship between internal and external activity, the disagreement of Rubinshtein with Leont'ev's idea of internalization, and different ideas of human abilities. However, on comparing the works of Rubinshtein and Leont'ev with those of Vygotsky one can note that both Rubinshtein and Leont'ev did not sufficiently consider the semiotic aspects of activity. The semiotic aspects of human activity include interaction with signs, symbols, and artifacts. Humans create these elements and are in turn influenced by them. According to Vygotsky signs and symbols were an important part of human development. They were internalized and played a critical role in cognition. Because signs and symbols are created and transferred to the individual by society, the social world was an important element of Vygotsky's theory of development. At the same time Vygotsky did not sufficiently consider individual, object-oriented activity. Development cannot be reduced to the internalization of social standards. Rather, activity is a creative process, and through this process the mind develops. Vygotsky's and Leont'ev's works overemphasized the problems of socialization of personality in comparison with individualization of personality. Rubinshtein's principle of personality in activity theory attempts to overcome these negative aspects of their work.

Sign-oriented activity and object-oriented activities are interdependent. According to this principle, during the thinking process, the subject, through manipulation of the sign system, interacts with the object's content which is expressed by this sign system. Therefore, theoretical sign activity always interacts with practical activity. However, it is important to note that not all signs are related to objects (Bedny and Karwowski, 2004). Brushlinsky (1979), Rubinshtein's student, argued that Vygotsky's sociocultural theory suffers from insufficient consideration of material and object-oriented activity. Studies demonstrate that when solving visual–practical tasks or problems, an individual can extract nonverbalized meaning (Pushkin, 1978) from the visual situation. The solution of problems is a function of thought, not speech per se. Speech and thinking emerge in unity, but they are not the same. For Vygotsky, the major units of analysis were meaning; for Rubinshtein, they were action that embodies cognitive as well as motivational components. Rubinshtein decomposes activity into major components, namely, motive, goal, action, and operation. Similar components were also described by Leont'ev (1947). The subject of psychology is, therefore, not an internal, psychological experience alone, but an "activity" that may take shape in either an external or internal form. Mental activity is considered a subcategory of "Activity." Not only does external activity depend on internal activity, it also governs it. This relates to the unity of internal and external activity stated above. External activity provides not only for the transformation of the environment, but also for subjective, explorative, and orientation functions in the environment. Practical actions constitute the basis of mental activity. Rubinshtein (1935) and Leont'ev (1947) defined mental and behavioral actions more precisely. With the help of external and internal actions, the individual can construct an image of the situation. Action is a tool for the creation of images. Similarly, concepts are formed through thought-actions. The actualization of images is treated as a reconstruction process through transitory, micro-operations. Thought and motor actions are also integrally related. For example, the thought about movement is conveyed by microelectrical activation of muscles, called ideomotor actions. Visual presentations of images are connected to micromotions of the eyes. Many studies show that motor activities are intimately involved in mental functions. This idea was first advanced in the 19th century by Sechenov (1968) who demonstrated that motor action was implicated, not only in the alteration of an object, but also in gathering information about an object. In this motor action, mental and motor components are interdependent. Leont'ev (1977, 1978) and Zinchenko (1961) claim that external and internal activities are to some extent isomorphic. At some point, they began to overestimate the significance of the motor function in the development of cognitive functions. Their enthusiasm for the motor domain resulted in a neglect of the semiotic functions that Vygotsky so successfully introduced. Of course, Vygotsky did not completely ignore the relationship between object, action, and meaning. However, meaning and speech dominated his explanations compromising his ability to distinguish between "speech" and "thought." More complicated experimental studies generated at the end of the 1980s suggested that we cannot always identify direct relationships between external, object-oriented activity and internal, mental activity (Pushkin, 1978). At the same time, studies demonstrate that absolute separation of external behavior and internal cognitive behavior is misleading, as may be evident when comparing cognitive and behavioral approaches.

According to our understanding of mental development semiotic mediation and external practical activity are interdependent and do not exist separately. For example, Vygotsky describes how gesture appears in a child's repertoire as a sign. During social interactions with others this gesture becomes a meaningful communicative act. However, in this example, not only social interaction but also subject–object interaction plays an important role. A child not only simply interacts with others, but also acts practically. Intersubjective aspects of activity can be observed even in individual activity. A child acquires speech through both social–verbal interaction and during practical manipulation of the external environment. Brushlinsky (1987), Rubinshtein's student, argued that Vygotsky's sociocultural theory suffers from insufficient consideration of material, object-oriented activity. At the same time

Rubinshtein's concept of activity underestimates semiotic aspects of mental development. Internalization is not a process of transformation of the external or practical components of activity into the internal or mental. Under internalization in Systemic-Structural Activity Theory (SSAT), one should understand the interdependence of external and internal components of activity and the ability at the first stage to perform mentally with the support of external activity and at a later stage to perform mentally independently. Internal or mental components of activity are a result of the active formation of internal and external components of activity during their interaction in the process of self-regulation. One should also pay attention to the fact that mental development is an important aspect in activity theory. However, one cannot reduce activity theory to this aspect alone. Rubinshtein's and Leont'ev's followers overemphasized this problem and were caught up in a controversy regarding the role of social interaction or object-oriented activity in mental development as a major aspect of activity theory. Social interaction and object-oriented activity are interdependent. Their consideration in mental development is only a theoretical question among many others in activity theory. Finally, it is well known that many Soviet psychologists deny Vygotsky's connection with activity theory. According to us Vygotsky, Rubinshtein, and Leont'ev are major contributors to general activity theory. At the same time, it is not correct to consider the work of Vygotsky and Leont'ev as a unitary school of psychology. Regarding this polemic, one of the leading specialists in the history of psychology in the former Soviet Union Yaroshevsky (1992) wrote "According to my opinion Leont'ev attempted at the later stage of his career to incorrectly sustain an idea about the unitary Leontév–Vygotsky school of psychology." He did this only for one reason — "to prove the connection between his theory and Vygotsky's theory" (Yaroshevsky, 1992).

Another important philosophical principle in activity theory is that the psychological processes perform reflective functions. For example, Leont'ev (1972) and Platonov (1982) wrote that the category of "reflection" is decisive for psychology. The notion of "reflection" in psychology cannot be adequately comprehended in isolation from the notion of "interaction." Reflection is a particular kind of interaction among phenomena in which the reflected object preserves its topological structure within a systematic reflective medium (Platonov, 1982). Psychological reflection is the complex process of capturing external reality. Later, reflection was analyzed in terms of information processing insofar as it transmitted information. Information reflection was studied from several vantage points — semantic, pragmatic, and quantitative. Semantic refers to the qualitative meanings, pragmatic to its utility, and quantitative to the density of information available.

In activity theory, emotion, sensation, memory, and thought are treated as distinct forms of psychological reflection. Reflection is considered a general category including the foregoing more specific vantage points. Psychological reflection is not a passive, mirror-like reflection; it possesses active features that imply some system of mental stages and operations. Therefore, activity includes two major levels of regulation. One level of regulation involves voluntary goal-directed actions. The other involves an automated operation which is triggered by external stimulation. It is a reflective process. These two levels are interdependent and influence each other. Moreover, these levels of regulation can in some degree be transformed from one to the other. The reflective process which is included into activity can be organized as a system of goal-directed actions.

The concept of internalization which was first advanced by Vygotsky was the basis for the development of the principle of genetic study in activity theory. According to this principle, psychological functions are studied as they are developed. At the end of his short career, Vygotsky began to address the absence of motivational issues in his social–cultural theory. In activity theory, motivation and goal-formation processes become central. Goal achievement is construed in terms of a concept of action. The activity consists of the conscious actions that are used to accomplish the goal of actions. According to activity theory, actions are composed of operations. The goal of actions is conscious and the operations are unconscious. With practice actions become automatic and are transformed into operations. The reverse process can also happen and operations can become actions. The interaction between subject, object, and community is mediated by tools. Activity is considered in the sociocultural context. Culture consists of shared social meanings that are internalized by individuals during their praxis. To understand human activity it is necessary to know how the subject perceives rules, roles, and social meanings. Activity is considered a historically developed phenomenon that evolved over time within a culture. In order to understand the dynamics of this development, it is necessary to comprehend the changes or evolution of human culture and related situations.

The aforementioned reveals that there may be two approaches to the study of "activity." One may be considered through individual psychological perspectives and the other in terms of cultural–historical perspectives. The first approach considers activity to be an attribute of individuals under which the individual is an agent of activity. The second approach points toward a formulation of activity not only as an individual trait but as normative standards for activity that transcend separate individuals (Schedrovitsky, 1995). In this latter perspective one emerges not so much as an agent but as a subject adjusting and adapting to the normative standards and requirements of activity. Activity captures individuals and engenders individuality as much as individuals create activity. The social environment and the surrounding reality determine how people behave. The social and physical environment prescribes "the space of possible actions" for individuals. To establish effective social interactions an individual must develop standardized actions. We form expectations and make predictions about how different people will act in different situations. Activity includes objects and sign tools as well as norms and procedures for attaining particular goals. This implies that individuals acquire the prescribed activity. When we study an individual's style of activity we should compare it with modal, normative activity. Individual–psychological and cultural–historical approaches do not conflict, but rather complement one another.

1.1.4 Concept of Self-Regulation in Activity Theory

Physiology and psychophysiology play a fundamental role in the developing activity theory. This area owes the most to Anokhin (1935, 1955) and Bernshtein (1935), who introduced the notions of feedback to the study of psychology and physiology from which they inaugurated a theory of self-regulation almost a decade before cybernetic thinkers such as Wiener (1958). Bernshtein's (1935, 1947) work was well known by American researchers in the field of motor learning. Anokhin's work was less well-known here. Anokhin (1962) developed a version of the systematic approach to the study of activity, called the theory of functional systems, that described the processes of self-regulation at the physiological and psychophysiological levels. He took as a point of departure the biological importance for an organism of the reflection of repetitiveness of certain external events. For successful adaptation to this environment, an organism must forecast different events and predict the consequences of its own reactions. At the same time, Anokhin introduced the idea that an organism functions on the basis of "polarity principles." It always evaluates the influences in its environment in terms of a dichotomous categorization and relates an event to a positive or negative pole. The assignment to a positive and negative pole is driven by emotional mechanisms. Anokhin (1955) developed a notion of "anticipatory reflection" based on these studies. In his research, he found that the brain has special mechanisms that reflect not only the representations of the current environment but also future possible events through which an organism can regulate its behavior. These physiological mechanisms are implicated in the formation of new goals, forecasting, expectations, and so on. He connected the goal aspects of behavior with the physiological apparatus that he called "Acceptor of Effect." He described a process that he called "Action Acceptance." This process involved the receipt of information about the result of actions. This is an important mechanism for the regulation and correction of human or animal acts. "Acceptor of Effect" enables one to compare the results of an act with requirements. These requirements are formulated in terms of a neural system mechanism that functions as a template with which actual, resulting states of neural systems are compared. Based on this, he developed a functional self-regulative model of conditioned reflex to provide an alternative explanation of conditioning. This system includes diverse mechanisms with feed-forward and feedback interconnections and a recursive, loop structure organization. Functional systems are dynamic entities that are mobilized, formed, and disappear upon consummatory activities.

Bernshtein was the other leading psychophysiologist who did fundamental research in the field of self-regulation. In the early 1920s, he worked in the Central Research Institute of Labor and then in the All Soviet Union Institute of Safety. He had a background in both medicine and mathematics. In the West he was known as a leader in the study of psychophysiological mechanisms of movement. He is the founder of the physiology of activity. He invested a great deal of thought to the control of behavior through a feedback mechanism similar to that of Anokhin. Bernshtein's (1966) ideas were extremely influential in the development of theories of motor learning in the West. He asserted that motor functions constitute a group of basic processes through which the organism not only interacts with its context, but also acts upon it in accordance with its needs. Each motor act is treated as an attempted solution to a problem of action. The performance of any act implies the creation of functional mechanisms or encoded states in the neural system called "Required Future." Performance and the consummation of action conveyed by continuous comparison of the process of execution of action is the result of execution with its required future state previously encoded in the nervous system. The performance of actions is also implicated in the prognosis and programming of actions. Probabilistic prognosis is based on the evaluation and exploration of a current situation and the forecast of the near future. Programming components of movement leads to the mobilization of movement in accordance with the "required future." Programming and realization of an action usually takes place under conflict between the developed program of performance and continual unpredictable changes in internal and external forces of performance of movement. These unpredictable forces include uncontrolled external forces of resistance, unexpected events, and the counterreactions of antagonistic muscle systems. Accordingly, feedback and correction are critically important for performance. Therefore, it is not possible to develop in advance a program that precisely executes an adaptive response without such feedback. The principle of self-regulation is a fundamental principle for Bernshtein and Anokhin in the analysis and explanation of behavior.

Bernshtein's self-regulative process contains external and internal contours of self-regulation. External contours include feed-forward and external feedback from external receptors. External feedback provides meaningful interpretation of events. An internal contour of regulation includes feed-forward and feedback in proprioceptive systems that are typically unconscious. The interrelation between these two contours has a dynamic character. Some internal components of regulation can be transformed into external contours enabling more exact conscious control of behavior. This becomes particularly important in engineering psychology because the transformation from one contour of regulation to the other is critically important to training and proficiency of performance.

Bernshtein also introduced the concept of the levels of regulation of movements and actions. High levels of regulation perform a governing role to which lower levels are subordinated. Low levels are performed unconsciously, while the higher levels are performed consciously. During the training process the relationship may be altered. For example, during *automatization* of the actions we can observe the transitions from a higher level to a lower level of actions regulation. Anokhin developed a model of the self-regulation of conditioned reflex. Bernshtein developed a model of the self-regulation of movement. It is worth noting that these models were developed prior to the existence of the discipline of cybernetics and therefore avoided the pitfall of post hoc analogizing of functions to independently developed computer simulations. Rather they are based on operational and functional constructs of behavior and their causal interrelationships. From about 1935 to 1955 the work of Anokhin and Bernshtein was officially suppressed and eclipsed by Pavlov's approach. Their research was conducted prior to the 1960s and assumed prominence in the former Soviet Union only after the death of Stalin and the publication in the U.S.S.R. of the works of cybernetics of Norbert Wiener (1958). Activity theory assimilates the theoretical principles and concepts developed by Anokhin and Bernshtein. They elevated the explanation of action and activity as a recursive feedback system, not as a linear sequence of mechanisms. Historically, activity theory evolved through the overlapping interests and efforts of its many founders and cannot readily be equated with any one of them (Zinchenko, 1961; Pushkin, 1965; Zaporozhets and El'konin, 1971; Kotik, 1974; Luria, 1975, 1979; Zinchenko, 1978; Brushlinsky,1979; Konopkin, 1980; Tikhomirov, 1984; Smirnov, 1985, and others).

Vygotsky, who was an unusual genius, of course, played a pioneering role, but he is not the "author" of the activity theory, nor is the sociocultural theory of the development of mind the same as activity theory, as some writers in the West imply (Kozulin, 1986). Since activity theory is the result of a nexus of schools and disciplines, there is a daunting complexity to the interpretation of the theoretical concepts and terms, its empirical results, and its organization into a comprehensive holistic model. An initial attempt was made in Bedny and Meister (1997), the current effort expands upon this from a historical perspective.

1.1.5 Activity Theory and German Action Theory

Earlier, at the turn of the last century, the sociologist, Max Weber (1947) introduced the concept of action. American sociologist George Mead (1944) introduced a slightly different concept of action on the basis of which he treated both social and psychological phenomena. Parsons (1937), the great systematizer of American sociology, developed yet another aspect of the notion of social action that ignited intense theoretical controversy over its application to diverse disciplines. A few years back German psychologists Heckhausen (1991) and Gollwitzer (1996) introduced the motivational concept of action. Frese and Zapf (1994) also presented other concepts of action as the basic approach in German psychology. In this section, we attempt to briefly describe the major differences between the concept of action in German psychology and that derived from the Russian activity theory. Action theory, like activity theory, focuses on goal-oriented human behavior and the concept of action. It has three major directions of development. One was developed in the United States and two in Germany. In the United States, Norman (1986) advanced what he calls the "Approximate theory of action." We will not discuss this theory here since English-speaking readers are familiar with it. According to Frese and Zapf (1994) German psychologists developed two different concepts. One was developed within the framework of the theory of motivation (Heckhausen and Gollwitzer, 1987; Gollwitzer, 1996) and the other was the action regulation concept (Wehner et al., 2000). According to Frese and Zapf, only the latter has been applied to the study of human work; we therefore do not consider the motivational concept of action in this study. Here, we concentrate on the concept of action theory described by Frese and Zapf (1994). They stated that action theory has been greatly influenced by activity theory, especially by the works of Rubinshtein (1973), Leont'ev (1978), Luria (1979), Vygotsky (1962), Oshanin (1977), Gal'perin (1969), and the Polish psychologist, Tomazevski (1978). They also mention the work of Miller et al. (1960). We note that these references neglect the advances within activity theory during the past 30 years. It is easy to understand why activity theory was particularly influential in East Germany. We need to distinguish the German term "Tatikeit," which translates into Russian as *Deyetel'nost'* (Activity) from the German word "Handlung," which translates into Russian as "Action." Psychologists in former East Germany utilized activity theory for a long time. Nevertheless, action regulation theory has been attributed to them. For example Hacker (Hacker et al., 1983; Hacker, 1985a, 1985b) who was previously considered an activity theorist, is now recognized as one of the leaders in the area of action regulation theory (Frese and Zapf, 1994). Further, some psychologists use the concept of action and activity interchangeably (Wehner et al., 2000). Despite the fact that action regulation theory as presented by Frese and Zapf does possess some features similar to activity theory, it differs with respect to their major concepts such as action, self-regulation, goal, and their respective understanding of human activity in general. First of all, we underline that in action theory, as described by Frese and Zapf, we cannot find differences between action and activity. According to the definition offered by Frese and Zapf (1994). "...action is a goal-oriented behavior that is organized by goals, plans, and feedback and can be regulated consciously or via routines." This is approximately the same as activity. Suchman's (1987) concept of the situated action has similar meaning. In activity theory actions are the major building blocks of activity and the major units of analysis (Bedny and Meister, 1997; Bedny et al., 2000). Therefore, the concept of action in the West has a different meaning in comparison with action in activity theory.

Suchman emphasizes the dependence on action from a situation and introduces the concept of a situated action. According to functional analysis, activity is always situated because it develops according to the principle of self-regulation. Shuchman argues against cognitive psychology. However she needs to enter into the polemics of activity theory, which has a long tradition, to study situated aspects of activity. Bernshtein (1947), one of the leading founders of the psychophysiological theory of self-regulation, in his book *About Constriction of Movements* wrote "repetition without repetition." By this expression he demonstrated that if a subject repeats the same action multiple times, the action will never be the same.

Actions and activity in general are constructed or adapted to situations according to the mechanisms of self-regulation. Description of activity as a self-regulative system is the purpose of functional analysis of activity. The failure to differentiate major concepts such as action and activity results in authors confusing concepts of task with concepts of action. For example, Frese and Zapf (1994) exhibit an example of the hierarchic structure of actions in the process of replacing a tree on a street. Their figure presents not so much a hierarchy of actions as a hierarchic description of tasks or subtasks to be performed. Similar examples can be found in a book edited by Kirwan and Ainsworth (1992). This approach is well known to human factor specialists as a hierarchical task analysis.

Further, the theory of activity formulates the notion of a goal in a different manner. A goal, according to Frese and Zapf (1994) has motivational and cognitive components, and the action is pulled by the goal. In activity theory the goal is a cognitive component connected with a motive. The motive-goal creates a vector that lends an activity a goal-directed character. Motives push people to reach goals; goals are cognitive representations of an imagined future result of an action. Goals do not exist without a motive. However, energetic (motivational) components of activity should be distinguished from cognitive (informational) components such as a goal. In activity theory we distinguish the final goals of a task from the intermediate goals of actions. Activity is a logically organized system of actions that are mental and behavioral. They are usually organized into separate tasks.

In activity theory, a theoretical concept of self-regulation of activity has been developed that is fundamentally different from the concept of self-regulation in action theory. For example, six action steps described by Frese and Zapf are presented as a single-loop cycle. Action proceeds from the goal to orientation, then to a plan, decision, and execution, and then to feedback and finally back to the goal. Action steps are not precisely defined. Feedback is presented only as a final step of action regulation. In reality, it is involved at different steps of action or activity regulation. Authors mix concepts of feedback with the evaluative stages of self-regulation. Self-regulation cannot be presented as a single loop and should be described as a multiloop model. It includes a number of steps and feed-forward and feedback connections omitted in this model. In activity theory these steps are defined as functional mechanisms and functional blocks. The existence of each block must be proved experimentally and theoretically (Bedny et al., 2000). In the action process described by Frese and Zapf action stages always include motor (external) behavior. However, mental actions can sometimes be performed without external behavior. Lacking a concept of mental actions, action theory cannot adequately describe such mental activity. It is important to note that the action process scheme presented by Frese and Zapf ignores motivational aspects of self-regulation.

Frese and Zapf (1994) discuss the operative image and consider it to be an important aspect of action theory. Oshanin (1976) introduced the concept of the operative image in activity theory in the late 1960s. He and his colleagues published multiple studies on this subject but they are beyond the scope of this work. The operative image is closely connected with the conceptual model (Welford, 1960; Zinchenko et al., 1974) and reflects only those components of the conceptual model that are significant for the operator during a particular period of time and emphasizes important action-oriented characteristics of activity. The operative image has orientational and regulative functions and we consider it to be an important component of the functional mechanism of self-regulation. This mechanism is called "subjectively relevant task conditions" (Bedny and Meister, 1999) and we consider it in greater detail in later sections during functional analysis of activity.

As in activity theory, Frese and Zapf discuss the relationship between personality and action. A basic principle of activity is that personality develops through activity and social interactions (Kovalev, 1965; Rubinshtein, 1973; Leont'ev, 1978; Petrovsky, 1982, etc.). Frese and Zapf (1994) state that a person is developed by acting or " ... people change the world and thereby change themselves." All these ideas were carefully developed in activity theory. Moreover, in activity theory social interaction is critically important (Vygotsky, 1962).

Frese and Zapf (1994) use the term *action style* where as activity theory uses the concept of the individual style of activity (Merlin, 1964; Klimov, 1969). Individual style of activity is critically important when studying human performance from the personality perspective (Bedny and Seglin, 1999a, 1999b).

Even very short comparative analyses of the concept of action theory described by Frese and Zapf and activity theory demonstrate not only similarities but also significant differences between them. Activity theory is considered one of the most important theoretical paradigms in psychology and is internationally recognized (Wertsch, 1981; Nardi, 1997; Engestrom et al., 1999). Nevertheless this theory has a number of less-known and critically important aspects. One such aspect is the principle of unity of consciousness and behavior or cognition and behavior. This aspect is central to this study.

1.2 SYSTEMIC-STRUCTURAL THEORY OF ACTIVITY

1.2.1 Introduction to Systemic-Structural Theory of Activity

One of the leading Soviet philosophers, Shchedrovitsky (1995), divided contemporary epistemology into two contrasting, nonexclusive, approaches, terming them the activity approach and the naturalistic approach. In the naturalistic paradigm, individuals confront various objects of nature that are independent of their activity. On the one hand, in the naturalistic approach the unmediated experience is transformed directly into knowledge about the existence of objects and phenomena. On the other hand, according to the activity approach the meaning of human life — things and events, features of those things and events, relationships among those things and events, etc. — takes shape through the process of human activity. The purpose of the existential context and its meaning is revealed through activity.

The activity approach and the naturalistic approach are not mutually exclusive. Rather, the two approaches constitute complementary frames of reference. Many important studies in the social sciences and humanities have been limited by reducing all methods to the naturalistic approach. Psychology, which in some ways bridges the humanities and the social sciences, has been particularly constrained by naturalism. Human information processing, the dominant paradigm in cognitive psychology, may be seen as an instance of the naturalistic approach. The cognitive approach is typically formulated in terms of artificial mentalistic assumptions removed from the concrete study of the human mind in interconnection with the real world through mediated activity. A major shortcoming of the naturalistic approach in general is its failure to appreciate the extent to which our knowledge about the external world is intersubjective in nature (Vygotsky, 1978) and mediated by human activity. This does not, of course, denigrate the fundamental data derived from the framework of cognitive psychology. Indeed, activity theory seeks to integrate the activity approach and the diverse naturalistic formulations into a coherent framework. Since human labor is a fundamental kind of activity, activity theory is of particular utility, not only for theoretical work, but in applied research and practical interventions as well.

In our work we utilize a systemic-structural approach for the study of human activity. This approach distinguishes between two kinds of systems. One system is organizational whereas the other is structural (Shchedrovitsky, 1995). The organizational system consists of different elements that have no relation to each other. Any change in one element of the system can change the system but does not change the other elements of the system. The structural system consists of different elements that are interrelated. Any change in one element of the system can change the system, its elements and the relationship between them. The latest kind of systems can be dynamic and develops over time. The systemic-structural approach studies structural systems. In studying these systems, particular attention should be paid to the units of analysis, the relationship between elements of the system, the stages and levels of analysis, and the relationship and transition between them. The genesis of these systems is also an important aspect of systemic-structural analysis.

Activity theory has a long history of development in the former Soviet Union. This theory may be considered a new paradigm for psychology, which is attracting ever-greater attention from professionals in the West. However, the exciting attempt to interpret and translate into English suffers from certain limitations attributable not only to the problems of translating terminology, but also to activity theory itself emerging from diverse, conflicting schools of thought. Thus, activity theory cannot be reduced to either Vygotsky's sociocultural theory of mind or Leont'ev's version of activity theory. Activity theory has only recently been used in ergonomics. Practitioners confront a number of difficulties in the translation and interpretation of different concepts and principles of activity theory. Objects of study get confused with units of analyses or objectives, actions get confused with tasks, body organs get confused with tools, and so on. We can consider Engestrom's (2000) study of the medical care of children. He described different actions performed by a junior physician. However, what he describes as actions are really tasks in the framework of activity theory. For example, examination and diagnosis of patients is not an action as was stated by Engestrom, but rather a diagnostic task. This task includes distinct actions, and not only subject–object interaction, but subject–subject interrelationships as well. Engestrom, in this example, considers a physician as the subject and the patient and his father as the object. However, in the rubric of activity theory the patient and his father are subjects; the object of the physician's activity is the health condition of his patients. Moreover, social interaction is also critically important. Therefore, in the physician's diagnostic task, the subject–object relationship is transformed into a subject–subject relationship, and vice versa. When a physician evaluates a patient's health, we refer to the subject–object aspects of the task; when a physician speaks with a patient and his father we refer to that as subject–subject aspects of the task. Further, we differ from Engestrom in our understanding of a model. What Engestrom presents as a model fails to represent the physician's activity as a formalized description, so we would consider his descriptions of procedures as simply a policy proposal. Further, when Engestrom (Engestrom and Escalante, 1997) refers to the use of another person to help read an instruction from the screen as the use of a tool, we would rather consider this situation as a subject-oriented activity or social interaction. Others in the West criticize from the activity perspective the concept of "task." For example, Nardi (1997) wrote that a task is something automatic, neat, and pure, and ignores the variability of human activity. She further argues that the notion of task ignores motivational forces. With due respect to her we disagree with this statement. The concept of task is fundamental in activity theory and is the major object of study from the activity point of view (Bedny and Meister, 1997; Bedny et al., 2000). The task in activity theory is inherently a problem-solving endeavor with an underlying subjective mental representation of the task. We briefly address this topic in what follows. We also respectfully disagree with some authors' interpretation of the concept of activity and action. For example, Kuutti (1997) defines "building a house" as an activity and defines "fixing the roofing" as an action. However, both examples are more properly construed as part of the production process, divided into task sequences in which each task constitutes an activity. Tasks may in turn be divided into actions, which can further be decomposed into psychological operations or into psychological acts, and so on.

We admire Engestrom's and other colleagues' vigorous promotion of the value of activity theory but wish to urge more fidelity to the carefully thought-out concepts and approaches of the established activity theory during the last 30 years. The current work wishes to clarify some of the specifics of activity theory in order to facilitate its introduction to and adaptation by more Western readers.

According to the general activity theory, the human mind develops from historically contextualized, object-practical activity. This object-oriented activity determines the genesis and structure of human psychology (Rubinshtein, 1935, 1959; Leont'ev, 1947, 1977). In Vygotsky's theory, the sign system is to some extent distinct from the object-practical activity (Yaroshevsky, 1985).

Vygotsky's theory of the sociocultural development of the human mind offers an ontology and history of the human mind. The development of human consciousness was always the major object of Vygotsky's research. The human mind is considered above all from the intersubjective perspective. In activity theory, developmental, genetic principles and social interactions are also important; however, activity theory is not focused only on this question.

In activity theory, individual human activity is the principal object of study. The principle of historicism is fundamental in Vygotsky's theory and, of course, is important for activity theory. Frequent assertions to the contrary by Western scientists notwithstanding (Engestrom, 2000), activity theory should not be limited to the cultural–historical paradigm. One of the founders of activity theory, Rubinshtein, never belonged to Vygotsky's school of psychology and many of his views diverged from those of Vygotsky. It is worth noting that Rubinshtein, like Leont'ev, presented activity theory as an alternative to Vygotsky's theory. Other founders such as Anokhin (1962) and Bernshtein (1966) established self-regulation as a theoretical foundation for activity theory. Bernshtein also demonstrated that motor action emerges as a psychological problem because motor actions inherently embody cognitive mechanisms. Based on this, in psychology, motor action emerges as an object of psychological analysis for researching cognitive regulation. It should be noted that Anokhin and Bernshtein developed psychophysiological concepts of self-regulation.

Shchedrovitsky (1995), points out that Vygotsky developed a sociocultural determinism of mind, but not of object-oriented, socially mediated, individual activity. This view, which expresses a general consensus in the former Soviet Union (Petrovsky, 1986; Yaroshevsky, 1985; Brushlinsky, 1987) differs from the Western understanding of the identity of sociocultural theory and activity theory. Thus, in the former Soviet Union, sociocultural theory and activity theory share some features but are not regarded as the same. Vygotsky's work has fundamental implications and influence on psychology in general with particular relevance to activity theory insofar as he inaugurated the sociocultural theory of development of the human mind, but this is distinct from activity theory (Brushlinsky, 1987).

Contrary to much of Western writing, while influential in its development, Vygotsky, himself, did not use the term "activity" as a basic concept of psychology. Both Vygotsky and activity theorists were responding to the challenge of developing a psychological theory aligned with Marxist philosophy in the early revolutionary culture. Accordingly, both Vygotsky's concept of sociocultural psychology and activity theory properly embody different aspects of history and culture in their study of individual psychology. Each of these approaches has its advantages and limitations.

1.2.2 Activity as a Multidimensional System

According to the systemic-structural approach, activity is a complex, multidimensional system, requiring the use of systemic principles. One can extract from the same activity different structures as independent objects of study, depending on the purposes of the study. Each of these objects of study can be represented as an independent system. Consequently, we may have different representations of the same activity.

Dividing activity into distinct elements and components, and *mutatis mutandis* from component to holistic activity is an important part of the system-structural analysis of activity. Morphological criteria entail representing activity as activity-action-operation. According, to the structural–functional criteria, activity may be subsumed under a tri-fold rubric: motive-goal-conditions (Rubinshtein, 1959; Leont'ev, 1977). Platonov (1982) outlined in activity elements such as goal-motive-methods-results. Shchedrovitsky (1995) expanded this to six major elements of activity: goal, task, initial material, methods, and product. Motor actions may be divided into motions and mental actions composed of discrete mental acts. Thus, the general structure of activity adduced by various authors converges, forming a basis for our formulation of activity (human) and the major elements (Figure 1.1)

Activity may be presented as a system that consists of heterogeneous, structural elements, composed of different units that allow for the representation of activity in terms of different models describing the same object of study. The description of activity as a multidimensional system significantly increases the applicability of this pproach to the study of human work. We shall briefly consider the subject of work activity and the elements of activity in Figure 1.1.

The subject of an activity is an individual who performs in accordance with conscious goals and tasks embedded in the goals. The subject is an agent with accumulated

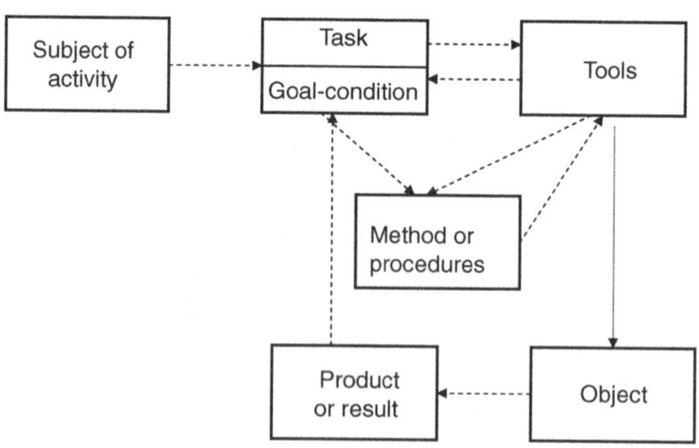

FIGURE 1.1 Major elements of activity.

historical and social experience. Through the objects that he/she transforms or changes according to the goal of activity, the acting individual emerges as a subject who reflects transformed reality in his/her consciousness, and based on this reflection regulates his/her activity in relation to others for whom he/she is a persona (Rubinshtein, 1935).

Tasks may be defined as a logically organized system of mental and behavioral actions, directed toward an ultimate task-goal. The task is the basic component of activity and human lives can be conceptualized as an ongoing attempt to solve tasks or problems. Typically tasks are organized in a logical sequence the performance of which enables attainment of the final system objectives. Sometimes such tasks are organized in accordance with technological requirements and are called production operations. Production operations may be studied from a technological frame, or from a behavioral or activity perspective. These two are, of course, interdependent. In the first case, the leading figure is a production engineer or related professional. In the second case, a human factors specialist is called for. However, in certain situations tasks are not well delineated. In accordance with prescribed rules and restrictions, as well as contextual purposes, operators formulate the goal of the task and the task itself. Changes in the situation, conditions, objectives, etc. may lead to reformulation of the task, rejection of the task, shifting attention to new tasks, and so on. In some cases, the performance of separate tasks entails different subjects requiring coordination of activity among them — including their informal, social interactions.

In order to understand what is task or action, it is essential to understand the goal of activity. In cognitive psychology goal is considered a combination of cognitive and motivational components (Pervin, 1989). In activity theory a goal is a conscious mental representation of humans' own activity in conjunction with a motive. Goals are considered cognitive, informational components of activity. In contrast motives or motivation in general, are treated as energetic components of activity. The more intense the motive is, the greater the effort to reach the conscious goal. Motive-goals create a vector that lends goal-directed activity its directness. Methods of task performance are determined not only by the goal but also by conditions in which the goal is presented. An analysis of the data gathered in psychology finds four approaches to understand the goal (Tikhomirov, 1984):

1. The goal is not a scientific notion. For example, Skinner (1974) described a person with the following terms: stimulus, reaction, and reinforcement. Here the goal is not considered a psychological concept.
2. The goal is the end state toward which the motivated behavior is directed and by which it is completed. This corresponds to the direction in psychology that studies purposeful behavior (Tolman, 1932).
3. The goal is considered the physical location of an object or a formal description of the final situation, which can be achieved during the functioning of the technical or biological systems. This corresponds to the cybernetic understanding of the goal.
4. The goal is considered a conscious mental representation of a future result connected with a motive. Only this last approach is associated with understanding of the goal in activity theory.

Although in activity theory informational (including goal) and energetic components (motive) are distinguishable, these components are closely interconnected. Vekker and Paley (1971) described several types of informational–energetic interconnections. The first type is described in psychophysical studies. The intensity of external stimuli results in increasing the experience of senses. Another type of informational–energetic interconnection is associated with the functioning of the reticular activating system of the brain. The third group of interconnection involves emotionally-motivational components of activity. The informational–energetic relationship in human performance is often overlooked in cognitive psychology.

The object of an activity refers to an object that has been modified by the subject according to the required goal of activity. This modification includes not only the physical transformation but also, for example, classification of objects according to the required goal and existing criteria. Objects may also include elements of the context within which the subject performs his or her task. People create artificial objects as a means of regulating their interactions with the external world and others. These objects are called artifacts, which are seen to hold a central place in the development of the human mind. Not every natural or artificial object is modified by humans in order to achieve a required goal. Subjects can change their own behavior or activity according to their objective environment. In order to discern objects that were modified during the achievement of a goal from objects that remain constant, but constrain or affect performance in activity theory, the notion of object of activity is used. Specifically, this object is modified and transformed during the subject's performance. Objects that are not transformed but affect a subject's activity can be referred to as task conditions. The environment possesses social and cultural properties that are considered to be as objective as physical ones. Cultural and social properties of an environment determine the manner in which people perform. Objects, which may be either material or ideal, determine the nature of human actions. Ideal objects refer to signs and symbols, and their constitution as an entity transformed by the subject in accordance with a required goal. These ideal objects exist in the form of special knowledge about external objects — particularly as images, concept, mental plane, etc. Depending on the character of the objects transformed, the performed actions can be practical or external and mental or internal.

Another important activity component is product. Product is a result of the transformation of an object of activity. Product may be material, spiritual, aesthetic, etc. Indeed, the subjects themselves may be the objects of change as a result of activity. This is why in activity theory, instead of the term product one may find the notion of result. The result does not always match the goal of activity.

The next important element of activity is the means or tools. They are divided into two types — external or internal tools. With the help of external tools, an individual may transform initial material or object of activity. Internal tools are internalized or acquired signs and symbols that are used during their internal mental activity. Through the manipulation of signs and symbols, subjects internally transform ideal objects of activity into their requisite product or result. The preceding elements of activity, methods or procedures include a logically organized system of external behavioral or internal mental actions through which external objects or mental situations are transformed to specification. The method of performance entails a plan of activity within which all components of activity — goals, conditions, tools, etc. — are integrated. Cognition is not merely a process or mental picture of the world but also a system of mental actions and operations intimately related to external actions (Bedny et al., 2000). As in physics, where light has both wave and particle characteristics, in the systemic-structural activity theory cognition is understood both as a process and as a system of actions or other functional information processing units. Thus, cognition incorporates both process and structure. Hence, cognitive task analysis used in ergonomics invites blending with activity principles (Bedny and Karwowski, 2000). The basic elements of activity do not exist in isolation, rather they function as a system.

1.2.3 Different Aspects of Activity Study

Because activity is a multifaceted system, it is studied from different perspectives. There are several approaches to the study of activity: social–historical, objectively logical, and individually psychological. The activity of a person is determined not only by individual development but also by the sociohistorical development of society (Vygotsky, 1978). A person learns norms, standards, and rules of behavior that were developed by society during the course of its development. Consequently, culture is an important element of activity. In activity theory, culture is viewed as a set of shared social meanings that are internalized by individuals during cooperative activity. A culture includes norms, attitudes, beliefs, values, philosophies, and ideology shared by individuals, which belong to the same community, where a community is an organization of people involved in the spheres of production, consumption, and culture each with its respective personal interrelationships (Shchedrovitsky, 1995). For example, the community of the production sphere includes such aspects as organizational relations, norms of behavior, and production discipline. The interrelationships of people that are formed in the spheres of production, consumption, and culture are interdependent but not the same. For example, interrelationships formed in the production sphere affect the interrelationships in the consumption sphere. The interrelationships in the production sphere are controlled more rigidly by external norms than interrelationships formed in the cultural sphere. Culture can pertain not only to society in general but also to a particular organization. Kotter and Heskett (1992) describe two levels of culture. At the less visible and deeper level, culture refers to values and norms that are shared by people in a particular group and have a tendency to persist over time, even when the group members change. By contrast, the more visible level of culture represents the behavior patterns of an organization that new employees are encouraged to follow by their fellow employees.

The concept of culture is connected to the historical aspects of activity development because activity evolved over time within the culture. Over the course of history the tools and methods of activity developed gradually, accumulated, and were selected. The understanding of changes, and evolution of human work over time, and its dependence on culture is important in the study of human performance. The principle that historicity is at the base of the genetic method of activity study was first proposed by Vygotsky (1978).

The genetic method examines how the cultural means used by society at different stages of historical development influence cognitive processes. Historical development of human culture is so important in Vygotsky's approach to the study of human cognition that his theory is often called the "cultural–historical" theory of mind. An understanding of the genesis of the development of activity, and its interconnection with culture and historical analysis, allows for a deeper grasp of activity as a whole. The structure of the developing activity can be better described with an understanding of the genesis of activity structure and the laws of its development. Therefore developmental or genetic principles of the study of activity are associated first of all with the name of Vygotsky.

Such concepts as community, culture, and historicity are important to the sociocultural analysis of activity. In this analysis activity is described in the context of the community. The community mediates the rules that describe how subjects should perform and the subject's beliefs, which in turn influence their performance. These rules are culturally accepted norms for human performance. The subject must acquire norms fixed by society that include tools of production and sign systems. Signs are words, mathematical symbols, gestures, etc. that carry a particular meaning. The sign systems can be interpreted in the same way only by those who share the same culture. In general, the social and cultural properties of the environment are important aspects of human work. Activity takes into account the cultural and developmental aspects of human life. Thus, activity should be considered to be historically developed phenomena, which are culturally mediated.

The sociocultural analysis of activity is tightly connected to the objectively logical analysis of activity. The objectively logical analysis of activity is related to those sets of activity, which the subject must perform according to his/her duties. To analyze activity from the objectively logical point of view is to determine the tasks performed by the subject in accordance with his/her position in the community. This analysis also determines, and describes, those objects, tools, and signs of activity which pertain to the task being performed. Of particular importance are the study of the product (result) of activity and the logic of the transformation process from the initial object to the required product. Tasks, tools, processes, and results are the basic elements of the objectively logical analysis of activity. Methods for the normative description of activity are central to the researcher. Through them the researcher reveals the material and symbolic tools of activity as well as the norms and procedures that must be performed to get the set product or result. In the process of studying human work, material components of activity are correlated with revealed components of activity. The analysis of activity conditions, and the description of activity norms that the individual must acquire independently of his/her individual characteristics, is particularly important to the objectively logical analysis of activity. At this stage of the research process, the subjectively psychological mechanisms of activity are not considered in detail. The researcher studies activity without revealing its internal mechanisms and psychological structure.

In order to acquire the socially fixed norms of activity, subjects must perform actions and operations with objects and symbolic systems, which pose fixed meaning. The subject must also acquire holistic individual strategies of performance in accordance with objective norm requirements and his/her personal characteristics.

This characteristic of activity requires not only sociohistorical but also an individually psychological stage of analysis, which involves a more detailed study of activity. In this stage of analysis, the structure of individual activity is the primary object of study. This structure is then coordinated with the objects, the tools, and the symbols that were revealed at the objectively logical stage of activity analysis. The subject–object relationship is major at this stage of analysis. This stage allows the researcher to address practical problems such as the designing of equipment, increasing the efficiency of performance, and training. The study of activity structure at the individual–psychological stage of analysis presupposes the study of mental and practical actions of the individuals, their logical organization, and the study of the self-regulation mechanism. The more detailed and individualized this stage of analysis, the more complicated the procedures for studying activity. In those cases where the activity of the subject is accomplished in collaboration with others, the researcher studies the individual structure of activity of each subject and their relationships. In this case we study social interaction.

The sociocultural and objectively logical approaches to activity study are used not only in psychology but also in sociological and philosophical research. In the latter cases these approaches to activity study acquire a specific character that distinguishes them from the purely psychological study of activity. Notably, the different approaches to activity study are separate in the theoretical rather than in the practical plane, because all of these approaches are tightly interconnected.

Currently, most of the works within activity theory in the west, are restricted to the sociocultural approach to activity study. The individual–psychological approaches to activity study, which are basic to the study of human work, are usually not discussed. Consequently, these aspects of activity study are not well known in the west. The individual–psychological analysis of activity includes the informational (cognitive), the morphological, the functional, and the parametrical methods of activity analyses (Bedny et al., 2001). All of the above methods of research in SSAT are considered to be interdependent and are logically organized according to stages and levels of activity analysis. This allows the researcher to tie together the obtained data into a holistic system. In order to fully capture its nature, activity is described as a multidimensional system. Consequently, the systemic-structural description of activity (not to be confused with the system analysis of men–machine system) is central to this work.

1.3 GENERAL CHARACTERISTICS OF ACTIVITY FROM THE SYSTEMIC-STRUCTURAL PERSPECTIVE

1.3.1 Object-Oriented and Subject-Oriented Activities

Activity determines the specificity interaction of conscious subjects with the external world. During this interaction, human mental processes evolve. From this follows the unity of consciousness and behavior. Cognitive mental processes evolved as a result of external activity of subjects mediated by intersubjective relations. Activity is an object-oriented, artifact-mediated and socially formed system. During activity humans create artificial objects that are a necessary precondition for the development of internal cognitive processes. The inner mental world of human beings is not naturally given, but mediated by artificial objects produced from human activity (Rubinshtein, 1935; Leont'ev, 1947). According to activity theory, external behavior is not the sum of reactions to external stimuli, but a complex external system of actions connected with internal cognitive activity (Bernshtein, 1966). Behaviorism formulates behavior in terms of stimulus and response reactions; activity theory interprets cognition and external behavior in terms of actions, the specificity of which is determined by the object and goal of activity.

A comparison with Piaget is also instructive. In Piaget's groundbreaking work, the interaction of subjects with the external world is similarly fundamental (Piaget, 1952). However, Piaget does not address the sociohistorical dimensions of this interaction in his studies. Rather, the development of the human mind is treated as the isolated interactions of subjects with surrounding objects. Since activity is culturally and historically shaped even when a subject privately and individually interacts with different objects, object-related activity is embedded in socially determined procedures for the manipulation of objects, which is especially true for artificial objects. People live in a world of stable things grounded in particular schemes of action with discrete meanings and purposes. Their internal activity utilizes a historically developed system of symbols and signs such as words, numbers, and icons, so that objects are not only confronted physically but are also encountered in defining intersubjective contexts.

Social–historical analysis reveals two closely related types of activity; "object-oriented" and "subject-oriented." Object-oriented activity is performed by a subject using tools on a material object. The simplest scheme of activity may be presented below as the following three components:

$$\text{Subject} \rightarrow \text{Tools} \rightarrow \text{Object}.$$

Through the use of tools the object is modified in accordance with the required goal. The content of activity progresses through determinate stages (1) the setting and acceptance of the goal, (2) the orientation in the situation in accordance with the goal, (3) the formulation of the task, (4) the evaluation of one's ability in comparison with the requirements (i.e., evaluation of the difficulty of the task), (5) development of strategies, and so on. Activity is completed only when subjects evaluate the results in accordance with the established goal and criteria of success (Bedny and Meister, 1999).

Subject-oriented activity refers to what is commonly called social interaction (obschenie). Social interaction may be presented as follows:

$$\text{Subject} \leftrightarrow \text{Tools} \leftrightarrow \text{Subject}.$$

Social interaction, or subject-oriented interactions, involves two or more subjects. Like object-oriented interaction, social interaction begins with a subject's goals, orientation in the situation, and so on. However, social interaction entails understanding of partners, predictions of their activity, evaluation of partners' goals, their abilities, past experience, personal features, possible strategies and actions in response to one's own. Social interactions are constituted by three sets of phenomena — exchange of information, personal interactions, and mutual understanding. The first of these includes both verbal and nonverbal communication. The second group refers largely to the coordination of actions among individuals, role definitions, development of social norms, standards, values, etc. The third set subsumes mutual understanding, comprehension of one another's inner experience, motives, goals, feelings, etc. In general, many aspects of social interaction are distinct from object-oriented activity. Object-oriented activity and subject-oriented activity during job performance continually transform into one another. They may be studied with common procedures as well as through distinct methods of study.

Intersubjective interactions may be found even subject–object activity. Intersubjective relationships arise from the observation of others even without direct contact with them or from the use of socially developed informal instructions. The intersubjective features of human individual activity (i.e., subject–object interactions) may be grounded in the work of the renowned Russian philosopher and literary theorist, Bakhtin (1982). His career began at the same time as Vygotsky's, but continued through the 1970s. He elaborated the interdependence of subject–object and subject–subject relationships. In those cases when we talk about subject–object relationships, subjects incorporate consideration of others through "inner dialogue." In this dialogue self-concept obtains its meaning, as well as the "image of me by others." Thus, in the study of object-oriented activity, intersubjective relationships must always be incorporated. Social interactions developed in a surrounding world of objects. Similarly, interactions with various objects arise on the basis of social norms and standards. Thus, we can eliminate the presumptive opposition regarding the primacy of either subject–object or subject–subject interrelationship between Vygotsky's sociocultural theory, on the one hand, and object-oriented activity theory on the other (Bedny et al., 2000).

Any activity has a recursive loop structure, organized according to the principles of self-regulation in which feedback mechanisms that evaluate performance are decisive (Anokhin, 1962; Bernshtein, 1966; Bedny and Meister, 1997). Subjects not only change their own strategies, based on self-regulation, but also provide scope for their external environment. Through mechanisms of self-regulation, internal activity is formed. Internal activity, which at first was performed with the support of external activity, is subsequently executed internally. The gradual transition from external, object-oriented actions to internal mental actions is called internalization. We consider this problem in Section 1.3.2 in relation to the basic principle of unity cognition and behavior. In our work, internalization is treated as an active process of the formation of internal actions and operations based on the mechanisms of self-regulation (Bedny, 1981). This is a formulation of internalization significantly different from the widely known ones of Piaget (1952), Leont'ev (1977), or Gal'perin (1969). Internalization is described as the creative process, which involves different self-regulated mechanisms. In our work the term internalization is used only for the designation of interdependences between external and internal activity but not as a process of transformation of external into internal. The opposite of the internalization process is the externalization process. Externalization is the transition of internal mental actions into the external plane. The processes of externalization and internalization demonstrate that mental or cognitive activity is tightly interconnected with external object-practical activity and that these two types of activity must be considered in unity. A more detailed discussion of this process is presented in the following sections.

During the description of activity, we discern various material and symbolic tools employed by the subjects, as well as standard procedures used to achieve the required product or result. The notion of description of standardized performance is vital to ergonomic design, which in activity theory is called a standardized description of activity. Activity is very dynamic and varies continuously even during repeated performances by the same subject. Accordingly, when we develop a template for activity as a standardized method of performance, it is merely an approximation that approaches the real activity. Analyses and description of activity must account for natural fuzziness and nonlinear dynamics (chaos) in the regulation of human activity (Karwowski, 1991, 1992, 2000). Since activity is variable, its performance must be modeled probabilistically, as well as deterministically. This enables the researcher to uncover how an operator's activity corresponds to constraints imposed for the purposes of particular tasks and designs.

1.3.2 The Principle of Unity of Cognition and Behavior

The idea of the interaction of cognitive and motor components of activity can be very useful but has not received enough attention in cognitive psychology. Let us briefly consider this problem from the cultural–historical and activity points of view. Vygotsky (1926) developed a social–historical theory of mental development, an important principle of which was the unity of consciousness and culture. In this theory, social experience is central to human mental development. Mental development is characterized by the use of tools, which mediate the subject's relation to reality, while various signs and symbols serve as tools that mediate the mental processes of humans. Tools used in human behavior are directed externally and change objects in the surrounding environment while tool-signs change mental processes internally. Signs begin with an external and material form and then according to Vygotsky become internalized and idealized. The major system of signs that mediates psychic activity is language, which also goes through a process of internalization during its development. In the beginning language is used to communicate with people and then is used by individuals for internal speech. The most important aspect in this process of internalization is not practical activity (labor) but the process of social interaction. Like external tools internal tool-signs are always a reflection of something in the environment and have a particular meaning for a person. According to Vygotsky (1978), this meaning is defined as a unit of individual consciousness. Therefore, cognitive development is in fact the assignment and development of meanings.

Rubinshtein (1973), Leont'ev (1977), and Gal'perin (1969) recognized the positive aspects of this theory but pointed out its limitations. While focusing on social interaction, Vygotsky's theory does not recognize the primacy of real work activity (labor) and a person's interconnection with real objects through labor. According to Leont'ev (1977) and Rubinshtein (1973), not only social interaction but also material activity and labor allow the formulation of concepts and meanings. Activity connects the person with the real world and leads to cognitive development of each individual and human kind in general. Therefore, while for Vygotsky, consciousness was mediated by culture, for Rubinshtein and Leont'ev it was mediated by material tools and objects. Unity of cognition and behavior becomes particularly obvious in the study of kinesthetic touch. People can perceive different features of objects without vision. Multiple studies were conducted by different authors in the former Soviet Union in this field (Ananév et al., 1959; Zaporozhets, 1969; Zaporozhets and Zinchenko, 1982, etc.) It was discovered that different kinds of hand and finger movements (including micromovements) perform cognitive functions. Several groups of movements were discovered. The first group of movements is called executive. The purpose of these movements is the transformation of an object or changing its position and orientation. The second group of actions is called gnostic. These actions are directed toward perceiving different properties of an object. The third group consists of adaptive movements like adjusting, correcting, and others. The relationship between the groups of this movement is changed during skill acquisition. Gnostic movements usually dominate at the first stage of skill acquisition. According to these studies and others Leont'ev (1977) and Gal'perin (1969) introduced the idea that external practical activity is internalized and becomes internal cognitive activity through human material activity. In contrast to Vygotsky they emphasized the primacy of external or motor activity in the process of internalization. As a consequence, internal mental activity is similar to external behavior in that it is composed of actions and operations. Not only does external, practical activity depend on cognition, but cognition also depends on behavior. Using this idea Gal'perin developed the "stage-by-stage formation of mental acts" theory, which is considered to be one of the important concepts of learning. According to this theory the development and acquisition of mental actions by the learner consists of the following stages. First the learner manipulates real objects (material actions) or draws symbols and pictures (materialized action). In the next stage actions are performed with the help of external speech, until the third stage where the actions can be internalized and mentally performed. In the verbal stage the student uses language not for communication purposes but to perform verbal actions upon the environment with which he interacts. Verbal action can be understood as a single meaningful output of verbal expression. Bakhtin's theory of utterance provides a theoretical basis for understanding how meaning is constructed in an utterance (Bakhtin and Voloshinov, 1973). In order to be effective the training process should be organized in accordance with these stages.

Piaget studied the individual explorative activity of an infant which is critical for its intellectual development. However the experience associated with social interaction with others cannot be totally eliminated in this situation. Moreover, objects with which infants interact are things that are produced by society. Therefore object-oriented activity and social interaction cannot be totally separated. They are interdependent and their opposition is incorrect.

Not all Russian psychologists fully accept the idea of internalization. According to the concept of self-regulation, learning is the process of constant transformation of the structure of activity. External, material activity contains cognitive components and serves as the basis for the formulation of internal mental actions. In the process of practical activity which involves manipulations with real objects, complex relationships between external and internal actions and operations are formed. Internal and external components of activity regulate and check each other based on the mechanisms of self-regulation (Bedny, 1987). Internal activity is shaped with the help of external behavior, and can then be performed independently. A person can perform mental actions immediately only if he/she is prepared for mental activity by previous experience. Very often in training, internal mental actions can be performed only with the support of external motor and verbal actions and only later can they be performed independently. In the beginning the learner manipulates external objects, signs and symbols using different written instructions and schemes, which facilitates externalization of internal mental activity. Mental activity is guided by external orientating components of activity. In activity theory an orientating component is viewed as a component of activity that precedes decision-making (associated with executive actions), performance, and evaluation of result (Bedny and Meister, 1999) and plays an important role in human performance. More specifically, because of the orientating component of activity, the operator develops a subjective model of reality from which he/she actively extracts distinct representations (Bedny, 1987). With the help of external objects and motor actions orientating components emerge as external tools for mental actions. They enable the learner to plan, regulate, and control not only external but also internal mental actions.

Speech improves the regulation of not only internal but also external activity. Verbalization of motor activity allows greater concentration of motor activity, fixation of attention on its separate elements, memorization of the methods of performance, and its more effective control. As motor activity is acquired the need for verbalization decreases and it becomes less conscious.

In some cases verbalization surfaces as self-instruction, important in the programming of actions and the integration of volitional processes into performance which enables the subject to voluntarily regulate his/her activity. At the same time it should be noted that not all components of motor activity can be verbalized because some have no verbal equivalent. In such cases conscious control of motor action can be achieved on the basis of nonverbalized visual or acoustical information. Studies demonstrate that speech is significant in the regulation of both internal and external activity. Thus speech can be seen not only as a tool for internalization, as noted above, but also as the means to a more effective control of internal mental and external motor activity. Regardless of whether the relationship of external behavior and internal psychological functions is viewed as the process of internalization of motor activity or as a process of their mutual regulation, this relationship is central to the theory of activity. Cognition is the regulator of external activity (behavior), and at the same time cognition is internal mental activity that has a great deal in common with external behavior. In the theory of activity, cognition is a system of perceptual, imaginative, mnemonic, decision-making, and other mental actions. These mental actions have been developed through practice (labor) and social interaction.

According to our understanding of mental development semiotic mediation and external practical activity are interdependent and do not exist separately. Intersubjective aspects of activity can be observed even in individual activity. A child acquires speech through both social–verbal interaction and during practical manipulation of the external environment.

An analysis of the presented material permits an outline of three approaches to the problem of internalization. Two of them are most widespread. Vygotsky (1978) formed the semiotic concept of internalization. According to this concept people internalize various sign systems, particularly the verbal system, during social interaction. What is most important here is not the internalization of the sign forms but rather the sign meanings.

The approach to internalization opposed to that of Vygotsky can be referred to as the object-practical concept of internalization (Gal'perin, 1969; Leont'ev, 1977). This approach emphasizes the internalization of external actions and operations performed on material objects, rather than the internalization of social interactions and verbal sign systems emphasized by Vygotsky. The object-practical activity determines the genesis and content of the human mind. Until recently these two approaches to internalization were in contrast to each other.

There is a third approach, which holds that these two concepts of internalization are interdependent (Bedny, 1981). From the first day after birth the infant is involved in object-practical and symbolic activities, both of which are social in nature. These two types of activity do not occur in isolation. The specificity of human activity lies in the fact that practical and symbolical activities (particularly verbal activity) are interdependent and are constantly transforming one another. Any practical action with an object, in contrast to a mechanical reaction to a stimulus, presupposes semiotic mediation. This semiotic mediation occurs because every object-practical action has a conscious goal, including planning and understanding of the possible outcomes. All these presuppose the presence of a symbolic representation of reality, where signs and meanings are of central importance.

Leont'ev (1977) argued that the structure of practical activity is similar to that of internal mental activity. From this idea Leont'ev and his colleagues formed the object-practical concept of internalization. However, the presented material demonstrated that the internal, gnostic, and dynamic activity is significantly different from external activity. Furthermore, internal activity cannot be reduced to conscious mental actions; rather it, includes unconscious mental operations. Such operations are not always elements of the conscious actions as held by Leont'ev. Internalization is not the transformation of the external into the internal, but rather the changing of the interrelationship between the internal and external in human activity (Bedny, 1981). At first, internal mental activity is supported by external behavioral activity. During the latter stages of activity acquisition, internal activity can be performed without the support of external activity. External activity never completely determines internal, mental activity. Internal mental activity is not determined by external reasons directly but rather through the mediation of internal conditions (Rubinshtein, 1973). These internal conditions include hereditary predispositions, abilities, past experience, and the temporal state of the subject.

In general, internal and external activities are interdependent and they are shaped based on the mechanisms of self-regulation. At the first stage internal activity is performed with the support of external activity and then internal activity can be performed independently. Therefore, in this case, the term internalization does not mean that external can be transformed into internal. Internalization is the process of mutual influences of external and internal activity through feed-forward and feedback interconnection and the gradual development of mental components via this process. Internal activity is constructed based on the mechanisms of self-regulation.

From the practical point of view the interrelationship of internal and external activity determines what must be the practical-external activity and what should be the external tools for the successful course of mental activity. Furthermore, the study of this interrelationship is important in the formulation of teaching and training methods.

1.3.3 Motivational Components of Activity

In this section we attempt briefly to describe the intentional or inducing component of activity. Interactions among such components as needs, motives, goals, and objects constitute the inducing aspects of activity. Inducing components begin with human needs. Needs are treated as states of individuals which they feel as desire for some objects that are required for survival and growth that becomes the ground for activity. Human needs are a function of activity itself. Natural things cease to be objects with merely biological meaning. Human needs are the result of acquired experience in conjunction with human culture. Through tool use mankind changes objects and modifies them in accordance with his needs and goals. During the satisfaction of human needs they change and develop. For example, meaning in work and spiritual expression are culturally formed human needs.

Needs become motives for activity when they motivate an individual toward a goal. Motives are defined as the inducing force that catalyzes the person's desire to reach the goal. Satisfaction or nonsatisfaction of diverse needs is conveyed by affects and emotions that in turn may induce activity. Such needs become capable of being sublimated into enduring interests, ideals, attitudes, and values which in themselves can become motivators. Thus, motives in activity theory include needs, affects, interests, etc., from which activity and goal striving emerge. The same motives under varying conditions can precipitate or influence diverse forms of activity.

Activity can be initiated by complexes of motives with varying weights or priorities assigned to each influencing factor. According to Maslow (1954), inducing components have a hierarchical organization. However, in his theory this organization is static and depends on qualitative features of motive. In activity theory the relationship among these inducing forces is typically dynamic and subject to modification during activity. Some motives may be salient in consciousness, others may be unconscious. The totality of these motives determines motivation of human activity. Motivation, therefore, encompasses more than the traditional study of motives. As noted, the goal of activity is a conscious future result of an individual's own actions or of activity in general. The relationship between motive and goal determines the directedness of activity. Motives are the energetic component, while goals are a cognitive element. More generally, activity theory requires that information and energy be treated as distinct but interrelated factors in accounting for behavior. In contrast to goals, which are always conscious, motives may be conscious or unconscious. Subjective awareness of the motive may affect the motives involved. The interrelationship between goals and motives is dynamic and complex, and may vary over the course of activity. For example, greater difficulty in attaining a goal generally requires greater motivation for goal achievement.

The specificity of the cognitive process such as perception, memory, and thinking, involved in task performance to a great extent depend on the vector "motive–goal," which mobilizes activity into a coherent structure. For example, a memorization task is dependent not so much on the nature of the material to be memorized as on how it is utilized in accordance with the goal (Zinchenko, 1961). In his experiment, subjects classified cards with pictures and numbers on them. The subjects were instructed to organize the cards either by pictures or by numbers. Those instructed to organize by pictures were unable to recall the numbers. In fact, some insisted that there were no numbers on the cards. Those instructed to organize the cards by numbers could not recall the pictures. According to activity theory, this experiment demonstrates that memorization was dependent not only on the particular features of the stimulus, but also on the way the material was used. In other words, a memorization task is stipulated by motives, goals, and the method of performing the activity.

Leont'ev (1977) sometimes talks about "removing the motive to the goal." However, here we are referring to functional coincidence between the motive and the goal and should not be confused with identifying motives and goals. When motives are "removed to the goal," the result of activity satisfies goal striving and motivation simultaneously. For example, if a person is very hungry he has a psychophysiological drive to reduce his hunger as well as a cognitive representation of the food through which goal he can consummate his hunger. In this case his attempts to obtain food to satisfy his hunger, the goal of his activity is to obtain food; the purpose of the motive is also contained in obtaining the food. According to its functional purpose the motive and goal coincide. On the other hand, the motivation of a starving person will differ significantly in quality from the motivation of an individual who is not particularly hungry. Often the motives of activity do not, however, coincide with the goal, because the goal may not gratify the motives. For example, if a subject produces something that constitutes the goal of activity, but this product does not satisfy the person's hunger, the individual must then exchange the product to satisfy his or her needs. When motives and goals are disparate, the final products or results of an activity are always mediated by the process of exchange. Thus, needs and motives may deviate from goals. This exchange process is unique to humans. Even when goals and motives are functionally matched, we should distinguish cognitive or representational aspects of the goal from the motivational or energetic aspects. When we study motivation of activity, the notion of "will" assumes importance. "Will" may be treated as a functional mechanism that sustains motivation to achieve a goal in the face of obstacles. Such obstacles may emerge during the formation of a goal or during the period when a person decides on the method of performance related to the goal. "Will" may also emerge in the face of contradictory motives or when they need to suppress some motives.

In activity theory motivation includes two basic functional mechanisms: one is the evaluative or "sense formative" mechanism of motivation. It embeds within itself cognitive–emotional components based on which the subject evaluates the personal significance of performed actions or activity. Sense refers to the emotional colorings an action has for a subject. The subjective personal sense should be distinguished from the objective meaning. Meaning is a form of presentation of reality to consciousness. Commonly accepted meanings are translated into an idiosyncratic sense for each individual. Idiosyncratic sense through interactions with another functional mechanism called "assessment of difficulty" catalyzes inducing components of motivation

(Bedny and Meister, 1997). For example, a very difficult task with very low personal significance results in the reduction of the inducing components of motivation. At the same time, very difficult tasks with high significance for subjects results in an increase in the inducing components of motivation. Significance, which derives from personal sense, influences the selection of specific information by an operator, developing strategies and criteria for the evaluation of task performance. From this it follows that the factor of significance introduces motivational factors into ergonomics research and practice. Motivation can be applied in ergonomic design by considering what information and which means of presentation of the information is most important for task performance. This way of dealing with motivation represents a distinctively original approach to motivation in design in ergonomics (Bedny, 1987).

1.4 OPERATIONAL SYSTEM OF ACTIVITY

1.4.1 Operational Components of Activity

During the study of activity we may extract such notions as object of study (do not mix with object of activity) and subject of study. The object of a study is the phenomenon or object that calls for the use of some theoretical or empirical methods. The subject of study is extracted from the perspective of framing the problem solving, in terms of a particular aspect of study. Thus we can have a single object of study that should be distinguished from diverse perspectives or subjects of study.

We also distinguish the object of study from units of analysis of activity (Bedny, 2000). Units of analysis are unified components into which we divide the whole for the purposes of studying the components and their integration into a dynamic whole. Distinct units of the whole are employed by distinct approaches. For example, behaviorism utilized S-R; gestalt psychology utilized figure-ground; Piaget (1952) utilized operations. Vygotsky (1956) marked out requirements for the units of analysis in psychology. However, he did not develop such units by himself. Rubinshtein (1935) and Leont'ev (1947) inaugurated the first general ideas of such units. Among these units the primary ones are internal, mental and external, behavioral actions and operations. The task is a specific kind of activity, which comprises different actions and operations and is present by itself as a complicated system. Accordingly, we consider the task as an object of study.

Operational components of activity represent logically organized system of actions through which an individual transforms activity, or initial material in accordance with the required goal. When we study the procedural components of activity, the selection of proper units of activity is decisive. Rubinshtein (1958), as Vygotsky (1962), wrote that units of analysis must retain the features of the whole. Analyzing into smaller units eclipses the quality of the whole. Thus, the primary unit of analysis is action. According to Rubinshtein, actions are basic to both external behavior and internal mental activity. Actions derive from particular motives and are directed toward a specific goal of action. Motives of each action and overall motivation for the activity must be distinguished. They may or may not coincide (Platonov, 1982). The goal of action should be distinguished from the goal of a task or an activity in general. By performing logically organized sequences of actions, subjects achieve intermediate goals of actions and then goals of task or activity in general. Action is a relatively bounded element of activity that fulfills an intermediate, conscious subgoal of activity. Rubinshtein and Leont'ev were the first to introduce into psychology the concept of mental action and describe the relationship between motives and goals of activity. The selection of actions as basic units of activity does not compromise the significance of images or meaning in psychology — a perceptual image is the result of perceptual actions of very short duration. Through perceptual actions, subjects develop images of perceived reality. Through thinking actions the perceived phenomena acquire conceptual formulation or meaningfulness. The relationship among images and concepts, or meaning in general, and action is complex. On the one hand, an image is the result of an action. On the other hand, how the image is developed affects the regulation of action.

Vygotsky, who first attended to the units of analysis of mental processes, lacked the time to develop and deploy such units as he adumbrated. In his studies he used meaning as a basic unit. Rubinshtein and Leont'ev argued that meaning and concepts are the result of mental actions, so that meaning cannot be used as a universal law of the genetically fundamental unit of mind. Moreover, the concept of meaning, developed by Vygotsky, marginalizes the motivational aspects of the thinking processes by emphasizing the cognitive aspects embedded in his notion of meaning (Gordeeva and Zinchenko, 1982). Meaning entails not only thinking but also other psychological processes. Meaning calls for the integration of diverse psychological processes. Meaning and signs should be treated as psychological tools of mental actions but not as units of analysis. These meanings are themselves products of action which, in turn, become tools of action in a continuous iterative process. Meaning is embedded within an ongoing loop structure of activity. Broadly stated, this loop structure of activity consists of discrete actions that include feedback and evaluations of results of performance. This loop-structure process for the formation of meaning includes interaction between internal mental and external tool-mediated actions. Cognition is not merely a mental picture of the world, it is also a system of mental actions and operations, intimately related to external actions. Thus, cognitive task analysis used in the field of ergonomics invites blending with activity principles, thereby overcoming the purely mentalistic approach to the study of human performance.

In the study of operational components of activity the concept of task is decisive. However, some psychologists in the West argue that the concept of "task" is useless and is not used in activity theory. "According to activity theory the task is suggestive of something automatic, net and pure." Similarly, Nardi (1997) writes that, according to activity theory, task does not indicate motives and she criticizes the concept of task used in conventional task analysis. She states that the concept of task is not in fact well elaborated in cognitive psychology. However it is a critically important concept not only in contemporary task analysis but in activity theory as well. The task in activity theory is the basic object of study. In contrast to cognitive psychology in activity theory the task always involves goal achievements that require motivational forces. Hence the task in activity theory includes motivational components. It contradicts Nardi's comments about the irrelevance of motivation to task. In activity theory tasks always include more or less problem-solving aspects. According to activity theory the task is a situation which requires achievement of a goal in specific conditions

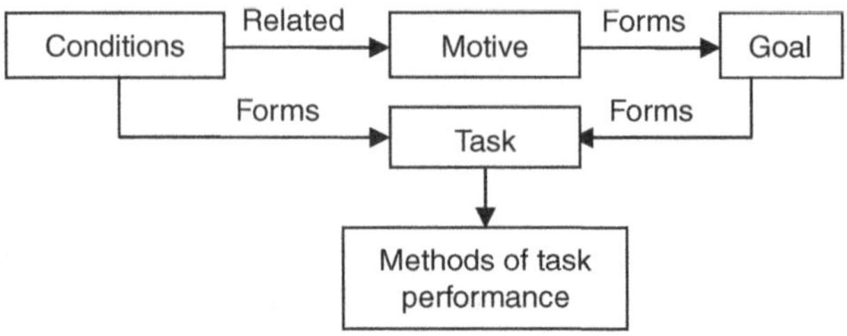

FIGURE 1.2 Scheme of task formation in structure of activity.

(Leont'ev 1977; Rubinshtein, 1973). Therefore the relationship between the goal and the conditions determines the task to be performed. The structure of the task defines the method of task performance (see Figure 1.2).

A general scheme of the components of activity that includes the task as an important element is shown as follows:

$$\text{Activity} \rightarrow \text{Task} \rightarrow \text{Action} \rightarrow \text{Operation} \rightarrow \text{Function Block.}$$

The last element refers to the functional analysis of activity at the microstructural level from the position of self-regulation, which will be considered in later sections. From this scheme we can see that activity may be decomposed into task, which may be subdivided into actions and further into operations and function blocks.

Activity may be crudely represented in linear form as

$$\text{Motive} \rightarrow \text{Method} \rightarrow \text{Goal} \rightarrow \text{Result.}$$

1.4.2 Triadic Scheme of Activity[1]

In reality, goal-oriented activity has a nonlinear organization composed of feed-forward and feedback loops.

On examination of the figures, it will be noted that the bottom line of the scheme (Figure 1.3) and the central axis of the scheme (Figure 1.4) present the relationship:

$$\text{Subject} \leftrightarrow \text{Object} \rightarrow \text{Outcome.}$$

In recent years, a number of researchers have interpreted the term "object" in this scheme as being synonymous with "objectives." In our view, this interpretation will always engender difficulties when attempting to apply activity theory in practice. To take one of the many possible examples, Bellamy asserts "...The individual's actions toward the object (objective) of the activity will be affected by three factors..." (Bellamy, 1997).

Most Western psychologists interpret the object in this scheme as objectives. However, it is an object of activity. In activity theory it may be a physical or mental one (sign, symbol, and image). Subjects in accordance with the required goal should

[1] This section has been prepared together with Steve Robert Harris, University of Glamorgan.

Engeström (1987) developed his basic triangle scheme based on the three-component linear scheme Subject $\rightarrow$ Tool $\rightarrow$ Object which is used in activity theory, and called it a triadic model of activity (see Figure 1.3). Later he (Engeström, 1999) developed a more complicated scheme (Figure 1.4).

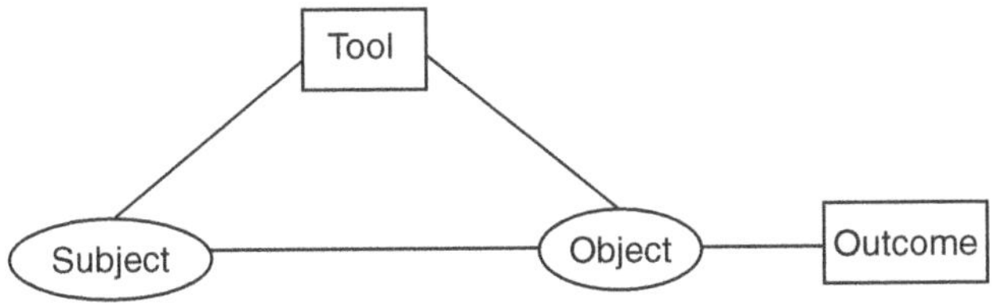

FIGURE 1.3 Simple triadic scheme of activity system according to Engeström.

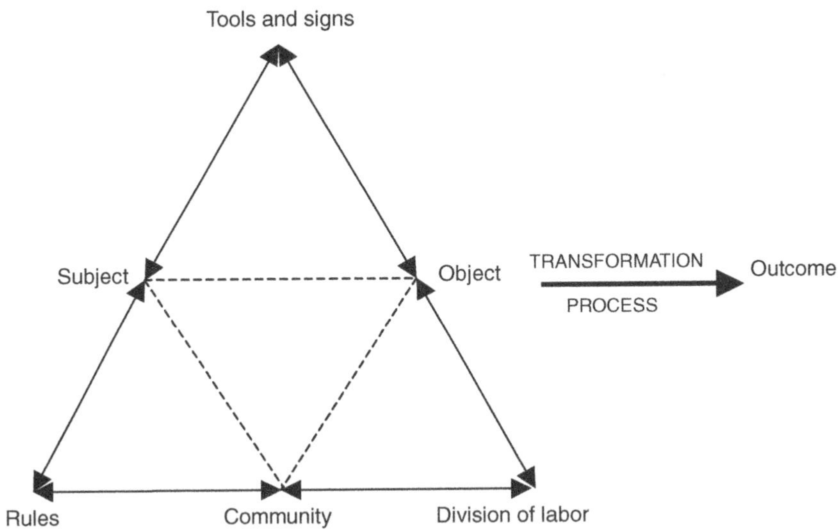

FIGURE 1.4 Complex triadic scheme of an activity system according to Engeström.

transform or modify the object of activity. Subjects can also perform explorative actions for discovering different features of the object. Very often the goal of explorative actions is not precise and can be modified during the performance of explorative actions. It is a goal formation process when the goal is formed based on an evaluation of the result of actions. Therefore the goal of activity can be modified during activity.

A somewhat more elaborate scheme is also presented in Figure 1.5 (Bedny et al., 2001). In this scheme we exhibit not only subject → object interaction but also subject ↔ subject interaction.

This scheme is intended to emphasize our point that the notion of "objectives" relates to the goal, rather than the object of activity. The broken circles in the figure indicate that subject–object interaction may be either direct or through the use of external mediating instruments. By the same token, intersubjective interaction may be direct (speech, gesture) or instrumentally mediated (e.g., telephone, email). In both object- and subject-oriented actions, direct interaction should not be taken as implying a complete absence of mediating instruments; rather; in such cases the

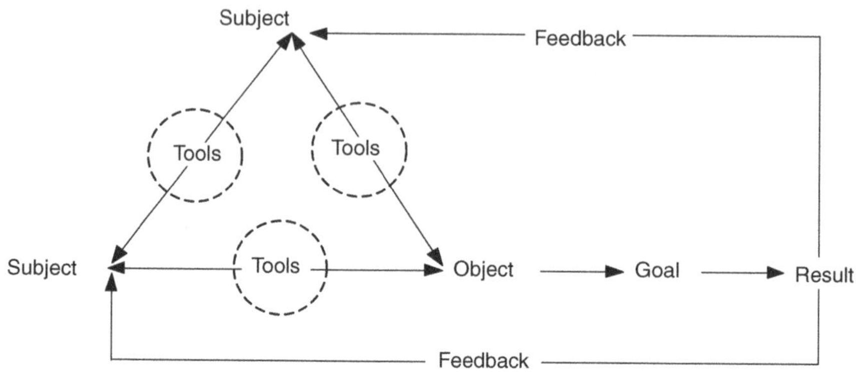

FIGURE 1.5 Triadic schema of activity.

subject employs "internal" tools. In activity theory, the subject is always understood to be socially constituted individual, who is in possession of internal, psychological tools acquired during ontogeny.

Further, in this scheme, the object and goal are treated as distinct components. From this scheme it may be inferred that the notion of "objective" is relevant to the goals and not to the object. Finally, this scheme possesses feedback influences, implying that an activity is organized according to the principles of self-regulation. Circles in this figure exhibit that subjects can interact with an object either through instruments or directly. By the same token, subjects can interact with each other either directly or indirectly through instruments. This brings us to the concept of collective activity. Under collective activity we understand the system of actions or tasks coordinated in space and time by diverse subjects toward achieving a common goal. In these cases, individual actions of subjects may be formulated as elements of collective activity. Collective activity emerges as a complicated system of individual actions. Practical significance attaches to the study of these systems themselves. Even without direct verbal interaction or direct visual contact with other subjects, powerful social and collective activity occurs. As classical economists like to point out in their discussion of abstract market forces, individual actions only require adequate and meaningful information about other subjects engaged in these interactions for coordinated social action to occur._

Figure 1.5 demonstrates one example of collective activity. In this example social interaction can be transferred into object-oriented activity and vice versa. The example demonstrates the situation when both subjects have the same object and a common goal of activity. Through feedback they coordinate their activity. However, there is another possibility for performing collective activity. For example, the object of activity for each subject may be different. Each subject can pursue his particular goal during transformation of his object. At the same time each subject should coordinate this transformative process in time with other subjects. This is the common goal of joint or collective activity. Therefore, the major criteria for common activity are the requirements to coordinate joint activity performed by different subjects and the existence of a common goal. Each subject in collective activity evaluates his result according to the following criteria (1) how an object is progressively transformed according to the goal of activity and (2) how the process of the transformation is coordinated with the other subject. From self-regulation perspectives collective activity requires coordination of activity strategies of different subjects. In Engestrom's model of interacting activity the major concepts are interaction and contradictions. They are important for the study of any system. However, without the concept of self-regulation and feedback individual and collective activity cannot be understood. Therefore, one cannot agree with Engestrom (1999) which states that in collective activity subjects always must share the same object of activity. Further, joint activity is not a unit of analysis but object of study.

1.4.3 Psychological Characteristics of Actions

An action is defined as a discrete element of activity that fulfills an intermediate, conscious goal of activity. The performance of all the actions required by a task leads to the achievement of the goal of the task. The structure of activity during task performance is formed by a logically organized system of motor and mental actions; action emerges as the primary unit for the morphological analysis of activity. Actions can be further divided into unconscious operations, the actual nature of which is determined by the concrete conditions under which activity takes place. In activity theory, cognition is considered not only as the storage of images, concepts, or propositions, but also as the system of mental actions and operations carried out with and upon them. All actions have a temporal dimension. The initiation of a conscious goal (goal acceptance or goal formulation) constitutes the starting point of an action; it concludes when the actual result of the action is evaluated in relation to the goal. This understanding allows for the depiction of a continual flow of activity, divided into individual units. Actions can be described in terms of a recursive loop structure, with multiple forward and backward interconnections. Figure 1.6 presents a simplified model of action as a one-loop system.

The systemic-structural theory of activity should respond to the following requirements (1) Psychological units of analysis should be expressed in such a way that will permit identification in real work processes; (2) The qualitative description of an activity should be combined with quantitative measures for the purpose of prognosis of the efficiency of the performance. (3) The description of activity should be performed in such a way as to allow us to make an inference or prediction as to how we can increase the efficiency of performance.

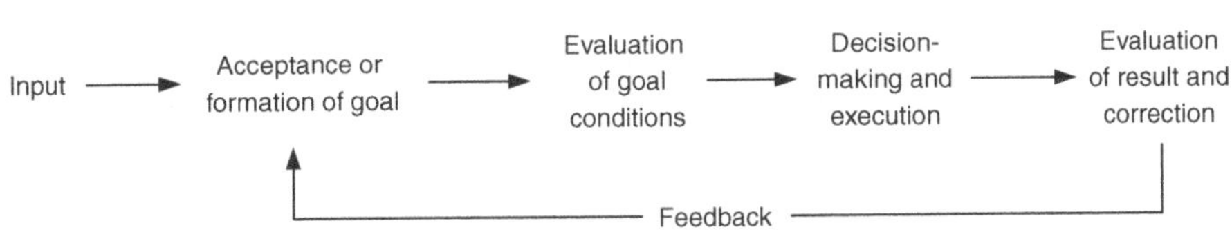

FIGURE 1.6 Simplified model of action as a one-loop system.

Actions as units of analysis satisfy these requirements. Action may be formulated in terms of the object of action, the tools and the subject of an action. Actions are the result of social–historical development. They are socially mandated prior to subjective realization. Subjects are taught to perform basic socially required actions. Each object has specific associated actions, governed by social norms and values. Actions are facilitated by tools that similarly possess a history and a cultural context. They imply the existence of an object of action. They are not isolated but are typically related to a class of similar actions. Individuals can extract principles of performance of particular actions from these classes because actions from the same class share general functions and purposes.

A similarity exists between action and words. Actions possess semantic, syntactic, and pragmatic features analogous to words. Syntactic features of actions are determined by their rules of organization into a system. Semantic features of action may be discovered through the relationship of an action to its object or to other actions. Pragmatic features of actions can be determined by their role for the subject and particularly in their relation to motivation. Verbal activity may also be presented as a system of actions possessing syntactical, semantic, and pragmatic features. Verbal actions may be considered a coherent organization of words around conscious goals integrated into a unified expression (Bedny et al., 2000). Verbal actions are more often used as a tool for communication that may also be used as a tool for self-regulation in a dialogic process. Nonverbal actions are typically object actions or may be mental actions involved with the manipulation of mental signs and images. We can outline two methods of action description. One method is based on the description of changes with objects that are performed by actions. Typically the names of action and changes performed are formulated as instructions analogous to software code. For example, "turn on the engine," "move the lever," "read display," and so on. These kinds of actions are conveyed by instruction and are classified according to specific features of an object. However, actions may also be classified according to their psychological characteristics, that is, by psychological processes and mechanisms implicated in their performance. For example, "memorize," "detect," "move arm," and so on. Based on these criteria we can infer two methods of describing actions. The first consists of actions classified as typical elements of a task, based on technological principles or on the nature of modifying the object. The second method is based on psychological principles that involve the description of typical elements of activity (Bedny, 1987). Usually, at the first stage, actions are described according to technological principles and are then transformed into typical elements of activity. For example, an action "move a lever into a particular position" is a technological description of the action. At the second stage the same actions may be described as "move arm into exact position with a force of 2 lb and a distance of 30 cm." This last is much more precise. Later, exact descriptions of the actions, unrelated to technological aspects of the situation, were developed. From these descriptions one can infer that this is a motor action requiring a high level of attention (third level of complexity) and performed over a distance of 30 cm with muscular effort equal to 2 lb. This gives us a precise picture of motor action even without the knowledge of the specifics of equipment and technology used (Bedny, 1987).

Since action is organized as a self-regulated system, the starting point of any action is the moment when the goal of the action is formulated or accepted. The terminus of an action occurs when the result is evaluated, thereby engendering a continuous flow of activity, divided into individual units, delimited by intermediate and terminal goals subject to the evaluation of the outcomes of the action. A simplified scheme of action is presented in the figure.

According to Leont'ev (1977), actions performed repetitively during training become automated and unconscious. During training, these actions are then abbreviated and become elements in more complex actions anchored in conscious goals. Leont'ev called these unconscious actions embedded in more complex ones "operations." Operations that are included in particular actions determine the method of performing actions. The notion of operation in psychological meaning should be distinguished from production operation. Dividing actions into small units is part of the consensual paradigm of activity theory. In the case of motor actions, instead of notions of operations, they consist of motions; in the case of mental actions these may be seen as composed of psychic acts. Psychic acts are cognitive actions automated during the training of such action. They lose their quality of consciousness of goal and are thereby assimilated to more complex cognitive actions.

According to Leont'ev, mental and motor operations always begin consciously; later, during automatization they become unconscious operations. We contend, however, that other motor and mental operations exist that are never conscious but are acquired unconsciously and remain unconscious elements of activity (Bedny, 1981, 1987). In order for these elements to become conscious, special methods of training and teaching are required. Frequently special training is called for to elevate these operations to consciousness and transform them into consciously regulated actions. Much of the work not only in ergonomics but also in clinical psychology seems to consist of this process.

There are different levels of regulation of activity that are a function of the extent to which an activity is voluntary and conscious. The more complicated levels of self-regulation of activity calls for orientation to the situation, development of goals, deliberate planning, etc. Highly automated activity entails goals involuntarily triggered by stimuli, which, in turn, guide subsequent cognitive operations and actions. Planning and the evaluation of results are extremely abridged. The lowest levels of regulation guide reactive behavior. In some cases activity can start from unconscious, automatized operations that can be raised to consciously performed actions at subsequent stages. This process was elucidated in the study of activity of pilots during emergencies (Ponomarenko, 1998). Thus, we respectfully disagree with Suchman (1987) that plans or goals have more to do with reasoning about the action after it has already taken place. Suchman (1987) ignores the notions of levels of regulation of activity and fails adequately to distinguish activity from reactive behavior. For example, in rule-based behavior according to Rasmussen's (1986) terminology we always have components of activity associated with preliminary planning which combine with mechanisms of situational adjustment and constructions. At the same time, thinking or creative activity (knowledge-based behavior) in general by definition cannot be fully anticipated or planned but develops as a process requiring direction and shaping in accordance with information obtained about results (Brushlinsky, 1987).

However some components of preliminary planning also can be found. Planning is an important anticipatory mechanism of activity. A plan cannot be considered to be "retrospective reconstruction" (Suchman terminology). It always precedes activity and includes conscious and unconscious components. Their relationship during planning depends on the level of self-regulation. As a result of self-regulation, the same task may be performed in various ways. In response to the external conditions and the internal state of the operator, goal directness, anticipation, and planning combine with flexible reconstruction of strategies of activity. The plan which can be adapted or changed depending on the situation is called "strategy." This understanding of activity, on the one hand, contradicts the construal of activity as a rigid, preplanned sequence of actions. On the other hand, the theory of self-regulation of activity contradicts the concept of situated action insofar as activity theory assumes flexible regulation of activity in accordance with a voluntary goal in response to varying situational requirements.

During activity, subjects process information at two major levels. The first level, derived from voluntarily regulated actions, provides conscious, goal-directed transformation of information. The second level consists of automatically performed operations in which goal-directed actions are obscured. These operations may precede goal-directed actions or may be performed in parallel to them. This second level is largely automatic with minimal conscious components. This level is important because it provides rapid shifting of attention from conscious activity to automatic activity performed at an operational level. In the last case, automatized activity becomes goal-directed and is performed in conjunction with voluntarily regulated actions. While in cognitive psychology, cognitive processes are fundamental, in activity theory psychic or mental processes and object-oriented activity occupy this conceptual role. This challenges the understanding of cognition as a continuous, uninterrupted process and adduces object-oriented cognitive activity as a discontinuous, interrupted activity. By contrast, we contend that cognition, when treated as a process, is continuous, but at the same time is organized into a recursive, loop structure of discontinuous, discrete units that transform from one into another. As light is both wave and particle in modern physics, activity theory treats cognition as a process and as a recursive system of actions or other functional units of processing of information.

Delineation of the basic components of activity and units of analysis empowers the design of man–machine systems informed by the alignment and coordination of external and internal means and conditions of activity. External means of activity include components of equipment and external tools with which a subject interacts during the process of work. External tools of activity refer to presentational controls, displays, screens, instructions, diagrams, and other media for conveying information to an operator. Internal tools of activity are conceptual models, images of the external world, skills, knowledge, etc. used by an operator during activity. These interactions must, of course, be responsive to external conditions and constraints. Effective alignment of external and internal tools of activity allows for the transformation of the object of work into a required product or result with maximum psychological and physiological efficiency. Individuals in this frame are not construed as a reactive organism but as a subject whose actions are guided by voluntary, established goals.

Therefore, the man–machine interface, or human–computer interaction, is treated as an interaction of the subject, tools, and objects.

1.4.4 Microstructure of Actions

Even short mental acts can be subdivided chronometrically into a series of very short stages. Each stage performs a particular function in the processing of information. Influenced by studies performed in cognitive psychology by Sternberg (1969a,b), Zinchenko et al. (1971) introduced the notion of a function block as a component of cognitive action. Each function block performs a discrete function. Anokhin (1962) during his studies of reflex as a functional system calls these stages "functional mechanisms." These stages are typically depicted graphically as a block or circles inspiring the appellation "functional block." On the other hand, outside activity theory, it is common when developing homeostatic models to analogize psychological processes to models of physics and engineering without any evidence supporting the existence of these as mechanisms. However valuable these analogies may be for stimulating thought, no one considers them to be representative of the actual units of action or their causal dynamics. On the other hand, activity theory aspires to and achieves a full-bodied definition of the psychological units and their causal interconnections. Such models are painstakingly developed, first, by empirical studies to identify a bounded unit of psychological functioning or functional mechanisms and, second, by empirical studies of the causal linkages among them. These criteria should be strictly applied when positing "function blocks" (Bedny and Meister, 1997). Thus, functional mechanisms or stages of information-processing may properly be conceptualized as "function blocks" only when they can be formulated within the context of a well-defined, nomothetic set of feed-forward and feedback relationships with other functional mechanisms. Activity theory is hierarchically organized so that more complex units may incorporate simpler ones. Thus, a concept of function block is appropriate for the representation of complex holistic activity as well as separate actions.

The foregoing enables us to delineate two stages of model building for either separate actions or holistic activity. The first stage suggests theoretical and empirical demonstration of discrete functional mechanisms. The second stage entails uncovering the causal relationships among these mechanisms from which we can develop a functional model with theoretically and empirically defined and grounded function blocks.

Since function blocks are stages of information-processing that sometimes occupies a very short duration, measured in milliseconds, this method of studying the content of different actions is called "microstructural analysis" and will be discussed later. The outcome of this analysis is a functional model of cognitive action. In cognitive psychology these models are usually presented as a chain of linear sequences (Sperling, 1960; Sternberg, 1969a, 1969b) In activity theory these models have a recursive structure of interconnecting feed-forward and feedback loops. The relative importance of different function blocks is contingent upon the situation. In some cases, particular function blocks drop out altogether. At present, the attempt to create a universal model of cognitive action cannot succeed. However, the importance of

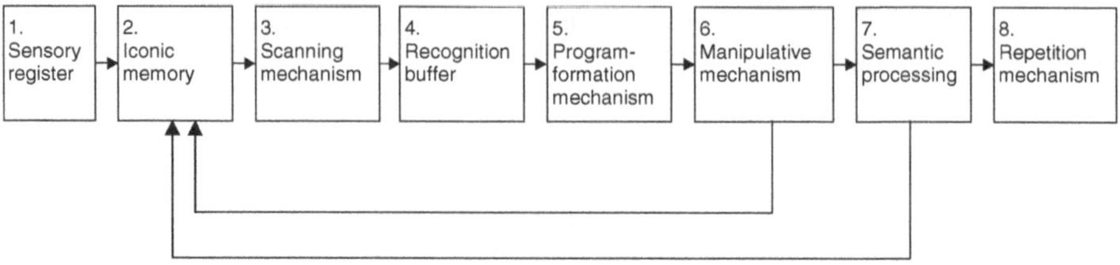

FIGURE 1.7 Microstructural model of perceptual action.

this study is that mental action emerges as a complex structure subject to self-tuning to a particular situation.

When we develop the functional structure of cognitive action, we base it on the following requirements:

1. Apparently, instantaneous cognitive actions actually occur over a short but measurable duration that can be represented as a series of subprocesses. Cognitive actions are implemented through a series of functional blocks that are stages of processing. The function block is a construct inferred from certain chronometrical (time measurement) experiments and qualitative analysis and should not be conceived of as a physical or observable process. Precise specification of events within the function block is not always possible (cf. cybernetic notion of black box). These function blocks have a very short duration. Sometimes they are called functional microblocks. The description of functional microblocks is always involved in chronometrical studies.
2. The list of function blocks and their content is not always the same and how it is utilized will vary according to the situation.

We offer as an example a microstructural model of perceptual action involved in visual recognition (see Figure 1.7) developed by Zinchenko and his colleagues (Zinchenko et al., 1971).

This is not offered as a universal model because the structure of cognitive actions is subject to change. Nevertheless, the structure of perceptual actions may be presented as follows:

1. Sensory-register is the first stage in the formation of visual images and constitutes the reflection of all the properties that are available to visual receptors. The content of sensory memory depends upon the physical characteristics of the stimuli such as intensity, duration, and contrast. Here information is preserved for 250–300 msec.
2. Iconic memory refers to the next block of information-processing. It is a "trace" of the stimulus that Neisser (1967) claims is a copying phase. Sensory and iconic memory have similar content. However, the duration of maintaining the content is longer — 1000 msec or more. In iconic memory

information is stabilized. Sensory and iconic memory stabilize and limit the information preserved to that required for further transformation.

3. Scanning mechanisms refer to the transformation of information from iconic memory. This mechanism can determine the sequence in which information can proceed to the next stage of processing. The sequence of scanning is contingent upon the external feature of stimulus kept in the memory as well as upon higher-level influences. This is the last "block" where the information is still largely determined by the stimulus features as they enter the visual system.
4. In the Recognition Buffer, information is transformed in accordance with internal psychological means. Some information is extracted from long-term memory and is implicated in framing "perceptual hypotheses" regarding categorization of incoming information. At this stage, incoming information is translated into the subjects' operational units. In the first three blocks perceptual processes depend upon external features extracted from the external stimulus. At this block subjects rely more on internal means — especially goals and expectations. Information irrelevant to goals is not passed on to higher levels of processing.
5. In the block "Program Formation for Motor Instructions," information is transformed into a usable form. It is transformed into some motor, verbal, or other form of speech. Rehearsal is one possible function at this stage. At this stage of information-processing, we no longer have traces of stimulus. However, we note that sometimes the distinction between the Recognition Block and the Block of Motor Instruction is not always well defined.
6. In the "Functional Manipulative" block nonverbal information may be transformed in different ways and new information may emerge during this manipulation. Information within this block may successively enter and register here following the transformation of preexisting information. The rate at which it operates must be commensurate with the speed of the recognition block.
7. The semantic processing block provides semantic processing of nonverbal information. The meaning is abstracted from the situation rather than from input information. A conceptual imagery model is thereby created.
8. In the block of Repetition of Information, information is transmitted to auditory memory. This block provides the interrelationship between visual and auditory memory in the perceptual process, as well as articulating with primary reflective data.

Due to the feedback influences, function blocks 6 and 7 can regulate the tuning of peripheral processes during perception. Therefore, perceptual action functions as a self-regulative system. During training and automatization of perceptual action, information can bypass some functional blocks or these blocks may be eliminated altogether. If the first blocks are involved more in reproductive transformation, they are later involved more in productive transformations of information. Consequently, a perceptual-imagery model is created. This model is the ultimate problem product of perceptual action. The same situation may be reflected through different operative, perceptual, and memory units. This implies that perceptual imagery models provide multidimensional reflection of reality that can be described in different symbolic, perceptual, and verbal languages. We can see that even very abbreviated perceptual actions may be represented as a complex self-regulative system involving symbolic manipulation and perceptual and high-level cognitive functions. Perceptual actions emerge as self-adapted to a particular task and situation. The microstructure of more complex thinking actions is not studied well.

Let us consider the microstructure of motor action. Motor and cognitive activity are not only closely interconnected but motor actions also include cognitive components within itself.

The study of the interaction between cognition and external behavior has taken two distinct directions. One comes from the works of Vygotsky and is connected with semiotic functions in psychology and the concept of internalization. Another direction, founded by Bernshtein (1947), is based on the study of the regulation of motor movements and actions. According to Bernshtein, motor actions include as an essential component, a continuous stream of afferent signals that are vital for the control and correction of movements. This means that the principles of self-regulation like the concept of feedback and correction are particularly important in the construction of movements. The principle of self-regulation implies the existence of cognitive mechanisms of regulation of motor actions. For example, Bernshtein introduced the concept of the "image of motor action" as an important regulative mechanism of cognitive action. When we discuss the development of a cognitive image of the situation the same self-regulative principles are applied. The continuous streams of feed-forward and feedback influences provide development and construction of cognitive images of a situation and motor action. The work of Zaporozhets (1960), Gordeeva and Zinchenko (1982) also demonstrates that motions possess not only reactive features but also some abilities to reflect a situation. This means that motor actions or motions include in themselves cognitive components.

The process of visual perception and sense by touch is important in the study of the relationship between external and internal activity. Touch perception is accompanied by complicated hand and finger macro- and micromovements (Anan'ev et al., 1959). Using these movements, the subject determines the position of the object in relation to the body as well as the shape of the object. This process is known outside the United States as the "dynamic touch" (Gibson, 1966; Turvey, 1996). The dynamics of a sense by touch, the number of fingers involved, and the speed of finger movements are significant in the Haptic perception of the object's shape. Haptic perception determines constantly changing movement "points of reference" in relation to which the mental image is constructed. Roze (1963) showed that a hand exploring an object performs a multitude of micromotions that have a cognitive function. In her study, Roze discovered that the fingers of the right hand executed more than 60 micromovements in 0.75 sec when pressing the button. This experimental study demonstrated that micromovements produce multiple kinesthetic signals required to regulate total motor action. As motor action becomes more complicated, the number of micromotions that also perform a gnostic function increases. The analysis of visual perception showed that the eye also performs a multitude of similar motions.

In visual perception, micro- and macro-eye movements depend on the specificity of the object and the strategy of perception (Yarbus, 1965). The analysis of the hand and eye movement in the process of perception revealed actions that perform the function of measurement. In visual tasks these movements are "tracking actions." These tracking actions are composed of discrete micromotions. The more precise the visual tasks, the more distinct are these micromotions. Corrective and control movements are present both in touch perception and visual perception. Sensory consequences of movement and micromotions are important in the perceptual process. As the subject becomes more familiar with the object there is a reduction in the amount of these explorative motor actions. The actions of the hand are decreased and replaced by eye movements, which are later also reduced. The successive perceptual actions become simultaneous. These studies point to the similarities between the functions of the hand and eye movements in the process of perception and the presence of motor components in perceptual activity (Zinchenko and Vergiles, 1969).

The study of more complicated cognitive processes presents a greater challenge in identifying mental actions correlated with external behavioral actions. Internal mental activity seems to be independent of external activity, and the presence or absence of motor actions does not indicate the presence or absence of mental actions. Cognitive actions can occur without external motor components. How do we reveal the commonality of external and internal activity and avoid their separation in this case? If it is not possible to find external manifestations of internal actions we must look for the internal components of external behavior based on this commonality.

Gordeeva and Zinchenko (1982) performed a microstructural analysis of motor actions and motions. The microstructural analysis was tied to the study of processes that occur in very short periods of time. The experiment involved moving a lever along three dimensions (X — left–right; Y — up–down; Z — away from oneself–toward oneself), in accordance with a particular requirement. The use of the lever allowed the subject to move a spot on the screen in accordance with the presented target. The movement of the lever along the Z-axis (depth movement) changed the size of the moving spot. The trajectory of the spot was determined by the positions and sizes of the targets on the screen. The movement of the lever was recorded on a multichannel self-recording tape. The tape presented information about the movement of the lever along all three dimensions graphically. The subject was given a signal that she/he should start the movements. Altogether there were three targets on the screen. After the spot reached the required target and corresponded to the target's location and size the subject pushed a button on the lever. The experiment involved multiple trials. In Figure 1.8, we present an example when the spot was moved from the first target to the second target. Figure 1.8a demonstrates the patterns of movement along all three dimensions in the beginning of the motor skill formation. Figure 1.8b demonstrates the data obtained at the end of the formation of motor skills.

The three stages of movement mentioned above were discovered along all three axes: the program formation stage (latent stage), the executive stage (motor stage), and the evaluative stage (evaluation of the result of movement and correction of movement).

The first and third stages are cognitive components of motor actions. The first stage is involved in the formation of a motor program of the action. The third stage

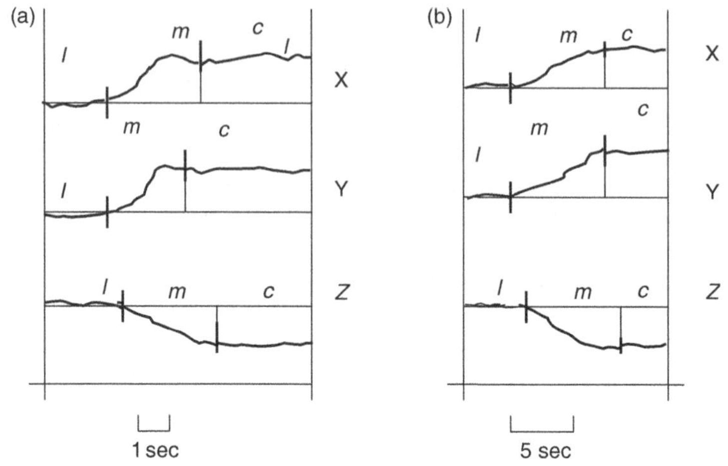

FIGURE 1.8 Microstructure of motor action.

is involved in the evaluation and correction of motor action. The second, executive stage, is directed to the transformation of the situation. (The stages are designated by letters *l*, *m*, and *c*, respectively). Through motor actions a subject not only transforms the situation but also explores it by building a mental image of the situation and of his/her own actions. Hence a motor action includes in itself a cognitive component. Furthermore, not only is the time of movement shortened but the interrelationship between cognitive and motor (executive) component changes unevenly along three coordinates during skill acquisition. The greater the duration of the cognitive components of motor actions, the more complex motor actions or motions are and the more concentrated attention they require.

Sometimes the operator is involved in performance of the tracking tasks when the operator has to coordinate her/his response with the constantly changing input signals. The existing response models attempt to explain and predict tracking performance. In Zabrodin and Chernishov (1981) study it was discovered that during performance of these tasks operator's movements can not be compared with the technical tracking system. The study discovered that the operator's responses in this kind of tasks contain micro-motions with additional harmonics that has not been anticipated as per tracking theory. Discovery of micro-motions gives the subject additional information that is important in regulation of movements. These additional micro-motions, that are considered by mathematical models as tracking errors, perform explorative cognitive functions. The obtained data demonstrates that during performance of the tracking tasks the operator works in a self-regulative mode.

2 Systemic-Structural Theory of Activity and Design

2.1 DESIGN PRINCIPLES AND CLASSIFICATION OF WORK ACTIVITY

2.1.1 WHAT IS DESIGN?

Ergonomics is an interdisciplinary approach that integrates the efforts of psychologists, physiologists, anthropologists, engineers, computer scientists, and other professionals. The major aim of ergonomics is to design human tasks, technical systems, informational components, including software, in relation to the psychological and physiological capabilities and limitations of humans.

The term "design" emerged from engineering practice. It determines the ultimate outcome of engineering activities the purpose of which is the creation of new products, software, manufacturing goods, etc. Objects of design in ergonomics and in engineering are very often the same. However, the specifics of ergonomic design is that design should take into account the capabilities and limitations of a human being. Ergonomic design draws its knowledge from various fields including psychology, physiology, and engineering. In order to obtain a better solution professionals involved in ergonomics design require fundamental principles and methodologies to guide the design process. There is no unified concept of design even in engineering where the first principles of design were stated. Some professionals even state that design, unlike the natural sciences, cannot stand on a scientific basis.

Due to the lack of a valuable theory of design, many professionals in ergonomics use the term "design" in many different ways and even incorrectly. Consequently, there are many definitions of design and of ergonomic design in particular. In the last decade the term "design" has been used in reference not only to material systems, but also to algorithms for human activities and related technological processes, including, for example, development of computer programs. In this book design is referred to as "the creation and description of an ideal image of artificial objects, in accordance with previously set properties and characteristics, with the ultimate goal of materializing these objects" (Neumin, 1984).

Engineering design involves four distinct aspects of engineering and scientific endeavor. They are similar to ergonomic design (Suh, 1990):

1. The problem of definition from a "fuzzy" array of facts and myths into a coherent statement of the question. As with any creative problem, definition

is one of the most important steps in design. They are often done through an interactive process involving different design cycles.

2. The creative processes of synthesizing a design solution in the form of physical embodiment. This is an ideation process, which is very subjective. The creative ideas and synthesis depend on specific knowledge and the designers' ability to integrate this knowledge.
3. The analytical process of determining whether the proposed solution is correct or rational with the ultimate check of the fidelity of the design product to the original needs.

From this follow the creative steps of design that should be checked by formalized steps of design. Therefore the design process can be depicted as a feedback control loop. This permits to correct a design solution. Hence creative and formalized or analytical stages are interdependent in the design process. The creative process depends on designer knowledge and creativity. This process is subjective. At this stage the designer can produce a number of possible creative solutions. In contrast, the analytic process is deterministic and based on a finite set of basic principles. In the absence of basic analytical principles of design this process is reduced to purely intuitive procedures. In this situation design becomes more of an art than a science. Analytical procedures are eliminated in this situation and design is performed through trial and error using different experimental procedures. This situation is encountered in ergonomics when psychological aspects of design are considered. Typically, an ergonomist attempts to use observation or questionnaire and then, bypassing analytic procedures (formalized description and analysis) he or she tries to develop a physical model of the design object. At the next step the ergonomist conducts experiments and analyzes the obtained result. In contrast, in engineering design analytical procedures precede experimentation. Moreover, in engineering design analytical procedures are also used at the experimental stage of design.

The following is another issue with psychological aspects of design in ergonomics. Specialist in cognitive psychology have shown a tendency to reduce design problems to the mentalistic orientated modeling methods. There are two kinds of models. One of them describes information-processing system of brain, rather than real activity during task performance. Such models sometimes are erroneously considered as design models. The other kind of models are design models. They describe the object of design. Design models in ergonomics are always task specific. They describe activity during task performance and tool or equipment utilized by subject during task performance, computer interface, etc. Analysis of relationship between these models is critically important in design process. The process of developing design models should be unified and standardized.

2.1.2 General Characteristics of the Design Process

The specificity of design lies in the fact that in developing the model of an object being designed, the object itself does not yet exist as a materialized system. In the design process the stage of developing of models precede to the stage of object creation.

The design process can be viewed in terms of stages of sequential refinement of design models. At the initial stage a designer has only the ideal's image or mental model of the object being designed. During the subsequent stages, the conceptual or mental model is externally described using symbols and signs. This makes the model available to other specialists involved in the design process.

There are three basic types of models: conceptual (thinking), symbolic (verbal, graphical, mathematical), and physical. Conceptual models are mental images about systems, events, objects, and phenomena that can be created in the mind. Symbolic models describe the object or phenomena verbally and graphically. One type of symbolic model is the mathematical model, with equations or algorithms representing relationships that occur in real objects. Finally, there are physical models that reflect some functional features of an object. In engineering design usually symbolic models precede physical modeling. In contrast, in ergonomic design, where there are no analytical procedures of the design, only physical models of the design object are used. Moreover only a physical model of the design object is created. Human behavior or activity is not modeled because there are no units of analysis or language of description for this purpose.

Any model is only an approximation of a real object or event. Because the real object or event has many features and properties, it is almost impossible to include all of them in any one model. This is why the design process requires the creation of different models of the same object. When addressing a problem experimentally, the choice of the model depends on the experimenter. However, when we address design problems, such freedom in the choice of models is not available. In technical design, for example, in contrast to experimentation, the model is used not only by the designer but also by those who attempt to produce or materialize an object based on the existing symbolic models.

As a general principle, there must be concordance between the specialists who create the models and those who must interpret it for production purposes. This may be attained in part by standardizing procedures and language of description of the model during the design process. For example, without a knowledge of specific drawing principles design engineers and production personnel cannot understand each other. Therefore in design process task analysis and task description are tightly interconnected procedures. This interdependence is very often overlooked by those who conduct task analysis. Usually designers have very limited sources for the collection of data. The obtained data should be described in a standardized and formalized manner and then these data become the source of subsequent analysis. Formalized description is considered to be a creation of models of the design object. Based on an analysis of the created models designers perform their modification (new models) and this cycle is repeated.

Design can also be defined as a mapping process from the functional space to the physical domain. The functional domain is associated with what we want or what a system must do.

In this domain one must determine the design objectives by defining it in terms of specific requirements, which can be called functional requirements. The physical domain is the solution to the problem. In order to satisfy functional requirements, a physical embodiment characterized in terms of the design parameters must be created

(Suh, 1990). More precisely, design can be imagined as a mapping between the functional requirements in the functional domain and the design parameters in the physical domain. The design process begins with the recognition of a societal need. Therefore, the first step of design is the definition of the problem in terms of functional requirements. The needs are formalized, transformed, and conceptualized in a set of functional requirements. At the next stage the object that satisfies these requirements should be created. Then the developed product should be compared with the original set of functional requirements. If the product does not satisfy the functional requirements then new ideas should be generated. This iterative process continues until an acceptable result can be achieved.

Functional requirements and design parameters have hierarchical organization. From this it follows that each level of functional requirements and design parameters can be decomposed. However, professionals can go to the next level of the functional requirements only after going on to the physical domain and developing a solution that satisfies the corresponding level of functional requirements. Having briefly described the general principles of the design process, we now consider the ergonomic principles of design that are derived from systemic-structural activity theory.

We can outline two aspects of design. One is involved in the design of totally new systems. The other is involved in the redesign or modification of an already existing system. Modification is required in any situation when the operational characteristics of the system do not match the new requirements. In order to meet the new requirements it is necessary to describe the advantages and deficiencies of the system and to develop new drawings and documentation. Based on the new documentation an experimental sample of the system can be developed. The system can then be evaluated through experimental procedures. New changes can be introduced into the developed documentation if required. Therefore during the design process one can distinguish the following general stages: analysis and observation of functions of the already existing system, development of new documentation and drawings that reflect the improved solution, and experimentation evaluation of the system. The second stage is analytical. Symbolic models of the modified system should be developed at this stage.

Ergonomics is often applied to the redesign of the various systems. All the stages listed above can be used in anthropometric or biomechanic aspects of ergonomic redesign. However ergonomists cannot use analytic procedures when they are dealing with psychological aspects of redesign. After observation of the existing equipment practitioners immediately go to the third stage of design, which involves experimentation. During anthropometric or biomechanic design practitioners can develop drawings of the human body and compare them with drawings of equipment. They can also perform the required calculations. In cognitive psychology, the language of the description that would allow creating models of activity does not exist. We do not consider mentalistic models in cognitive psychology as design models. Behavior or activity is a process that is very difficult to present as an analytical model. There are no required units of analysis and language of the description of human activity in cognitive psychology. This reduces the effectiveness of psychological aspects of design in ergonomics. The analytical stage of design becomes particularly important in the design of a new system. Systemic-structural activity theory that creates a hierarchical system of units of analysis permits professionals to use the analytical stage of design during the system modification or in the design of a new system. It is impractical to make any changes to the system after the system is actually produced. Thus, the initial design must be as efficient as possible at the analytical stage. It is particularly relevant to the design of a new system when the analytical stage is the major focus of ergonomic design. Analytical aspects of psychological design in ergonomics are relevant to product safety. They can substantially reduce potential hazards of the system. This stage of design is critical in the study of computer-based tasks.

The ergonomic principles of design should be based not only on the study of separate cognitive processes but also on the study of activity as a whole where there is a complicated interrelationship between these components of activity. In this book we will consider the ergonomic principle of design from this point of view.

2.1.3 Structure of the Production Process[1]

A production process can be defined as a sequence of transformations of raw material into finished product. The process usually begins with the entry of the raw material and proceeds through various steps until the material becomes a finished product, or result, in accordance with the purpose of the production process. Any production process can be seen to contain three basic elements: the human work activity, or work process; the means of work; and the product. The means of work are the tools, equipment, and instruments used by subjects in the production process. While, in general, the purpose of any production process is the transformation of raw material into finished product, it is possible to distinguish various types of production process. For example, in manufacturing there are mechanical production processes, physical–chemical processes, transportation, and control production processes (Gal'sev, 1973). The structure of the production process is presented in Figure 2.1.

The left side of the figure represents the work activity, or *work process*. Here, the term process is used to emphasize that activity is performed according to some prescription or order. The work process contains a substructure of basic components: *motive-goal* as a vector which demonstrates the directional and energetic aspects of work activity; *knowledge and skills* which demonstrate the relevance of pastexperience to the work process; *abilities* related to the tasks to be performed; and *work actions* which are organized into a structure, and together present the *method of work*. In this scheme, "action" refers to both cognitive and motor action. The presence of the concepts of knowledge and action in the structure of work process also implies the existence of mental tools.

The right side of Figure 2.1 represents the technological aspects of the production process, which are also performed in accordance with some prescriptive order. The technological process includes the instructions according to which the worker performs the job and the means of production; equipment; the raw material or initial material object; and the finished product or result.

The production process can be described as a sequence of separate steps or production operations. By production operation one understands the isolated part of the

[1] This section was prepared together with Steve Robert Harris, University of Glamorgan.

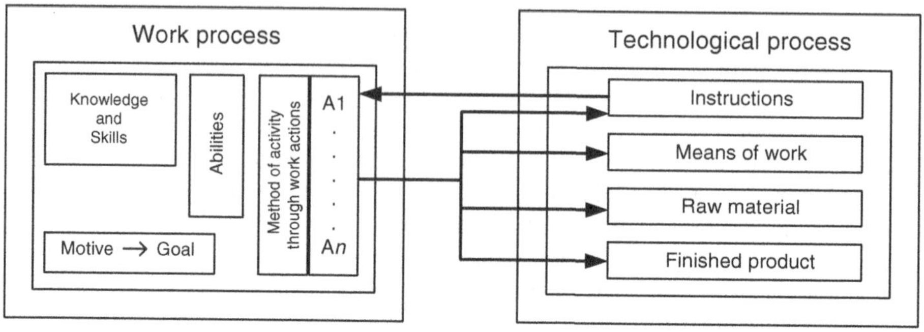

FIGURE 2.1 The structure of the production process.

production process which is performed upon the work object in one work place by one or several workers. The existence of the same equipment and tools, object of work and technological completeness characterizes a production operation. Production operations according to the technological principle can be divided into smaller standardized technological units.

According to Figure 2.1 production operation includes human activity, technological components or tools, and the object that is transformed. The nature of the object will change depending on both human activity and the specificity of the technological process. Production operations may be studied within the technological frame, or from the activity perspective. In the first case, the leading figure is the production engineer or related professionals. In the second case, a human factors specialist is called for. If one studies production operation from the activity perspective the term tasks can be used. Such tasks are organized in accordance with technological requirements.

Task or production operation from the activity point of view can be divided into a logically organized system of cognitive and motor actions that can be divided into smaller units or integrated into a combination of actions. We consider this later in more detail. The presented scheme demonstrates that work process should be considered as a specific type of activity. One of the most important specifics of the work process is that the worker knows about material, tools, equipment, and the final result of the activity in advance. The worker should possess the required abilities and the necessary professional background and perform in accordance with the given instructions. The tasks can be with different degrees of freedom of performance. It can be a deterministic or skill-based task. This task requires the same sequence of actions. The task can be deterministic-algorithmic. This is performed according to deterministic algorithmic prescription. Sometimes the task can also be probabilistic-algorithmic. This is performed according to probabilistic-algorithmic prescription. Algorithmic tasks are performed according to rule-based principles. In general, a work process is always performed according to some prescriptions and constraints. A subject should know a sequence of actions and technological stages that cannot be violated. This kind of work can be design. Creative work cannot be design. Some additional principles of categorizations are given in Section 2.1.4.

2.1.4 Structure of the Operational-Monitoring Process[2]

One of the specific characteristics of work with the automated and semiautomated systems is that the operator is involved not merely with changes to physical material but also with the transformation of information. In many cases, it is the operator alone who must determine what has to be done. Although the work is based on certain procedures and rules, it also involves creativity and problem solving. In these cases, the work is organized as probabilistic algorithms and quasi-algorithms, with the operator forming the performance rules based on his or her experience.

Another characteristic of an operator's performance with automated systems is that he or she is required to perceive information from a variety of displays and instrumentation. Rather than controlling the power sources directly, the operator uses intermediary control devices. In this type of work process, inspection, controlling, and monitoring functions predominate, and the motor components of activity are significantly reduced. Work activity in its external appearance (as motor actions) loses continuity and acquires an episodic character, while at the same time, the role played by the sensory–perceptual and thinking components of activity increases.

An operational-monitoring process is defined as a combination of duties essential to accomplish some automated or semiautomated system function. Recognition of the specific characteristics of the operational-monitoring process requires some reconsideration of our description of the work process. The notion of the production operation is no longer appropriate. Rather, the task is seen as the basic component of operational-monitoring processes. In this case, the task goal is not always reduced to the transformation of the material or physical object but often involves changes in the state of a control object, or the transformation of information. The structure of an operational-monitoring process is illustrated in Figure 2.2.

In this figure, "technological process" is replaced by the notion of *control process*. It is the transformation of information, rather than the material transformation of the object, that now plays a major role. Changes in the physical state of a controlled object become possible mainly on the basis of the transformation of information. In the control process, the components "raw material" and "finished product" are replaced by *input* and *output*. In most cases, the input is the information received by the operator about the initial state of the control object. The output is the information about the control object following the completion of the task. In this type of process, it is very often the case that a sufficiently detailed description of how to accomplish the task goal is lacking. The task then becomes one of problem solving. The operational-monitoring process can thus also be described as a system of logically organized problem-solving tasks.

In a production process, the tasks or production operations are prescribed in advance. By contrast, in an operational-monitoring process the tasks are often self-initiating. For example, while a pilot may have as her final goal the completion of a flight from one city to another, she will be required to formulate a variety of ongoing tasks herself, according to the situation presented by flight conditions. When

[2] This section was prepared together with Steve Robert Harris, University of Glamorgan.

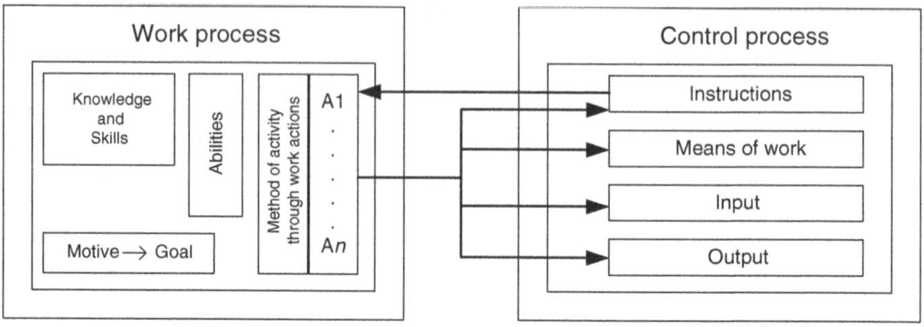

FIGURE 2.2 The structure of an operational-monitoring process.

analyzing operating-monitoring processes, the functional analysis of activity where the concepts of *objectively given* task and *subjectively accepted or formulated* task are critical become particularly important.

Acceptance or formulation of a task is closely associated with the subject's representation of the task. The subjective, or mental representation of a task, is characterized by the following features (Kozeleski, 1979):

1. It is dependent on the objectively presented structure of the task.
2. It is a dynamic phenomenon that can change during task performance.
3. The mental representation determines task performance.
4. Success in solving a task is dependent on the performer's personal representation.

Investigating the influence on task performance of the relationship between the objective presentation of the task and its subjective representation is a major aspect of the functional analysis of activity.

Computerization significantly alters the specifics of an operational-monitoring process, often leading to greater demands on the task performer. When a subject directly interacts with a computer during task performance, this becomes a specifically human–computer interaction (HCI) process, which includes logically organized computer-based tasks. The schema of this process is basically the same as that for operational-monitoring processes, the only major difference being that the computer is now the dominant means of work.

The material we have presented above demonstrate how, from the perspective of activity theory, a work process may be understood as a combination of tasks performed by subjects to accomplish the objectives of the system. We now briefly summarize the main features of this approach.

Each task in the work process is regarded as a situation-bounded activity which is directed to achieve a goal under given conditions. Any task includes both the subject's activity and the material components of the task, with all the elements of activity during task performance being organized by the task goal. It is only when the objectively given or subjectively formulated requirements of the task are accepted by the subject as a desired future result that they become the goal of the task. Whatever is presented to the subject for the performance of the required actions constitutes the conditions of the task. Task conditions include both the subject's past experience and such material components as instructions, means of work in given conditions, raw material, and input information. These conditions also determine the possible constraints on activity performance. The raw material, or input information, is considered to be the object of activity. What is actually achieved (finished product, output) is the result of activity. The vector motive→goal determines the directedness of activity during task performance.

Any task includes an initial situation, intermittent situations, and a final situation. By associating the notion of a *situation* with the stages of task performance, it becomes possible to study how the structure of a task changes during different stages of performance and how many basic transformational stages are required.

In the schema presented in Figure 2.1 and Figure 2.2, *knowledge* includes images, concept, propositions, and nonverbal sign systems. According to Landa (1976), knowledge includes not only data about objects and their attributes and relations, but also knowledge about (motor or cognitive) actions on objects. When a subject is able to perform mental actions on images, concepts, propositions, and other sign systems, those sign systems become internal, psychological tools for action. This is essential for the practical application of knowledge.

2.1.5 Classification of Work Processes[3]

On the basis of the material presented earlier we can now sketch out an activity-theoretical classification of different types of work processes. The criteria for classification include the features of the object of activity (i.e., the material components of activity); the extent to which the subject is involved in the transformation of the work object (which may be raw material or input data); and the functions performed by the subject. A classification of work processes according to these criteria is presented in Table 2.1.

Of course, other types of work activity exist in addition to the work processes described in the Table. There are numerous professions where the task performance methods are not precisely determined. For example, the work activity of engineers, medical doctors, teachers, etc., can be widely varied in order to respond to the conditions for achieving a particular goal. Strumilin (1983) designated this kind of work as "independent work within the given set of the requirements." It is also necessary to distinguish the kinds of creative work performed by scientists, artists, inventors, etc. In this type of work, at times it is not only the method of achieving the goal that is unknown but also the goal itself. It can be seen that the activity-theoretical classification of work activity based on the criterion of predetermination must have a relative character. Work processes, independent work within the set of requirements, and creative work may be combined in different proportions in different professions.

Different specialists focus on different aspects of the work activity. For both ergonomists and engineers, the task as a component of a work process is a major object of study. However, while for the ergonomist the main focus is on the work process (depicted on the left side of Figure 2.1 and Figure 2.2), he or she only considers the

---

[3] This section was prepared together with Steve Robert Harris, University of Glamorgan.

TABLE 2.1
Classification of Work Processes

Classification criteria	Type of work process
Character of object	Substance-energetic informational mixed
Extent of subject's involvement in transformation of object	Manual mechanical-manual automated
Functions performed by the subject	White-collar workers Blue-collar workers Operators (technologists, controllers, inspectors, dispatchers, etc.) Workers involved in the performance of computer-based tasks

technological process and control system (depicted on the right side of Figure 2.1 and Figure 2.2) as they relate to the work process. On the other hand, for the engineer-technologist the main object of study is the technological process itself, which can often be considered exclusively from the technological point of view without relating it to human activity. To return to a point made earlier, we can say that these different specialists share the same object of study, namely, the work process, but have different subjects of study, namely, the different aspects of those processes.

An analysis of the work process has demonstrated that the work activity of the subject need not be rigid. Work activity contains both anticipated and unanticipated elements (Vicente, 1999), and the proportion of predictable to unpredictable factors will vary across work processes. We can conclude that the more predictable the work process, the easier it will be to precisely design the work activity. It is also important to recognize that activity-theoretical models created during the process of ergonomic design represent idealized versions of activity and can only approximate the real activity of the work subject. However, a very useful comparison of different versions of the equipment configurations can be accomplished based on an analysis of these idealized models of activity.

The physical characteristics of the equipment impose different strategies of activity, in a probabilistic manner. The space of possible strategies for activity is defined by the totality of these strategies, taking into account the constraints on performance. Given the same performance constraints, changes in the equipment configuration will result in a new space of possible strategies for activity. In order to analyze the basic characteristics of the space of possible activity strategies, the ergonomist need not analyze all possible strategies of performance but rather must select and analyze those more representative and critical strategies of activity that correspond to the basic trajectories of activity within the space of all possible strategies. This allows him or her to gain an understanding of the space of activity strategies under the given performance constraints.

We have described some ways in which the systemic-structural activity theory approach can be used to solve design problems. While this approach is based on the analysis of theoretical models of activity which do not always describe all the possibilities for action, it does take into consideration the constraint-based principles of design proposed by Rasmussen (1986) and Vicente (1999). Our contention is that without principles for the development of theoretical models of activity, there can be no scientific basis for ergonomic design. This follows from the assertion that any design process requires models to represent the object being designed. Finally, we would like to distinguish three major aspects of ergonomic design: the design of equipment, the design of human performance, and the design of human–computer interaction. In Section 2.2, we consider some practical examples connected with the latter category.

2.1.6 Classification of Tasks

Any work process includes a number of different tasks. There are skill-based tasks on the one hand and problem-solving tasks on the other hand. These two types of tasks require different levels of automaticity with which the tasks are performed. Skill-based tasks require standardized methods of performance without logical decisions about the sequence of possible actions. Skill-based tasks are performed in a rapid automatic way with minimum concentration. The simple production operations are an example of this kind of tasks. The simplest type of work task is rigid and deterministic, when workers produce identical products in large quantities. Such tasks are often highly repetitive and carried out at great speed. We define such tasks as skill-based tasks. These tasks require automatically performed actions. According to Russian terminology automatically performed actions are called "naviky." However, there is a second level of skills. These kinds of skills consist of an individual's ability to organize knowledge and the first level of skills into a system and efficiently use it to perform a particular class of tasks or solve a particular class of problems. For more detailed information see Bedny and Meister (1997a).

Any task includes some problem-solving aspects. When a task requires conscious deliberation on how to accomplish the goal of the task, it becomes one of problem solving. Hence the tasks can be presented as a continuum (Figure 2.3).

On the left-hand side there are skill-based tasks and on the right-hand side there are problem-solving tasks. Problem-solving tasks can be divided into two major groups: algorithmic and nonalgorithmic. Algorithmic tasks are performed according to some logic and rules. Algorithmic tasks can be divided into deterministic and probabilistic types. In deterministic–algorithmic tasks workers perform simple "if-then" decisions based on familiar perceptual signals. Each decision usually has only two outputs. For example, "if the red bulb is lit then perform action A, if the green bulb is lit then perform action B." Algorithmic tasks completely define the rules and logic of actions to be performed and guarantee successful performance if the subject

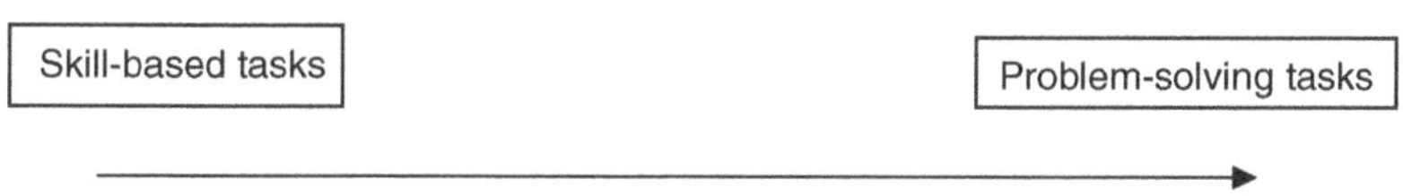

FIGURE 2.3 Continuum of possible tasks.

follows the prescribed instructions. Deterministic–algorithmic tasks can be compared with rule-based tasks according to Rasmussen terminology. Probabilistic–algorithmic tasks involve logical conditions with the possibility of three or more outputs, each of which possesses a different probability of occurrence. This probabilistic element significantly increases the operator's memory workload and complexity of task performance in general. Probabilistic–algorithmic tasks can also include nonalgorithmic problem-solving components.

More complex tasks are nonalgorithmic. The latter class, according to Landa (1976), should be divided into three subgroups: semialgorithmic, semiheuristic, and heuristic. The distinction between the types of tasks is relative, not absolute. We recommend the following major criteria for classification of tasks (a) indeterminacy of initial data; (b) indeterminacy of goal of task; (c) existence of redundant and unnecessary data for task performance; (d) contradictions in task conditions, and complexity or difficulty of task; (e) time restrictions in task performance; (f) specifics of instructions, and their ability to describe adequate performance and restrictions; (g) adequacy of subject's past experience for task requirements.

If the situation or instructions contain some uncertainty resulting from vagueness of criteria that determine the logical sequence of actions and, therefore, require the subject not only to perform actions based on prescribed rules but also to create his or her own independent cognitive actions which should be performed in order to achieve the required goal, such tasks are semialgorithmic. Because of the inability to remember all possible rules for the performance of probabilistic–algorithmic tasks and because of insufficient familiarity with the probabilistic characteristics of such tasks these tasks very often become semialgorithmic and even semiheuristic.

If uncertainty is even greater, and includes some independent solutions without precise criteria, the tasks are semiheuristic. This class of tasks does not only fully determine executive actions but also requires explorative actions for analyses and comprehension of the situation. Semiheuristic problems may include algorithmic and semialgorithmic subproblems. The purpose of the ergonomic design of such tasks is to reduce the degree of objective and subjective uncertainty in problem solving. A significant part of probabilistic-algorithmic tasks and all nonalgorithmic tasks can be regarded as knowledge-based tasks according to Rasmussen terminology(1986).

Purely creative tasks are a heuristic task problem. The major criteria for categorization as a heuristic task problem are an undefined field of solution, indeterminacy of initial data, and indeterminacy of goal of task. Task categorization or taxonomy of task is presented in Figure 2.4.

Therefore in activity theory there is a more detailed system of task classification than in cognitive psychology.

The basic characteristics of tasks are structure, complexity, difficulty, and degree of physical effort. The structure of a task is the spatiotemporal organization of its elements and the actions to be performed by the subject. The structure of a task can be externally determined by or depends on the mental representation of the task. In the last situation task performance is more complex. The structure of the task influences strategies of task performance.

Complexity of a task depends on the number of static and dynamic components of the task and specificity of their relationship. A degree of uncertainty or

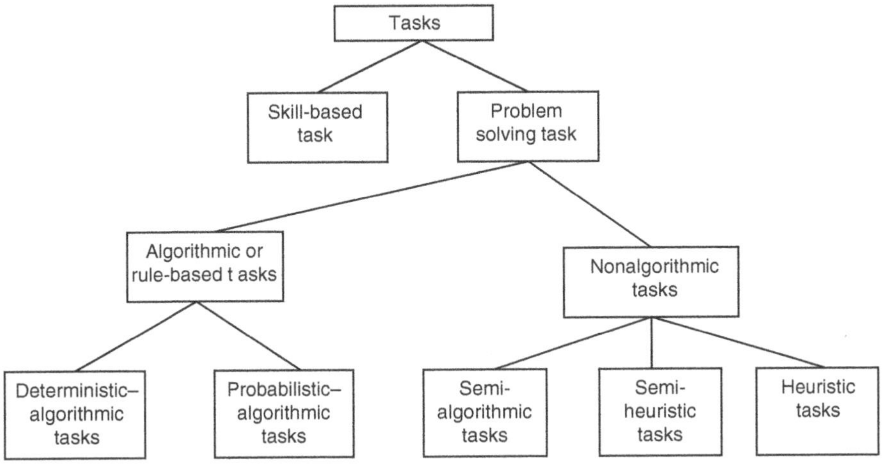

FIGURE 2.4 Tasks' taxonomy in SSAT.

unpredictability of a task is also an important component of complexity. The number of interactions among components of a task, specificity of instructions, indeterminism of the task, concentration of attention, and excessive cognitive requirements, etc. are some characteristics of task complexity. In general the more complex the task, the more mental efforts required for its performance. Complexity is an objective characteristic of a task.

Task difficulty is a subjective characteristic of a task. Task complexity evaluation is based on the assumption that the more complex a task is, the higher is the probability that it will be difficult for a subject. However, the same task complexity can be subjectively perceived as a task with different difficulty levels for different subjects. Task complexity and difficulty are associated with usability of equipment. Complexity of a task influences precision and reliability of task performance. The lower the task complexity, the easier the process of skill acquisition, and the easier to perform the task, the less mental fatigue.

The degree of physical effort is a particularly important characteristic of manual task. Physical efforts are easier to measure. However physical efforts also have subjective components (a feeling of physical stress) that are affected by the individual's physical conditions. Physical and mental efforts can influence each other.

2.2 INTRODUCTION TO GENERAL PRINCIPLES OF SYSTEMIC-STRUCTURAL ANALYSIS

2.2.1 Basic Principles of Systemic-Structural Analysis

Systemic-structural activity theory is distinguished by the careful delineation of the structure of the activity describing both the basic components of the activity and their interrelationships. Here, one is not speaking of the system analysis of the man–machine systems, but rather is describing human activity and behavior as a system, which, of course, is in a dynamic interaction with machinery. A system is a set of interdependent elements that is organized and mobilized around a specific purpose or goal. Systemic-structural analyses entail extracting the relevant elements of the system and their dynamic interactions. Systemic-structural analyses not only differentiate elements in terms of their functionality but also describe their systematic interrelationship and organization. Whether or not there is a systemic-structural approach depends not so much on the qualities and specificity of the object under consideration but rather on the perspective and methods of analysis.

In the former Soviet Union, Shchedrovitsky first wrote about this in 1974 (Shchedrovitsky 1995), describing two systemic approaches. The first he termed object-naturalistic; the second, theoretical–methodological. In the first case, the system approach is determined by the specificity of the object being studied and emphasizes the objective nature of the object as a system. Object-naturalistic approaches overlook that the consideration of the object as a system depends on the abilities and perspectives of scientists. Thus, Shchedrovitsky encourages a second approach, which he calls theoretical–methodological method, which emphasizes the specificity of procedures and methods for considering objects as a system. While not rejecting objective reality, the theoretical–methodological approach enables the articulation of systematic particularity and specificity of objects of concern in operational terms. The particularity and specificity of the object as a system is, of course, closed to us until the requisite tools and methods of study are developed. We believe that a major contribution of activity theory to current endeavors in psychological studies is the systemic tools and methods that it has crafted to study behavior as a system. During these procedures one can open or discover totally new features of an object as a system. All of this is obscured in the absence of the appropriate units of analysis of activity. In activity theory, under which the major units of analysis are actions, operations and function blocks that have specific organizations, the systemic-structural approach is fundamental. In this new approach of activity study, one uses the term systemic-structural approach because it elucidates the structure of activity.

Cognitive psychology treats activity and behavior as a process, making it difficult to study activity and behavior from a systemic-structural perspective. When considering cognition merely as a process, systemic-structural description of activity becomes impossible. In the sequential process of activity at any particular moment, only a particular slice of activity can be realized. In spite of this, those slices that represent our past and future influence what we do at present. All of this leads to the conclusion that despite the unfolding of activity over time it is a holistic activity with its own structure, much like a symphony. However, the notion of a process does not allow us to capture this structure, nor does it allow integration into a holistic structure of past, present, and future (Shchedrovitsky, 1995). By introducing special units of analysis, the AT engenders a structure of activity that unfolds over time. Temporal models of activity are particularly effective methods of representing this structure. One important aspect of this problem is the further development of methods of description for standardized cognitive actions and their duration.

Systemic-structural approaches invite and empower the study of the same object from different points of view and distinct aspects, thereby legitimating the use of multiple approaches to the description of a single object. This implies that in applied

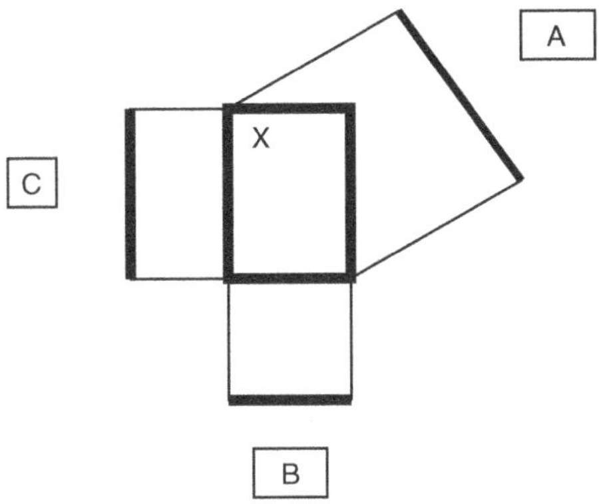

FIGURE 2.5 Systemic representation of object.

research, adequate descriptions of the same object of study can be represented by multiple, interrelated and supplemental models and languages of description. In Figure 2.5 we diagram this in an accessible manner. "X" is the object under consideration; "A," "B," and "C" are different interdependent and supplementary presentations of the same object which constitute a systemic approach. This calls for different stages and levels for the description of activity. We consider this problem in the next section.

One can outline the following approaches in the study of activity: parametric and systemic. Parametric analysis entails the study of distinct components of activity. For example, we can measure the time taken for the performance of the task. The cognitive approach is also an example of parametric analysis. The systemic approach includes a morphological and functional analysis of the activity. Each of them comprises different methods. Morphological analysis is involved in the description of constructive features of activity. Actions and operations are used as units of analysis at this stage. Researchers of these stages attempt to describe logical and spatio-temporal organization of cognitive and behavioral actions. This is the description of the structure of activity on a morphological level. In the functional analysis of activity major units of analysis are function blocks. Activity in this case is regarded as a self-regulative system (Bedny and Meister, 1997; Bedny and Karwowski, 2003). This allows the identification of potential strategies of activity performance. Activity in this case is regarded as an adoptive system that actively interacts with the situation. Functional analysis can be regarded as an application of systemic principles to a qualitative stage of analysis. We discuss this approach in Chapter 3.

In a morphological analysis the structure of activity is a logical and spatio-temporal organization of actions and operations performed to achieve the goal of a task. To describe the structure of activity it is necessary to subdivide the work process into tasks that should then be individually described in terms of mental and motor actions and operations. Each action has a separate, intermediate goal, which must be reached to attain the goal of the task. Therefore, objects of study in this case are work process and tasks. Units of analyses are cognitive and motor actions and operations.

Systemic-structural analysis of work activity entails four stages of analysis organized according to the loop-structure principle (Figure 2.6). One important feature of

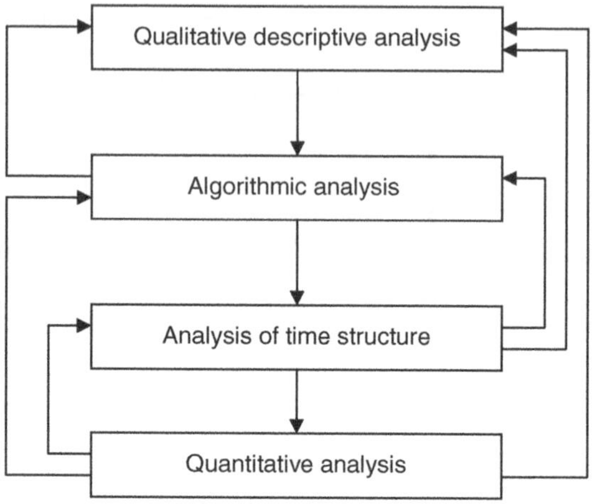

FIGURE 2.6 Four stages of systemic-structural design.

systemic-structural analysis is the hierarchical description of activity. According to this, activity is organized in a hierarchy. Thus, activity needs to be described at different levels of decomposition, and therefore calls not only for different stages but also for different levels of analysis. Transition from one stage and level of the description of activity to another has a loop structure of organization, implying that the result of analysis from one stage or level may require reconsideration of preliminary stages and levels of analysis.

Zinchenko and his colleagues introduced into the Theory of Activity the notion of the "microanalysis" and "macroanalysis" of activity. Macroanalysis of activity involves studies of motives, goals, actions, skills, etc. This level of analysis suggests that the processes under study possess significant duration that may be indexed with traditional chronometric methods. Microstructural analysis is concerned with the internal structure of actions and psychological operations that are sometimes of very short duration. Typically, these processes cannot be observed externally or reported introspectively. One of the ways of studying microstructural analysis is the use of function blocks that are components of cognitive actions and motions (cf. Figure 1.7 and Figure 1.8, see Section 1.4.4), performed over very short periods of time and sometimes measured in milli-seconds. Later, we will show that a function block as a unit of analysis may be used at a macroanalysis level. In such cases, function blocks reflect more complicated psychological phenomena, not closely related to chronometrical procedures. In other situations microstructural analysis uses psychological operations or cognitive actions as units of analysis that also have a brief duration. Systemic-structural analysis is classified in the following way:

Parametrical method — concentrated on the study of different parameters of activity that are treated as relatively independent. Experimental methods in cognitive psychology are related to the parametrical method of study.

Morphological analysis — in which the major units are actions and operation, based on which one may describe the structure of activity, in terms of logical and temporal-spatial organization of actions.

Functional analysis — in which the major unit of analysis is the function block. Analysis of the structure of activity is based on different functional models of self-regulation of activity. Describing the specificity of the operation of different functional mechanisms and their influence on strategies of task performance becomes possible at this stage.

Macrostructural and microstructural analysis determines the level of analysis. Macrostructural analysis includes larger units of analysis. Microstructural analysis suggests more detailed ones. These levels of analysis can be used in morphological as well as in functional analysis.

All methods of analysis of activity are intimately related and mutually interactive, so that according to the systemic-structural method of analysis all methods are a unity.

Formation experiment — based on a genetic explanation of activity development derived from the work of Vygotsky is also important. This method of study emphasizes that activity is a constantly developing system. Existing norms of human activity are historically shaped. Analysis of activity development and its association with the material world helps us to understand how activity can be developed in future. This is important for the prediction of the development of a design object. Concepts such as the zone of proximal development, task complexity and difficulty, and guided and independent activity help us to understand the dynamics of the various stages of activity transformation during the design process. In general, the genetic method can be formulated in the following manner. We need to describe the activity structure while subjects are acquiring competence in task performance. If it is known how the structure of activity changes and what the final structure of such activity is during skill acquisition when a subject uses a different kind of equipment, then we can evaluate the usability of this equipment more efficiently. Very often it is impossible to discover differences in the design solution when the activity is acquired and perfect. Only through analysis of the acquisition stages of activity can the design solution be found. The formation experiment can be formulated in different ways. Subjects learn to perform tasks from precise instructions; they learn to perform a task using different prompts and finally do so in self-learning conditions. The structure of activity at different stages of skill acquisition is examined.

We do not consider more specific methods of study. We only mention that in activity theory physiological methods are widely used, especially electrophysiological methods, including the physiology of higher neural functions. Here we also make note of the "poly-effector" methods, a complex approach that simultaneously registers different indices such as the electroencephalographical, electro-oculographic (i.e., eye movements) and electromyelogram.

2.2.2 Qualitative Stage of Systemic-Structural Analysis

A qualitative analysis of activity starts with objectively logical analysis. It may be reduced to providing a short verbal description of job performance, analysis of related production operations or tasks, and determination of their logical organization in space and time. Furthermore, a short description of technological processes is provided including a description of major equipment, tools, raw materials, and the sequence of basic technological procedures. Conditions of work such as temperatures, noise, illumination, and the potential for extreme situations should also be described. The relationship between the computerized and noncomputerized components of work and the calculated proportion of time in computer-based work should be developed.

Activity theory is distinctive in the attention it devotes to sociocultural aspects of activity and the role of external and internal tools of activity in task performance (Vygotsky, 1962). Culture is regarded as a mediator between the user and technology. It is an aggregation of beliefs, attitudes, values, social norms, and standards. Culture comprises shared social meaning. Individuals associated with a culture internalize this shared meaning. The social context under which a task is performed should be noted, including the social dynamics of the group involved in the job performance. This stage of analysis studies the relationship and coordination between several interdependent activities, including an understanding of the relationship between communities, culturally accepted norms of activity, and roles that are templates of activity developed in a community. All these questions may be addressed in detail.

Another aspect of qualitative activity analysis is individual–psychological method of study. Personal requirements for job performance such as aspects of the individual's personality, educational background, motivational aspects of work, needs and desires, and wishes should be considered at this stage of analysis. The background and training of the individuals in the use of computers and their subjective relationship to computerization are analyzed along with the relative job satisfaction derived therefrom.

In the design of activity one can outline two styles of performance, one that is standard and the other, individual. The first is a method of performance prescribed by instructions, the second depends upon individual features of personality (Bedny and Seglin, 1999a). As in manufacturing, the design of activity requires a "standard description of activity" to deal with the inherent variability of an individual's performance. In manufacturing, the actual size and shape of every part is unique and differs at the microscopic level from all others of the same kind. If the actual size of the part deviates from the nominal size and shape but does not exceed the range of tolerance, then this part is acceptable.

A standardized style of performance may be considered to be the psychological equivalent of the nominal size of the particular parts. Actual performance is subject to variation, similarly to the actual size of the parts. An individual style of performance is a kind of "central tendency" of actual performance. If the individual style of performance is sufficiently efficient and does not exceed tolerance as determined by an ergonomist, then the performance is acceptable. This problem is considered in more detail later.

During the next step of qualitative analysis, researchers perform a detailed task analysis. Specialists become involved in the detailed description of task performance. This stage of analysis can be performed based on a parametric or functional analysis of work performance. In simple cases, when general qualitative analysis of a task performance is discussed, scientists use parametrical methods of study that allow concentration on distinct aspects of activity. Objectively logical methods of analysis of a particular task and the cognitive task analysis, etc can be used at this stage. In some cases, qualitative methods of analysis suggest the use or development of symbolic process models. These models are distinctive insofar as each symbol refers to a unit of analysis in activity theory. Depending on the degree of detail, this analysis may be at either a micro-structural or macro-structural level (Bedny, 2000).

A more complicated qualitative analysis, based on systemic principles of study is functional analysis. The description of the structure of activity becomes important at this stage. This method of analysis is closely associated with cognitive analysis. Functional analysis regards activity a self-regulative system. Major units of analysis at this stage are function blocks. Experts as well as novices are involved in experimental study and observation. Changes in the structure of activity during skill acquisition are an important method of study at this stage. One important aspect of functional analysis is that at this stage experts use various complex qualitative methods that supplement each other.

During functional and general qualitative analyses of activity, widely used methods of comparative analysis, such as contrasts between effective and substandard performers, workers' error analysis, and definition of difficulties and obstacles, are used. The workers' strategies of performance may also be compared. Observation, experimentation, verbal protocols, etc. are combined in such studies. Methods for changing strategies of performance during the acquisition of skills and experience are fundamental to functional analysis. Individual styles of performance, derived from personal features, are also relevant (Bedny and Seglin, 1999). Error analysis and variability of time of task performance, etc. are some of the essential components of this approach. We consider functional analysis further in Chapter 5 and Chapter 6.

One should also distinguish between macro-structural and microstructural analyses which determine the level of analysis of activity. The macrostructural level of analysis includes larger units of analysis, where as the microstructural level is more detailed. These levels can be used at different stages of analysis. An example of microstructural analysis is given in Section 1.4.3. *Microstructure of actions*. In general, microstructural analysis is associated with the description of short cognitive actions and the decomposition of motor and cognitive actions into smaller units. Finally, task analysis is closely connected to task description. True task analysis always includes task description. At the qualitative stage of analysis the language of description is not sufficiently precise. This stage of description is then transferred into a more precise description at the later stages of analysis.

2.2.3 Application of Principle of Unity Cognition and Behavior in Practice

Before we start to consider this principle we present a brief, general description of cognition as done in activity theory. In activity theory there are three levels of information processing: sensory–perceptual, imaginative, and verbal–logical (Lomov, 1982). The sensory–perceptual level forms images of objects directly influencing the sense organ, resulting in sensory–perceptual images during perception ("primary" images). On the second imaginative level, the image emerges without any direct influence on the sense organ ("secondary" images). This results from an imaginative memory and processes that enable one to combine past images. Although this level is less precise and focused than the sensory–perceptual images, it has some advantages. Essential features of different objects in the same category can be extracted from multiple perceptions of those objects. Accidental features are sifted, leaving the more important features in the memory, and the images are reconstructed. At this level the ability to operate with images evolves into the ability to undertake the imaginative actions. Secondary images can be divided into reproductive (*predstavlenie*) and productive or creative images (*voobrazhenie*). Creative images are the result of the imaginative process that includes thinking. Here we also use the concept of "imaginative thinking." During the transition from perception to imagining the structure of the image changes. Some features are strengthened whereas others begin to weaken and fade away. The specific transformation of a perceived image into a reproductive one (pzedstavlenie) depends on the characteristics of the tasks. For example, when a subject was asked to compare similar objects in his imagination process differences between the objects emerged. The conscious part of the image changes during task performance. When the requirement was to compare different objects, similarity in features was emphasized (Lomov and Surkov, 1980). The image can be compared to an iceberg — at any given moment only the tip is visible above the surface of the water (Zavalova et al., 1986). Images are critically important in developing dynamic models of a situation.

The third level of information processing is verbal–logical thinking. At this level the individual learns to operate with concepts or propositions using signs and other symbols. Efficiency in verbal–logical thinking depends on the ability of the subject to manipulate concepts and propositions.

These three levels of processing of information are closely interconnected and are continuously transformed into each other. All the three levels of processing of information involve conscious and unconscious processes. We can reflect on the past, present, and future using different levels of processing of information. For example, the anticipation on the sensory–perceptual level is limited by the actions that are being performed. On the secondary imaginative level there emerges the ability to anticipate potential actions. On the level of verbal–logical thinking, anticipation supplies the ability to plan activities as a whole entity.

The following are aspects of the application of the principle of unity of cognition and behavior: the study of motor components of activity in cognitive tasks; the study of cognitive components of activity in motor tasks; and the relationship of cognition and behavior in the training process. We consider the role of cognition in regulation of the motor components of activity in a different section of this book. Therefore, in this section we briefly consider the last two aspects (Bedny et al., 2001).

Motor components of activity are deeply implicated in the cognitive tasks. For example, the evaluation of ocular movement assumes importance in the study of attention and perception. The characteristics of the visual field influence the strategies of visual search. This has significance to the method of presentation of information on a computer screen. In one study it has been discovered that when the users scanned through web sites seeking particular information to identify the object of their interest, their attention shifts involuntarily to other areas (Nielsen, 1994) if they cannot find what they seek within 5–10 sec. Accordingly, the method of organizing information on a screen is decisive in human–computer interaction. Mit'kin's (1974) study is of relevance to this issue. Mit'kin used what is called an electro-oculographic technique to study how different forms of screens affect strategies of visual search. Information display on the screens was presented within diverse geometric figures — square, rectangle, triangle, circle, and ellipse. Each figure was presented to a subject 60 times. All the figures were black. The subjects were required to locate a white dot subtending a visual angle of 9° as quickly as possible. This white dot was small and difficult to discover. The task did not require any adaptation to darkness. The subjects had to perform the search utilizing their central, as opposed to their peripheral, vision. It was found that subjects used specific strategies of visual search. They usually started with an exploration of the visual contour of the figure. Based on this, subjects first determined the areas of restriction for visual search. Mit'kin discovered that for each figure specific saccadic movement was required. When a rectangle or square was used, horizontal and vertical saccadic movements dominated. When a circle or ellipse was presented, inclined or radial saccadic movements were dominant. Each figure had areas of preferred frequency of fixation (Figure 2.7). Vertical columns in the figure reflect the frequency of fixation. Black squares in Figure 2.7 reflect the frequent area of fixation within the figures. In general, the study demonstrates that the motor eye-movement is a fundamental component in perceptual and attention processes. The specific nature of this eye-movement depends on how the information is organized on the screen. Mit'kin's findings may be applied to the study of HCI.

So far, we have discussed examples related to perceptual tasks and the regulation of motor actions. Activity theory also includes analyses of motor activity during thinking processes, such as problem-solving and decision-making. Sokolov (1960) discovered that electrophysiological activation of the lip and throat muscles is heightened during problem-solving. This activation changed depending on the specificity of the problem, and in some cases the activation of the lip muscles was correlated with the EEG of the brain.

Zinchenko and Vergiles (1969) studied the process of visual thinking that is significant in solving visual problems. They discovered complicated micromotions of the eyes during visual problem solving. These micromotions are considered to be an externalization of the thinking process. While performing these micromotions the eye can be relatively stationary around some point. During this period the subject does not perceive external information but manipulates visual images and performs mental transformations of the situation. Kamishov (1968), in his study of a pilot's eye movements, discovered "blind fixations," when the eyes seem to be on the display panel, but the pilot does not perceive any information. Kamishov assigns these micromovements to periods of mental manipulation of flight images. Zinchenko and Vergiles (1969) called them "vicarious actions." Pushkin (1965) studied eye movement in the process of solving chess problems and described the strategies of eye movement during problem solving. Later Tikhomirov (1984) and Telegina (1975) did similar studies. These researchers discovered complicated "gnostic dynamics" directed to the analysis of various elements of the situation, mental transformation of the situation, and other aspects. Such "dynamics" include a system of thinking and exploratory motor actions directed to the extraction of the dynamic meaning of the situation. The difference between visual actions that are components of the thinking processes and visual perceptual actions is that in the thinking process a person perceives the function of an object and its functional relationship with other objects rather than its perceptual

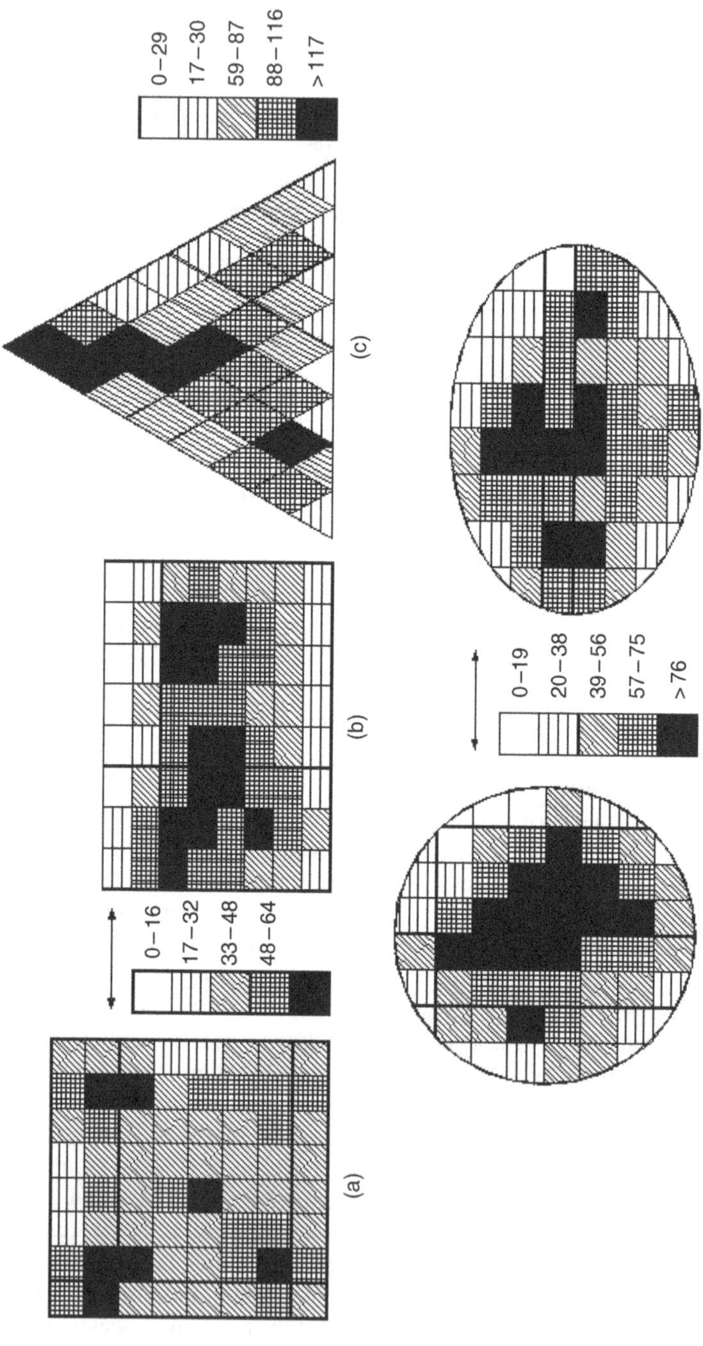

FIGURE 2.7 Pattern of eye fixation corresponding to different figures.

qualities such as color and form. When a chess player visually receives information about the interaction between the pieces, the actions involved are not perceptual but rather actions that are included in the thinking processes that occur while the person is discerning the functional relationship between the figures.

Recent research in neuropsychology also provides evidence of the union of motor and mental activity. The cerebellum and basal ganglia, which are known to sequence and time muscle movements in human beings, are also involved in the timing and sequencing of mental processes (Allen et al., 1997).

The technique of registration of macro and micromovements of the eyes during the solving of visual problems is widely used in activity theory. Usually this method is combined with observations, analysis of errors, thinking aloud, verbal protocol, changing the methods of problem solving, etc. For example, Mit'kin (1974), in his study of different versions of graphic technological control mnemonic scheme (GTCMS), used an electrooculographic recording method to register eye movements. He studied two versions of GTCMS during the solving of different mental problems and called them version A and version B. The subject was presented with problems of different levels of complexity. Registration of eye movements' was combined with observation, verbal protocol, and analysis of errors and performance time. This allowed scientists to separate perceptual visual actions from thinking visual actions. The study was performed with operators who had different skill levels in order to compare how experienced and inexperienced operators solved different problems during knowledge and skill acquisition. Particular attention was paid to the change in the strategies of task performance using eye movement registration. Figure 2.8 shows how eye fixation changed when the operator worked with two versions of GTCMS.

The study demonstrates that the number of eye fixations in version B was less than that in version A. There was a sharp increase in the number of eye fixations when operators shifted from task 4 to task 5 due to the complexity of the latter. The duration of eye fixation also changed depending on the task complexity. In this experiment registration of eye fixation was combined with the qualitative analysis of task performance. This experiment showed that version B of information presentation was better than version A and helped to develop a more efficient training method. As is seen, motor eye movements are involved in the thinking process. It was recognized in activity theory that motor movements are also important in the study of cognition.

The above discussion provides evidence of the close interconnection of motor actions with cognitive operations and actions. The movements and motor actions themselves have not only transformative functions but also cognitive functions and there is a complicated transition from the external material activity to internal cognitive activity. The interrelation between motor behavior and cognitive functions has also been demonstrated in the study of nonvisual (dynamic touch) perception (Turvey, 1996). This suggests that cognitive processes should not be studied separately from external behavior; rather these two types of activity are interconnected and form a unity. Cognition is not only a process but also an internal activity comprising discrete mental actions and operations. Both internal and external actions have a logical organization and form the structure of activity. This structure can change based on the mechanism of self-regulation and the strategies of activity connected to it.

The principle of the unity of cognition and behavior plays an important role in the development of training methods in activity theory. The mastering of different skills

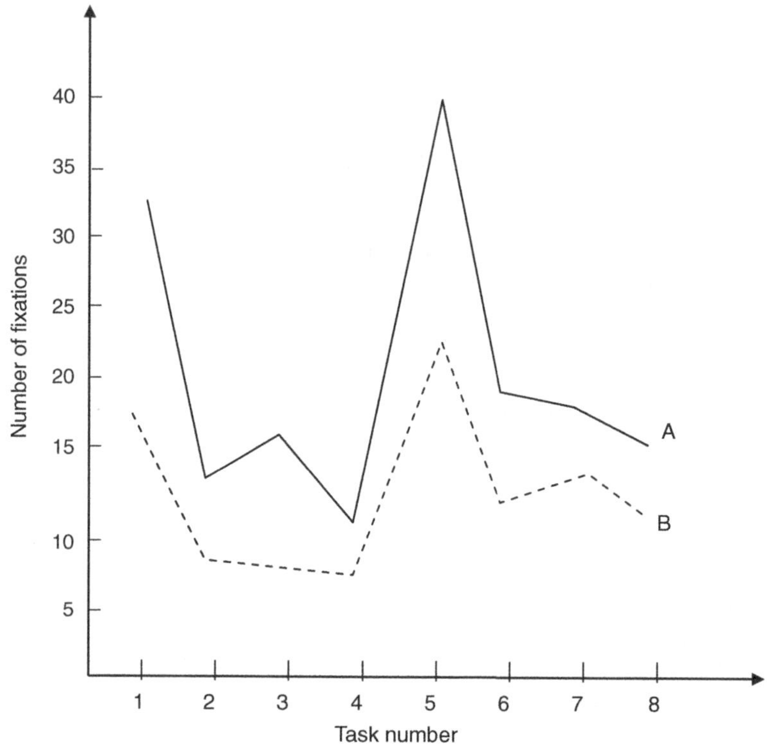

FIGURE 2.8 Frequency of eye fixations during performance of different versions of a task.

and knowledge is viewed as a process of forming cognitive and motor actions and organizing them into a structure. Action is the basic unit of learning activity. During different stages of the learning process the learner uses a different system of actions and the structure of activity gradually changes. Thus, learning can be considered a transformation of the strategies of performance. The more complicated the task, the more intermediate strategies used. And the developed strategies are evaluated and tested based on mechanisms of self-regulation (Bedny, 1981). Learning activity presupposes a dynamic orientation of the learner during the acquisition process, where object-practical actions are formed in close union with cognitive actions. Accordingly, the instructions in the learning process should be changed based on the various stages of skill and knowledge development.

In a study by Ponomarenko et al. (1993) this concept of learning was applied to pilot training. In order for the pilot to develop an image of the aircraft position in space, distinct indicative features (indices) or whole templates of different situations were presented to the pilot using pictures, slides, film, video, etc. Combinations of these different media played an important role in this experiment. Visual information was used along with acoustic information, vestibular information, and vibration. During the first stage the pilot learns distinct indicative features and then templates of the situation, while being involved in active manipulations. The pilots were given algorithmic prescriptions pointing out the actions to be performed. Verbalization and motor responses became important at this stage of learning. The pilots performed the required motor, verbal, perceptual, and other actions in accordance with set goals, while the information was presented and changed in response to the motor and verbal actions of the pilots. In the beginning it was observed that the pilots performed mental and motor actions in the most elaborate manner using a variety of actions to facilitate the formation of mental actions. Only gradually did the pilots transfer their activity onto a mental plane. Thus, activity progressed from a maximally externalized form to the abbreviated and internalized one. Some actions performed temporal, auxiliary functions and were not included in the final stage of the pilots' performance. This training method made it possible for the pilots to develop simultaneous perception of the situations and formation of the image of a flight. Similar training methods were used to develop the problem-solving strategies of the pilots. As a result of such training the precision of the pilot's performance increased by 15 to 20% and decision-making time decreased 2.5 times (Ponomarenko et al., 1993).

The same principles were used in the experiment during concept learning, when the major purpose of the training process is the formation of mental operations and actions. For example, during concept learning, teachers try to discover the essential features of the concepts being learned. Then the educator determines a logical organization of the actions that will be involved in identifying these features (conjunction and disjunction). In the following step the educator develops a human performance algorithm. Using this kind of human algorithm (identification algorithm) the student relates objects to a particular class or category. This algorithm is presented to students through written instructions. If the concept consists of several features organized in conjunction, all features must be discovered during the performance of the algorithm. Positive and negative instances of the concept are presented to the student. The student performs a search for positive indicative features of the concept in question based on algorithmic instructions and makes a conclusion on whether the object belongs to a particular concept or category. Only after a successful performance of different tasks externally will a student gradually start performing a similar task on the mental plane. After multiple performances of different tasks the conscious, deliberate, and consecutive identification process is transformed into a simultaneous and nonconscious mental process (Landa, 1976).

Training and teaching according to the self-regulation concept of learning (Bedny and Meister, 1997a) can be regarded as a sequence of particular stages. Very often, in the first stage of learning, mastering of mental skills and knowledge involves the use of different objects, schemes, and external signs. The learner's cognitive activity is, to a large degree, externalized. Only in the subsequent stages of learning can a student's actions be performed on mental internal plane. This is why motor and verbal actions are important in the training process. Mental activity is formed with the support of external activity and this relationship should be closely supervised at different stages of learning. The relationship between external and internal components of activity is considered a self-regulative process that provides formation of motor and mental actions based on their mutual comparison and corrections. Therefore, we do not consider internalization to be a simple process of transformation from the external to the internal plane. We will consider this problem in greater detail later.

The principle of unity, cognition, and behavior becomes critically important in an automatic system. At the same time there is a tendency of diminished interest in the study of motor components in human performance due to the increasing automatization and computerization of work that is accompanied by an increase in the mental components of human performance. This tendency is also cased by the methods of study of human work that derives from cognitive psychology. Cognitive components of activity are studied separately from motor components.

Here, we will demonstrate that motor components are important in highly automatic systems. We will also show that motor and cognitive components of activity should be studied in unity and this unity cannot be understood outside the study of temporal parameters of activity. The time factor always plays an important role in the study of work activity.

In automatic systems the operator functions as a monitor entering the control loop to override the automatic system in critical conditions. In this situation one can observe the transition from automatic to manual control. For example, the pilot begins to intervene in the control of the aircraft and performs emergency cognitive and motor actions. The more unexpected and unfamiliar the situation is, the more important it is to study behavior and cognition in their unity. It should also be taken into consideration that the less the operator is involved in the performance of emergency tasks, the more attention should be paid to the design of tasks. Special attention in these circumstances should be paid to motor components because automated control may be replaced by manual control.

It was discovered that automation has both positive and negative consequences. This can be explained by the fact that with manual control the operator has greater knowledge of the influences he exerts on the system, due to which he can easily predict changes in it. From this it follows that for quick recognition and correct interpretation of an unexpected situation motor actions are important. These motor actions both transform a situation and also perform cognitive functions. The motor actions and system responses preserve situation awareness. It was discovered that the more pilots engaged in manual rather than automatic flight, the more they were alert and hence more responsive to minor cues. Based on the advantages of manual control, Zavalova et al. (1986) introduced the principle of joint control by the pilot and the automatic system. This system combines the work of an automatic system with manual work and provides immediate transfer of control to the pilot. In an automatic control system, only those components are shut down that interfere with manual performance. Joint control of the flight decreases the pilot's workload, reduces fatigue, and permits urgent intervention by the pilot. Joint control is associated with the concept of the active operator (Zavalova et al., 1971). The essence of this principle lies in the fact that in automated systems the operator must be actively engaged in a goal-directed activity. The operator cannot simply act as a reserved component engaged in task performance only in the case of technological malfunctioning. Activity contains behavior and cognitive components that function in unity. Research based on activity theory shows that there is a complex relationship between these two types of activities. Without an analysis of this relationship it is not possible to understand cognition. Cognition is not only the regulator of external behavior but is itself formed based on motor activity. Cognition is "behavioral" in nature and is both a process and a system of mental actions. Such concepts as mental and motor actions, their interconnections and transitions, are central to activity theory. Internal and external activity is formed interdependently based on the process of self-regulation (Bedny et al., 2001).

2.2.4 Description and Classification of Motor Actions

Units of analysis of human activity and mind play a fundamental role in the development of psychology (Zinchenko, 1995). Formulation of normative requirements for the units of analysis in the study of human activity is also critically important for ergonomics. A scientific foundation for the design process requires a language of description that provides a fundamental basis for the creation of models to design any object. Standardized units of analysis are essential in creating a language of description for designing motor components of activity. We describe below units of analysis for the motor components of activity.

Actions performed by an individual through their skeletal–muscular system that can change the state of objects in the external world are called object-practical-actions or, more simply, motor actions or object-actions. These actions include distinct motions that, in activity theory, are called motor operations. Verbal actions are a particular class of motor actions that include motor components.

A motor operation (motion) is a relatively homogeneous act that lacks a conscious goal. Motor actions integrate a set of motor operations around a conscious goal. We can provide the following definition of standardized action and motion (Bedny, 1987). Standardized motor action is a complex of standardized motions (usually no less than two or three motions) performed by the human body, unified by a single goal and a constant set of objects and work tools. By standardized motion, or motor operation, we understand a single motion of body, legs, hand, wrist, and fingers that has a definite purpose in work processes and also corresponds to the rules of standardized description. One can clearly describe motor action only if one can define standardized motions embedded in motor actions. Motor actions performed by different parts of the body cannot be integrated into one action. For example, two motions simultaneously performed by left- and right-hand cannot be considered one motor action. Actions can be combined with supplemental motions. For example, actions performed by hand can be combined with supplementary motions of the body. Sometimes separate motions can be associated with conscious goals in the work process. On the one hand, they have only one homogeneous motor motion and thus should be regarded as a motion, where as on the other hand, they have a conscious goal and can also be regarded as a simple motor action. For the purpose of standardized task description we call them conscious motions. These simplest motion/action elements can be combined in sufficiently long sequences with one member of an algorithm that can be integrated with a higher-order goal. The member of the algorithm is discussed later. For descriptions of standardized operations or motions, we use the Methods-Time Measurement (MTM-1) system. For example, "move arm and grasp lever" is considered a standardized motor action that comprises two standardized motions, "move arm" and "grasp." MTM-I provides the most useable description of standardized motions. However, MTM-I ignores the concept of action. A specialist in MTM-1 system begins task analysis by dividing activity during task performance into discrete motions. In contrast, in AT motor components of a task are divided into actions and then each action is, in turn, divided into operations (motions).

Systemic-structural analysis of activity utilizes different principles for analysis of activity. When using MTM-I for activity theory, we do not need to build up a holistic activity out of the separate elements. Rather, holistic activity is the point of departure for analysis of the separate elements of activity into sequential decompositions of activity. If required for feedback, we can revisit the holistic activity, implying that analysis of design for performance has a recursive, loop structure (Bedny et al., 2001). We begin with a general analysis of task performance for which we must discover the goal of task strategies, mechanisms of self-regulation, and content of actions only after that do we employ the MTM-I system through which we can ultimately derive a holistic time-structure for the activity.

Another advantage of using the MTM-1 system is that it enables the determination of the duration of separate elements of activity. In social, philosophical, and scientific research time is an important criterion of the measurement of human labor (Marx, 1969). The factor of time can also be utilized for the measurement of human productivity, where labor productivity is defined as "output per units of time" (Barnes, 1980).

Analysis of MTM-1 system demonstrates that basic unites of analysis in this system are purposeful motions. A purposeful motion can be compared with the motor operation in activity theory. In contrast, Hick's law and Fitts' law describe a human operator as a reactive system that performs with maximum speed.

We can not agree with such psychologists as Frese and Zapf, (1994) who criticize Time and Motion Study approach for its behaviorist orientation. It is an important area of industrial engineering which offers very useful data for the study human work. SSAT utilizes MTM-1 system for the study of human performance.

One can differentiate between two kinds of units of analysis. One is called "typical elements of task" or "technological units"; the other is called "typical elements of activity" or "psychological units." At the first step of the analysis, one usually uses technological units. They can be transformed into psychological units at a later stage of the analysis. "Take part" and "press a button" are examples of motor actions that are described using typical elements of task or technological units. It is not a very precise description of actions. Depending on the conditions of performance, the same motor action can involve different motor activity elements. For example, distance to object and its precise direction can be changed during the performance of action "take part." This action can be performed automatically, or under precise conscious control. Different conditions from past experience require totally different elements of activity. Ignoring the differences between typical elements of task and activity can result in inadequate task analysis. The action "take part" can be more precisely described using a typical element of activity if one knows the distance, shape of the object, its position, etc. Suppose a subject performs this action under conscious careful control, such as where the distance of the movements is 30 cm and the object has an approximately cylindrical shape with a diameter of 15 mm, this action can be described using standardized description principles in the MTM-1 system. In the above example, action "take part" can be described as follows: "R30C+G1C1." According to MTM-1 this standardized description means "reach under careful control with distance 30 cm, and grasp approximately cylindrical object with diameter more than 12 mm." Following this description those who know the

MTM-1 system precisely understand this motor action. From this example, one can also see that motor action comprises two motions (operations) that are integrated into holistic units by the goal of action (grasp object). This future desired result is a conscious goal. However, how a person moves his arm, what the exact trajectory of the movement is, or how the subject moves his fingers is, in most cases, unconscious. As a first step, the specialist describes action using typical elements of task and then using typical elements of activity; in other words, technological units are transferred into psychological units of analysis. Typical elements of task and typical elements of activity as units of analysis are unfamiliar concepts for MTM-1 system. There is no also concept of cognitive and motor actions in MTM-1 system. Sometimes actions can be integrated by a higher hierarchical goal into some combination of actions. During algorithmic description of task, this combination of actions is called "member of algorithm." For example, a subject performs two actions "take part" simultaneously using right and left hands. Therefore, the left and right hands are included in the same motions (operations) "R30C+G1C1." Simultaneously performed actions ("take part") are the examples of members of algorithm. As example one can also consider action "reach and grasp mouth with right hand." This action is described in terms of typical element of task or technological units. However, one can transfer them into typical elements of activity or psychological units. The above action in the HCI study can be described as R30A+G1A. According to MTM-1, it means "move arm 30 centimeters and, with low concentration of attention, and grasp the easiest way." Hence, task can be described using a different hierarchically organized system of units. One starts task analysis from a description of the goal of the task, in other words, analysis of different strategies of performance. Then the task is divided into separate actions and their overall combination (members of algorithm of task performance). The activity specialist, in an approximate way at least, determines the logic or the sequence of the members of algorithm, or the overall actions. Each motor action is divided into motor operations (motions). The verbal description of the action can sometimes be translated into a symbolic one, using such techniques as decision-action diagrams. At this step, each symbol corresponds to particular units of analysis in activity theory. Symbolic models can describe a task with different levels of detail. All steps are not in a strictly sequential process. Owing to the transition from one step and level of analysis to the next, and the transition from technological units to psychological units of analysis and vice versa, more and more precise descriptions of performance are provided. Motor components of an activity are presented not merely as the sum of motor motions but as part of a logically organized system of motor actions. This aspect of design is considered in more detail in the algorithmic description of task performance.

Verbal action is another category of actions that performs not only communicative but also regulative functions within human activity. Speech can be both external and internal. The latter is sometimes manifested by the electrophysiological indices of the articulation muscles, even in the absence of audible speech. For example, it was discovered that increasing the complexity of a task increases the activity of the articulation muscles (Sokolov, 1960).

Verbal performance of a task differs from verbal communication or explanation. In verbal performance, speech clearly emerges as a system of verbal actions that correspond to the actually performed actions. Verbal actions may be the minimal verbal expression for the transmission of meaningful information aligned with the desired goal. If we segment verbal speech too discretely, the extracted segment loses all meaning and we fail to achieve the goal of expression. Segments of speech, used as verbal actions, should correspond to all requirements of actions. These requirements include such features as (a) the expression we produced intentionally implying that we were motivated to produce it; (b) prior awareness of what we wish to tell others, which implies that the expression is goal directed; (c) alteration of prior incorrect expressions that enables voluntary regulation of expression. At the same time, we do not realize how we produce separate verbal operations.

2.2.5 Description and Classification of Cognitive Actions

The development of units of analysis for the study of the human mind plays a critical role in the description of the structure of human cognition. AT distinguishes direct connection actions from transformational actions (Zarakovsky and Pavlov, 1987). Direct connection of mental action proceeds without distinct differentiated steps and requires less attention. It is less consciously directed and is subjectively experienced as instantaneous. For example, recognition of a familiar object may be seen as this kind of action. Because direct connection mental actions have a short duration, they are often called mental operations. Transformational mental actions involve more deliberate examination and analysis of stimulus as, for example, the perception of an unfamiliar object in a dimly lit environment. Mental actions may be classified based on the dominant action of the cognitive process and the ultimate purpose of the actions (Bedny and Meister, 1997):

1. The first group of actions are those of direct connections
   - 1.1 Sensory actions enable detection of signal from noise or require a decision about the signal at the threshold level. These actions enable us to detect objects from their background noise, obtain information of the distinct features of the objects such as color, shape, sound, etc.
   - 1.2 Simultaneous perceptual actions implicated in the identification of clearly distinguished stimuli that are well known to the operator and only call upon immediate recognition. Perceptual actions enable us to perceive whole qualities of objects or events — as, for example, the recognition of a familiar picture.
   - 1.3 Mnemonic (memory) actions involve memorization of units of information, recollection of names and events, and so on. A direct connection of mnemonic actions includes involuntary memorization without significant mental effort.
   - 1.4 Imaginative action as, for example, in mentally rotating the visual image of an object from one position to another according to a specific goal. This action involves manipulation of images based on perceptual processes and simple memory operations.
   - 1.5 Decision-making actions at a sensory–perceptual level that are involved in operating on sensory–perceptual data. For example, detection of a signal, comparative analysis of sensory or perceptual features of objects and their categorization, etc.
2. The next group of actions are transformational actions

2.1 Reproductive transformational actions

2.1.1 Successive perceptual actions are involved in the recognition of unfamiliar stimuli and the creation of a perceptual image of the object. They require more deliberate examination and analysis of stimuli. During skill acquisition, one may observe the transition from successive transformational action to simultaneous direct connection of perceptual actions.

2.1.2 Explorative-thinking actions are performed, based on sensory–perceptual information. They are involved in deliberate examination of different elements of tasks, extraction of subjectively significant elements, discovering specificity of their interaction, the interpretation of obtained information, and the creation of mental pictures of a situation.

Very often it is difficult to find the differences between successive perceptual and explorative-thinking actions. The major difference between them is as follows: The main purpose of successive perceptual actions is to develop a perceptual image of the object (percept), for example, categorization of objects based on their shape, color, size, etc. The purpose of explorative-thinking actions is to discover a functional relationship between the elements of a situation based on the presented sensory-perceptual data. Frequently, the functional property of objects can be discovered only through analysis of the relationship between different elements of a situation. Sometimes perceptual properties directly demonstrate the functions of the object. At the same time, the shape of an object and its function can deviate from each other. In the first case, the thinking process is almost entirely eliminated and we classify such actions as successive perceptual, while in the second case, the thinking process dominates and we classify them as explorative-thinking actions, based on sensory-perceptual data.

2.1.3 Decision-making actions at the verbal–thinking level, for example, if, after receiving some information, the operator must determine what steps are needed next, based on a logical analysis of the situation. These actions can be performed based on an algorithmic and heuristic level of regulation of the thinking processes.

2.1.4 Recording actions enabling an operator to transform one kind of information into another. For example, transformation of meaningful verbal expression from one language to another.

2.2 Following are higher-order transformational actions

2.2.1 Categorization actions that include information processing resulting in the division of some signals into a series of separate subjects or classes.

2.2.2 Deductive actions that refer to the application of general rules to draw novel conclusions from existing data. (Many mathematical tasks involve deductive actions)

2.2.3 Creative-imaginative actions empowering combinations of logical and intuitive operations on images.

2.2.4 Mnemonic actions entailing complex manipulation of information in working memory, extracting information from long-term memory, storage of requisite information, and maintaining information in working memory.

3. Creative actions

3.1 They include operations that generate new knowledge from old knowledge, either logically or intuitively. Creative actions are different from reproductive actions, which involve convergent thinking. Creative actions, on the other hand, involve divergent thinking.

In AT there are additional methods for classifying actions (Lomov, 1986; Zarakovsky and Pavlov, 1987; Bedny et al., 2000). The basic criteria for classification are both the nature of the object of actions and the method of their performance. This classification includes the following actions: (1) object-practical action that is performed with real objects; (2) object-mental action that is performed mentally with images of objects; (3) sign-practical action performed with real signs, such as receiving symbolic information from different devices, as well as its transformation; and (4) sign-mental action is performed mentally by manipulating symbols (Figure 2.9).

Cognitive psychologists have, for example, studied the chronometrics of the manipulation of mental images (Cooper and Shepard, 1973; Kosslyn, 1973). They found that the time of mental rotation of objects was similar to the time of actual external rotation. According to AT, if an individual intentionally turns the mental image of an object to a required position, it is an imaginative action or an object-mental action. We formulate analysis of this in terms of the individual's goal to rotate to a particular position.

Cognitive actions can also be described using technological (typical elements of task) or psychological (typical elements of activity) units. For example, taking a reading from a pointer or from digital displays are examples of perceptual actions that are described based on technological principles. Depending upon the distance of observation, illumination and constructive features of display, the content of mental operations and the time of action for performance can vary. At the same time, action "eye focus or fixation" (EF) in the MTM-1 system is a simultaneous perceptual action, or a simple decision-making action at the sensory–perceptual level. Actions such as "detection of signal," (performing decision "yes," "no") logical decision "if. . .then," etc. are examples of mental actions that are cognitive units of activity. The more standardized the conditions of the actions described according to technological principles, the more often they become similar to the standardized actions described

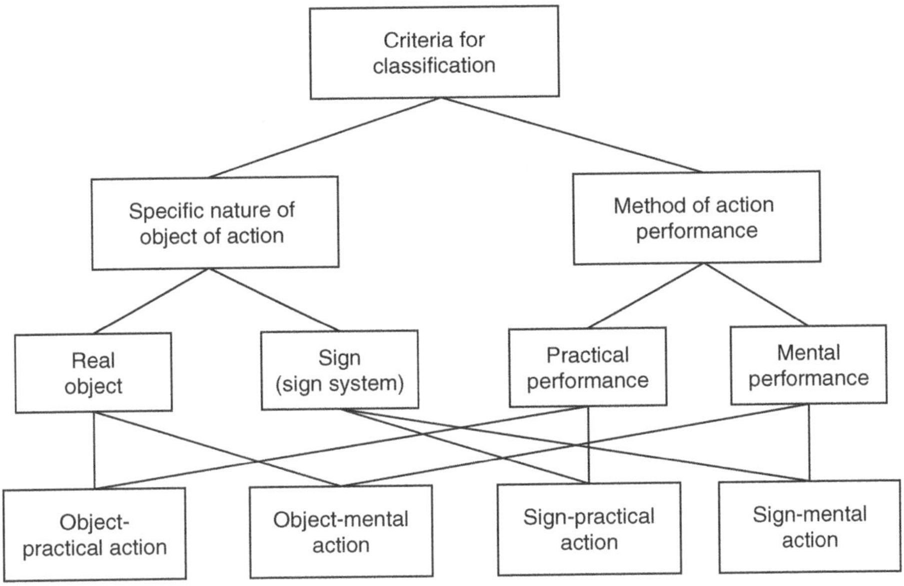

FIGURE 2.9 Classification of actions based on the nature of objects of actions and method of their performance.

according to psychological principles. It can be explained by the fact that the content of mental operations of these actions has also become similar.

The purpose of classifying actions is to present an activity as a structure with systemic organization. All actions are organized as a system because of the existence of the general goal of activity and of mechanisms of self-regulations (Bedny and Karwowski, 2004b).

Operative units of activity (OUA) are also important concepts in the study of activity. We understand OUA as contextually defined entities (image, concept, statement, comment, etc.) formed through training or experience that enables a subject to mentally manage semantically meaningful units at levels of specificity relevant to the execution of the indicated task (Zarkovosky et al., 1974; Zinchenko et al., 1974). Operative units of activity (not to be confused with units of analysis) are the symbolic entity that are used by cognitive actions during performance. Appropriate characterization of OUA provides a great deal of leverage in the analysis of cognitive actions.

2.3 ALGORITHMIC DESCRIPTION OF WORK ACTIVITY

2.3.1 Decomposition of Activity into Hierarchically Ordered Units during Morphological Analysis

Let us consider how one can decompose activity into hierarchically ordered units of analysis during morphological description of activity. A subtask of a production operation is used as an example. According to algorithmic analysis, a combination of actions that integrate higher hierarchical goals, we will, in our example, refer to

TABLE 2.2
Example of Decomposition of Activity during Task Performance

Members of algorithm	Actions	Operations (or motions)
Look at one bulb, then another	Simultaneous perceptual actions 1. Look at bulb #1 2. Look at bulb #2	1. The same (look at the bulb #1) 2. Move eyes to the second bulb 3. The same (look at the bulb #2
Decide which bin to choose	Decision-making actions at perceptual level 1. Decide to choose left bin or 2. Decide to choose right bin	1. The same 2. The same
Take part	Discrete motor actions 1. Move arm and grasp part	1. Move right arm to part (motion Reach-R) 2. Grasp part (motion Grasp-G)
Install part in air-operated clamping device	Discrete motor action 1. Move part to the clamping device and install it	1. Move part to air operated clamping device (motion Move-M) 2. Turn hand with part in required position (motion Turn-T) 3. Align part with clamps and insert part into fixed secure position (motion Position-P)

as "member of algorithm." Members of algorithm are important units of analysis that are used during algorithmic description of activity. We consider this to be a unit of analysis that is discussed in greater detail in Section 2.3.2. Let us consider an example of how we can decompose a subtask into discrete units. Suppose a worker performs the task in the following manner: if a red bulb and a white bulb are both lit, a worker should take a part from a bin on the right and install it into the air-operated clamping device. If both a green bulb and a white bulb are lit, then the worker should take the part from the bin on the left, and install it into the same clamping device. The method of decomposition of the subtask is presented in Table 2.2. Subtask includes four members of algorithm. In our example, a member of algorithm is composed of one to two closely interconnected actions.

From this table one can see that during the decomposition of the task we applied hierarchically organized units of analysis.

The second example what we consider below is a computer-based task. Prior to discussing this example, it is necessary to briefly consider the relationship between concepts of means of work and tools. In our discussion of the work process, we utilized the notion of means of work (see Figure 2.1 and Figure 2.2). The means of work is a broad concept that includes a variety of tools and equipment and cannot simply be used as a synonym for tools. In the systemic-structural theory of activity, the concept of tool is, from a psychological point of view, closely associated with the concept of action; outside a specific task, we cannot precisely determine what a particular tool is for. In this sense, the personal computer certainly cannot be classified simply as a tool. Rather, we must consider the computer first as a means of presentation or creation of different artificial tools for use by the subject during his performance of actions in a computer-based task. Moreover, the computer as a means not only presents subjects with tools for the performance of the required actions but also creates artificial objects toward which actions may be directed. In this sense, the computer is a special kind of means, mediating human interaction with the external world by creating artificial objects and tools required for the performance of computer-based tasks.

The concept of tool can be used differently depending on the function it serves in the analysis of human activity. In relation to the technological process, it is usually some kind of technical equipment. In the general analysis of the task, the term *tool* has a fixed meaning that is determined by the specificity of the task. In relation to mental and practical actions and operations, the term psychological tool or psychological instrument has a more dynamic meaning that changes depending on the functions served by actions in the structure of activity. The analysis of Vygotsky's concept of tool demonstrates that he used this concept in relation to mental actions and operations. More precisely, we can see that it is a tool for mental actions and operations. From this perspective, we will consider the concept of tool when we describe classification of actions according to the second principle described above (see Figure 2.9). We will also touch on the concept of object in relation to mental actions and operations.

Let us now consider a generalized example taken from the domain of human–computer interaction. The overall task activity involves a subject using basic word processing software running on a personal computer to produce some document. As text is entered, misspellings are underlined in red. At this particular stage of activity, subjects have to correct all the misspellings. The task is "Correct spelling." This is a deterministic task, requiring a well-defined sequence of actions. In order to correctly extract the required actions and develop their classification, we need to identify the object, tool, and goal of each action; the nature of an action is dependent on the interrelation between these components in any particular situation. In our analysis, we describe the following actions required during the performance of this task:

1. Reach and grasp the mouse with the right hand.
2. Move the cursor to the initial position preceding the misspelled word and depress the left mouse button with the index finger.
3. With the mouse button depressed, highlight the required word by dragging the cursor to the end of the word; release the mouse button.
4. Move the pointer to the spelling icon on the toolbar, depress the left mouse button with the index finger, then release.
5. Examine the list of options presented by the dialog box.
6. Decide on the most suitable spelling option.
7. Move the pointer to the desired spelling option, depress the left mouse button with the index finger, then release.
8. Move the pointer to the OK button, depress the left mouse button with the index finger, then release.

When a subject performs the first action, the mouse is the object engaged by the subject. The conscious goal of this action is to grasp the mouse, which is what the subject understands that he wishes to achieve as a result of taking action. As the mouse is a real, material object, this is classified as an object-practical action. In the second step, the mouse becomes a tool through which the subject implements the movement of an object, the cursor, to the start position. The transformation of the object (cursor) is performed according to the goal of the action, which is now to move the cursor to the required position. The pointer is a symbol on the screen. However, as the meaning of the sign is not especially important here, this is also regarded as an object-practical action.

When the subject performs the third step, the word to be highlighted becomes the object of the action; the subject transforms the quality of the object according to the goal of the action, and the background of the text becomes black, while the characters become white. It should be noted that at times the subject may not be aware of all the changes in the object. For example, sometimes the subject may not be aware that the characters have changed color. This illustrates that the subjectively formulated goal of action may not always coincide with the objectively required goal and result of action. This third action also includes several tools, such as the cursor, mouse, and the left mouse button. In the fourth action, the spelling icon is the object of the action and the mouse and pointer are the tools. Actions 3 and 4 are both object-practical actions.

In the fifth action, the list of options in the dialog box becomes the object. In executing this action, the subject does not employ any external tools. According to the type of tool use, this is classified as a sign-practical action in which the sign becomes the tool of internal, mental action; in terms of the dominant psychological process, this is a simultaneous perceptual action. In action 6, a particular item in the list of options is the object. This is also a sign-practical action; the dominant psychological process is decision making.

In action 7, the pointer, mouse, and button are the tools, and the selected spelling option chosen in the previous action is the object. In action 8, the OK button is the object; the mouse, mouse button, and pointer are the tools. Actions 7 and 8 are both object-practical actions. At times, it may be difficult to decide whether an action should be classified as object-practical or sign-practical. This is because the subject simultaneously manipulates different tools. When actions with the mouse are performed almost automatically, we can classify them as object-practical actions; when conscious manipulative effort involving meaning is required, the actions can be classified as sign-practical.

When attempting to solve HCI usability problems, a series of questions may be raised on the basis of this kind of analysis. For example, during the performance of the first action in our example, the usability of the mouse, its "graspability," and "clickability" are the issues of concern. In the second action, the ease and precision of directing the pointer are highlighted. Similarly, perceptual actions, such as action 5 and simple decision-making actions (6), present issues connected with the ease of performance of such mental actions. In this regard, systemic-structural activity theory supports the quantitative evaluation of the complexity of performance. For example, it is possible to estimate the time involved in decision-making processes during a task and define it as a ratio of the overall time taken for task performance. This approach gives a basis for considering methods of reducing task time and complexity.

This hypothetical example illustrates one stage in the morphological analysis of activity and is primarily intended to provide a simple demonstration of one way in which the activity-theoretical concepts of object, action, goals, etc. may be applied for detailed studies in HCI and other design-oriented research. When undertaking a morphological activity analysis, the activity under study is initially formulated in terms of tasks. Next, the structure of task performance is described. As in our example, this involves determining the content of tasks through the delineation of the actions implicated in task performance. Along with this kind of analysis, we may also need to precisely describe the actions in terms of typical elements of a task (technological units), or in terms of typical elements of an activity (psychological units). Additionally, the type or level of attention required for each action may be studied. However, further discussion of these aspects of action classification is outside the scope of this book.

The morphological analysis of activity itself constitutes only one aspect of a four-stage, multilevel, methodology for the systemic-structural analysis and design of work activity (Bedny and Meister, 1997; Bedny et al., 2000; Bedny et al., 2001). The systemic-structural approach offers an integrated framework for the iterative description and analysis in both qualitative and quantitative terms, supporting the stepwise development and testing of the models of human activity that are used as a basis for the design of equipment and work processes. Both the morphological and functional aspects of activity are studied and described from multiple perspectives and at varying levels of decomposition.

2.3.2 Basic Principles of Algorithmic Analysis

Algorithmic analysis of activity is a particularly powerful method of morphological approach. It consists of the subdivision of activity into qualitatively distinct psychological units and determination of the logic of their organization and sequence. These moments are formulated as elements of activity with a specific logical structure. Typically such elements, called members of algorithm, are made of actions with their associated subgoals, integrated through supervening goals. A member of algorithm consists of closely interdependent homogeneous actions (only motor, only perceptual, or only decision-making actions) that are integrated by a particular goal into a holistic system. Subjectively, a member of algorithm is perceived by a subject as a component of activity having logical completeness. Filing of logical completeness of actions, which are included in one member of algorithm, is usually associated with the subgoal of the task that integrates several actions. Sometimes sensing the logical completeness of homogeneous actions can be determined within the capacity of short-term memory. When actions are performed simultaneously or require keeping their order in working memory, due to limits on the capacity of working memory, each member of algorithm is limited to between one and four homogeneous actions. If these actions are performed in sequence, and their order need not be kept in working memory, then the quantity of these actions in one member of algorithm can be increased. While motor actions can be performed simultaneously, mental actions are usually performed sequentially. Subjects may also combine motor and cognitive actions according to rules described later. As units of activity, the members of algorithms are termed "operators" and "logical conditions." Operators consist of actions that transform objects, energy, and information. For example, we can describe operators implicated in receiving information, analysis of a situation and its comprehension, shifting of gears, levers, etc. Logical conditions are members of algorithm that determine the logic of selection and realization of different members of algorithm, and include decision-making process. What are called human algorithms are just such algorithms defined by associated units of analysis made up of human actions. Such algorithms represent the logic and sequence of actions and enable the development of instructions to guide enactment of that action, logic, and sequence. Actions as units of analysis constitute the distinctive features of human algorithm, such as those with flow charts widely used to represent human performance. There is also a functional analysis of activity. Analysis of the structure of activity is performed according to different models of self-regulation of activity. Here we outline micro and macro levels. In the first situation, one can use microblocks and in the second situation macroblocks as the units of analysis. We considered earlier the first method very briefly. We will consider the second method later during the functional analysis of activity. The material presented above allows us to outline the following hierarchically organized units of analysis of activity (Figure 2.10).

Sometimes, algorithmic description of activity may be represented as an iterative process sequentially approaching an optimal method of description (Bedny, 1987). It can be explained by the fact that in some cases members of algorithm can be divided into several smaller ones or be integrated together. The more complicated the task, the

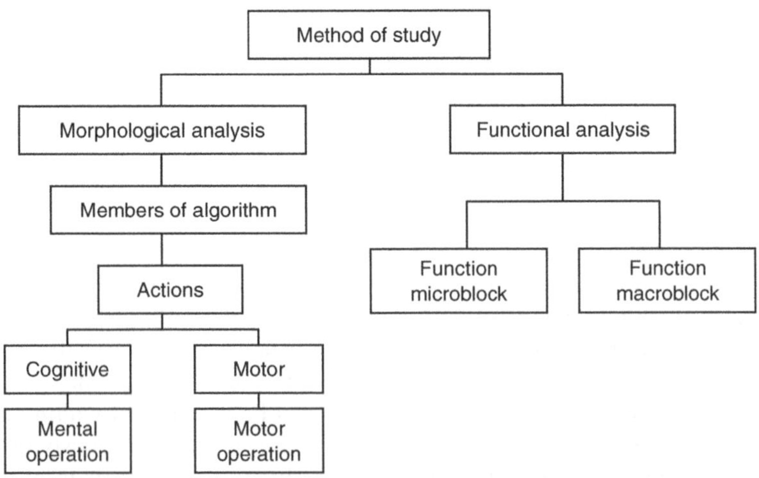

FIGURE 2.10 Major units of analysis in systemic-structural theory of activity.

smaller the members of algorithm. For example, Newtson (1976) presented subjects with a film depicting various tasks. They pressed a button to mark the beginning and the end of each separate element of the task. It was found that in a more complex task, subjects separated it into smaller units. Subjectively, members of algorithm are perceived by a performer as a holistic element of task. At the same time, qualitatively different components of task should be related to different members of algorithm. For example, cognitive and motor actions can never be integrated into one member of algorithm. In the same way, perceptual and decision-making actions should belong to different members of an algorithm.

Sometimes similar members of algorithm follow each other. For example, several members of algorithm related only to motor activity, or perceptual activity can follow each other. In this case, experts can use different approximate rules to extract different members of algorithms.

If the sequence of the performed actions can be kept in working memory, then the number of actions in one member of the algorithm should be no more than three to four.

If actions are simple and performed sequentially, and their order does not need to be kept in working memory, then their integration into separate members of algorithm is determined by logical completeness of parts of the activity. Such actions are integrated by higher-order goals and have a limited number of interdependent work tools and objects. The limited capacity of short-term memory can also influence strategies of the grouping of these actions.

Each member of algorithm is designated by special symbols. For example, operators can be designated by the symbol "*O*" and logical conditions by the symbol "*l*." All operators involved in the reception of information are categorized as afferent operators and are designated with superscripts α, as in "O^{α}." If an operator is involved in extracting information from long-term memory, the symbol μ is used as in O^{μ}. The symbol $O^{\mu w}$ is associated with keeping information in the working memory, and the symbol O^{ε} is associated with the executive components of activity, such as the movement of a gear. Operators with the symbol O^{ε} are designated as efferent operators. From this description of rules, one can see that, for example, O^{ε} cannot include any cognitive actions. In the same way "O^{α}" can include only perceptual actions and operator O^{μ} can include only mnemonic actions. The superscript of a member of algorithm determines what kind of actions can be included in this particular member of algorithm. There are two types of algorithms, deterministic and probabilistic. In deterministic algorithms, the logical conditions designated with "*l*" have either of two values, zero or one. The symbol "*l*" for a logical condition must include an associated arrow with a number on top that corresponds to the number of logical conditions. For example, logical condition l_1 is associated with a number on top of the arrow $\uparrow^1$. An arrow with the same number but a reversed position must be presented in front of another member of algorithm to which the arrow makes reference, $\downarrow^1$. Thus the syntax of a system is based on a semantic denotation of a system of errors and superscripted numbers. An upward pointing of the logical state of simple logical conditions, "*l*" when, "*l*" = 1, requires skipping the members of algorithm until the next appearance of the superscripted number with a downward arrow (e.g., $\uparrow^1\downarrow^1$). So the operation with the downward arrow with the same superscripted number in front of it is the next to be executed.

TABLE 2.3
Description of the Algorithm in Tabular Form

Members of algorithm	Description of members of algorithm
O_1^α	Take reading from display
l_1	If pointer demonstrates more than 100 go to O_1^ε. If less than 100 go to O_2^ε
O_1^ε	Press red button
O_2^ε	Press green button

Let us consider a simple example, in which the worker should check a corresponding display. If the pointer rises over 100, the worker must press a red button. If the pointer indicates less than 100, he must press a green button. A verbal and symbolic description of this algorithm is presented in Table 2.3.

The algorithm logical formula is as follows:

$$O_1^\alpha \, l_1 \uparrow^1 O_1^\varepsilon \downarrow^1 O_2^\varepsilon.$$

In this formula, the logical condition has only two states, 0 or 1. If logical condition equals 0, then the next member of algorithm is activated. If logical condition equals 1, that it requires skipping the following members of the algorithm, until the next appearance of the superscripted number with a downward arrow. The algorithmic formula is read from left to right. In some cases, logical conditions can be a combination of simpler ones. These simple logical conditions are connected through "and," "or," "if-then," etc. rules. Complex logical conditions are designated by a capital "*L*," while simple logical conditions are designated by a small "*l*." Logical connections between the simple ones are designated with standard symbols such as "&," g"∧," "→," etc. For example, complicated logical conditions, comprising simple ones, may be designated as $L_1(l_1^1\& \, l_1^2\& \, l_1^3)$. Symbol 1 as a subscript of *L* designates that it is the first complex logical condition. Symbol 1 as a subscript of "*l*" designates that it is a simple logical condition that belongs to L_1. The numbers 1 to 3 used as superscripts designate the number of logical conditions. Complicated logical conditions are particularly important in diagnostic tasks. In the above example complicated logical condition L_1 comprises 3 simple ones that are combined with a logical conjunction *and*. This complicated logical condition can be used to determine whether particular phenomena belong to a certain category, particularly when the attributes of the phenomena are connected through the conjunctions. Three simple logical conditions must give responses to the following questions: is feature 1 present? Is feature 2 present? Is feature 3 present? Only when all the three questions receive the response "YES," can one conclude that a phenomenon belongs to a particular category. In contrast, if, in our example, simple logical conditions are combined with logical disjunction *or*, it will be sufficient when any one simple logical condition has the required attribute. Sometimes a complicated logical condition includes different logical connections. The more complicated an aspect of algorithmic description, the more likely the situation in which homogeneous members of algorithm follow one after the another. For example, there are several afferent members of algorithm such as O_1^α,O_2^α, and O_3^α or several efferent members of algorithm such as O_1^ε, O_2^ε, and O_3^ε. Dividing them into separate members of algorithms should be performed according to the above described qualitative analysis.

A significant part of an operator's work is dedicated to waiting when they are not directly involved in task performance. These intervals of time can be encountered within or between specific tasks. For example, an operator performed one task and waited before starting to perform the second one. In spite of the absence of external observed behavior, this time requires concentration of attention. The operator was ready to become immediately involved in the next situation. Readiness to be immediately involved in performance, without an externally observed activity, is a difficult part of any operator's duties. This period of the operator's work is called the active waiting period. In an algorithmic description of the operator's performance, this period is designated by the symbol $O^{\alpha w}$.

In a probabilistic algorithm, logical conditions may have two or more outputs with a probability between zero and one (Bedny and Karwowski, 2003). As a simple example, these logical conditions may be represented in the following way. Suppose we have logical conditions with three outputs with distinct probabilities of occurrence. In such a case logical condition can be designated as $L_1 \uparrow^{1(1-3)}$ which possesses not two but three potential values. In this case we have three versions of output. $\uparrow^{1(1)}$, $\uparrow^{1(2)}$, $\uparrow^{1(3)}$. Each output, respectively, has its own probability. For example, the first output can have the probability 0.2, the second 0.3, and the third 0.5. Knowledge of the probability of the output may be taken into consideration for studying variables such as the probability of the performance of different members of algorithms, strategies of performance, calculation of performance time of the algorithm or components of the algorithm, analysis of errors, and evaluation of task complexity. Frequently, in algorithmic description we can use the always-false logical condition, defined by the symbol ω. This logical condition is introduced only to make it easier to write the algorithm. It does not designate real operations performed by the subject. It always defaults to the next member of algorithm, as indicated by the arrow included in the specification of this always-false logical condition.

An arrow designates the logic of transition from one member of algorithm to another. Thus an algorithm exhibits all the possible actions and their logical organization and, therefore, constitutes a precise description of human performance. It describes activity of a subject in terms of actions through which the subject attains the goal of activity. An algorithm can be presented in a tabular form or as a formula. However, we recommend combining the tabular form of an algorithm with the description of the algorithm as a formula. This significantly simplifies the algorithmic description of the task. This combination is carried out in the following way. On the left-hand side of the table, there is a column where the symbols are placed. It is a symbolic description of the algorithm or its formula. On the right-hand side, there is a verbal description of the members of algorithm. The symbols "*l* " or "*L*" for logical conditions in the left column include an associated arrow, with a number on top. An arrow with the same number, but in a reversed position, should be presented in front of another member of algorithm to which the arrow makes reference, as described above. In a probabilistic algorithm, we need to skip the next appearance of the superscripted numbers for all possible arrows of the logical condition. Each superscripted number is associated with a discrete probability that needs to be represented as a transition process from one member of the algorithm to another. In order to summarize this data, we present major symbols and their descriptions in Figure 2.11.

In the first stage the specialist presents a preliminary verbal description of the task. The verbal description is then transformed into an algorithmic description. Sometimes it is a sufficiently complex process. In the first step, the specialist presents the algorithm in a tabular form. A table that describes an algorithm is read from top to bottom. Progressive reading of the table from top to bottom and careful analysis of each member of algorithm, with information presented to the operator, physical characteristics of equipment, and analyses of relationship between members of algorithm, enable one to understand the structure of activity during task performance. Reading the algorithm also allows a specialist to understand the logic of transition from one action, or several actions of the algorithm, to others. The left column, with symbolic description, is called the formula of algorithm that is presented vertically. A separate formula can be set as horizontal lines only for the description of a short simple algorithm, or for the presentation of algorithm in an abbreviated manner. The right column of the table contains a verbal description of the algorithm. Sometimes the tabular form of algorithmic description can be combined with the symbolic description of algorithm. For this purpose, a specialist can utilize only standardized symbols for the description of the task. Then a tabular form of algorithm can be transferred into a symbolic description or a symbolic model. This model is particularly useful when probabilistic characteristics of the task have to be described. Symbolic model is also useful when one calculates the reliability of task performances or attempts to use analytic procedures for the analysis of the layout of indicators and controls (Bedny and Meister, 1997). As an illustration, let us consider a probabilistic and reliability analysis of a hypothetical task.

An operator performs the task of keeping technological parameters in the required range. The operator looks at the display and checks if the pressure is preserved and does not exceed particular restrictions. If the pressure moves out of the range between 100 and 60, he presses the red button. If after pressing the red button, the pressure does not return to the normal range, the operator calls the supervisor. This event is considered a failure. The algorithmic description of the task in tabular form is presented in Table 2.4.

Based on the tabular method of algorithm description, the specialist develops a symbolic description. To designate different members of an algorithm we utilize the symbols shown in Figure 2.12.

In order to illustrate how these symbols may be used, see the symbolic algorithmic description of the task shown in Figure 2.13.

Logical organization of symbols corresponds to the tabular algorithmic description of the task. A combination of these two descriptions can help a specialist to describe more precisely an algorithm of performance. This model also demonstrates probabilistic transitions from one member of algorithm to another. These data can be obtained based on either experimental studies or expert analysis. At the next step of analysis in this study, a symbolic model is used for reliability analysis (quantitative stage of analysis). Performance of O_6^{α} considered a failure. Because the probability

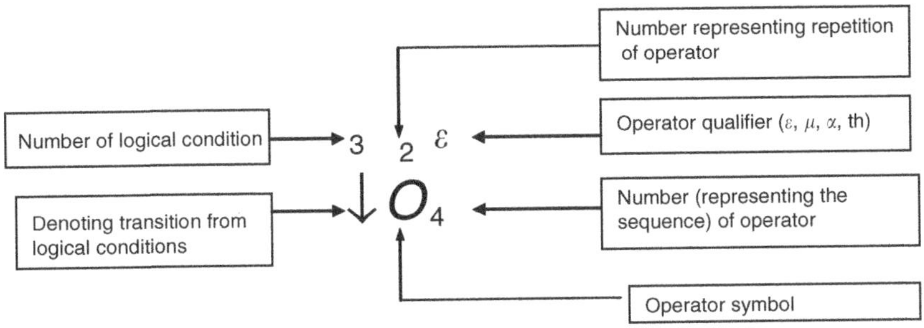

(a) Operator

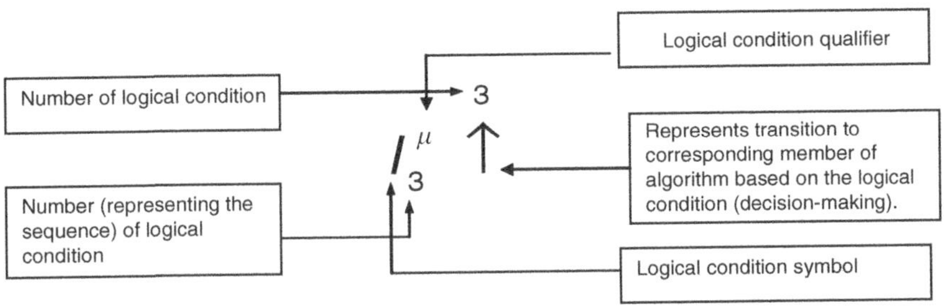

(b) Logical condition

Symbol	Denotes	Description
O	Operator	Consists of actions that transform objects energy and information
l, *L*	Logical condition	Determine the logic of selection and realization of different members of algorithm and include a decision making process. Can be designated as *l* or *L* *(based on combination of several logical conditions)*
α	Perceptual operator qualifier	Involvement of receipt of information for example O^{α}
μ	Memory involvement qualifier	Involvement of memory actions O^{α}
ε	Executive operator qualifier	Involvement of executive action in terms of motor performance O^{ε}
th	Thinking qualifier for different members of algorithm	Involvement of thinking action O^{th}

(c) Description of symbols

FIGURE 2.11 Explanation of major symbols consisting of a member of an algorithm.

TABLE 2.4
Algorithmic Description of Task

Members of algorithm	Description of members of algorithm
$\overset{2(1)}{\downarrow}\ O_1^{\alpha}$	Take a reading from display
$l_1 \overset{1}{\uparrow}$	If pointer demonstrates pressure between 100 and 60 perform $O_2^{\alpha w}$. If pointer demonstrates pressure more than 100 or less than 60 perform O_3^{ε}
$\overset{3}{\downarrow}\ \overset{2(2)}{\downarrow}\ O_2^{\alpha w}$	Wait up to 30 min
$l_2 \overset{2(1-2)}{\uparrow}$	If half an hour passed then go to O_1^{α}, if not return to $O_2^{\alpha w}$
$\overset{1}{\downarrow}\ O_3^{\varepsilon}$	Press red button
O_4^{α}	Take a reading from display
$l_3 \overset{3}{\uparrow}$	If pointer demonstrates pressure between 100 and 60 to perform $O_2^{\alpha w}$. If after performance O_3^{ε} the pointer again demonstrates pressure more than 100 or less than 60 call the supervisor
O_6^{ε}	Call the supervisor (this event is considered a failure)

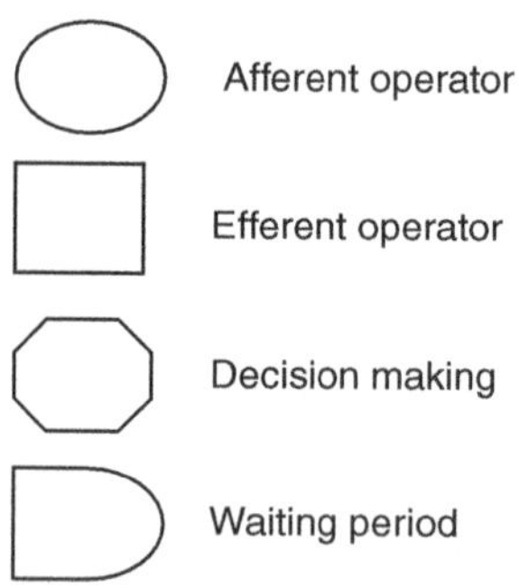

FIGURE 2.12 Basic symbols.

of transition from l_1 to O_3^{ε} is 0.2 and the probability of transition from l_3 to O_6^{ε} is 0.5, the probability of performance of O_6^{ε} is $P_1 = 0.2 \times 0.5 = 0.1$. This is the probability of failure according to the requirements of task performance. In the same way, one can, for example, calculate the probability of appearance (probability of reparative

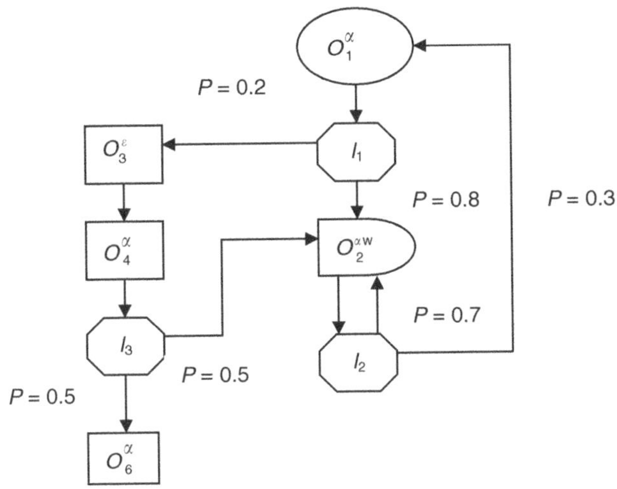

FIGURE 2.13 Symbolic algorithmic description of task.

performance) of O_1^α. This probability can be calculated in the following way:

$$P_2 = (0.8 \times 0.3) + (0.8 \times 0.7 \times 0.3) = 0.41.$$

This suggested method can be used not only for reliability analysis but also for other design purposes in general (Bedny and Meister, 1997). From this example, one can see that in systemic-structural activity theory isolated techniques or procedures are never used, Logically organized steps, procedures, and stages of analyses are used.

In some cases algorithms have such large realizations that experts extract only the critical ones to serve as markers for analysis. An algorithm enables an expert to describe human performance in a probabilistic manner and uncovers the constraints of the work process. As may be seen, algorithmic analysis of human performance can be an important method for the morphological description of activity wherein the units of analysis are human actions.

Following the development of the algorithm, experts perform psychological analysis of the algorithm, returning to a qualitative stage of analysis. Each member of algorithm can be evaluated as a whole, from both the qualitative and quantitative points of view. Members of algorithm and concomitant actions may be studied as functional subsystems of activity, making them a self-regulating organization directed toward achieving particular subgoals of activity. Qualitatively, each member of the algorithm may be approached in terms of either cognitive psychology or functional analysis of activity.

From a morphological perspective, each member of algorithm can be described at a more detailed level, in terms of cognitive and motor actions and operations performed by humans. Actions, in turn, may be described in terms of "typical elements of the task" (technological units) and typical elements of activity (psychological units).

In cases where a specialist uses only two stages of analysis (qualitative and algorithmic), he usually describes activity in terms of "typical elements of task" or "technological units." This approach simplifies analytic procedures. The use of typical elements of activity (psychological units) during the study of HCI should be carried out when researchers need a very detailed description of the structure of activity. Usually, these units of analysis are indicated in time-limited conditions or settings with a low tolerance for errors.

The relationship between qualitative and algorithmic analysis of activity is not strictly linear. It is possible to transfer from qualitative analysis to algorithmic and also in the reverse direction. This relationship, between stages of analysis, demonstrates principles of systemic-structural analysis of activity. When the resources of the qualitative stages of analysis are exhausted, the researcher switches to algorithmic analysis. Thereafter, a qualitative analysis of the human algorithm can be performed. This allows for the correction of algorithmic description. Hence, design is iterative in nature.

If necessary, more detailed stages of analysis may be pursued. A designer may even proceed to a third stage of systemic-structural analysis composed of a description of the time structure of activity using psychological units of analysis.Thereafter, the complexity of task performance can be evaluated. These two stages are not considered in this chapter. In Chapter 3, we will discuss computer-based tasks that is important in the study of HCI.

2.3.3 Algorithmic Analysis of Computer-Based Tasks

2.3.3.1 Qualitative Analysis

As the object of study we select an inventory receiving task. This is a computerized task that comprises a number of mental actions, including a high level of variability, and presents difficulties in observation and formal description. The first stage of study starts with qualitative analysis. In this particular case, it begins with objective logical analysis. It consists of a sequence of steps intimately related to algorithmic analysis. Each sequential step of the qualitative analysis is carried out in greater detail and for a distinct purpose. The first step is restricted to an analysis of what is currently being done, which in computer tasks is frequently quite vague or variable. The major emphasis is on identifying the content of the task under investigation and its relationship to other tasks. Discussion with workers or supervisors, observation, review of documents, available data compared with literature on similar work, etc. can be used at this stage. This stage enables the researcher to obtain a general understanding of the technological processes and methods of work. The result of such an analysis provides a model of inventory process for a manufacturing firm (Figure 2.14).

The obtained data form a platform for a more specific analysis of the inventory-receiving task, henceforth to be known as receiving, which constitutes the major focus of the present study. It may be seen from Figure 2.14 that the inventory process for any company consists of three subsystems: (1) Stocking; (2) Record keeping; (3) Work-in-Process (WIP).

The first subsystem (box 1) refers to the physical movement of items into and out of stock, providing a physical quantity on- hand. Raw materials, intermediate products, or finished goods are physically brought in or taken out of stock. What

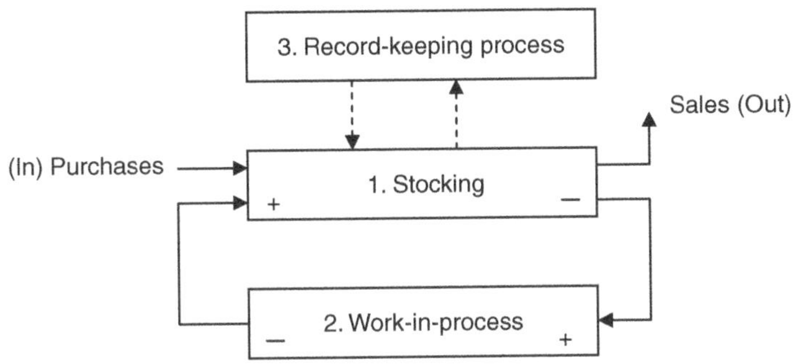

FIGURE 2.14 A model of inventory process for a manufacturing firm before improvement. Route 1 goes from (in) purchase, to stocking, then to sale (out), or WIP.

remains is the actual on-hand quantity — an "In" is an increase in stock and an "Out" is a decrease. An "In" occurs when some material is entered; an "Out" occurs when something leaves. Stock can be increased by either purchasing or returning items from manufacturing to stock. Stock is decreased by sale of products or component parts to customers, by putting intermediate products into manufacturing, or by scrapping. When purchases are brought into storage, stock increases. This is designated by a "plus" sign; a "minus" sign represents a reduction in stock.

The second subsystem (box 2) represents WIP. This is a value adding, manufacturing process in which diverse raw materials or intermediate products are transformed into a ready product. Movement "In" and "Out" is designated the same as in the stock process.

Whenever material moves physically into or out of stock, that movement is mirrored as a transaction in the record-keeping process, which is the third component of the process model. A properly designed inventory process is capable of producing a match between the physical events that occur in box 1 and box 2. The record keeping process is a complicated computerized system that has to track all physical movements of different parts, purchases, intermediate products, etc. The model of the inventory process depicted in Figure 2.12 facilitates an understanding of the specifics of different tasks involved in this process.

The first task is called Inventory-receiving. Four workers responsible for registration of all purchases and movement of intermediate and final product perform this task. The task includes two parts of the job performance. One part involves physical work, that is, when a worker (later receiver) receives a box with raw material or intermediate parts. The receiver can perform two similar tasks. One task entails reception of parts from different vendors to restock the warehouse and fulfill special and emergency orders. The second task entails receiving intermediate or finished product from WIP. The present study pertains to the first task.

Parts arrive at a plant in special boxes that are delivered to the reception area. Figure 2.15 depicts a view from above of the workplace for the receiving task.

The dashed lines designate equipment introduced following the improvement of this task, which will not be discussed at this point. Number 1 represents the receiver, who opens the boxes placed on base unit 5. For this purpose, he uses a special

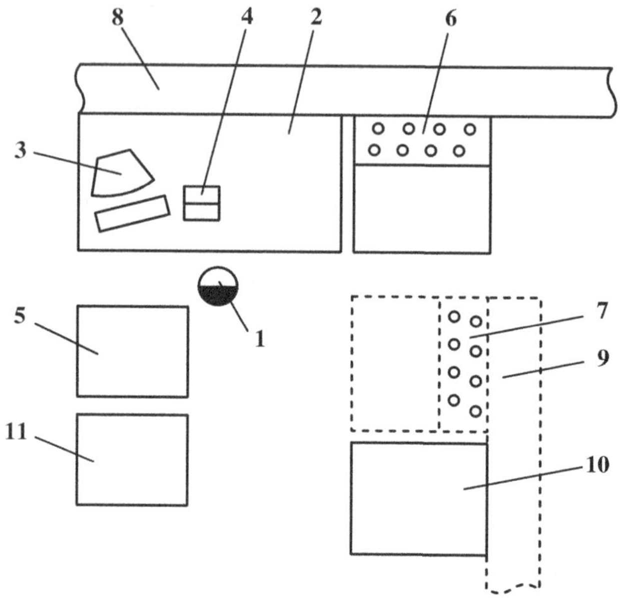

FIGURE 2.15 A view of the receiving task from above the workplace. 1 — Receiver; 2 — work table; 3 — computer; 4 — tag printer; 5 — base unit for unpacking; 6 — base unit for stock process; 7 — base unit for work-in-process (WIP); 8 — belt for staking; 9 — belt for WIP; 10 — put aside area; 11 — place for tote.

knife. After opening the box, the receiver removes a packing slip and reads it. Then the receiver uses a computer-based warehouse management system 3. He enters the purchase order (PO) number listed on the packing slip and hits the F3 key to check what is still open on the PO. The receiver takes the parts out of the box and compares the order quantity with the received quantity. He chooses the item from the PO, then changes or confirms the quantity and the price, and he assigns allocation if necessary. If allocation is already reserved for the item, the system will select it automatically. All the required information is shown on the screen and is later printed on the label. One can specify two kinds of subtasks; the first is the setup subtask and the second is the main subtask. The setup operation includes login, menu selection, key in PO number, etc. The main operation begins when an item is taken out of the box and ends when it is put in the tote. The receiver places each part from the vendor into a "tote." The tote thus filled with parts is placed in a "put-away" area by stock belt 8. The second task is "putting-away." The "put-away" operator takes parts from the "tote" and places them on the corresponding shelves.

The next task is called "pick-up." The "pick-up" operator takes the parts that have been ordered from the shelves and places them in the "tote." This "tote" is later delivered to the workshop for production. The "pick-up" operator also places ready for sale products into the tote. Delivering the required parts for production is related to

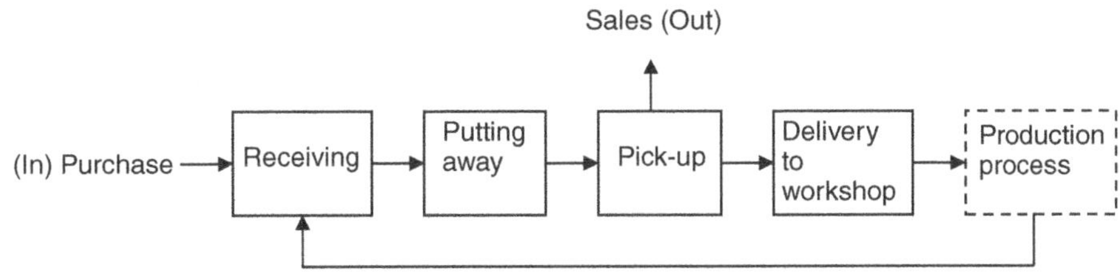

FIGURE 2.16 The sequence of tasks before the improvement.

the delivery task. Figure 2.16 designates the sequence of operations described above, before improvement. These operations are covered in relation to the receiving task.

The section above gives a brief description of the receiving task and those tasks that attend it. The qualitative analysis may be labeled as analysis with a technological orientation. The data gathered by qualitative–technological analysis (the first step of qualitative analysis) may now be used for the second stage of analysis called algorithmic analysis.

2.3.3.2 Algorithmic Description of the Receiving Task before Its Improvement

The data obtained in the previous section merely provides preliminary data related to qualitative analysis. In order to perform more in-depth qualitative analysis, more detailed information is needed. Accordingly, a second stage of analysis using algorithmic description is called for. Only after this second stage, may a qualitative analysis be revisited with greater elaboration and insight. The algorithmic model of activity during the performance of the receiving task is presented in Table 2.5 (only those actions intimately related to computer use are described).

In this table, the left column delineates the symbolic model of the activity. The right column contains a verbal description of the algorithm. A short description of the development of the algorithm is offered in the preliminary sections above. Table 2.5 reveals that for the description of human activity, a probabilistic algorithm, rather than a deterministic one was used. Actually, many logical conditions possess more than two outputs. Moreover, each output can possess different probabilities. Hence, this task has different degrees of uncertainty. Progressive reading of this algorithm, from top to bottom, and comparison of each member of the algorithm with information presented on the computer screen enable one to see a precise picture of how users carried out the computer-mediated task.

Reading each member of an algorithm in symbolic representation allows one to understand the logic of the transition from one member of algorithm to another. Algorithmic description also provides insight into the psychological peculiarities of each member of the algorithm. For example, O^{α} indicates that this member of algorithm refers to perceptual actions; $l_{15}^{\mu} \uparrow^{(1-10)}$ exhibits decision-making actions with ten outputs and shows its involvement in intensive utilization of memory.

TABLE 2.5
Algorithmic Description of Activity during Computer-Based Task Performance (before Improvement)

Members of algorithm	Description of members of human algorithm
O_1^α	Check for presence of inventory receiving screen
$\overset{1}{\downarrow} O_2^\varepsilon$	Type 1 and then press ENTER to choose ADD INVENTORY RECEIVING screen (see Figure 2.17)
O_3^α	Check to see if you are at the ADD TRANSACTION screen (cursor on the field 1)
$l_1 \overset{1}{\uparrow}$	If you are at the right screen, go to O_4^ε If the screen is wrong, hit F3 for exit and go back to O_2^ε.
$\overset{17(1)}{\downarrow}\ \overset{9(1)}{\downarrow}\ \overset{7(1)}{\downarrow}\ O_4^\varepsilon$	Take a packing slip from the box placed on base unit 5 (see Figure 2.15)
O_5^α	Find PO (purchase order) number on the slip
O_6^ε	Key in PO number and hit enter (Figure 2.17, field 1)
O_7^α	Look at the screen message
$l_2 \overset{2}{\uparrow}$	If the screen displays an error message, INVALID PO NUMBER, then go to operator O_8^α. If PO number is correct, the cursor moves to the second field RECEIVED –DATA (see Figure 2.17), go to O_{11}^ε
O_8^α	Compare PO number on the screen with the number on the packing slip
$l_3 \overset{3}{\uparrow}$	If PO number does not match go to O_9^ε. If the PO number is correct and error message persists (system can not find purchase order) go to O_{10}^ε
O_9^ε	Key in the correct number again
$\overset{3}{\downarrow} O_{10}^\varepsilon$	Call manager
$\overset{2}{\downarrow} O_{11}^\varepsilon$	Key in a current date or the date it has been received (the cursor moves to field 2, see Figure 2.17)
$\overset{17(2)}{\downarrow}\ \overset{11(2)}{\downarrow}\ \overset{7(2)}{\downarrow}\ O_{12}^\varepsilon$	Press F8 to look up items on the purchase order (Figure 2.17)
O_{13}^ε	Take an item from the box placed on base unit 5 (see Figure 2.15)
$^{a}\ O_{14}^\alpha$	Look at item number and compare it with item numbers (Figure 2.17, field 3) on the screen
$l_4 \overset{4}{\uparrow}$	If item number is on the first page, go to O_{16}^ε. If item number is not on the first page, go to O_{15}^ε
O_{15}^ε	Hit arrow key (repeat if required)

(Continued)

TABLE 2.5
Continued

Members of algorithm	Description of members of human algorithm
$\overset{4}{\downarrow} O^{\varepsilon}_{16}$	Put cursor on the selected line (Figure 2.17) and hit ENTER to go to the screen with detail item information (Figure 2.18)
O^{α}_{17}	Compare received quantity with PO (purchase order) quantity (Figure 2.18, field 4).
$^{b}l^{\alpha th}_{5} \overset{5}{\uparrow}$	If received quantity and ordered quantity are the same, press ENTER and go to O^{ε}_{24}. If received quantity is greater or less than ordered quantity, go to O^{ε}_{19}
O^{ε}_{19}	Type the received quantity and press ENTER to get a question at the bottom of the screen
O^{α}_{20}	Read the statement: THE RECEIVED QUANTITY AND ORDERED QUANTITY DO NOT MATCH. DO YOU ACCEPT? (YES/NO)
$l^{th}_{6} \overset{6}{\uparrow}$	If quantity is not accepted (computer defaults to "N") go to O^{ε}_{21}. Otherwise, go to O^{ε}_{23}
O^{ε}_{21}	Press ENTER
O^{α}_{22}	Check if there are other items on this PO to receive
$l_{7} \overset{7(1-2)}{\uparrow}$	If there are no more items in the box, go to O^{ε}_{4}, otherwise go to O^{ε}_{12}
$\overset{6}{\downarrow} O^{\varepsilon}_{23}$	Type "Y," press ENTER
$\overset{5}{\downarrow} O^{\alpha}_{24}$	Compare price of the item on the shipping list with price on the screen
$l_{8} \overset{8}{\uparrow}$	If the price on the screen and on the shipping list are different, go to O^{ε}_{25}. Otherwise, go to $^{2}O^{\varepsilon}_{31}$
O^{ε}_{25}	Key in the new price and hit ENTER
O^{α}_{26}	Read the message, THE PRICE YOU ENTERED DOES NOT MATCH INITIAL PRICE. DO YOU WANT TO ACCEPT? (Y/N) on the screen (Figure 2.18, field 5)
$^{c}O^{\alpha th}_{27}$	Compare new price with ordered price
$l_{9} \overset{9}{\uparrow}$	If new price is smaller or equal, go to $^{1}O^{\varepsilon}_{31}$. If new price is greater go to $O^{th\mu}_{28}$
$O^{th\mu}_{28}$	Mentally calculate the price difference
$l^{\mu}_{10} \overset{10}{\uparrow}$	If difference is less than 10%, go to $^{1}O^{\varepsilon}_{31}$. If difference is greater than 10%, go to O^{ε}_{29} (unless instructed otherwise)
O^{ε}_{29}	Type N and hit ENTER (the item is put aside and task is completed)
O^{α}_{30}	Check if there are other items in this box to be received

(Continued)

TABLE 2.5
Continued

Members of algorithm	Description of members of human algorithm
$\overset{11(1-2)}{l_{11} \uparrow}$	If there are no more items to be received, go to O^{ε}_{4}, otherwise go to O^{ε}_{12}
$\overset{10}{\downarrow}\ \overset{9}{\downarrow}\ {}^{1}O^{\varepsilon}_{31}$	Type "Y"
$\overset{8}{\downarrow}\ {}^{2}O^{\varepsilon}_{31}$	Hit ENTER to go to the Completion Flag field (Figure 2.18, field 6)
O^{α}_{32}	Check system default (Y/N) "flag." System gives default according to the rule "If received quantity $\geq$ ordered quantity system defaults to "Y," otherwise it defaults to "N"
$\overset{12}{l_{12} \uparrow}$	If you accept the system default (Y/N), go to O^{ε}_{34}, otherwise go to O^{ε}_{33}
O^{ε}_{33}	If the system defaults to "N," press "Y," and go to O^{ε}_{34}. If system defaults to "Y," press "N," and go to O^{ε}_{34}
$\overset{12}{\downarrow}\ O^{\varepsilon}_{34}$	Hit ENTER to go to the next field (Figure 2.18, field 7)
O^{α}_{35}	Check if there is a bin for this item
$\overset{13}{l_{13} \uparrow}$	If the bin is not assigned for this item, go to O^{μ}_{36}, otherwise the system will automatically assign the required bin, then go to O^{ε}_{40}
$\overset{15(1)}{\downarrow}\ O^{\mu}_{36}$	Depending on the size, shape and special features (HazMat) of item, recall required bin type
$\overset{14(1-10)}{l^{\mu}_{14} \uparrow}$	If bin type is 1, go to ${}^{1}O^{\varepsilon}_{37}$, up to "If bin type is 10" go to ${}^{10}O^{\varepsilon}_{37}$
$\overset{16(1)}{\downarrow}\ \overset{14(1)}{\downarrow}\ {}^{1}O^{\varepsilon}_{37}$	Press 1 and hit ENTER
$\omega_1 \uparrow_{\omega}1$	Always falls logical condition (see O^{α}_{38})
...	Choose the required bin type
$\overset{16(10)}{\downarrow}\ \overset{14(10)}{\downarrow}\ {}^{1}O^{\varepsilon}_{37}$	Press 10 and hit ENTER
$\downarrow_{\omega(1-9)}\ O^{\alpha}_{38}$	Check error message on the screen (O^{α}_{38} follow after every O^{ε}_{37})
$\overset{15(1-2)}{l_{15} \uparrow}$	If you get an error message THIS IS A WRONG BIN TYPE go to O^{α}_{39} or O^{μ}_{36}, otherwise go to O^{ε}_{40}
O^{α}_{39}	Look at the bin type chart
$\overset{16}{l_{16} \uparrow}$	If bin type is 1 go to ${}^{1}O^{\varepsilon}_{37}$. If bin type is 10 go to ${}^{10}O^{\varepsilon}_{37}$
$\overset{15(2)}{\downarrow}\ \overset{3}{\downarrow}\ O^{\varepsilon}_{40}$	Hit ENTER to print the label
O^{ε}_{41}	Peel the label off the printer and put it on the part
O^{ε}_{42}	Put the part in the tote
O^{α}_{43}	Check if there are other items in the box to receive

(Continued)

TABLE 2.5
Continued

Members of algorithm	Description of members of human algorithm
$l_{17} \overset{17(1-2)}{\uparrow}$	If there are no more items to receive, go to O^{ε}_{4}, otherwise go to O^{ε}_{12}. If there are no new boxes to work with, go to O^{ε}_{44}
O^{ε}_{44}	Hit F3 and go to the previous screen
$\overset{18}{\downarrow} O^{\varepsilon}_{45}$	Press 3 and then press ENTER to choose PRINT REPORT
O^{α}_{46}	Check to see if you are at the PRINT REPORT screen
$l_{18} \overset{18}{\uparrow}$	If you are at the right screen, go to O^{ε}_{47}. If the screen is wrong hit F3 for EXIT and go back to O^{ε}_{45}
O^{ε}_{47}	Key in start date, hit ENTER; key in end date, hit ENTER
O^{ε}_{48}	Wait for the report completion message

[a] One PO usually requires no more than 3 passes on the screen.

[b] $l^{\alpha th}$ stands for "thinking actions," involved in decision-making, that are performed based on visual information.

[c] $O^{\alpha th}_{27}$ stands for executive "thinking actions" that are performed based on visual information.

$\uparrow^{(a-b)}$ This means that the logical condition has an output from "a" to "b."

Always falls logical conditions. It is introduced to help understand the way the algorithm flows. It is artificial and does not require any action to take place.

Figure 2.17 and Figure 2.18 are examples of screens used in the algorithmic description of activity.

Algorithmic description of activity entails much effort and time. Nevertheless, it should be noted that many people repeat some tasks over periods of decades, which in the absence of a precise formulation of human performance results in user unfriendly software designs. Currently, improvement and innovation are gate ways to the user, who, lacking a paradigm for conceptualizing the process, is plunged into endless iterative cycles of trial and error. The precise algorithmic description of the task, in combination with qualitative analysis, can significantly reduce cycle time. The presented algorithmic description allows revisiting of the preliminary steps of analysis.

2.3.3.3 Second Step of the Qualitative Analysis of an Inventory Receiving Task

According to the loop-structure principle of design analysis, the qualitative stage in this practical example requires reconsideration of the preliminary qualitative stage of analysis. At this step, the receiving task was formulated at a more detailed level. While the first step focused on the present circumstances, the second step emphasized locating deficiencies in the existing task, identifying psychological difficulties of

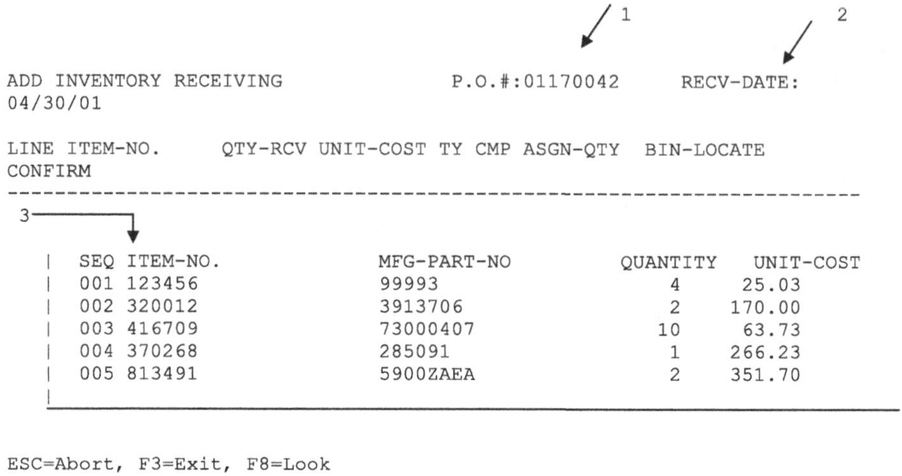

FIGURE 2.17 Add inventory receiving screen. 1 — Purchase order number; 2 — received date; 3 — item number.

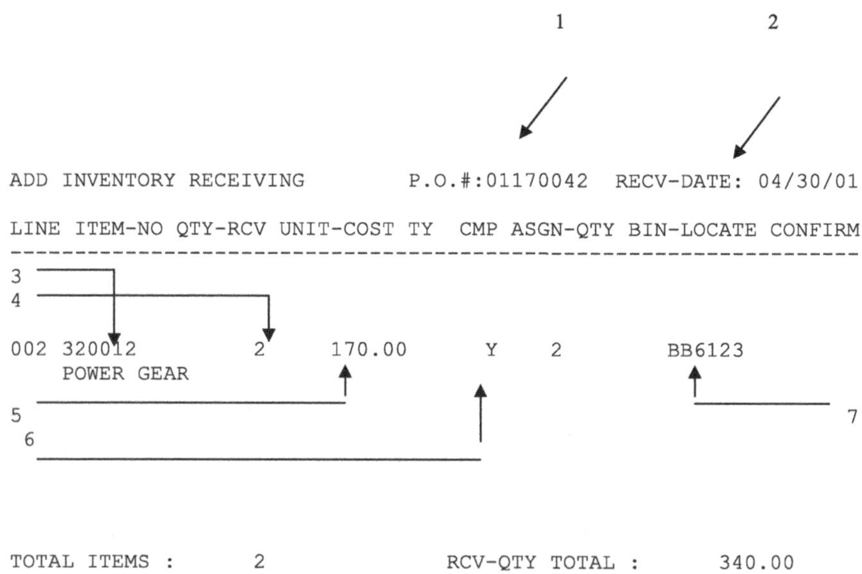

FIGURE 2.18 Add receiving screen with detailed item information. 1 — Purchase order number; 2 — received date; 3 — item number; 4 — received quantity; 5 — unit cost; 6 — completion flag; 7 — bin location.

performance, quality analysis, etc. It was discovered that tasks are multivariant and cannot be treated as deterministic. Subjective difficulties of performance, such as specificity of interactions with supervisors and others were noted. This means that the sociocultural context of work assumes importance. The objective of the study was not merely the task as a whole, rather each member of algorithm was studied as a quasi-system. The logic of the transition from one member of algorithm to another was also defined. Comparing this data with direct observation facilitates the discovery of many deficiencies in the task and in the inventory process, in general. The model of the inventory-receiving process (Figure 2.14) and algorithm of performance (Table 2.5) suggested that all parts received from vendors should be placed directly in storage on the warehouse shelves, from where they can be delivered upon request to the various workshops. However, there were instances in which observations or deliberations called for immediate deployment of items into the manufacturing process. These parts should be treated as special-order items. A model of an inventory process that encompasses this latter contingency is illustrated in Figure 2.19. This model shows that received parts have two possible routes: one proceeds to stock while the other is sent to WIP.

Work procedures prior to improvement were performed according to the model of the inventory process that is exhibited in Figure 2.14. According to this model, the task unfolds as shown in Figure 2.16. The task, "Delivery to Workshop," may be carried out only after the "receiving," "putting away," and "pick-up" tasks are performed. This results in a delay in the production process, unnecessary work, and performance under rushed, stressful circumstances. It has been empirically discovered that more than 20% of the parts should be sent immediately to the workshops. Thus, the model of inventory process depicted in Figure 2.14 fails to correspond to the real requirements of a production process. It ignores situations in which there is an emergency order and parts from vendors have to be delivered immediately to production, that is, WIP. Therefore, the second model in Figure 2.19 was recommended. With this process, roughly 20% of the parts go immediately to production, while the majority of the parts are sent to stock. Under this model, a different sequence of operations is depicted (Figure 2.20).

Figure 2.18 shows that the receiver is required to place some parts directly into the "delivery to workshop" tote, while the majority of the parts should be directed to the stock bins. Hence, the content of the receiving task should be changed. In addition,

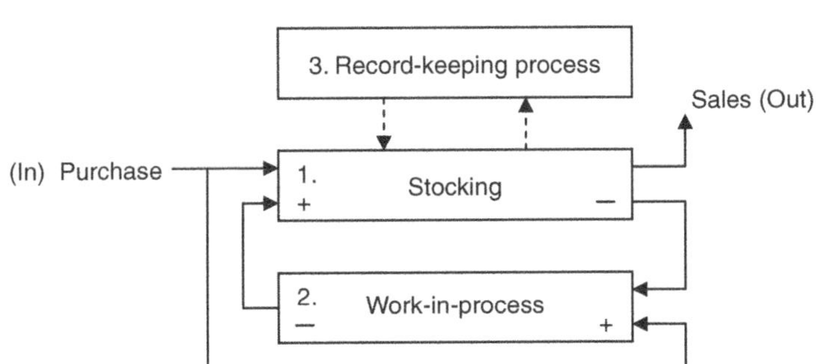

FIGURE 2.19 A model of the inventory process for a manufacturing firm after improvement. 1 — Route 1 goes from (in) purchase to stocking then to sales (out) or work-in-process (WIP); 2 — Route 2 goes from (in) purchase to WIP then to stocking and sales (out) or again to WIP.

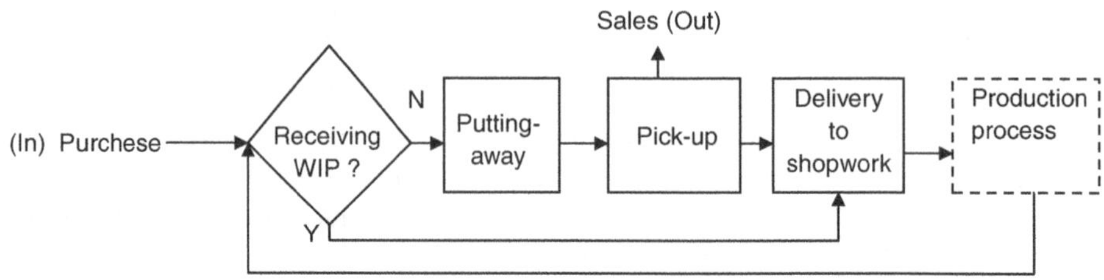

FIGURE 2.20 The sequence of tasks after the improvements. 1 — Route 1 receiving and question work-in-process (WIP; response "N") then putting away up to production process; 2 — Route 2 receiving and question WIP (response "Y") then delivery from workshop up to production process.

```
ADD INVENTORY RECEIVING          P.O.#:01170042           RECV-DATE:
4/30/01

LINE ITEM-NO. QTY-RCV UNIT-COST TY CMP ASGN-QTY  BIN-LOCATE  CONFIRM
----------------------------------------------------------------------
---------

002 320012         2         170.00     Y       2
    POWER GEAR



1   ____________________
                        |
WORK-IN-PROCESS (Y/N)?_v_
```

FIGURE 2.21 Screen with the work-in-process option. 1 — "yes" or "no" answer is required.

the receiver has to determine the category of the order he is processing by answering the question, work-in-process (Yes/No)? on the screen (see Figure 2.21).

The algorithm of performance facilitated the determination of the step needed to introduce decision rules regarding assignment of parts and new actions related to the WIP. The analysis of the logic of the performance algorithm revealed that these actions should be introduced following the checking of the PO and the item numbers, and comparing the received quantity and price. In terms of the algorithm, this meant that introduction of decision rules should be inserted after O^{ε}_{34} and before O^{α}_{35} in Table 2.5. In Table 2.6 (algorithmic description of task after improvement) members of the algorithm from O^{α}_{33} to O^{ε}_{36} describe components of activity related to WIP.

The method of developing an algorithm following improvement, in this case, was the same as described above. The difference here was that the researcher had the opportunity, at this stage, to compare preliminarily performed qualitative analysis with algorithmic description of the task in Table 2.5. Accordingly, comparative analysis follows immediately, bypassing the development of a second version of the algorithm presented in Table 2.6.

TABLE 2.6
Algorithmic Description of Activity during Computer-Based Task Performance (after Improvement)

Members of algorithm	Description of members of human algorithm
O_1^α	Check for presence of inventory receiving screen
$\overset{1}{\downarrow} O_2^\varepsilon$	Press 1 and then press ENTER to choose ADD INVENTORY RECEIVING screen
O_3^α	Check to see if you are at the ADD INVENTORY RECEIVING screen (cursor in field 1)
$l_1 \overset{1}{\uparrow}$	If you are at the right screen, go to operator O_4^ε. If the screen is wrong, hit F3 for EXIT and go back to O_2^ε
$\overset{16(1)}{\downarrow}\ \overset{10(1)}{\downarrow}\ \overset{7(1)}{\downarrow} O_4^\varepsilon$	Take a packing slip from the box placed on the base unit 5 (see Figure 2.18)
O_5^α	Find PO number on the slip
O_6^ε	Key in PO number and hit enter (Figure 2.17, field 1)
O_7^α	Look at the screen message
$l_2 \overset{2}{\uparrow}$	If the screen displays an error message, INVALID PO NUMBER, then go to operator 0_8^α. If PO number is correct (the cursor moves to the second field RECEIVED-DATA; see Figure 2.17) go to O_{11}^ε
O_8^α	Compare PO number on the screen with the number on the packing slip
$l_3 \overset{3}{\uparrow}$	If PO number does not match go to O_9^ε. If the PO number is correct and error message persists (system cannot find purchase order) go to O_{10}^ε
O_9^ε	Key in the correct number again
$\overset{3}{\downarrow} O_{10}^\varepsilon$	Call manager
$\overset{2}{\downarrow} O_{11}^\varepsilon$	Hit ENTER to get a current date (the cursor moves to field 2; see Figure 2.17)
$\overset{16(2)}{\downarrow}\ \overset{10(2)}{\downarrow}\ \overset{7(2)}{\downarrow} O_{12}^\varepsilon$	Press F8 to look up items on the purchase order (Figure 2.17, field 3)
O_{13}^ε	Take out item from box 5
O_{14}^α	Look at item number and compare it with item number (Figure 2.17, field 3) on page one on the screen
$l_4 \overset{4}{\uparrow}$	If item number is on the first page, go to O_{16}^ε. If item number is not on the first page, go to O_{15}^ε
O_{15}^ε	Hit arrow key (repeat if required)
$\overset{4}{\downarrow} O_{16}^\varepsilon$	Put cursor on the selected line (Figure 2.17) and hit ENTER to go to the screen with detail information (Figure 2.18)

(Continued)

TABLE 2.6
Continued

Members of algorithm	Description of members of human algorithm
O^{α}_{17}	Compare received quantity with purchase order quantity (Figure 2.18, field 4)
$l^{\alpha th}_{5}\ \overset{5}{\uparrow}$	If received quantity and ordered quantity are the same, press ENTER and go to O^{ε}_{24}. If received quantity is greater or less than ordered quantity, go to O^{ε}_{19}
O^{ε}_{19}	Type the received quantity and press enter to get a question on the bottom of the screen
$O^{\alpha th}_{20}$	Read the statement: THE RECEIVED QUANTITY DOES NOT MATCH ORDERED QUANTITY. Do you want to accept? (Yes/No)
$l_6 \overset{6}{\uparrow}$	If quantity is not accepted (computer defaults to N) go to O^{ε}_{21}. Otherwise, go to O^{ε}_{23}
O^{ε}_{21}	Press ENTER
O^{α}_{22}	Check if there are other items in the box to be received
$l_7\ \overset{7(1-2)}{\uparrow}$	If no more items in the box, go to O^{ε}_{4}, otherwise go to O^{ε}_{12}
$\overset{6}{\downarrow} O^{\varepsilon}_{23}$	Type "Y," press ENTER
$\overset{5}{\downarrow} O^{\alpha}_{24}$	Compare price of the item on the shipping list with price on the screen
$l_8 \overset{8}{\uparrow}$	If the price on the screen and shipping list are different, go to O^{ε}_{25}. Otherwise, go to $^{2}O^{\varepsilon}_{29}$
O^{ε}_{25}	Key in the new price and hit ENTER
O^{α}_{26}	Look at information on the screen. (Cursor can move to the next field, or the message, PRICE DIFFERENCE IS GREATER THAN 10%. DO YOU WISH TO PROCEED? (Y/N) can be presented)
$l_9\ \overset{9(1-2)}{\uparrow}$	If the message, PRICE DIFFERENCE IS GREATER THAN 10%. DO YOU WISH TO PROCEED? (Y/N) appears and the answer is "N," go to O^{ε}_{27}; if "Y" (special instruction) go to $^{1}O^{\varepsilon}_{29}$. If there is no message (cursor moved to the next field) go to O^{α}_{30}
O^{ε}_{27}	Type "N" and hit ENTER
$\underset{\omega 1}{\downarrow} \underset{\omega 2}{\downarrow} \underset{\omega 3}{\downarrow}\ O^{\alpha}_{28}$	Check if there are other items in this box to be received
$l_{10} \overset{10(1-2)}{\uparrow}$	If no more items to be received, go to O^{ε}_{4}; otherwise, go to O^{ε}_{12}
$\overset{9(1)}{\downarrow}\ {}^{1}O^{\varepsilon}_{29}$	Type "Y"

(Continued)

TABLE 2.6
Continued

Members of algorithm	Description of members of human algorithm
$\overset{8}{\downarrow}\ {}^{2}O^{\varepsilon}_{29}$	Hit ENTER to go to the Completion Flag field (Figure 2.18, field 6)
$\overset{9(2)}{\downarrow}\ O^{\alpha}_{30}$	Check system default (Y/N). (System gives a default according to the rule) "If received quantity $\geq$ ordered quantity, system defaults to 'Y', otherwise it defaults to 'N'")
$l_{11}\overset{11}{\uparrow}$	If you accept system default (Y/N), go to O^{ε}_{32}, otherwise, go to O^{ε}_{31}
O^{ε}_{31}	System defaults to "N," type "Y," and go to O^{ε}_{32}. System defaulted to "Y," type "N" and go to O^{ε}_{32}
$\overset{11}{\downarrow}\ O^{\varepsilon}_{32}$	Hit ENTER to go to the next field (Figure 2.18, field 7)
O^{α}_{33}	Look at the screen message
$l_{12}\overset{12}{\uparrow}$	If screen displays a message, WORK-IN-PROCESS? (Y/N) and the answer is "Y" go to O^{ε}_{34}; otherwise, go to O^{ε}_{37}. (Figure 2.21)
O^{ε}_{34}	Type "Y," hit ENTER to print out a label, and put label on the part
O^{α}_{35}	Look at the label to determine which department within the plant the item will be shipped to
$l_{13}\overset{13(1-3)}{\uparrow}$	If it goes to department 1, go to ${}_{1}O^{\varepsilon}_{36}$, if it goes to department 2, go to ${}_{2}O^{\varepsilon}_{36}$; otherwise, go to ${}_{3}O^{\varepsilon}_{36}$
$\overset{13(1)}{\downarrow}\ {}_{1}O^{\varepsilon}_{36}$	Put the part in box 1
$\omega_1 \uparrow_{\omega^1}$	Always falls logical condition (see O^{α}_{28})
$\overset{13(2)}{\downarrow}\ {}_{2}O^{\varepsilon}_{36}$	Put the part in box 2
$\omega_2 \uparrow_{\omega^2}$	Always falls logical condition (see O^{α}_{28})
$\overset{13(3)}{\downarrow_2}\ O^{\varepsilon}_{36}$	Put the part in box 3
$\omega_3 \uparrow_{\omega^3}$	Always falls logical condition (see O^{α}_{28})
$\overset{12}{\downarrow}\ O^{\alpha}_{37}$	Check if there is a bin for this item

(Continued)

TABLE 2.6
Continued

Members of algorithm	Description of members of human algorithm
$\overset{14}{l_{14}\uparrow}$	If the bin is not assigned for this item, go to O^{ε}_{38}; otherwise the system will automatically assign the required bin, then go to O^{ε}_{41}
O^{α}_{38}	Depending on the size, shape, and special features, (HazMat), choose the bin type from the list of bin types on the screen
$\overset{15(1\cdots10)}{l_{15}\uparrow}$	If bin is type 1, go to ${}^{1}O^{\varepsilon}_{39}$; up to if bin is type 10, go to ${}^{10}O^{\varepsilon}_{39}$
$\overset{15(1)}{\downarrow}\ {}^{1}O^{\varepsilon}_{39}$	Choose the bin type 1 (Move cursor to required position and hit ENTER)
$\omega_4 \uparrow^{\omega 4}$	Always falls logical condition (see O^{α}_{28})
⋮	Choose the required bin type
$\overset{15(10)}{\downarrow}\ {}^{10}O^{\varepsilon}_{39}$	Choose the bin type 10
$\downarrow^{\omega 4} O^{\varepsilon}_{40}$	Hit ENTER (system will assign the available bin of the chosen type)
$\overset{14}{\downarrow} O^{\varepsilon}_{41}$	Hit ENTER to print the label
O^{ε}_{42}	Peel the label off the printer and put it on the part
O^{ε}_{43}	Put part in the tote
O^{α}_{44}	Check if there are other items in the box to be received
$\overset{16(1-2)}{l_{16}\uparrow}$	If there are no more items to be received, go to O^{ε}_{4}; otherwise go to O^{ε}_{12}. If there are no new boxes to work with, go to O^{ε}_{45}
$\overset{17(2)}{\downarrow} O^{\varepsilon}_{45}$	Hit F3 and go to the previous screen
O^{α}_{46}	Check for presence of inventory receiving screen
$\overset{17}{\downarrow} O^{\varepsilon}_{47}$	Type 3 and then press ENTER to choose PRINT REPORT
O^{α}_{48}	Check to see if you are at the PRINT RECEIVING REPORT screen
$\overset{17(1-2)}{l_{17}\uparrow}$	If you are at the right screen and you choose current date, go to O^{ε}_{49}. If you choose a different date range (From–To) go to O^{ε}_{50}. If you are at the wrong screen, go to O^{ε}_{45}

(Continued)

TABLE 2.6
Continued

Members of algorithm	Description of members of human algorithm
O^{ε}_{49}	Hit ENTER twice and go to O^{ε}_{51}
17(1) $\downarrow O^{\varepsilon}_{50}$	Type the Start Date and End Date (see pattern — MM/DD/YY), hit ENTER, and go to O^{ε}_{51}
O^{ε}_{51}	Wait for the report completion message

2.3.3.4 Comparative Analysis of Activity Algorithms before and after Improvement of the Receiving Task

The comparative analysis began with a study of WIP improvement. Consider Table 2.6, in which WIP is presented algorithmically. Members of algorithm from O^{α}_{33} through O^{ε}_{36} define the content and logic of actions performed by a receiver, if the screen displays "work-in-process" and the answer is "Yes" (Y). If according to the received message, the answer is "No" (N), then the receiver bypasses all the above-mentioned members of the algorithm and performs the tasks in a regular order. It is clear that if computer programmers received a precise description of the actions performed by users related to WIP, they could introduce more efficient changes in the design of the software, providing a new way of task performance. After obtaining a clear and precise description of the actions performed by a user, a programmer could develop programs that require the minimum of corrections and "debugging." It is, of course, well known that since users are frequently unable to explain the task requirements to programmers, software design expands into a long sequence of improvements. Moreover, users often change their opinion after improvements. Thus, algorithmic description of human activity during task performance enables evaluation of the efficiency of user actions. Actions performed by users according to the algorithmic representation become clearly understandable if the algorithmic description of the task and workplace arrangements are compared. According to the algorithmic description in Table 2.6, and the work arrangement in Figure 2.15, if a receiver gives an answer on the screen "work-in-process"-"No," he/she uses base unit 6 for the stock process and belt 8. If the user's answer for "work-in-process" is "Yes," he uses base unit 7 for work-in-process and belt 9. Base unit 7 and belt 9 were introduced after improvement, as designated by the dashed line.

Thus, the fragment of algorithm introduced as improvement adds to "up-front" work but eliminates subsequent steps of the algorithm altogether, thereby reducing the overall task burden. It is also worth noting that in 20% of the cases that route directly to "work-in-process," the unnecessary tasks (putting away, pick-up) are completely eliminated. The following stages of analysis involving evaluation of task complexity before and after improvement, show that there is a negligible increase in task complexity for some subtasks. This is attributable to additional afferent operator O^{α}_{33} and logical condition l_{12}. Finally, some measures of variability of task performance may be slightly increased, which is offset by the elimination of unnecessary work and reduction in time constraints.

Let us consider the following steps involved in comparative analysis of different members of the algorithms, before and after improvement. For the purpose of discussion only, the more important members of algorithm that are germane to performance improvement against the baseline are selected. Let us compare a member of algorithm O^{ε}_{11} before and after improvement. Before improvement, it is revealed that the worker must recall the current date and key it in. After improvement, he must simply hit "enter" and the system defaults to the current date. Logical condition l^{th}_{6} includes reasoning actions (thinking action category) before a decision is made. The performance of actions complying with these logical conditions is not altered by the improvements. Analysis of these logical conditions demonstrates that special training is required for efficient performance of the above-described members of algorithm. In this case, training is not reducible to explanation and demonstration, rather a set of scenarios reflecting diverse contingencies and outcomes must be developed around l^{th}_{6} in combination with members O^{α}_{20}, O^{ε}_{21}, and O^{ε}_{23} of the algorithm. This shows that algorithmic formulation of a task is useful for training development.

Member of algorithm O^{α}_{26} is different before and after the improvement. Before improvement, the message, "PRICE DOES NOT MATCH INITIAL PRICE. Do You Wish To Proceed? (Y/N) " always emerges when the price is different from the price on the screen (order price). After improvement, this message appears only when a new price is more than 10% over the order price. This reduces the perceptual workload during the performance of this member of algorithm.

Members of algorithm $O^{\alpha\text{th}}_{27}$, l_9, and $O^{\text{th}\mu}_{28}$ (see Table 2.5) are performed only before improvement. Taking into consideration that these members of algorithm involve thinking and decision-making processes, their elimination is particularly potent, by virtue of a reduction of task complexity. Logical condition l_{10} before the improvement (Table 2.5) is carried out by maintaining information in working memory until these logical conditions are completed. The function performed by a logical condition l_{10} before improvement (see Table 2.5), is performed by l_9 after improvement (see Table 2.6). In the last case, decision making is carried out based on exteroceptive information presented on the screen. This significantly reduces the load on working memory and the complexity of task performance in general.

Comparison of the methods of performance implicated in the evaluation of the prices of the parts, before and after improvement, generally reveals that before improvement multiple diverse steps, including many behavioral and mental actions, are required. After improvement, all these actions were eliminated. There are also differences in the performance of logical condition l_{10} (before improvement) and l_9 (after improvement). After improvement, l_9 is performed not only on the basis of exteroceptive information but also partially automatically by the computer system. The message appears on the screen only in those cases where the price exceeds a threshold of 10% or more. In this case, the operator has a choice to answer "YES" (Y) or "No" (N). In all other cases (price is less, equal to or less than 10% difference), the decision is taken by the computer. Only in special cases, when a worker receives an instruction from his or her supervisor, can the worker supply the answer "Yes"

(Y), even when the price is more than 10% over order price. The system does not default to "N," if the price variance is more than 10% positive, since under particular circumstances it is possible for the worker to answer "Yes" (Y) if specific instructions are given by a supervisor. In this situation, a worker can hit "ENTER" prior to conscious decision making. Decision making connected with logical condition l_9 (Table 2.6) is only partially automated providing for flexibility of the worker's performance and his ability to decide what to do in any particular case. Moreover, this prevents mindless hitting of "ENTER."

If a worker's answer is "No" (N), then he transfers to O^{ε}_{27}, O^{α}_{28} and l_{10} see Table 2.4). The worker then progresses to a new item or a new box with another item inside. By the same token, if it branches to $^1O^{\varepsilon}_{29}$, the cursor moves to the next field and the worker goes to O^{α}_{30}. Members of algorithm from O^{α}_{30} to O^{ε}_{32} describe the worker's activity when he compared received and ordered quantities (see Table 2.6). This part of the task was not altered by intervention.

The following part of the algorithm describes WIP. This part of the task has already been discussed. Consequently, the part of the algorithm that begins with O^{α}_{38} up to O^{ε}_{40} (see Table 2.5) will be considered. In Table 2.5 (before improvement), this part of the task is described by members of algorithm from O^{α}_{35} through logical conditions l_{16}. Let us compare these members of algorithm with those mentioned in Table 2.6.

Under conditions where the bin exists, the tasks performed before and after improvement are the same. Before improvement, workers perform O^{ε}_{40} following l_{13} (see Table 2.5). Workers performed it the same way after improvement — $l_{14}$and then O^{ε}_{41} (see Table 2.6). However, in approximately 10% of the cases, bins are not assigned for particular items. Workers must categorize the item themselves. This part of the task is treated below. Prior to improvements, after l_{13}workers should perform O^{μ}_{36}, (see Table 2.5). Symbol μ designated the situation under which the worker must retrieve the required information from long-term memory ("recall required bin type"; see O^{μ}_{36}, Table 2.5), then follow l^{μ}_{14}. There are different bin types. Based on the information retained in working memory, the worker makes a decision on which bin to select. Analysis of the above-mentioned members of algorithm requires workers to continually maintain in memory the required information, causing an overload on working memory. The decision-making process is based, not on exteroceptive information, but on information extracted from memory. This is a complicated decision-making process that increases the probability of the computer presenting a warning on the screen, "Wrong Bin Type." To avoid the warning, monitoring and controlling actions O^{α}_{38} through l_{16} were introduced into the task algorithm, and if required, the worker should return to O^{μ}_{36} (see Table 2.5).

After improvement, the list of bin types is presented on the screen (see O^{α}_{38}, Table 2.6). This eliminates the necessity of retrieving information from long-term memory (instead of O^{μ}_{38} in Table 2.5, O^{α}_{38} in Table 2.6). Decision making is executed now on the basis of exteroceptive information presented on the screen, rather than from information extracted from memory. In this case, mnemonic actions are transformed into perceptual actions. This facilitates the decision-making process and reduces the probability of an erroneous decision. As a result, members of algorithm requiring control actions for correction can be eliminated. Moreover, if this correction is indicated, it can be performed with the assistance of information regarding the bin type from the screen.

The following members of algorithm O^{ε}_{41} through O^{α}_{48} are not affected by improvements (see Table 2.6). Let us consider the final step of the algorithm of activity. This part of the algorithm describes those portions of the task that are involved in producing reports. A worker may print what was received during one day or up to several days later. In most cases, reports are related to what was done during the day. Before improvement, workers in both the cases (report during one day or report during several days) must key in "Start Day," hit "ENTER" and key in "End Date" and hit "ENTER" (see O^{ε}_{47}, Table 2.5). Moreover, the computer expects the date to be entered in a specific way. If the keyed-in date pattern does not match the date pattern in the computer, the report will be empty. Thus, it is important to provide the worker the date pattern, so that he knows how to present the date to the computer. This pattern was not presented before improvement.

After improvement, if a report is produced for one day, the worker simply hits "ENTER" twice (see O^{ε}_{49}, Table 2.6). If a report covers several days, the worker carries out O^{ε}_{50} in Table 2.6, which corresponds to O^{ε}_{47} (Table 2.5). The difference is that the worker is always presented with the required MM/DD/YY date pattern. This reduces the probability of errors caused by the worker's preferred pattern of keying in date fields. Consider also that workers produce reports for only one day, meaning that they typically only have to hit "ENTER" twice. In general, it may be seen that the suggested method of morphological description of activity, which includes qualitative and algorithmic stages of analysis, is a powerful tool to be used for the study of HCI.

3 Time Study and Design

3.1 PRINCIPLES OF TIME MEASUREMENT IN ERGONOMICS

3.1.1 TRADITIONAL METHODS OF TIME MEASUREMENT IN ERGONOMICS

Time not only reflects the duration of human performance and the distinguishing features of external behavior but also specifies internal cognitive processes. For example, chronometrical studies play an important role in cognitive psychology (Sperling, 1960; Sternberg, 1969). Therefore, indices of time can be used not only as characteristics of productivity and efficiency but also as criteria for evaluation of internal cognitive processes and external behavior. The time factor becomes particularly important in those professions that have time restrictions. There is a traditional area that was known in the United States as time study. The founders of this field were Taylor (1911) and Gilbreth (1911). Further, this approach is sometimes called "work method design and work measurement" because the time of task performance depends on the method of performance (Karger and Bayha, 1977). This direction of time study is very useful for determining the standard time to perform a specific task. The preferred work method should be clearly defined (Barnes, 1980). However, according to the systemic–structural approach, these two steps have a loop structure organization. The preliminary preferred method should first be defined, and then the time performance can be determined. After that the method of performance can be reconsidered. Time study can also be used for evaluation of efficiency of performance.

Time study, as described above, is usually applied to traditional professions, such as those of blue-collar workers. Another direction studies temporal parameters/characteristics of operator activity in the man–machine system. The time during which the system is transferred from the initial to the required state is called the "cycle of regulation time." Task performance time very often constitutes a substantial part of the cycle of time regulation.

In general, the form of the cycle of time regulation may be presented as the sum of time delays that are produced, both by human information processing and by the physical responses of the equipment. Thus,

$$T_0 = t_0 + t_1 + t_2 + t_3 \tag{3.1}$$

where T_0 is the cycle of regulation time; t_0 — time during which the equipment presents information; t_1 — time required for the operator's perception and processing of the information; t_2 — time needed for the operator to perform his/her control action; t_3 — time required by the equipment to respond to the operator's control action.

Another important system characteristic is the reserved time it possesses (Siegal and Wolf, 1969; Kotik, 1974). Reserved time is defined as the surplus of time over the minimum that is required for the operator to detect and correct any deviations of system parameters from allowable limits, and to bring the system back into tolerance. Thus,

$$T_{\text{res}} = T - T_0 \qquad (3.2)$$

where T — time that cannot be exceeded without peril to the system; and T_0 — cycle of regulation time.

For example, when a ship is following a certain course and an obstacle (e.g., an iceberg or reefs) suddenly appears, the human-ship system needs, let us say, at least 8 min to effect a change of course and avoid the obstacle. If the ship can make the change in 5 min, it has a 3 min margin of reserved time. In an accident situation, parameters quickly shift to the minimum allowable value, and reserved time decreases sharply. Usually, the equipment responds more quickly then the fastest operator action. In such cases, the operator's delay, as a component of the system, can significantly affect the length of the reserved time. Nevertheless, there are systems in which the delays can exceed the delays of the human components of the system.

From the point of view of the functional analysis of activity, when activity is regarded as a self-regulation system, it is necessary to differentiate between objectively existing reserved time and the operator's subjective evaluation of that time. In many cases they are not the same. This may lead to an inadequate evaluation of the situation and, more importantly, to inadequate behavior of the operator in an accident situation. A person often roughly evaluates reserved time by making statements such as "I have plenty of time" or "I have a little time" and "I have no time." Such statements may reflect a sharply changed activity strategy, which suggests that the transition from one level of activity regulation to another has discrete features.

A decrease in reserve time can often produce different kinds of tension. In activity theory, one can distinguish two kinds of tension (Nayenko, 1976). One is called operational and the other emotional. Operational tension is determined by a combination of task complexity and lack of available task time. Emotional tension is determined by the personal significance of a task to the operator. The concept of significance serves an important functional purpose, which will be considered further in our discussion of the functional analysis of activity. It should be noted that both kinds of tension are closely interrelated and under certain conditions they can be transferred to each other.

Measurement of reaction is another example of studying the temporal characteristics of behavior in ergonomics. Many studies have been performed to determine the effect of stimulus information on response time. Hick (1952) and Hyman (1953) were the first to discover the linear relationship between average information (I) and reaction time (RT):

$$\text{RT} = a + b\text{I} \qquad (3.3)$$

where "a" and "b" are constant coefficients dependent on the conditions of the experiment. In our study of the interaction of complex choice reactions (Bedny, 1987), it was discovered that the more complex first and second reactions are, the more they influence each other, and the more time required for the second reaction. This means that information theory cannot be used for prediction of operator performance time.

During the process of extracting information from long-term memory, the alphabet used by a subject (the number of memory storage units of information associated with required response) constantly changed; that is, alphabet is dynamic. As a result, information-processing changes too. This means that information theory can be used only in simpler situations when the amount of information used by an operator is not larger than the capacity of short-term memory. Further, an operator does not react to isolated stimulus with maximum speed during actual task performance. Therefore, we cannot ignore the concept of pace of performance.

There are two aspects to time measurements. The first is involved in determining the time performance of motor components, and the second, with cognitive components of work activity. There are less precise standardized data for cognitive elements of activity. Usually, they are described in terms of typical elements of tasks (technological units). As an example, we present abbreviated data in Table 3.1 and Table 3.2 from a handbook of engineering psychology (Lomov, 1982).

Such tables do not give accurate activity description because the elements of activity are not clearly defined. The lack of a standardized description for activity elements is a common problem in ergonomics and psychology. We can roughly interpret them as different kinds of perceptual actions. The time required for detecting a target in an informational field can be determined according to the following formula (Lomov, 1982):

$$M(T) = 1 + \frac{(N/\eta_i) + 1}{N_i + 1} \times 0.3 \tag{3.4}$$

where, $M(T)$ — mathematical mean of time performance searching or detecting a task; N — general number of elements in the informational field; N_i — number of elements related to solving the present step of the task (an element of search that possesses as the required features); and η — number of simultaneously perceived search elements. The value of η is restricted by the capacity of the working memory (4 to 6 elements) and by the angle of the operative visual field $\alpha \approx 10°$ (see Figure 3.1).

TABLE 3.1
Duration of the Visual Fixation

Task	Average time of fixation T_f
Search for simple geometrical figures	0.18–0.20
Search for letters and numbers in tables	0.30
Search for letter–number lists	0.31
Search for the target on the locator screen	0.37
Orientation and navigation with the help of a locator	0.64
Acquaintance with the situation denoted by conventional symbols	0.63
Search for conventional symbols	0.25–0.33
Detection of changes in a familiar situation denoted by conventional symbols	0.55
Counting of the conventional symbols	0.52

TABLE 3.2
Temporal Characteristics of Operator Work-Activity

Action performed	Average duration (sec)
Reading of digital indicators:	
Fluorescent light IN-1	0.73
Optical projection board	0.58
Seven-segment luminescent	0.58
Eight-segment luminescent	0.63
Open-window scale	0.20
Reading of pointer indicators:	
Damper	0.4
Average damper	1.0
Small damper	1.5

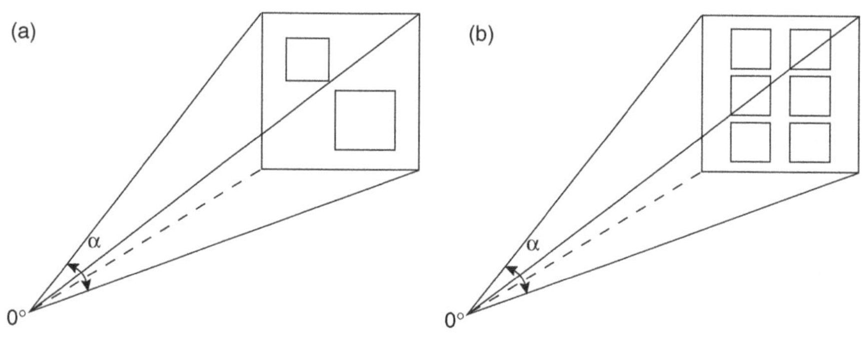

FIGURE 3.1 Determination of capacity of visual field: (a) two elements in visual field ($\eta = 2$); (b) six elements in visual field ($\eta = 6$).

Some other temporal data can be found in the work of Zarakovsky (2004) or other cognitive psychology sources.

Another approach useful for determining temporal data for cognitive components of activity is an experimental one. Donders (1862) method, widely used in cognitive psychology, is called the method of subtraction. A person performs two mental tasks, X and Y, where Y = X + K. Psychologists measure the time of performing X and Y, and the subtract time Tx from Ty to derive Tk. This procedure permits one to determine the duration of the mental process, even if this process cannot be directly observed and has a very short duration. Another method is derived from the work of Sternberg (1969, 1975). The intent of this method is to define the existence of different stages of information processing. From the sequence of these stages and their duration the duration of cognitive actions can be calculated.

If visual information is important, eye tracking data can be used to determine mental strategies and the duration of cognitive actions or operations. Gaze and eye movement data are important in this kind of study. In activity theory, multiple studies were conducted which led to the discovery of eye movement activation (macro- and micromotions) during problem solving and decision making. The micromotions of eyes during visual problem solving are considered to be an externalization of thinking actions. There is a lot of evidence of the close interconnection of motor actions with cognitive actions (Bedny et al., 2001).

When eye movements are recorded in studying cognitive task performance, it is necessary to distinguish between perceptual and thinking eye movement. The latter is often unconscious. Such actions are explorative in nature and directed at extracting the meaning of a situation. Distinction can be made based on a comparison of the physical eye movement data with qualitative analysis. Eye movements, verbal protocols, thinking aloud, subject self-report, and examination of task characteristics under different conditions of performance, all complement each other. Subjects may sometimes explain the rationale behind scanning strategies. Combinations of these methods can help, with some approximation, to describe hidden or unobservable cognitive actions, or operations, and their duration.

Eye movement recording indicated that while solving various problems, the operator often looked at an area of the control panel not linked with the information required by the problem. This was not always an erroneous search for information. The operator often observed instruments unrelated to the particular problem they were trying to solve for the purpose of determining that the parameter values displayed in these instruments were, in fact, unrelated to the problem. At the same time these explorative unrelated eye movements can be triggered automatically. The relationship between related and unrelated eye movements can be used as one criterion for evaluation of visually presented information. Sometimes, while a subject performs automatic micromotions of the eye, a subject's eyes can be relatively stationary around some point. During this period of time, the subject does not perceive external information but manipulates visual images and performs mental transformation of the situation.

There have been attempts to use Fitts's Law in ergonomic study (Fitts, 1954; Fitts and Patterson, 1964). Some scientists have tried to apply this law to the determination of the performance time of motor components of activity. Fitts investigated the time performance of positing actions. Usually, these actions are involved in moving small objects from one position to another, such as moving a cursor on the computer screen. All these results of positioning movements are widely used to differentiate operator tasks. The study of positioning actions described the relationship between such factors as time, accuracy, and distance. Fitts's Law stated that when movement amplitude (A) and target width (W) were manipulated, the time of performance of the movement can be determined with the following equation:

$$M(T) = a + b\log_2(2A/W) \tag{3.5}$$

where "a" and "b" are constants; $\log_2(2A/W)$ is called the *Index of Difficulty*.

The index of difficulty integrates two characteristics: amplitude and precision. One specific aspect of Fitts's experiment was that subjects must move metal sticks between two targets with maximum speed, or with maximum speed move sticks from the starting position to a particular target. Trying to transfer this result to a work environment, one can assume that each operator's action is performed at the maximum pace and that each performed action does not depend on either previous or subsequent actions. However, actions cannot be considered independent and isolated from each other. The subject cannot move objects or control them only with maximum speed during an entire shift. Therefore, this formula ignores the pace of performance. We conducted a special experiment which proved our statement (Bedny, 1987). In one series of experiments, the subjects were required to hit two targets with metal sticks, while in another series of experiments, four targets were required to be hit. It was discovered that when the subjects hit four targets, they changed the strategies of performance and slowed down its pace. Studies have shown that positioning actions in the context of the entire activity cannot be seen as independent. This allows us to conclude that Fitts's Law cannot be used to determine the time for motor components of a task.

A comparison of traditional work in time study and the study of temporal parameters of operator performance demonstrate that they have both common and distinctive features. The first approach concentrates on studying production operations and efficiency. The second approach is important for studying performance strategies in time-restricted situations, evaluation of safety, etc. However, in both cases, temporal parameters of human performance must be considered. From the above material, it became evident that there are limited data that can be used for the prediction of time performance in ergonomics. This is particularly relevant for the time study of cognitive components of activity. Therefore, some experimental procedures for prediction of time performance of these components of activity can be used.

Analysis of the obtained data demonstrates that there is a restricted number of methods that are utilized for the analysis of temporal parameters of human performance in Work Psychology and Ergonomics. Here one can mention such methods as timeline analysis (Kirwan, Ainisworth, Ed. 1992), reaction time measurement, (Hick-Hyman law), Fitts' law that is applied to the study of positioning actions, study of transfer functions when a time-varying input is compared with a time-varying response (Wickence, 1992). All these methods have significant limitations. According to SSAT these are parametric methods of analysis when a specialist pays attention to separate parameters of activity. These methods consider human activity as a summation of independent responses that are performed with the maximum speed. Suggested methods can not be used to analyze an activity as a system. The subtractive method suggested by Donders (1862) and additive method develop by Sternberg (1969a) can be very useful for the development of a predetermined time system of the cognitive components of activity. Usually these methods are utilized for the measurement of qualitatively different stages of a reaction time. However these methods can be adapted to the measurement of the performance time of the goal directed cognitive actions. If we have a taxonomy of such actions it becomes possible to create predetermined time system for the cognitive components of activity. Finally, the eye movement analysis is useful for determining temporal data of the cognitive components of activity. This method will be considered later.

3.1.2 Time Study Using MTM-1 System

The most significant achievements in the study of the temporal characteristics of work activity have been made by specialists in time–motion study. Here, we can refer first of all to the work of Frank B. Gilbreth and his wife Lillian M. Gilbreth. (Gilbreth, 1911; Gilbreth and Gilbreth, 1920). The fundamental principles and techniques that they developed many years ago are still being adopted by industry today. One of the more widely known systems, at present, for the study of behavioral components of activity is Methods Time and Measurements (MTM-1) (Karger and Bayha, 1977). Some scientists criticized this method because they believed that according to system MTM-1, the total time for task performance equals the sum of the time for elements of activity. However, this is incorrect. The MTM-1 system has rules for combining motor components of activity. Other researchers connected this system with the principles of simplification of tasks. This is also incorrect. This is particularly important in ergonomic study. The major purpose of the MTM-1 system in ergonomics is the description of the structure of activity and optimization of task performance. The weakness of this approach is the fact that the MTM-1 system ignores the concept of motor and cognitive actions and the existence of hierarchically organized units of analysis of activity. In the following section, we will demonstrate how we can overcome this weakness. According to systemic–structural analysis, the MTM-1 system uses units of analysis that are related to typical elements of activity. The other important feature of this system is the fact that MTM-1 has very precise descriptions for units of analysis, and all data are integrated into a holistic system. For example, after performing "reach," one can perform "grasp"; then "move," "release," and so on. It is an example of a well-developed taxonomy, or principles of classification of human behavior. Development of a unified system of classification of units of analysis, or taxonomy, is recognized as an important aspect of the design process. This allows one, in standardized ways, to interpret and describe activity during the performance of different tasks. Temporal data in an MTM-1 system can be corrected through comparison with experimental data, if this is possible. Finally, MTM-1 does not ignore the pace of performance.

MTM-1 has a predetermined time standard for each behavioral unit of analysis. Time standards for each one of the elements takes into consideration the specificity of cognitive regulation of movement. For example, the more concentration is required to perform a particular movement, the more time is assigned for its performance. MTM-1 also includes the time data for simple recognition and decision making. For this purpose, MTM-1 uses eye focus (EF) time as the time required to focus the eye on an object, and look at it long enough to recognize readily distinguishable characteristics. We will describe this microelement later in this section. A drawback of the MTM-1 system is the fact that this system does not sufficiently take into consideration the specificity of the holistic structure of activity and possible strategies for its performance. For example, Bedny (1981) conducted a study in which subjects performed a sequence of motor, perceptual, and decision-making actions. The duration of each action and the entire task was measured. These data were then compared with MTM-1 estimates. It was discovered that the duration of particular actions depends not only on their specific characteristics but also on how they influence each other and the strategies a subject used during task performance.

From this study, it was concluded that predetermined time systems are used incorrectly. Such systems start by breaking down the task into individual motions. As a result, the strategies used by the performers are ignored. The use of predetermined time systems must begin with a qualitative analysis that involves consideration of the task goal, the significance of the task, and possible strategies for goal attainment. Sometimes it requires functional analysis of activity which we will consider later. After that, algorithmic analysis should be performed. Holistic activity during task performance is divided into cognitive and behavioral actions. These actions are then integrated into members of algorithms and the logic of their organization can be determined. Only after these steps and stages of analysis can each action be broken down into behavioral or cognitive operations (motions or cognitive acts).

At this time, there are many different sources that give an abbreviated description of the MTM-1 system. Such an abbreviated description of this system is rather misleading. We will present here, as an example, a few microelements to give a better idea about the above described system, MTM-1. In Chapter 4, we will show how this system can be applied in ergonomic design. We start our description with microelement EF, which was briefly presented before.

Microelement EF is itself included in the following type of activity. It is used for the detection and recognition of an object in a visual field (simple sensory action and simultaneous perceptual action). EF can also be used to determine the duration of a simple decision-making action (decision-making action at the sensory–perceptual level) according to "yes–no," "if–then" rules (e.g., pushing a red button if the stimulus is red, pushing the green button if the stimulus is green). Subjects can concentrate on one or several points in a normal visual field. Normal visual field is about 10° in diameter. This field provides precise and simultaneous detection and recognition of targets. According to the data presented in Figure 3.1, the capacity of working memory should also be taken into consideration during visual recognition. While in one gaze a subject can detect or recognize no more than five to six objects, even when they are inside the normal visual field, during this analysis, it is important to determine what kind of objects should be detected or recognized by the subject. According to Zarakovsky (Zarakovsky and Pavlov, 1987), for practical purposes, it is recommended that there be no more than three to four targets in a normal visual field. When a subject performs an activity associated with EF, the eyes are usually stationary. However, the head can be turned for tracking the target. One encounters this situation when a worker needs to detect labels that are glued to boxes moving on a conveyor. EF can also be performed without head movement. It is in this kind of situation that an operator should turn on a particular kind of equipment. For example, a red bulb lights up when the equipment works. The subject looks at the bulb and decides "the bulb is red, then the equipment is working." EF can be overlapped by motor components of activity or it can be performed independently. The time for making an EF is 0.43 of time measurement units, or 0.26 sec. Microelement EF is completed when the subject finishes a decision-making action. EF is closely connected with eye travel time ET (we do not consider this microelement in our discussion). Unfortunately, no more data are available in this predetermined time system for more complex cognitive activities.

Sometimes, mental actions require more time for their performance than is allocated by one EF time. In order to overcome this limitation, we have developed additional rules about EF. These rules will be used during activity time structure analysis:

1. If stimulus characteristics are amorphous and cannot be easily distinguished (e.g., radar, sonar), additional time for analysis of the stimuli is required (two EF instead of one). One EF element is devoted to detection of stimulus and decision making (what does this stimulus mean). The second one belongs to decision making (what one should do according to obtained data).
2. In cases when signals are easily recognized, only one EF is required. Half of this EF time involves recognition and the other half decision making.

Let us consider another example with a second microelement called "Reach" (R), which will be described shortly. The R element is used when the predominant purpose is to move a hand or finger to a destination. The time for performance of a reach depends upon the following factors (1) class of Reach (nature of destination), (2) length of the motion, and (3) type of Reach. The time to perform a Reach is, of course, directly affected by these factors.

There are five classes of Reach:

1. Reach to object at a fixed location, or to an object in the other hand or on any other hand on which it rests.
2. Reach to a single object at a location which may vary slightly from cycle to cycle.
3. Reach to an object jumbled with other objects in a group so that search and select occur.
4. Reach to a very small object or where accurate grasp is required.
5. Reach to an indefinite location to get a hand in position for body balance or for the next motion or for getting out of the way.

The more complex a microelement is in the MTM-1 system, the greater the level of attention it requires. For example, class A requires a low level of attention. Class B requires an average level of attention. Classes C and D require high levels of attention.

The length of a motion is the true path, not just the straight-line distance between two terminal points (Figure 3.2). It is recommended that one measure distance 3 first and then calculate data multiple on 1.3. This result is the real movement distance 2. In the table, there are distances up to 80 cm. If a movement is longer then one should use extrapolation. If the actual distance is within any data interval, then one should use interpolation.

There are four types of Reach (1) the hand does not move at the beginning and at the end of Reach, (2) the hand moves at the beginning of Reach, (3) the hand moves at the end of Reach, (4) the hand is in motion at both beginning and end of reach. For a description of these types of movements, one should use the symbol "m." If Reach is performed according to type 1, the letter "m" need not be used. If Reach is performed according to type 2, the letter "m" must be placed before the symbol R (mR) and if it is performed according to type 3 the letter "m" is placed after R (Rm). Finally, if it

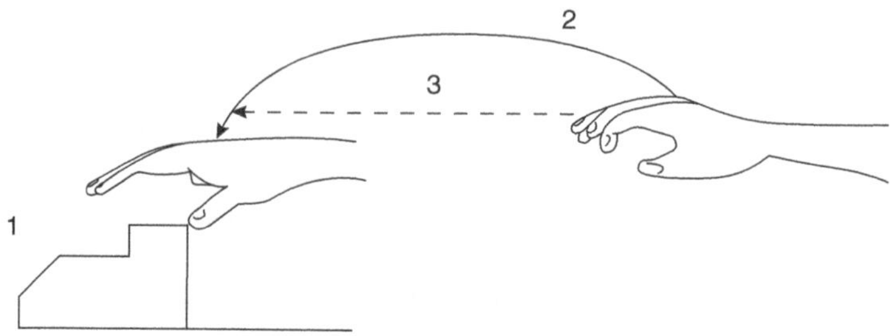

FIGURE 3.2 Measuring of length of the motion. 1—Object; 2—true path; 3—straight-line distance.

is type 4 the letter "m" should be placed before and after symbol R (mRm). For the microelement "Reach," the distance of movement is always placed immediately after the symbol R. For example, mR30m means that the hand performs microelement Reach, the distance of movement is 30 cm, and the type of Reach is 4. For another example, "hand performs Reach, the distance is 40 cm, and the type of movement is 3." This can be described as R40m. Time of performance for different kinds of Reach can be found in Table 3.3 (fragment).

There are time standards only for classes "A" and "B" combined with types 2 and 3 (mR and Rm) in the table. For other methods of combination, classes, and types, time of performance is determined by using the following formulas:

$$mR\,\#\,C = R\,\#\,C - [(R\,\#\,B) - (R\,\#\,Bm)]$$

$$mR\,\#\,D = R\,\#\,D - [(R\,\#\,B) - (R\,\#\,Bm)]$$

$$mR\,\#\,E = R\,\#\,E - [(R\,\#\,B) - (R\,\#\,Bm)]$$

$$R\,\#\,Em = R\,\#\,E - [(R\,\#\,B) - (R\,\#\,Bm)]$$

Symbol # means distance of movements.

Let us consider as an example, the time for element mR50C, which can be determined according to the following formula:

$$mR50C = R50C - [(R50B) - (R50Bm)] = 1.18 - [(1.1 - 0.94)] = 1.02 \text{ sec}$$

The third type of performance is impossible for microelement Reach using classes "C" and "D." This can be explained by the fact that these two classes require unattainably high levels of concentration during performance. The time performance for the fourth type (mRm) can be calculated in the following way:

$$mR\,\#\,Am = R\,\#\,A - 2[(R\,\#\,B) - (R\,\#\,Bm)]$$

$$mR\,\#\,Bm = R\,\#\,B - 2[(R\,\#\,B) - (R\,\#\,Bm)]$$

$$mR\,\#\,Em = R\,\#\,E - 2[(R\,\#\,B - (R\,\#\,Bm)]$$

Time standards in the table are given without taking into consideration movements of other parts of the body. For example, if a worker makes an auxiliary torso movement,

TABLE 3.3
Time of Performance of Element Reach (R) (Extraction from Representative MTM-1 Table)

Distance in cm	RA	RB	RC and RD	RE	mRA and RAm	mRB and RBm
2	0.12	0.12	0.12	0.10	0.10	0.10
4	0.20	0.20	0.31	0.20	0.18	0.15
6	0.27	0.27	0.39	0.27	0.23	0.18
8	0.32	0.34	0.45	0.33	0.27	0.22
10	0.36	0.40	0.50	0.38	0.29	0.25
⋮	⋮	⋮	⋮	⋮	⋮	⋮
60	0.88	1.28	1.34	1.14	0.76	1.11
65	0.94	1.36	1.46	1.21	0.81	1.20
70	0.99	1.45	1.50	1.28	0.86	1.28
75	1.04	1.53	1.58	1.36	0.91	1.37
80	1.09	1.61	1.67	1.43	0.95	1.46
For each 5 cm more than 80	0.05	0.08	0.08	0.07	0.05	0.08

they shorten the hand movement by the distance of the auxiliary movement. Auxiliary torso movement during performance of Reach is described by the symbol "BA." For example, when a hand performs an R115 B the torso moves at 35 mm. In this case, the hand movement (115 − 35) = 80 mm. The symbolic description of this movement is as follows:

R80B BA
(AS35)

The symbol AS35 describes the reduction of the hand movement. Other auxiliary movements of the hand and changes in direction of the hand movement are also taken into consideration. In these examples we would like to demonstrate that the MTM-1 system has a very precise and detailed description of the elements of activity and this is important for ergonomic design. The MTM-1 system should be regarded first of all as a language of description of possible activity elements and then can also be regarded as a tool for determining the time of performance. Such a system is very useful for description of motor and simple cognitive components of activity.

3.1.3 Pace of Human Performance

The above described methods to study the temporal parameters of activity in ergonomics demonstrate that there are two different major approaches, both of which are derived from cognitive psychology. In those cases when motor response is relatively simple, and perceptual or central-processing elements of activity are the major components of human performance, reaction time is used. For evaluation of the speed of processing of information, specialists use the Hick–Hyman Law. In situations when manual control is more important, Fitts's Law is applied. These two approaches suggest that different components of a task are performed with maximum speed and the operator's actions are independent and they do not influence each other. However, data presented in previous chapters refute these assumptions. Even in emergency conditions, when an operator attempts to perform a task with maximum speed, the pace of performance is slower in comparison with the total time of separate reactions or motor actions. Moreover, an operator very often produces errors, not because of delay, but by being in a hurry. One cannot determine time performance, or design time structure of an activity during task performance, without understanding the concept of work pace.

Unfortunately, a precise definition of work pace does not exist. Barnes (1980) defines work pace as the rate of speed of an operator's motions. However, this definition is unsatisfactory because it ignores cognitive components of activity and their logical organization. The operator's ability to sustain a specific speed (below maximum) of holistic activity during task performance can be defined as pace of performance. The pace can be imagined as the speed of flow of different components of activity whose structure is organized in time. It was discovered that a blue-collar worker's pace of performance can vary from one to two units (Barnes, 1980). This means that in a large group of workers who perform the same task using the same method, the fastest operator would produce approximately twice as much as the slowest operator. Usually, during pace evaluation the specialist uses experimental methods or expert analysis or a combination of both.

3.1.3.1 Evaluations of Pace Based on Subjective Judgment

The subjective evaluation of performance pace is called rating. We will consider this method very briefly. Rating is a process during which a specialist compares the pace of the operator's work with the observer's own concept of normal or standard pace. The last can be understood as an average worker's pace that can be maintained during a shift, without excessive mental and physical effort, assuming that the quality of work performance will correspond to assigned requirements.

An average person walking on a level grade at 3 mi (4.8 km)/h along a straight road is used to represent normal walking pace. This criterion has been supported by physiological studies. It is a traditional type of activity that is also easy to compare with subjective feelings and psychophysiological measurements. Physiological studies demonstrate that energy expenditure per unit of covered distance is minimal if the speed of walking is between 4 and 5 km/h (Frolov, 1976). As can be seen from Figure 3.3, increase in energy expenditure is not proportional to the speed of walking. Experimental data demonstrate that there is a clearly defined minimal energy expenditure in walking within the range 1–1.5 m/sec. This is why a walking speed of 4–5 km/h is considered optimal. Professionals should receive special training to evaluate walking speed. During this training, comparison of visual information with verbal description becomes important.

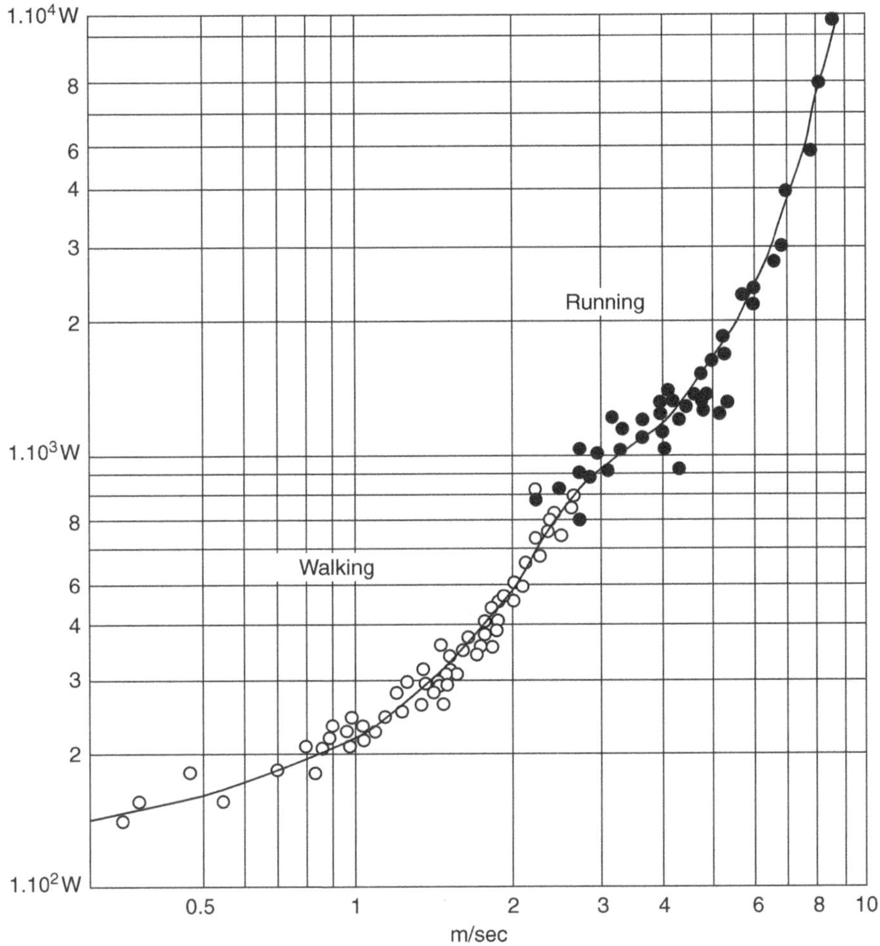

FIGURE 3.3 Relationship between energy expenditure and speed of walking.

There are several different scales for evaluation of work pace. For example, there is a scale where standard or normal pace is assigned the number 100. If the actual pace of performance is less than normal, it receives a number less than 100. If the actual pace is more than the standard, it receives a number above 100. These kinds of scales are developed based on psychophysical methods. The last number that is assigned to the real pace of performance should be either "0" or "5." For example, those numbers can be 85, 90, 95, or 105, 110, 115, etc. If in a particular case, the real pace of performance would be evaluated as 96, the specialist would assign this pace number 95. Pace evaluation is produced for separate elements of a task. Pace of performance should be determined over each half-minute for those elements of task that have significant duration.

A ratio can be established between standardized time value for an element when pace is equal; for example, 100 and the actual time value for this element that was obtained during chronometrical study. Therefore, the standardized time for a particular element can be determined based on the following formula:

$$S = T \times P, \tag{3.6}$$

where S is the standardized time that demonstrates time performance of a particular element if it is performed at a standard pace, T — time obtained during chronometrical study; and P — coefficient of pace performance (it depends on the relationship between the real pace of performance determined by an expert and the standardized pace of performance).

Assume that average time performance of an element (real-time performance) $T = 0.25$ min; then the real pace of performance determined by an expert is 80 units. Therefore, standardized time performance would be $S = 0.25 \times 80/100 = 0.20$ min. This value 0.20-min represents time of performance when the standard pace assigned number is 100.

The standardized time does not contain any special allowances. However, it is not expected that an operator can work during shifts without some interruptions. The operator may take time to rest and for other personal needs. The final value of time for task performance contains these allowances. In ergonomics, this problem is known as break/work time schedules.

The pace of performance in an MTM-1 system corresponds to a level of walking 5.8 km/h. Highly skilled workers who perform the same tasks or similar tasks during a shift can maintain this pace. It is considered the average, or standardized pace, for workers according to the MTM-1 system. In the automation process, the worker becomes a machine monitor rather than a task performer. The worker is involved in task performance only sporadically, particularly in emergency situations. The normal pace of operator performance is less under these conditions. Moreover, some have considered the pace of performance in MTM-1 to be too high (Gal'sev, 1973). Gal'sev recommends that the data that is obtained based on the MTM-1 system be multiplied by a coefficient of 1.2 for calculation of the time required for task performance. In engineering psychology, or ergonomics, to determine time of performance for different elements of activity, mean time of performance and standard deviation time are used (Zarakovsky, 2003). All the following calculations are performed based on these data. This method significantly complicates procedures that are used for calculation of time of task performance. Standard deviation should be used because the concept of "standard pace" does not exist in ergonomics. It is much easier to calculate time performance when a specialist uses a concept like "standard pace" for an ideal user. The obtained data can be easily corrected by introducing special coefficients. For example, for those operators who work faster, this coefficient can be 0.9; for those who work slower, it can be 1.1, etc. These coefficients can be determined by comparison of time performance for different operators. We will show that the MTM-1 system can be used not only, for analyses of repetitive production operations, but also for activity time structure design and, therefore, for designing complicated man–machine systems.

3.1.3.2 Physiological Evaluation of Pace of Human Performance

There are different subsystems of activity and their associated relationships (Zarakovsky, 2003). One important subsystem of activity is called a supplying physiological subsystem, which is an integration of the physiological mechanisms for providing functionality of the organism. The functional state of an organism is an important concept in the study of the supplying physiological subsystem of activity. The functional state of the organism refers to the organization and integration of physiological and psychological processes over time with qualitative characteristics that serve specific activity outcomes. An example of functional states of human organisms is work capacity and fatigue. These states change with the flow of time and depend on the character of the job being performed as well as the subjective characteristics of the workers. During the work shift, these states change according to certain stages. One of the important characteristics of human functional states is the ability of the organism to adapt to work demands. The subject can voluntarily or involuntarily mobilize resources and adapt to the work. But these resources are limited, and the speed of their recovery can lag behind the speed of their expenditure. These resources contain both physiological and psychological components. A supplying physiological subsystem provides the selective activation of different physiological processes and their integration into a system, which is directed toward achieving the particular goal of the task. The self-regulation process ensures that activation should not be lower or higher than a required level. Physiological self-regulation involves stabilizing and compensatory factors. It works according to homeostatic principles. This system also provides relocation of the organism's energetic resources depending on the nature of task activity and environmental conditions and is associated to a significant degree with an unconscious level of self-regulation. In contrast, a psychological system of self-regulation provides conscious reallocation of physiological resources. For example, when a person notices significant fatigue, he can consciously reduce the pace of work or sustain it by increasing effort. In this section, we will pay attention to the physiological subsystem and its functional state. A subject can demonstrate a high level of pace and quality of performance that cost a high physiological price. Reducing the pace of performance or introducing break time is an important intervention that prevents overloading the supplying physiological subsystem of activity.

Physiological evaluation of pace of performance is related to experimental methods. In those cases when a practitioner should evaluate medium and heavy physical tasks, physiological evaluation of pace of performance is possible. Oxygen consumption in calories per minute and heart rate in beats per minute can be used for evaluation of pace of performance. One of the basic questions is how changes in pace influence the strenuousness of the work. Presently, there are a number of approaches for classification of strenuousness (Marchenko et al., 1972). We present in Table 3.4 the classification suggested by Rozenblat (1975).

Expenditure of energy at 4.17 kcal/min is equivalent to a pulse rate of 100 beats/min. Based on analysis of the existing literature (Christensen, 1953; Lehmann, 1962) and his own studies, Rozenblat demonstrated that a pulse rate of 100 beats/min, or 4.17 kcal/min, should be used as the benchmark for the boundary

TABLE 3.4
Classification of Strenuousness of Work According to Rozenblat

Easy work	Average physical intensity of work	Heavy physical intensity of work	Very heavy physical intensity of work
Below 90 beats/min	90–99 beats/min	100–119 beats/min	Over 120 beats/min

between acceptable and unacceptable strenuousness of work. This corresponds to the boundary between easier and heavy physical intensity of work according to Rosenblat's classification. In any work condition in which the pulse rate increases beyond this standard, an additional standard break time is indicated.

At present, energy expenditure measures are used for calculation of break time (Lehmann, 1962; Murrell, 1965). Evaluation of energy expenditure in a work situation is a very difficult task. Therefore, Rozenblat suggests that a simple method for calculation should be acceptable. This method is not based on energy expenditure but is associated with the pulse rate calculation procedure. In any work conditions in which pulse rate increases beyond 100 beats/min an additional break time is indicated. In theoretical studies that use a laboratory setting, pulse can be monitored continuously. However in practice, discontinuous measurement procedures are typically used. Rosenblat suggests a sampling pulse rate at specific times during an 8-h work shift. In these cases it is necessary to measure periods that represent typical operations and typical rest periods. If the work is repetitive, the measurement is conducted (1) at the end of the first hour of work, (2) at the end of the third hour of work, (3) one hour following meal time, and (4) during the last hour of the work shift. Two hours between measurements is the prevailing rule.

When it is difficult to use special devices for measuring the pulse rate, the subject can be palpated. This procedure is important for specialists working in practical settings. Palpation takes place at the carotid artery in the neck or the radial artery in the wrist. The measurements should be taken immediately after a task is completed, preferably during the first few minutes of a break. If there is no opportunity to the measurements for an entire minute, a sample of 15–45 sec or 20–40 sec can be used. Table 3.5 provides the relationships between average pulse rate during work and the pulse rate during break periods at low or moderate temperatures.

Table 3.6 provides the same data for high temperature environments (temperature higher than 25°C and intensity of heat radiation >1.5 kcal/cm^2/min).

This approach fails to provide information regarding maximum pulse rate during the shift, but rather provides average data for different periods of work. Table 3.5 and Table 3.6 were developed based on correlation studies between pulse rate during work and rest periods.

Based on the obtained data the average pulse rate (PR_w) during work can be calculated according to an approach introduced by Rozenblat (1975):

$$PR_w = (P_1T_1 + P_2T_2 + \cdots + P_nT_n)/T_s, \tag{3.7}$$

TABLE 3.5
Evaluating Work Pulse Rate during Rest Time (Environment Not Characterized by High Temperatures)

Level of pulse rate (beats/min)			
Average pulse rate work (PR_W)	Rest time (min)		
	1	2	3
84	76	74	72
86	78	76	74
88	79	77	75
90	81	79	77
92	83	80	78
94	84	81	79
96	86	83	81
98	88	84	82
100	89	85	83
102	91	86	84
104	92	87	85
106	94	89	86
108	96	90	87
110	97	91	88
112	98	92	89
114	100	93	90
116	102	94	91
118	104	95	92
120	105	96	93
122	107	97	94
124	109	99	95
126	111	100	96
128	113	101	97
130	115	103	98
132	117	104	99
134	119	106	100
136	121	107	101
138	123	108	102
140	126	111	104
142	130	115	107
144	134	118	101
146	137	121	113
148	141	125	116
150	145	128	118
152	148	131	121
154	152	135	124

TABLE 3.6
Evaluating Work Pulse Rate during Rest Time (High Temperature Environment)

Level of pulse rate (beats/min)			
Average pulse rate work (PR_W)	Rest time (min)		
	1	2	3
86	77	74	74
88	79	76	75
90	81	77	76
92	83	79	78
94	85	81	79
96	87	82	80
98	89	84	82
100	91	86	83
102	93	87	84
104	94	88	85
106	96	89	86
108	98	91	87
110	100	92	88
112	102	94	89
114	104	95	90
116	106	96	90
118	107	97	91
120	108	98	92
122	109	99	93
124	111	100	94
126	112	101	95
128	114	102	96
130	116	104	97
132	118	105	98
134	120	107	100
136	122	108	101
138	125	110	103
140	127	112	105
142	130	114	107
144	132	116	109
146	135	119	111
148	138	122	113
150	141	124	115
152	144	127	118
154	147	130	121
156	150	133	123
158	154	137	126
160	158	140	128
162	162	144	132
164	165	147	134
166	169	151	138
168	173	155	142

where, $P_1, P_2, \ldots, P_n$ are the pulse rates of the first, second operations, etc.; $T_1, T_2, \ldots, T_n$ are the time performances of the first, second operation, etc.; T_s is the overall duration of the actual work performance during the shift.

$$T_s = T_1 + T_2 + \cdots + T_n. \tag{3.8}$$

The average pulse rate during break time PA_{br} is calculated similarly to PR_w. The next step involves the calculation of the pulse rate during the shift PA_{sh} according to the following formula:

$$PA_{sh} = (PR_w T_s + PA_{br} T_{br})/(T_s + T_{br}). \tag{3.9}$$

PA_{sh} is the major criterion for evaluating the intensity of work and for the estimation of break time. If the value of PA_{sh} is less than 100 beats/min, one needs no additional time to rest. If PA_{sh} is more than 100 beats/min, then additional break time should be allotted (Rozenblat, 1975). This method of evaluating break time based on pulse evaluation procedures is much easier to follow than those which require evaluation of energy expenditures (Lehmann, 1962; Murrell, 1965). Calculation of break time based on pulse rate criterion can thus be performed according to the following steps.

The first step calls for calculating a theoretical break time, which is designated as calculating break time (BT_{cal}), which is a percentage of the shift time. Using the theoretical break time and the real break time (BT_{rl}), the required break time (BT_{rq}) is calculated. The criterion BT_{cal} requires that the average pulse rate during the shift be less than 100 beats/min. BT_{cal} is determined by the following formula:

$$BT_{cal} = 100(PR_w - 100)/(PR_w - PA_{br})\% \tag{3.10}$$

PR_w is calculated for an 8-h shift when the duration of the average work week is 41 h, including overtime. For a shorter work week (e.g., a 6-h workday), 17.88%, or 1 h of break time, needs to be subtracted when using this formula. In cases when PA_{br} equals or is greater than 100 beats/min, the following stages are used. First, 10% of shift time is added to BT_{rl}. Then new measurements of pulse rate are taken until PA_{br} is less than 95 beats/min. Next BT_{cal} is calculated using Equation 3.10. The next stage entails calculating BT_{rq} as follows:

$$BT_{rq} = (BT_{rl} + BT_{cal})/2 \tag{3.11}$$

The average is used in this formula because PR_w and PA_{br} are not constant, but change when new work regimes and rest schedules are introduced. Let us consider a practical application. Assume that in a foundry with a 6-h shift, the following pulse rate is obtained: $PR_w = 115$ beats/min, $PA_{br} = 92$ beats/min, $BT_{rl} = 30\%$. According to Equation 3.11, $BT_{cal} = 65\%$. Because the shift is of 6 h, 1 h (or 17.88%) must be deducted from BT_{cal}, resulting in the adjusted $BT_{cal} = 48\%$. According to Equation 3.11, $BT_{rq} = 39\%$, or 2 h and 20 min.

We suggest extending this method in order to evaluate the cost-effectiveness of mechanization for the improvement of a shop's environment and similar situations (Bedny and Seglin, 1997; Bedny et al., 2001). We will consider this problem further in the next section. This method is a combination of theoretical and experimental procedures of design.

3.1.4 Physiological Evaluation of Cost-Effectiveness in Ergonomic Interventions

The data presented in the previous section can be used for evaluation of cost-effectiveness in ergonomic interventions. We must consider how we can study the physiological subsystem of activity to increase efficiency in functioning of the energetic components of activity. Practitioners currently appraise the effectiveness of work environments and performance enhancement interventions on the basis of productivity and apparent gains through empirical data. In this section we introduce a method developed to establish the benefits of ergonomic interventions based on heart-rate evaluation procedures as described above (Bedny et al., 2001).

We suggest the aforementioned method of heart rate evaluation for totally different purposes such as for calculating the economic efficiency of some ergonomic interventions and pace evaluation. In this section, we will consider an example of estimating the cost-effectiveness of environmental improvements, particularly cost-effectiveness of air-conditioning in the cabin of a large excavator. At a given level of productivity we determine the amount of break time for existing conditions using the previously described pulse rate technique. Following this we can calculate break time under the projected improvements with appropriate mechanization, automation, improvement of microclimate, etc. The savings created by the new work environment are indexed by the differences before and after implementing the break time changes. There are well-established data that describe cost of equipment and man work as measured by labor hours. Based on this data, one can infer effective gains in terms of work time, thereby indexing the effective gains from the intervention. This approach may be used in situations when average pulse rate is higher than 100 beats/min during the shift. The task was to determine the cost-effectiveness of air-conditioning the cabin of a large earth remover and excavator by creating a simulation of the physical and psychological activity in the following way.

An excavator's cabin was placed in a laboratory (Figure 3.4). The subjects moved the control levers of a simulator so that the vertical rod would not touch the walls of the slots in which they were mounted. If contact occurred between the rods and the slot, a lamp flashed. If contact was made against the right slot by the right hand, a red lamp flashed; if contact was made with the left slot by the left hand, a green light flashed. The errors are counted by a special device, that allows for continuous self-monitoring. The ambient temperature was set to approximate normal working conditions. The independent variable was the presence or absence of air-conditioning in the cabin. Ten subjects, all trained in the use of the excavator simulator, took part in the study. The experimental trials each lasted two hours. Before the tasks began, each subject spent 25 min in the cabin adapting to the existing climate.

During the experimental simulation, we counted the pulse rate and breathing rate. The pulse rate was determined by a photopleismograph. A pulse rate detector was fixed to the ear lobes of the subject. The breathing rate was determined with a gauge fixed to the operators' nostrils. We also registered the blood pressure using standard medical methods.

The preintervention portion of work break time (existing break time rest BT_{rl}) equals the production pauses during work shift t_{pp}, that can be determined from a

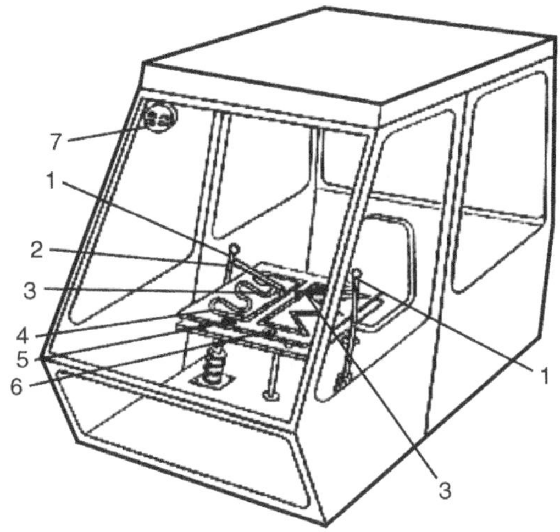

FIGURE 3.4 Excavator's cabin with imitator. 1 — Control levers; 2 — handles; 3 — vertical rod; 4 — board; 5 — red lamp; 6 — grean lamp; 7 — error counter.

handbook for the utilization of construction machinery (Kantorer, 1977). According to the handbook data, t_{pp} is 17.88% or BT_{rl} for the shift. It equals 88 min for the shift of 492 min. Accordingly, for 1 h of work, the rest period equals 11 min and work time equals 49 min. Therefore, the subjects in our experiment have the opportunity to rest twice, first for 5 min and a second time for 6 min. Thus, we attempted to simulate an operator's load closely approximating actual work conditions.

Preliminary experiments were conducted in order to establish a baseline functional level of the subject's physiological state (cf. Table 3.7), later to be called the relaxed state.

Following initial measurements, the subjects began to work in the excavator's cabin; each subject was exposed to both control and experimental conditions. Under control conditions, subjects worked without air-conditioning. The conditions inside the cabin with no air-conditioning were made to approximate summer temperatures (air temperature 40 to 41°C, wind speed 1 m/sec, relative humidity 55%). These values for the microclimate were selected because they simulate conditions in some of the warm climates of the former Asian Soviet Republics.

According to the Russian classification system, the results of pulse rate under control conditions (without air-conditioning) place the strenuousness of this job into category 3. The mean pulse rate is substantially increased over that of the relaxed state. A substantial pulse rate increase combined with the absence of increased blood pressure suggests that working in hot climate conditions produces substantial stress. The high load on circulatory and respiratory systems correlates with the dilation of blood vessels in the skin. This, in turn, causes a reduction in blood pressure. Thus, concurrent substantial shifts in pulse and breathing rates with small shifts in arterial blood pressure is a negative indicator.

An analysis of individuals #5, #7, #9, and #10 in Table 3.8 reveals that their arterial systolic pressures in relaxation state (Table 3.8) was even higher than at work

TABLE 3.7
Pretest Functional Measures: Pulse, Breath Rate, Arterial Blood Pressure

Subject	Pulse rate (beats/min)	Breath rate (beats/min)	Arterial blood pressure	
			Systolic	Diastolic
1	80	12	125	85
2	78	13	120	80
3	70	16	115	75
4	67	10	110	74
5	70	12	120	75
6	65	11	117	68
7	68	10	120	67
8	66	14	110	69
9	65	13	119	70
10	68	10	117	70
Mean	69.7	12.1	117.3	73.3

time. PR_W and breathing rate (BR) substantially increased, so we infer that the activity of the heart and cardiovascular system of these individuals was most affected.

At break time their breathing and pulse rates were maintained at high levels. This means that the operator's work in the excavator's cabin, with specified microclimatic conditions, produces significant stress on their bodies.

The next series of experiments (experimental conditions) were conducted with air-conditioning (air temperature 24°C; wind speed 1 m/sec; relative humidity 45%). The results of the trials under air-conditioning conditions are presented in Table 3.8. The average pulse rate was 83 beats/min. The breathing rate increased insignificantly and systolic blood pressure increased negligibly in comparison with the relaxed state (compare Table 3.7 and Table 3.8 with air-conditioning). Further, during work breaks the indices approached their initial values. In terms of Russian classification, this work with air-conditioning would be assigned to the second level of strenuousness, in which the pulse rate is over 80 beats/min. Individuals performing excavation work with air-conditioning are assigned to this intermediate level of stress. A Student's t for interdependent data was calculated to determine the significance of the differences between average values in air-conditioning vs. values in its absence as presented in Table 3.8. In this experiment Student's t was used to compare the pulse rate in the work period without air-conditioning vs. the pulse rate in the work period with air-conditioning; the pulse rate during the rest period without air-conditioning was also compared with the pulse rate rest period with air-conditioning. In the same way other physiological data were statistically evaluated.

The differences in the pulse rate between the two situations, with air-conditioning and without air-conditioning, were statistically significant according to Student's t index ($p < .01$). Differences in the breathing rate were also statistically significant at ($p < .05$). Differences in blood pressure were not statistically significant. This

TABLE 3.8
Functional State of Subjects during Work Process

	Functional state of the subjects without air conditioning						Functional state of the subjects with air conditioning					
	Pulse rate (beats/min)		Breath rate (beats/min)		Blood pressure		Pulse rate (beats/min)		Breath rate (beats/min)		Blood pressure	
Subject	Work	Rest	Work	Rest	Syst.	Diast.	Work	Rest	Work	Rest	Syst.	Diast.
1	108	97	24	19	138	97	88	82	19	14	130	92
2	110	98	23	18	135	90	92	80	18	15	125	85
3	107	96	25	17	130	88	85	82	16	15	120	80
4	108	98	21	14	120	82	80	65	19	14	110	75
5	105	95	22	16	115	80	82	77	17	13	117	77
6	102	93	18	16	125	80	84	70	15	15	110	70
7	104	91	20	14	110	70	80	74	18	13	115	77
8	106	96	21	15	118	64	77	70	16	12	118	75
9	98	89	19	14	116	69	79	69	17	14	120	80
10	103	93	22	17	110	75	82	72	18	13	110	67
Average	105.1	94.6	21.5	16.0	121.4	79.5	82.9	74.1	17.3	13.8	117.5	76.9

experiment shows that air-conditioning changes the excavator's job from the third to the second category of strenuousness without any loss of productivity.

Next, let us utilize the pulse rate to calculate the necessary break times when working without the air-conditioning. The average pulse rate (PR_w) during work time is 105.1 beats/min; during break time PA_{br} drops to 94.6.

According to formula 3.10:

$$BT_{cal} = 100(105.1 - 100)/(105.1 - 94.6) = 48.57\%$$

Taking into consideration that the preintervention proportion of work break time $BT_{rl} = t_{pp}$ that has been determined from the handbook of Kantorer (1977) to be 17.88%, the required proportion of rest time can be determined from Formula (3.11) for the calculation of the required break time:

$$(BT_{rq}) = (17.88 + 48.57)/2 = 33.22\%$$

In other words, to obtain an average shift of pulse rate (PR_{sh}) less than 100 beats/min, 33.22% break time must be provided instead of the prescribed 17.88% working time. This means that we need to increase break time by 15.34% of the overall work time on the excavator ($T_{ex\ w}$). The time worked on the excavators during a shift includes breaks for maintenance and other technical reasons t_{tr}. According to standard requirements, $t_{tr} = 39$ min [1] (Kantorer, 1979). In the case at hand, the time of work ($T_{ex\ w}$) when the excavator operated during shift becomes

$$T_{ex\ w} = T_{sh} - (t_{pp} + t_{tr}) \tag{3.12}$$

Thus, $T_{ex\ w} = 492 - (88 + 39) = 365$ min. The required increase for work breaks (BT_{rq}) should be 15.34% from 365 min. So, in the absence of air-conditioning we need an additional 56 min (15.34%) of break time per shift.

Working in an air-conditioned environment generates pulse rates much lower than 100 beats/min. This reduces the need for additional work breaks. In the present case, air-conditioning resulted in a saving of 56 min of work time per shift. Knowing the hourly expense of running the excavator, it is easy to calculate the economic gain from the introduction of air-conditioning into the cabin of the excavator. When we take into account that the average pulse rate is 83 beats/min under air-conditioning, the value added following air-conditioning can be formulated as consisting of a given level of productivity being reached at a lower functional state of the organism.

The results of these experiments allow us to draw the following conclusions.

1. This study shows that without air-conditioning, operators must have an additional 56 min of break time per shift in order to avoid overloading the cardiovascular system. With air-conditioning additional rest time was not required. This means that air-conditioning yields 56 min of additional excavator use per shift.
2. When evaluating physical work — especially in adverse microclimates — overloading often depends less on energy expenditure than on cardiovascular strain. Under these conditions, an analysis of energy expenditure is not

[1] Time for production pauses and breaks for maintenance or other technical reasons may vary from nation to nation.

sufficient. The evaluation of loads imposed on the cardiovascular functional system is more appropriate.
3. Evaluation of the cost-effectiveness of interventions reducing the physical workload and stress through environmental improvements and other technical innovations can be made on the basis of the pulse rate recorded during work periods. The method contrasts the required work break-time with and without the implementation of work improvements.
4. The proposed method of cost-effectiveness calculations of different interventions can be applied both when work pulse rate exceeds 100 beats/min and when it is much lower. When the pulse rate exceeds 100 beats/min, some intervention to reduce the workload by using break time evaluation is indicated. When the pulse rate is lower than 100 beats/min it implies that the pace and work load can be safely increased.
5. Evaluation of the cost efficiency of reducing workload and stress by different innovations can be achieved for the evaluation of the required break time, both before and after the introduction of stress reducing innovations.

Pulse rate criteria can also be used during evaluation of the pace of performance. After calculation of the pulse rate during a shift, one can increase or decrease the pace of performance. Pulse rate criteria can be used to increase the pace of work when pulse rates are much less than 100 beats/min. When the pulse rate exceeds 100 beats/min the same method may be used to recommend decreasing the pace of work performance. Therefore, this method may be used to regulate the pace of dynamic physical work.

In this section, we considered an example which demonstrates how a supplying physiological subsystem of activity could be studied during design performance, by using a combination of experimental and analytical procedures.

3.2 SYSTEMIC PRINCIPLES OF TIME STUDY

3.2.1 Systemic Organization of Activity and Temporal Characteristics of Operator's Performance

In highly automated human–machine systems the human being provides monitoring, diagnosis, and planning activities. Automated monitoring can often be effective when a system functions in a standardized work environment that meets established requirements. However, the system can fail and needs to be monitored by a human operator. Hence, cognitive analysis of situations and manual control over automated tasks in emergency conditions will always be critical, even in automated human–machine systems. In an automated system the operator functions as a monitor usually entering only under emergency conditions as the control loop to override the automatic system. Therefore, even in a highly automated system, manual control cannot be totally eliminated. The development of principles of activity design of cognitive and manual tasks will always be important in any kind of man–machine system.

Activity is a structure that unfolds over time as a process. We cannot design a process without evaluation of the temporal parameters of activity and analysis of its time structure.

Often within the field of ergonomics, during analysis of the temporal characteristics of activity, researchers ignore the systemic principles of its organization. It is assumed that isolated elements of activity performed with maximal speed can be performed with the same speed in holistic activity. The total task performance time is erroneously viewed as the sum of the duration of the separate elements of activity. This ideology is derived from the behaviorist approach, which considers behavior as a sum of independent reactions. This contradicts the systemic organization of activity. The interaction of the elements during their sequential performance, as well as the possibility of their being performed simultaneously, is generally not considered. An exception is the system of MTM-1, which to some extent takes into consideration the specifics of the interaction of motor activity elements and the possibility of their parallel performance. As demonstrated in the previous chapters, the informational approach is widespread in the study of time characteristics of activity within ergonomics. This approach typically utilizes the probabilistic structure of signals. It is well known that the probabilistic structure of signals will affect the speed of information processing (Hick, 1952; Hyman, 1953). However, the probabilistic characteristics of activity are not sufficient to determine the duration of operator activity, because other factors impact its duration. In a real work situation, it is very difficult to calculate the "amount" of information being analyzed by the operator. For example, in the process of perception there is an interaction of two streams of information. One stream of information stems from the environment while the other comes from memory. For a valid utilization of the theory of information, one needs to know the statistical structure of these two information sources. At this point, it is practically impossible to determine the statistical structure of the information that flows from memory. In the process of extracting information from long-term memory, the alphabet utilized by the operator is constantly changing. As a result, information theory can only be used in those cases when the amount of information used by the operator does not exceed the capacity of short-term memory.

Actions that are performed in a holistic activity are different from those that are performed independently. For example, it was discovered that the time for complex choice reactions to be performed by the right hand depends on the choice reaction times performed by the left hand. The more complex reactions performed by the first (left) hand and the more complex reactions performed by the second (right) hand, the more time required for the second reaction (Bedny, 1987). Therefore, reactions that are performed in a holistic activity cannot be considered independent. They are organized as a system and influence each other. For example, in a study (Bedny, 1987) of positioning actions it was discovered that the pace of performance significantly changed when subjects hit not just two but four targets. From this, it can be concluded that the same elements of task will be performed with different speeds if they are combined in different ways with other elements of tasks. Therefore, one cannot ignore the systemic organization of activity during time study.

The systemic organization of activity is apparent in the study of the skill acquisition process. Changes in the duration of one of the components of a skill lead to changes in the duration of performance of the other components. A reordering of elements of task performance can also lead to changes in the duration of the whole task performance, even if the duration of those elements in isolation did not change (Bedny, 1981).

Chebisheva (1969), in laboratory studies of manual tasks, described several levels of work pace:

1. Too slow and uncomfortable pace
2. Optimal pace subjectively evaluated as a comfortable pace
3. Stressful pace
4. Difficult to achieve pace
5. Unachievable pace

Studies demonstrated that the relationship of pace and performance quality could change with skill acquisition. Furthermore, with skill acquisition the subjective evaluation of the work pace changes (Bedny, 1981). It is important to note that in man–machine systems changes in the pace of performance affects the reliability of performance. The reliability of performance, to some extent at least, depends on the subjective evaluation of the pace of performance as either comfortable or uncomfortable. A pace which is uncomfortable for the performer cannot ensure reliability of performance. In this way during the analysis of the pace of work activity, it is important to use experimental methods which change the pace of performance and relate these changes with the subjective opinions of operators and experts. Therefore subjective evaluation of the pace of performance by an operator cannot be ignored. Training performed at a pace slightly above the optimal gives positive results. It was demonstrated that the benefits of training at a pace slightly above the optimal are transferred to all lower levels of pace and even, but to a lesser extent, to higher levels of pace. Research has demonstrated that skills of performing at a given pace must be through specialized training methods. Variation in the pace of activity during the training process and evaluation of this process by a performer gives the subject a feeling of time (Gellershtein, 1966).

In the United States, the issue of training for performance at a particular pace has been considered under the title of above real-time training (Miller et al., 1997). Training in conditions above real time reduces the perceived workload when the pilot is tested in real-time performance. This method is especially valuable in those cases when the task must be performed in a restricted time, such as under emergency conditions. Usually training at an elevated pace is done on simulators.

3.2.2 Analysis of Temporal Parameters of Human Activity in Man–Machine Systems

The pace of operator performance during work with semiautomatic and automatic types of man–machine systems has not been sufficiently studied. We consider below some aspects of this problem. For the study of temporal parameters of activity and, in particular, the pace of performance, we constructed a specialized control board. On one side of the control board there was a panel for the participant and on the other side a panel for the experimenter. The experimenter's panel allowed him to set the program, which would present the participant with different versions of the task. The duration of the performance of the various versions of the task was registered automatically through timers (exact to 0.01 sec). In addition to the main experimental panel, the participant was also instructed to use another panel, which he believed to be part of the experiment, but which simply served to increase task complexity and did not register any measurements. In one series of experiments, the participant worked only on the main panel; in the other series the participant worked simultaneously on two panels. Each experimental trial lasted 1 h and 30 min. During breaks between tasks performed on the panels, the participants performed supplementary activities, such as copying texts, performing simple arithmetic tasks, or performing attention tests. A sound signal indicated to the participant that they should cease any supplementary activities and switch to the major task with the panels. The work with the panels was presented to the operator as "emergency conditions." The participants were told that they needed to perform quickly and reliably. As a result, the participants could not completely concentrate on the work with the main panel. This manipulation of introducing a second panel and supplementary activities resulted in a more naturalistic simulation of work conditions required for evaluation of the work pace.

The participants were shown various versions of the task performance. After this the participants were trained to work on the panels. The panel of the participant had (1) a signaling bulb on which lit up either number 1 or 2; (2) a pointer indicator; (3) a digital indicator, which showed one of 10 possible numbers; (4) a signaling bulb which lit up as green; and (5) a signaling bulb, which lit up as red. All the indicators were located on a vertical panel, which was slanted according to ergonomic standards. Under these indicators on another horizontal panel were located the controls. All the indicators and controls were situated in order from left to right and had linear organization. The controls were as follows (6) a four-position switch; (7) a hinged lever, which could be moved to four perpendicular positions (up, down, left, and right); on the top of the lever there was (8) a button which could be pressed with the thumb. Only following the depression of the button (8) could the hinged lever (7) be moved. The next control was (9) a ten-position switch. Furthermore, the panel had (10) a red button and (11) a green button. In sum, there were five indicators and six controls. Each organ of control was located under the corresponding instrument. Consequently, movement of the eyes from instrument to instrument and the movement of the right hand, which was used to manipulate the controls, had linear organization. In order to make the task more complicated, green button 11 had been installed under red bulb 5. Red button 10 had been installed under green bulb 4. Therefore, color was used as an interfering factor. In real work conditions, when an operator performs a variety of tasks this kind of interfering becomes critical. The work on the panel was an imitation of a logically organized system of mental and motor actions the completion of which was done under conditions of constrained time. The incorrect sequence of actions or exceeding the required time of performance of a task (4.5 sec) was followed by an unpleasant sound. After the completion of a task on the panel, the participant returned to the interrupted task.

The system of signals presented to the participant using 5 indicators allowed the presentation of 110 different versions of an algorithm. According to the classification of tasks assumed by the systemic–structural activity theory, this task can be considered a deterministic algorithmic task (Bedny et al., 2005). According to the typology of Rasmussen (1986), this is a rule-based task.

The work on the two panels had the following logic. The operator receives information from instrument 1 at the main panel that can demonstrate number one or two. If this instrument presents number "1," the subject is to turn a switch down; if number "2" is presented the switch is to be turned up. Then he uses a lever that can be moved into one of four directions, depending on information that was presented on the pointer display. However, before the subject can do it he depresses the button at the top of the handle with his thumb. After the hinged four-position lever (7) is moved in the required position digital indicator (3) presents the corresponding number. Depending on the presented number, the operator can turn the 10-position switch in the required position. Green or red bulbs may be illuminated. Depending on which indicator is lit, the subject presses the green or red button. After pressing the corresponding button, the task on the main panel is completed. We can describe a more general version of task performance. In this version of task all instruments and controls are used during task performance. In other versions of task, some instruments are not activated and they, together with associated controls, were not involved in the performance of this particular version of the task.

Seven male subjects were involved in an experimental study. A warning signal that conveys information about emergency conditions and the necessity of working on the main control panels had been presented to the subjects in a random fashion. In this situation, the subjects had to drop their supplementary activities and switch to the major task with the panels. During experimental trials, performance time was registered for only three preselected versions of the task on the main panel. These versions of the task had been combined with other versions. The time performance of the other versions of the task was not measured. All versions of the task had been presented in a random fashion. The subjects did not know that only the time performance of three selected versions of the task was recorded. All versions of the task (selected as basic and supplementary) were equally important subjectively for the subjects.

The first versions of the task included the following steps. After signal bulb (1) went on, the subject was to turn on the four-position switch (6) to the required position. After checking the instruments and making sure that none of the 1's were turned on, the subject moved this hand to the hinged four-position lever (7), grasped the handle and pressed the button on the handle using the thumb. As a result, the pointer of indicator (2) assumed one of the four possible positions. The subject then moved the four-position hinged lever (7) to one of the four possible positions. Then digital indicator (3) displayed number "5" (possible numbers which can be presented by this indicator varied from 0 up to 9). After that, red bulb (5) was turned on and the subject moved his hand toward the green button (11) and pressed it.

The second version consisted of the following steps. After signal bulb (1) was turned on, the subject was to turn on the four positioning switch (6) to the required position. After making sure that none of the instruments was turned on, the subject moved his hand to the four-position hinged lever (7) and in the same way pressed button (8) the subject then moved the four-position hinged lever (7) to one of four possible positions. Then digital indicator (3) displayed number "0" and simultaneously the red bulb was turned on. After that, the subject was to press green button (11).

The third version of the task consisted of the following steps. The subject switched position switch (1) after the bulb went on. The digital indicator showed number "0" and simultaneously the green bulb was turned on. In response to this signal, the subject was to press the red button. We measured the time performance for only these three versions of the task, which were presented randomly between others versions of the task.

In the first series of experiments, the stopwatch on the experimenter panel and signal bulb (1) on the subject panel had been turned on simultaneously (synchronous regime). In the second series of the experiments (asynchronous regime), the stopwatch had been started only after the four-position switch was turned on in the required position. In the third series of experiments, the subjects worked on two control panels in stressful conditions. Introducing an 85-dB noise created this condition. In these series of experiments two control boards were used instead of one, as in the other series of experiments. Therefore, in this series of experiments stressful conditions were combined with more complicated tasks.

Both control boards were placed next to each other. The subjects did not know which control board they were going to work on in each trial. As in the previous series, the subjects used their right hand when working on the control panels. In the third series of experiments, the stopwatch was started only after the four-position switch (6) was turned to the required position (asynchronous regime).

The first group of experiments focus on performance in synchronous stopwatch regime and one panel work under normal conditions (see Table 3.9). One of our goals in conducting this experiment was to compare the experimentally derived data with analytically derived estimates of task performance time utilizing the MTM-1 system (see section 3.3.1). A comparison of experimental, and analytically derived estimates illustrates that the MTM-1 can be utilized for analytical evaluation of task time performance. A comparison of the experimental data, and analytical estimates illustrates that subjects pace of performance similar to pace of performance in MTM-1 system. However, in a post experimental questionnaire subjects evaluated the pace of performance as above optimal for extended periods of work. At the same time they judge this pace as reliable for emergency conditions (over short intervals of time). Thus while MTM-1 was able to accurately predict the pace of performance of manual, skill based tasks, this was not a sustainable pace for rule based tasks. The pace of manual work in mass production process is different from the working pace of operator on a semiautomatic system. More over, some scientists have argued that the pace estimated with the MTM-1 system is also above optimal for manual work (Gal'sev, 1973).

Let us compare the pace of performance while working on one control board with the asynchronous regime of starting the stopwatch (second series Table 3.9) with the pace of performance while subjects work on two control boards in noise conditions when the asynchronous regime was also used (the third series Table 3.10). The time of performance in normal conditions with a synchronous regime (series one) and an asynchronous regime (series two) of starting of the stopwatch is presented in Table 3.9. The time of performance in stressful conditions with an asynchronous stopwatch working regime (series three) is presented in Table 3.10. Our hypothesis was that in stressful conditions the pace of performance would be slower. However, the pace of performance increased in the third series of experiments in comparison with that in the second series of experiments (see Table 3.9 and Table 3.10, asynchronous stopwatch working regime in normal and stressful conditions).

TABLE 3.9
Work on One Control Panel in Normal Conditions

	Stopwatch working regime					
	First version of task		Second version of task		Third version of task	
Subject	Synchronous	Asynchronous	Synchronous	Asynchronous	Synchronous	Asynchronous
1	3.8	4.0	2.7	2.2	1.1	0.8
2	4.0	3.6	3.0	2.6	1.3	0.9
3	3.8	3.2	2.8	1.7	1.2	0.8
4	3.5	3.3	2.7	2.2	1.0	0.7
5	4.6	4.1	3.4	2.7	1.8	1.2
6	3.9	3.5	3.1	2.3	1.3	1.0
7	3.4	3.0	2.9	2.3	1.2	0.8
Average	3.85	3.52	2.93	2.3	1.27	0.9

TABLE 3.10
Work on Two Control Panels in Noise Interference Conditions

	Asynchronous stopwatch working regime		
Subject	First version of task	Second version of task	Third version of task
1	3.3	2.1	0.8
2	3.2	1.9	0.7
3	3.4	2.0	0.6
4	2.8	2.2	0.6
5	3.6	2.3	1.0
6	3.3	2.2	0.8
7	3.1	2.1	0.7
Average	3.2	2.11	0.74

It should be remembered that subjects were not instructed to work faster in the stressful conditions. The transition to the faster pace happened involuntarily. For the first version of the task, the difference in time performance was statistically significant ($p < .01$). The difference in time performance for the second version of the task was also statistically significant ($p < .05$). The difference in time performance for the third version of the task (the simpler one) was not statistically significant. Therefore, the more complex the version of the task, the more the pace of performance increases under stressful conditions. The simpler the version of the task is, the lesser the differences in the pace of performance. This can be explained by the fact that in stressful conditions, when a subject performs a more complex activity, activation of the neural system increases and the physiological resources of the organism are mobilized involuntarily. This can cause involuntary and unconscious acceleration of performance. The same phenomena can sometimes be observed when a driver involuntarily increases the speed of a car under a stressful functional state. As a result of excessive activation of the neural system, the operator often increases his error rate in stressful conditions, not because the work slows but because of his being rushed. The subjects perceive work in stressful conditions as requiring more effort and convey this by feeling tense. From this, it follows that the pace of performance should be appropriate to the complexity of the task and of the possibility of stressful conditions under which this task may be performed. Moreover, when the subject performs the task in time-restricted conditions, additional time should be introduced for checking actions. These kinds of actions can often be triggered involuntarily and are not necessarily under conscious control.

Another group of experiments examined how the pace of performance of the same elements of activity changes depending on the complexity of tasks of which they were a part. A group of four subjects worked on an experimental control board along with the asynchronous regime of using a stopwatch. Different versions of the task were presented to the subjects randomly. Each version was performed ten times by each subject. The first version of task performance prescribed the use of all instruments and controls (five instruments and five controls). In the second version of the task, one

TABLE 3.11
Time Performance of the Simplest Version of Task When Performed Randomly between Other Versions of Task and When Performed in Isolation

Subject	Presentation of different versions of task	Presentation of only one version of task
1	1.0	0.6
2	0.8	0.5
3	1.1	0.7
4	1.4	0.9
Average	1.0	0.65

instrument and one control were eliminated. In the third version, two instruments and two controls were eliminated. In the simplest version, three controls and three instruments were eliminated. The simplest version of the task consisted of turning switch (6) to the right, moving the right hand to button (10) and pressing this button. Therefore, the hand moved from the left most control to the next to the right button (10) which was located before the last control (button 11). The subjects did not know that the time of performance was measured only for the simplest version of the task. Hence the study compared the time performance of the simplest version of the task when performed randomly between other versions of the task and when performed in isolation. The average time performance of the version of the task is presented in Table 3.10.

After this series of trials was over, the subjects moved to performing the same simple version of the task 10 times. The average time performance of the same version of the task is presented in Table 3.11.

The average time performance of the simplest version of the task performed randomly along with the other versions is 1.0 sec. The average time performance of the same task when only one version of the task is involved is 0.65 sec. This difference is statistically significant according to student's criteria ($p < .01$).

This difference can be explained by the fact that all actions of the task that are performed are closely interconnected and influence each other. When the same version of the task is performed the uncertainty associated with the presented version is eliminated. The task is performed based on automatic processing mechanisms (Shneider and Shiffrin, 1977). The increase in the stereotype of cognitive activity and automaticity leads to an increase in the speed of switching attention to bulb (4). The stereotype of the motor components of activity also increases. The program of performance is developed in advance for only one motor action. The time of right-hand movement from the start position to the red button (10) and pushing it is close to the motor reaction time. When different versions are presented randomly such speed of cognitive and motor actions is impossible to reach. This experiment confirms that the time of separate reactions to a limited number of signals cannot be used to determine the operator's performance time in time-restricted conditions. The informational method of the operator's time performance does not take into consideration the above discussed factors.

The task performed on the main control board has 110 versions of realization. All versions can be with the same approximation divided into three groups of complexity. In the first, more complicated version of the task, all controls were used. In average group complexity, one of the controls was not used (hinged lever 7). In the simple version of the task, two controls were not used (additionally digital indicator 3 was not used). It was discovered that the complexity of the version of the same task influences the pace of performance. The simpler the version of the task, the higher the pace of performance under the same working conditions. In different versions of the task, some actions are eliminated while some actions are performed in all versions. From this one can conclude that the pace of action performance depends on logical organization and mutual influences. In simple versions of the task, actions are performed with a high level of automatism and as a result, the pace of performance increases. When a specialist attempts to determine time of task performance he should select a pace of performance based on analysis of the different versions of the task and conditions under which it may be performed. Very often he should choose a pace of performance that is adequate to more complex versions of the task.

3.2.3 Activity Time Structure and Human Performance

Traditional methods of time study in industry, as well as chronometric methods of study in ergonomics, according to systemic–structural theory of activity are related to "parametrical methods of study." In contrast to the parametric methods, this chapter proposes a method that belongs to systemic principles of time study. Here, rather than consider separate parametrical characteristics of activity, such as time of task performance, reaction time, reserve time, pace of performance, etc., we attempt to develop a holistic time structure of an activity. It requires the development of certain methods to describe the time structure of work activity. Design of a holistic time structure of activity is related to the third stage of analysis. At this stage, all activity elements are translated into temporal data that demonstrate the duration of standardized elements of activity. The third and the fourth stages of analysis are used in all cases when the most precise description of the structure of activity is required. The description of activity time structure is important in the study of HCI and in the design of tools and equipment for the operator. The main idea is that changes in equipment configurations probabilistically change the time structure of activity. The specialist can evaluate and change the equipment characteristics based on time structure analysis. The time structure of activity helps the specialist to evaluate the efficiency of the performance of production operations; thus it can be used in the evaluation of safety and training. Design of the time structure of activity is, accordingly, a necessary stage for evaluation of task complexity.

There is no clear understanding of time structure of activity in ergonomics. Sometimes, a specialist confuses the time-line chart with the time structure of activity during task performance. We define time structure of activity as the logical sequence of activity elements, their duration, and the possibility of their being performed simultaneously or sequentially. The appropriate design of an activity time structure is possible only after a preliminary qualitative and algorithmic description of activity. This is critically important in differentiating between activity elements (psychological units) and task elements (technological elements). A time structure of activity can be developed based only on psychological units of analysis. This means that all units of analysis at this stage should be transferred into psychological units. At the next step, the duration of each element of activity should be determined. A time structure of activity cannot be developed until it is determined what elements of activity can be performed simultaneously and what elements sequentially. It is important also preliminarily to determine the logical organization of the elements of activity. Therefore, before developing a time structure of activity it is necessary to describe the task algorithmically.

The following are the stages of developing a time structure of activity:

1. Determine the content of activity with the required level of decomposition for defining their elements (psychological units of analysis)
2. Determine the duration of elements while considering their mutual influence on each other
3. Define the distribution of activity elements over time, taking into account their sequential and simultaneous performance
4. Specify the preferable strategy of activity performance and its influence on the duration of separate elements and the total activity
5. Determine the logic and probability of transition from one temporal substructure to another
6. Calculate the duration and variability of activity during task performance
7. Define how strategies of activity change during skill acquisition, and estimate what is intermediate and final about the time structure

Suppose we wish to develop a time structure of activity during the performance of a particular task, such as the following. An operator is simultaneously presented with a weak acoustic signal and a well-defined red light. He must detect the acoustical signal and perceive the red light at the same time. In response to this information he must grasp a lever on his right with his right hand and move it forward to a specific position. The movement of the lever is complicated because the operator has to press a pedal with his right foot at the same time. On the other hand, if the acoustical signal is accompanied by a green light, the operator must move the same lever backwards, similarly accompanied by the same pedal movement with the right leg. If we know the duration of each action, then based on this the time structure of activity may be developed (Figure 3.5). Behavior actions can also be described using system MTM-1. In order to make the description simpler, the hand movement distance has not been considered in this example.

In the above case, we can extract the following actions to describe tasks (Figure 3.5).

According to MTM-1 rules RA represents reaching the object at a fixed location. G1A is to grasp the object when it can be easily done. MC means "move object to exact position." FM means "press and release pedal with the foot." In this example, cognitive actions are described in terms of typical elements of activity. Motor actions are first described as typical elements of task (technological units). Then they are transformed into standardized motions (typical elements of activity) using the MTM-1

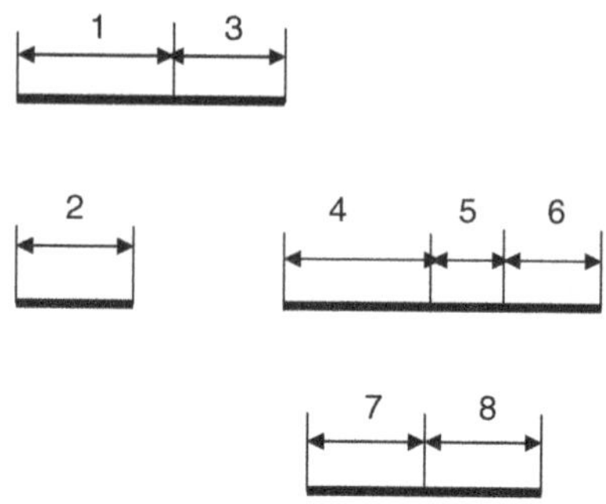

FIGURE 3.5 Time structure of activity during performance of hypothetical task. 1 — Sensory action of detection of acoustic stimulus in threshold area; 2 — perceptual action of recognition of visual stimulus in optimal visual conditions; 3 — mental action of decision making having two alternatives; 4 and 5 — motor action "reach for lever and grasp it" (action includes two motions or behavioral operations; 4-method RA and 5-method G1A); 6 — Motor action that includes "move lever to exact location without releasing it" (includes one motion-method MC); 7 and 8 — foot action "press pedal and release" (action includes two motions: FM-foot motion down 7 and FM-foot motion up 8).

system. Such a description allows a clear understanding of what the operator is doing. For instance, "reach lever and grasp" can be performed in different ways. The hand can move carefully or automatically. The lever can be easy or difficult to grasp, and so forth. MTM-1 allows transferring ambiguous descriptions of motor actions into a standardized language of description (typical elements of activity).

Another problem in the design of the time structure of activity is associated with the classification of cognitive actions and determining their duration. The principles of classification of cognitive actions were described before. The duration of their performance can be determined based on methods developed in cognitive psychology or in activity theory. However, some additional requirements should be mentioned. Chronometrical measurements of performance of different cognitive actions should be performed in different conditions. In one situation, during chronometrical study cognitive actions should be performed independently. In other conditions, they should be performed in combination with other actions (cognitive and motor). Some simple rules can be recommended. Time performance of actions included in a nonrepetitive task should be increased approximately by 20 to 30% in comparison with actions performed in isolation with maximum speed (Bedny, 1981; Zarakovsky, 2004). In different tasks, the same action can include different content operations. Therefore, we also need to pay attention to the fact that the time performance of different actions can be changed because the subject can decrease or increase speed in the performance of actions and because strategies of performance of a holistic task can be changed (Bedny, 1987). These two factors are interdependent. In rule-based tasks, the operator often performs evaluative actions for checking the correctness of a preliminary action performance. This is an evaluative stage in task performance. This stage includes perceptual and simple thinking actions. The time requirement for performing evaluative actions can be estimated as time for performing perceptual action during the receiving of one operative unit of information which ranges approximately from 0.3 to 0.4 sec

(Zarakovsky, 2004). More operative units of information should be evaluated as more time is required. The reliability of actions or components of a task, or tasks in general, depends on the reliability of the last evaluative action. Sometimes evaluative actions can be externalized when the operator performs different forms of writing or manipulations with instruments, etc. There are other strategies to check performance. For example, the operator can repeat the same actions, if this is possible. In rule-based tasks, the operator can perform a sequence of the same actions. In this situation, the time for performing subsequent similar actions should be decreased by approximately 40%. If a stereotype sequence of actions is expected to be performed, interruption of performance and involvement in unexpected tasks will increase the time of performance of the first starting action of the preliminary task by approximately 25% in comparison with regular conditions (Zarakovsky, 2004). Increasing time performance can be much more significant if the interrupted task is very complex. From this it follows that the concept of self-regulation is important for chronometrical studies. The study of self-regulation of activity is related to functional analysis of activity. It helps us to evaluate possible strategies of performance.

3.2.4 Basic Principles of Combining Different Elements of Activity

The time structure of activity cannot be developed until it is determined what elements of activity can be performed simultaneously and what elements can be performed only sequentially. The determination of which activity elements can be performed simultaneously, and which ones must be performed sequentially, is an integral part of the design of time structure of activity. This stage of design depends upon the goals, motives, significance of action and tasks, and, eventually, a strategy of task performance. For example, in a dangerous situation, when actions have a high level of significance, an operator performs them sequentially, even if they are simple. However, in a normal situation, where the consequences of error are not severe, the same simple actions will be performed simultaneously. The strategies of activity also depend on the logical components of the work process and the complexity of separate elements of activity. During the design of time structure one should distinguish cognitive elements that are independent components of activity (cognitive actions) and those cognitive elements that are components of motor actions (microblock of programming and correction of motor motions of activity). The last cognitive components are not independent elements and will be related to motor activity. We will discuss this problem during the analysis of the microstructure of motor actions. Therefore, we will consider further the following situations. (1) The possibility of a combination of motor components of activity. (2) The possibility of a combination of cognitive components. (3) The possibility of a combination of motor and cognitive components of activity.

Let us consider the possibility of a combination of motor components of activity. The level of concentration of attention during the performance of different motor actions and motions can be used for evaluation of their complexity. According to system MTM-1, there are three groups of motions depending on the level of control of motions during their performance. These are low, average, and high levels of control. In accordance with our rules, we relate those motions or actions that require a low level of control, to a group of the elements of activity associated with a low level of attention. Those elements of activity that require average and high levels of control are associated with average and high levels of concentration of attention. There are two types of control of motor actions and motions. One type is motor control. This type of control is based on the evaluation of muscle effort and kinesthetic information. The second type of control is cognitive. These two kinds of control are interdependent and can be explained in a more detailed manner based on the mechanisms of self-regulation. The level of concentration of attention or the level of control is also important for the evaluation of the complexity of cognitive actions. The level of concentration of attention or the level of control influences the time of performance, such as the ability to perform different elements of activity simultaneously or in sequence. The higher the level of concentration of attention, the greater the associated complexity of actions. We will consider this problem in more detail later.

According to the MTM-1 system (Karger and Bahya, 1977), simultaneous motions can be performed if they require low or average control. Two motions of a high level of control (or complexity which require a high level of attention, according to our rules) can be performed simultaneously only under specific conditions where the point at which the motion is terminated is in the normal field of view. If the point of motion termination is out of the normal field of view its final steps must be performed sequentially. The more experienced a performer, the greater the ability to simultaneously combine certain motions. This statement was proven experimentally. It was also discovered that two motions performed in parallel require more time than one, particularly when these motions are complex and require more attention. If two motions are simple, the duration of their performance does not significantly increase. If two motions with low and average levels of attention should be performed at the same time, their performance duration is identical to that of the two motions performed separately.

MTM-1 rules are limited because they consider only visual control. However, actions also require motor, mental, and other types of control. For example, motor actions requiring a high level of examination, such as by touch (motor control), can sometimes be performed simultaneously regardless of the field of view (Bedny, 1987).

A more detailed analysis of the possibility of combining motor actions was performed by Gordeeva and Zinchenko (1982). They found that if two motions are performed together their simultaneous combination depends on the phases of performing these actions. If the cognitive components of one motion coincide with the motor components of the other motion, they can be performed simultaneously. If the motor phases of both motions coincide, they can also be performed simultaneously. However, the cognitive phase of two motions cannot be performed simultaneously.

Let us consider the ability to perform simultaneously cognitive actions. The MTM-1 system does not consider the ability to combine mental actions, nor mental with motor actions. However, in this system there is data about the ability to combine eye fixations (EF). This element is associated with the process of perceiving stimuli and of simple decision making. According MTM-1, two-element EF should be performed sequentially because they require a high level of control. From this it follows that mental operations involving recognition and simple decision making cannot be performed simultaneously. The above described microstructural analysis of motor actions (Gordeeva and Zinchenko, 1987) also demonstrates that cognitive components of activity cannot be performed simultaneously.

This conclusion was supported by studies in cognitive psychology. In Broadbent's (1958) study, when unskilled subjects attempted to listen to one message while answering another, accuracy deteriorated. According to Glezer and Nevskaya (1964), if two objects appear in the visual field, recognition of the second object starts after a 0.7 probability of recognition of the first. Smirnov (1985) discovered that the ability to perceive stimuli significantly deteriorates when a person simultaneously attempts to perform thinking operations. He also discovered that if stimuli always have the same meaning and are encountered in similar contexts, information processing becomes almost automatic. However, if the same signal is encountered in different situations and requires different interpretations, verbalization is involved and recognition requires a high level of attention and conscious information processing. On the other hand, automatic recognition processes can be performed simultaneously. Control recognition processing should be performed in consecutive order. According to Lindsay and Norman (1992), automatic attention processes can proceed concurrently. Conceptually driven processing requires switching from one channel of information to another because the conscious part of the attention process is restricted. This also requires recognition of different stimuli in consecutive order. According to the model of attention described by Bedny (Bedny and Meister, 1997), the more involved long-term memory and the regulative integrator in information processing, the more difficult simultaneous performance of actions becomes. The function block "Regulative Integrator," which is responsible for coordinating energetic and informational components of attention, becomes overloaded as the task becomes more complex. This means that it is difficult for a person to allocate his attention between two conscious goals of activity. In this situation, the individual prefers to switch attention from one goal to another. Kahneman's (1973) model of attention similarly suggests that increased task complexity requires more attention resources and restricts human ability from simultaneously performing different components of activity. Zarakovsky (2004) also demonstrates that simultaneous performance of thinking actions is almost impossible.

Based on analysis of existing data and our own studies we developed formal rules for determining the possibility of combining mental or mental plus motor components of activity. These rules should be taken into consideration during design of the time structure of activity.

- Motor components of activity
    1. Two actions that require high levels of concentration of attention and visual control can be performed simultaneously only after the development of high-level automatic skills and if the two motor actions are performed in normal visual field.
    2. Two motor actions that require high levels of concentration of attention and visual control along all trajectories of actions and are performed outside the normal visual field can be performed only in sequence.

3. Two motor actions that require low and average levels of concentration of attention can be performed simultaneously.
4. Two motor actions when one of them requires a high level of concentration of attention and the other requires a low or average level of concentration of attention can be performed simultaneously.

- Cognitive actions
  1. The simultaneous recognition of different stimuli is possible if they are well structured and the number of stimuli is not greater than 3 to 4, (based on working memory capacity) and the stimuli are familiar. In other cases, input information should be received sequentially.
  2. If an operator recognizes well-known stimuli in a familiar situation, mental actions can be simultaneously combined with motor actions whatever the level of attention required.
  3. If an operator recognizes unfamiliar stimuli in unfamiliar situations and the system of expectation does not coincide with ongoing information, motor actions that require only lower and average levels of concentration of attention can be performed simultaneously with perceptual actions.
  4. The decision-making process and motor actions that require a high level of attention should be performed sequentially.
  5. Simple decision making (e.g., choosing between alternatives) can be simultaneously performed with motor actions that require a lower or average level of attention.
  6. Cognitive components should be performed sequentially.
  7. Simultaneous performance of activity elements that might result in working memory overload (e.g., requiring simultaneously keeping different data items in memory) should be performed sequentially.
  8. In stressful situations, or when personnel are not highly skilled, all activity elements requiring high levels of attention should be performed sequentially.

In the final stage of design, when analytical models are tested experimentally, some correction is possible. These are common steps both for ergonomic and for engineering design. It is important to keep in mind that all the above discussed models are idealized models of human performance. Between these models and real performance are probabilistic relationships. The more subjects perform the same tasks, the more closely their activity approaches developed models. Sometimes a practitioner can develop more individualized models. These models take into consideration individual strategies of the performer. Based on idealized models of performance, a practitioner can evaluate design solution, HCI, safety, efficiency of performance, etc. Experimental procedures significantly reduce and become simpler or are sometimes eliminated altogether.

3.3 DESIGN OF WORK ACTIVITY TIME STRUCTURE

3.3.1 DESIGN OF WORK ACTIVITY TIME STRUCTURE IN SEMIAUTOMATIC AND AUTOMATIC SYSTEMS

In semiautomatic systems an operator receives information directly from the environment or indirectly through devices that were developed for receiving information from the environment or the machine. Based on this information, the operator is manipulated by different controls. The machine provides the power and the human provides control. The physical workload in this system is significantly reduced. In this system, cognitive components of activity are closely connected with manual control. Therefore, cognitive and motor components of activity are equally important in these systems.

Even in automatic systems manual control is important. In emergency situations the operator functions as a monitor entering the control loop to override the automatic system. In such cases, the operator manually controls the system. Usually the operator is involved in an emergency situation very rarely, and at the same time such situations are very critical. Therefore, these rare, unexpected, and at the same time crucial, situations should be carefully designed. Moreover, as discussed before, manual, semiautomatic, and automatic regimes of control should be carefully balanced.

Here, as an example, we will discuss the task that was considered in section 3.2.2. In this section, we give an algorithmic description of this task and then describe the time structure of activity during task performance. We also compare experimental data obtained before with data that we received through analytical procedures. An algorithmic description of the task is presented in Table 3.12. This description starts after switch (1) is turned into vertical position, only after that is number 1 or 2 lit. Therefore, algorithmic description begins when the right hand is in the start position.

After the algorithmic description of the task, one can develop time structure of activity during task performance. We will describe three versions of realization of task. They correspond to three versions of task performance that was studied using chronometrical methods in section 3.2.2. The first version corresponds to the situation when one uses all instruments and controls including hinged lever (7) and multipositioning switch (9). The hinged lever should be turned in one of four positions (up–down and right–left). However, before the subject can do it he must press button (8) with the thumb (the button was installed directly into the lever's handle). Movements in one to four positions were similar and required the same time for performance. The multipositioning switch, according to this version of the algorithm, should be turned to position 5 according to the information presented from the digital display (3). After that the green bulb turned on and the subject moved the right arm to the right direction and pressed the green button (11). Here we want to mention that this version of algorithm realization is presented between other versions of algorithms in a chance manner.

In the second version of algorithm realization after moving hinged lever (7) to the required position, digital indicator (3) presents number "0" and simultaneously red bulb (4) turned on. After that the subject moves the arm to the red button (10) and presses it. Therefore, in this particular situation, multipositioning switch (9) is eliminated.

TABLE 3.12
Algorithmic Description of Task on Experimental Control Board (the General Description)

Members of algorithm	Description of member of algorithm
O_1^{α}	Look at first digital indicator
$l_1 \overset{1}{\uparrow}$	If number 1 is lit, perform ${}_1O_2^{\varepsilon}$; if number 2 is lit, perform ${}_2O_2^{\varepsilon}$
${}_1O_2^{\varepsilon}$	Move the two-switch 6 to the right
$\overset{1}{\downarrow}{}_2O_2^{\varepsilon}$	Move the two-switch 6 to the left
O_3^{α}	Determine whether to turn on the digital indicator 3 or the signal bulbs 4 or 5
$L_2 \overset{2(1-3)}{\uparrow}$	If neither the digital indicator 3 nor the signal bulbs 4 or 5 are turned on ($L_l = 0$) perform O_4^{ε}; if digital indicator 3 presents numbers 1–9 ($L = 1$) perform L_4; if digital indicator 3 presents number 0 and bulbs 4 or 5 turn on perform l_5
$\overset{2(1)}{\downarrow} O_4^{\varepsilon}$	Move right arm to the fourth positions hinged lever 7, grasp the handle and press the button 8 with the thumb
$O_5^{\alpha w}$	Wait for 3 sec
O_6^{α}	Determine the pointer's position on the pointer indicator 2
$L_3 \overset{3(1-4)}{\uparrow}$	If the pointer position is 1 perform ${}_1O_7^{\varepsilon}$; if 2 perform ${}_2O_7^{\varepsilon}$; if 4 perform ${}_4O_7^{\varepsilon}$
$\overset{3(1)}{\downarrow} {}_1O_7^{\varepsilon}$	Move the hinged lever 7 to the position that corresponds to the number 1
⋮	
$\overset{3(4)}{\downarrow} {}_4O_7^{\varepsilon}$	Move the hinged lever 7 to the position that corresponds to the number 4
O_8^{α}	Determine whether the second digital indicator 3 or the signal bulb 4 or 5 is turned on
$\overset{2(2)}{\downarrow} L_4 \overset{4(1-10)}{\uparrow}$	If the digital indicator 3 displays a number 1 perform ${}_1O_9^{\varepsilon}$; if number 2 perform ${}_2O_9^{\varepsilon}$; if number 9 perform ${}_9O_9^{\varepsilon}$; if digital indicator 3 presents number 0 and bulbs 4 or 5 turn on perform O_{10}^{α}
$\overset{4(1)}{\downarrow} {}_1O_9^{\varepsilon}$	Turn multipositioning switch 9 to position 1
⋮	
$\overset{4(9)}{\downarrow} {}_9O_9^{\varepsilon}$	Turn multipositioning switch 9 to position 9
O_{10}^{α}	Determine which one of the two bulbs 4 or 5 (red or green) turns on
$\overset{2(3)}{\downarrow} \overset{4(10)}{\downarrow} l_5 \overset{5}{\uparrow}$	If the red bulb 4 turns on ($l_5 = 0$) perform O_{11}^{ε}; if the green bulb 5 turns on perform O_{12}^{ε}
O_{11}^{ε}	Move the arm to the red button 10 and press it
$\overset{5}{\downarrow} O_{12}^{\varepsilon}$	Move the arm to the green button 11 and press it

In the third version of algorithm realization after the two-positioned switch (6) was moved to the required position, the digital indicator (3) demonstrated that number "0" and red bulb (4) simultaneously turned on. The subject then moves the arm to the red button (10) and presses it. Therefore, in this version of task performance hinged lever (7) and multipositioning switch (9) were eliminated. The purpose of our study was to demonstrate how we could develop a general time structure of activity during task performance that was almost the same as the first version of algorithm realization. We also present the time structure of activity that corresponds to the other two versions of algorithm realization. The time structure of activity helps us to not only precisely describe the structure of activity but also to determine the time of task performance. If time determined activity based on analytical procedures is similar to experimental data described before, it means that our time structures are correct.

There are 110 versions of task realization. However, the number of studied versions should be much less. This can be explained by the fact that a significant number of versions of task performance are close in their content and temporal parameters. For example, the difference in realization time of two different versions of task (versions of realization algorithm) that differ only in switch position (9) on one–two positions is not essential and this difference can be ignored. Therefore, we need to select the most representative versions of task algorithm realization. This design principle does not contradict the multivariable features of operator performance. If designers consider spacing possible strategies of performance, they should describe first of all the most representative strategies of task performance, or strategies that are the most critical.

In the case when our goal is to design a time structure of activity without using experiments, it is necessary to know the performance time of separate cognitive or motor actions or operations. In the considered task, the subjects received signals, evaluated them, made decisions, and performed actions. Mental actions consisted of simple recognition and decision making, and retrieving from long-term memory well-known information. This kind of activity is common for an operator's activity in semiautomatic systems when they perform familiar tasks. Therefore, in this study we limit ourselves to using the MTM-1 system or data from engineering psychology handbooks.

The duration of perceptual and mental actions or operations can be determined based on a technique presented in a handbook of engineering psychology (Lomov, 1982). The duration of mental actions can also be determined based on microelement EF. The MTM-1 system also takes into consideration cognitive processes during the performance of different motions. For our example, we selected the following data: reading pointer display 0.4 sec; recognition of simple signal 0.4 sec; decision making at sensory–perceptual level 0.29; retrieval of information and simple, well-known signal 0.25 to 0.35 sec; EF recognition and making decision "yes–no" "if–then" 0.27 sec.

It was discovered that in complex decision-making actions, when wrong decisions have undesired consequences and the operator checks actions, the duration of the decision-making process increases. In time restricted conditions the existence of repetitive checking actions was discovered during a study of astronaut tasks (Hrunove et al., 1974). For the determination of the duration of motor actions we used system MTM-1.

The decision making involved in this task is the simplest. It merely requires the operator to recognize the initiating stimuli and to remember the (usually) binary rules associated with it. Because of this, stimulus recognition is conflated with decision making. According to our classification it is decision making at the sensory–perceptual level. MTM-1 system assigns 0.27 sec for this type of action. In our task there are also more complicated decision-making actions at the sensory–perceptual level. They require simultaneous receiving of information from several instruments and making decisions based on it (two to three stimuli can be perceived as one operative unit of information that requires no more than one fixation of eye). For this kind of decision making at the sensory–perceptual level we can assign 0.3 sec. According to the rules described in Section 4.1.2, the time for the simplest recognition and decision-making operations at the sensory–perceptual level can be determined by dividing element EF by two, 1/2 EF being related to sensory–perceptual operation and the other 1/2 EF being related to the decision-making operation. This rule will be applied in our example. However, one should understand that this is simply a conventional rule that helps us pay attention to perceptual and decision-making mental operations.

Eye travel time in our study had not been taken into consideration due to the fact that the control board was relatively small. The first version of the algorithm of realization was almost the same as the general algorithm of task performance. The only difference was that the arm was always turned to multiposition switch position 5 and at the final stage the subject always pressed button (11).

The graphic presentation of time structure described the first version of algorithm realization. When describing motor actions one should always consider motor operations as components of these actions. In a study of cognitive actions, cognitive operations are described only for complex cognitive actions. The exception should be made for decision-making actions at the sensory–perceptual level when perceptual and decision operations are considered separately.

In the tables presented, time is given in most cases in units that correspond to 0.01 min. In those cases where time is given in seconds, it is shown in Table 3.13 to Table 3.15.

We presented above the time structure of activity in tabular form. However, the most informative and comprehensive method is a graphical form of time structure. This method of presentation can usually be done after a tabular form is developed. Sometimes the graphical method helps us to correct a preliminarily developed table of time structure. Both tabular and graphical forms of description of time structure are regarded as informative models that present qualitative and temporal aspects of activity during task performance.

We present in Figure 3.6 a general graphical model of the time structure of activity during the performance of the above described task (first version).

In the presented graphical model of the time structure of activity, individual elements of activity are presented as a horizontal line. The elements are specified by symbols above the segments. Microelement EF describes the perceiving of signals as simple decision making at the sensory–perceptual level that includes "yes–no" or "if–then" decisions, etc. A segment under EF designates the duration of this kind of mental action. In the same way, duration of any other segment designates the duration

TABLE 3.13
Time Structure of Activity during Performance of Task According to the First Version of Algorithm

Members of algorithm	Description of elements of tasks (technological units of analysis)	Description of elements of activity (psychological units of analysis)	Time
O_1^{α}	Look at first digital indicator	Simultaneous perceptual operation	0.15 sec
l_1	If the number 1 is lit, turn switch left (perform ${}_1O_2^{\varepsilon}$;) if the number 2 is lit turn switch right (perform ${}_2O_2^{\varepsilon}$)	Simultaneous perceptual operation	0.15 sec
${}_1O_2^{\varepsilon}$ or ${}_2O_2^{\varepsilon}$	Move the two-positioned switch 6 to the right or move switch 6 to the left	M2,5A	0.14
O_3^{α}	Determine that the digital indicator 3 or the signal bulbs 4 or 5 did not turn on	Simultaneous perceptual operation	0.15 sec
L_2	Decide to move an arm to the hinged lever 7 and press button 8 (perform O_4^{ε})	Decision-making operation at a sensory–perceptual level	0.15 sec
O_4^{ε}	Move right arm to the four-position hinged lever 7, grasp the handle and press the button 8 with the thumb	RL1 + R13A + AP2 ~~G1A~~	1.15
$O_5^{\alpha w}$	Wait for 3 sec	Waiting time	3.00 sec
O_6^{α}	Determine the pointer's position on the pointer indicator 2	Simultaneous perceptual operation	0.15 sec
L_3	Decide how to move hinged lever 7 (if the pointer position is 1 perform ${}_1O_7^{\varepsilon}$; if 2 perform ${}_2O_7^{\varepsilon}$; if 4 perform ${}_4O_7^{\varepsilon}$)	Decision-making operation at a sensory–perceptual level	0.15 sec
${}_1O_7^{\varepsilon}$ ⋮ ${}_4O_7^{\varepsilon}$	Move the four-position hinged lever 7 to the position that corresponds to the number of pointer indicator	M5B	0.27
O_8^{α}	Determine that digital indicator 3 demonstrates number 5	Simultaneous perceptual operation	0.15

(Continued)

TABLE 3.13
Continued

Members of algorithm	Description of elements of tasks (technological units of analysis)	Description of elements of activity (psychological units of analysis)	Time
L_4	Decide to move multipositional switch to position 5 (if the digital indicator 3 displays a number 5 perform ${}_1O_5^{\varepsilon}$)	Decision-making operation at a sensory–perceptual level	0.15
${}_5O_9^{\varepsilon}$	Turn multipositioning switch 9 to required position 5	RL1 + R13A + G1A + T150S	1.12
O_{10}^{α}	Determine that bulb 5 (green) turns on	Simultaneous perceptual operation	Overlapped by motor activity (0.15 sec)
l_5	Decide to press green button 11 (if the green bulb 5 turns on ($l_5 = 1$) perform 0_{11}^{ε})	Decision-making operation at a sensory–perceptual level	Overlapped by motor activity (0.15 sec)
O_{11}^{ε}	Move an arm to green button 11 and press it	RL1 + R26B + G5 + AP2	1.46
Total	Working time — 3.73 sec; waiting time — 3 sec		6.73 sec

TABLE 3.14
Time Structure of Activity during Performance of Task According to the Second Version

Members of algorithm	Description of elements of tasks (technological units of analysis)	Description of elements of activity (psychological units of analysis)	Time
O_1^{α}–O_7^{ε}	From operator O_1^{α} to O_7^{ε} the same as in Table 3.13		1.83 sec
O_{10}^{α}	Determine that red bulb 4 turns on	Simultaneous perceptual operation	Overlapped by motor activity (0.15 sec)
l_5	Decide to press red button 10	Decision-making operation at a sensory–perceptual level	Overlapped by motor activity (0.15 sec)
O_{11}^{ε}	Press red button 10	RL1 + R26B + G5 + AP2	1.46
Total	Working time — 2.7 sec; waiting time — 3 sec		5.7 sec

TABLE 3.15
Time Structure of Activity during Performance of Task According to the Third Version

Members of algorithm	Description of elements of tasks (technological units of analysis)	Description of elements of activity (psychological units of analysis)	Time
O_1^α	Determine that digital indicator 1 demonstrates number 2	Simultaneous perceptual operation	0.15
l_1	Decide to turn two-positioning switch 6 right (perform ${}_2O_2^\varepsilon$)	Decision-making operation at a sensory–perceptual level	0.15
${}_2O_2^\varepsilon$	Move switch 6 to right	M2,5A	0.15
O_3^α	Determine that red bulb 10 turns on	Simultaneous perceptual operation	Overlapped by motor activity (0.15 sec)
L_2	Decide to press button 10 (perform O_{11}^ε)	Decision-making operation at a sensory–perceptual level	Overlapped by motor activity (0.15 sec)
O_{11}^ε	Move arm to the red button 10 and press it	RL1 + R39B + G5 + AP2	1.68
Total	Working time — 1.38 sec; waiting time — 3 sec		4.38 sec

of other elements of activity. According to the introduced rule (rule 2, page 122) in order to distinguish simple perceptual actions from decision-making actions at the sensory–perceptual level, one can sometimes divide EF into perceptual and decision-making mental operations. In more complicated situations the duration of decision-making actions can be evaluated experimentally or the required data can be taken from other sources. For example, O_6^α and L_3 are also involved in the decision-making process at the sensory–perceptual level. We defined duration of this element based on data from a handbook of engineering psychology (Lomov, 1982) and divided this action in the same way as EF. Those elements of activity overlapped by other longer elements are designated by a dashed line. For example, O_{10}^α and l_5 overlapped by O_{11}^ε.

Because of that, for O_{10}^α and l_5 we did not assign time for performance. One can make conclusions about the possibilities of performing actions simultaneously or sequentially, based not only on analyses of separate actions or operations, but also on the analysis of possible strategies of task performance. For example, if a performer is very skilled and the consequences of a wrong action are not important, then actions can be performed simultaneously. If actions are not automated and errors are undesirable, they should be performed sequentially.

Some symbols used in Figure 3.6 require additional explanation. M2.5A means, "move object against stop" when the distance is 2.5 cm. Letters P and D over the segment mean "perception" and "decision making." RL1 means "normal release

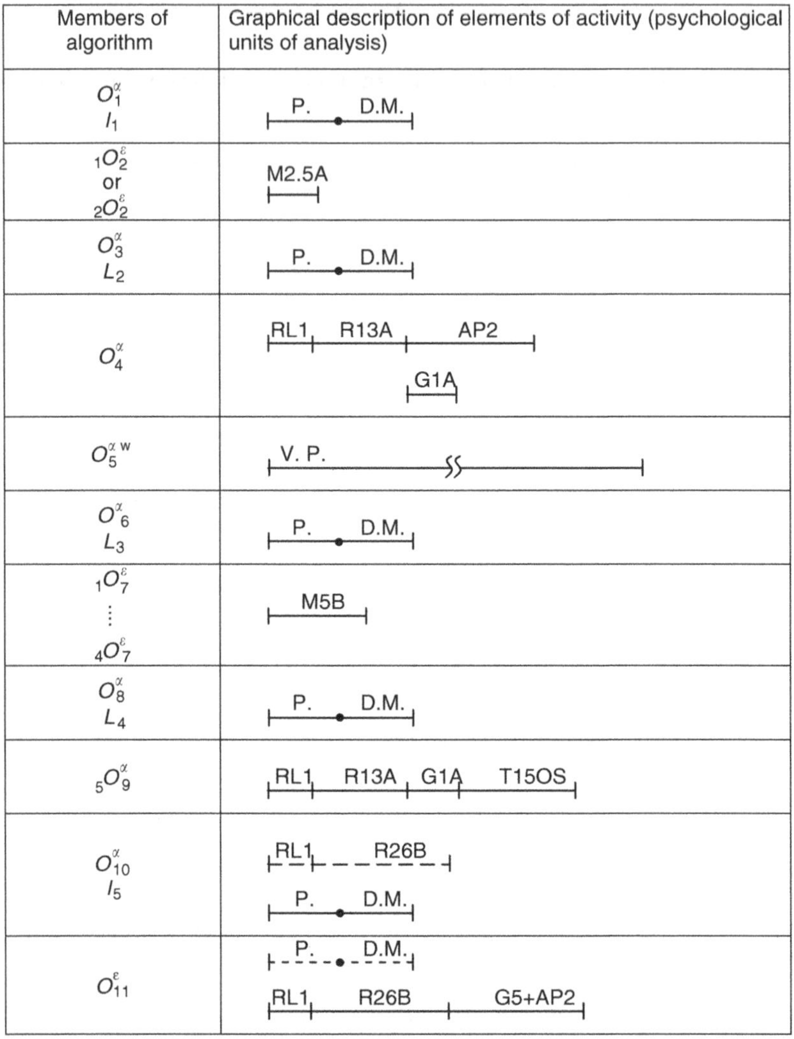

FIGURE 3.6 Graphical representation of time structure of task performance on experimental control board (the first version of algorithm).

performed by opening fingers." R13A means "reach to the object in fixed location, when distance is 13 cm." AP2 designates "apply pressure with effort less than 15 kg." G1A means "easily grasped." This element is overlapped by AP2. The letters W and P mean "waiting period." M5B designates "move an object 5 cm to approximate location (requires an average level of concentration of attention)." T180 S designates "turn 180° with small effort (from 0 to 1 kg)." In a similar manner other elements of activity are designated.

Let us consider units of analysis that are used during algorithmic description of task and temporal analysis of activity. The first two members of the algorithm

(O_1^{α} and l_1) are the result of artificially dividing element EF into two separate mental operations, which are related to different members of the algorithm. This was performed for purposes of distinguishing, in future analysis, members of the algorithm associated with decision making at a sensory–perceptual level from members of algorithm that comprise simultaneous perceptual actions (operation). This is why in Section 2.3.2, we introduced an artificial rule according to which in some situations we can divide decision-making actions at a sensory–perceptual level into operations and relate them to different members of algorithms, when one member of the algorithm is associated with the perceptual stage and others the with decision-making stage. In all other situations we divide tasks into separate members of algorithms according to the recommendations described in Section 2.3.2. For example, a member of algorithm O_2^{ε} contains one motor action "move arm to the lever, grasp it, and simultaneously press the button with the thumb." This action, in turn, consists of the following motor operations (motions) "move arm," "grasp the handle," and "press the button with the thumb." All these operations are integrated by the goal of action. In a similar manner other members of the algorithm are described.

Here we compare the experimental data presented in section 3.2.2 with the analytical estimate derived in this section. We examine the average time of task performance for three version of task, working on one control panel under normal conditions, with a synchronous stopwatch regime (see Table 3.9). The analytical estimates are presented in Tables 3.13–3.15 (working time). Experimental times of task performance were 3.85 sec, 2.93 sec, and 1.27 sec on versions one through three of the task respectively. The analytically derive estimates were 3.73 sec, 2.7 sec, and 1.38 sec. A comparison of the experimental data, and analytical estimates illustrates that the analytical estimates are sufficiently precise.

3.3.2 Time Structure of Work Activity during Manual Task Performance

The following is an example of developing a time structure of production operation under laboratory conditions. We select, as the first step, a laboratory experiment, because this gives us an opportunity to carefully control different versions of task performance without any violation of safety requirements. In order to conduct this experiment a special physical model of production operation was developed (Figure 3.7).

This model consisted of a pin board that contained 30 holes for metal pins. On the front of the board were two hand push buttons. Behind the pin board was a box containing pins. On the left side, there was a panel with 30 cells that lit up when each space was filled. This panel also contained a stopwatch. The stopwatch turned on when the subject pressed two buttons and turned off when the last space was filled in. The left side panel allowed measurement of time performance and also registered the sequence of performance. In this particular experiment, we measured only the time of performance. Therefore, the lit up cells were not used in this experiment. There are regular (without a flute) pins and fluted pins (see pins inserted into the space).

In the first version of the task, all the pins were without a flute. In the second and third versions of the task, we used fluted pins for the purpose of manipulating the

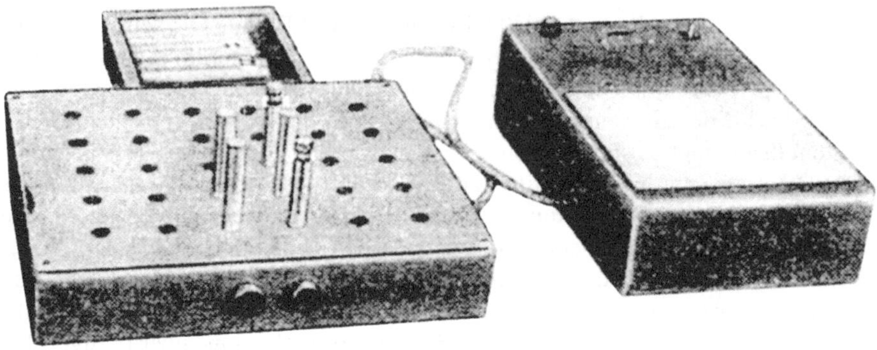

FIGURE 3.7 Physical model of production operation for pins installation.

TABLE 3.16
Algorithmic Description of the Task "Installation of Pins" (First Version)

Members of algorithm	Description of members of algorithm
$\omega\downarrow O_1^{\varepsilon}$	Move both hands to the pin box and grasp two pins
O_2^{ε}	Put pins in a hole in any position
$\omega\uparrow_{\omega}$	Always false logical condition (return to O_1^{ε} and repeat 15 times until task is finished)

complexity of task performance. In the experimental example, personnel had to fill a pin board with thirty pins in a particular order and according to specific rules. In order to manipulate the complexity of performance, the rules were also changed according to which the subjects were to install pins into the holes in the second and third versions of the task. In all three versions of task performance, the subjects used both arms to fill in the center row with pins, starting with the space closest to them and then working upwards and outwards.

Let us consider the first version of the task (pins without flutes). This task consists of two members of the algorithm: "Move both hands to the pin box and grasp two pins" and "Put pins in a hole in any position." Each member of algorithm consists of two simultaneously performed motor actions. The algorithmic description of this version of the task is presented in Table 3.16.

From this example, one can see that there are no separate cognitive actions, particularly those that are related to decision making and described as logical conditions. It does not mean that cognitive components of activity were ignored during task analysis. They were taken into consideration during qualitative analysis of motor actions, but were not considered apart from motor activity.

In Table 3.17 we present a time structure of activity for the first version of the task. The same members of algorithm are repeated 15 times. The time structure represents just one occurrence of each member of the algorithm.

TABLE 3.17
Time Structure of Activity for First Version of Task "Installation of Pins" (Table Form of Time Structure)

	Left hand			Right hand		
Members of algorithm	**Description of elements of task**	**Description of elements of activity**	**Time**	**Description of elements of task**	**Description of elements of activity**	**Time**
O_1^{ε}	Move left hand to the pin box and grasp pin	R32B + G1C1	1.23	Move right hand to the pin box and grasp pin	R32B + G1C1	
O_2^{ε}	Put pin in a hole in any position	M22B + mM10C ~~T90S~~ + + P2SE + RL1	2.08	Move pin in a hole in any position	M22B + mM10C ~~T90S~~ + + P2SE + RL1	
Total in units (0.01 min)		O_1^{ε} and O_1^{ε} performed one time				3.31 units
Total for task in sec		O_1^{ε} and O_1^{ε} performed fifteen times				29.88 sec

We used "technological units of analysis" to describe each member of the algorithm. At the next stage of time structure design, we transformed them into "psychological units of analysis."

For this purpose in this example, we used system MTM-1. The cognitive components of motor activity at this stage were considered during the selection of microelements from system MTM-1. Let us explain the symbols used in Table 3.17.

In the first action "Reach pin and grasp" 32 cm is the average distance of hand movement. Therefore R32B means "Reach single object in location" which may vary slightly from cycle to cycle, when the distance of movement is 32 cm. G1C1 is the microelement (motor operation) when the subject grasps a nearly cylindrical object with a diameter greater than 12 mm. According to MTM-1 rules, when hand movements cover significant distance only approximately, the last 10 cm before the final destination usually should be described by microelement MC (which requires a high level of concentration of attention). The first stage of action is usually related to RA (requiring a minimum level of concentration of attention). However, in our example, this stage of movement requires some anticipative control (how to grasp a pin). Hence for this stage we select method RB.

In the second action "Move pin, position it into the hole and release it," when the hand moves a pin M22B + mM10C means "move object in approximate position with in a distance of 22 cm and then without interruption move hand in exact position with in a distance of 10 cm." During movement of the pin to the hole the hand simultaneously turns the pin 90°. This movement is described by microelement T90S (turn object 90° with small effort). P2S means positioning of a symmetrical object, second class of fit, and easy to handle. RL1 stands for normal release (open fingers).

This simple example allowed us to describe what kinds of units of analysis are used in the study of this version of the task. The task consists of two different members of an algorithm. Each member of the algorithm, in turn, consists of two simultaneously performed actions. Motor actions include motions or motor operations according to activity theory. Therefore, this example demonstrates a hierarchically organized system of units. At the first stage we use technological (typical elements of task) units of analysis. Later we transfer them into psychological units (typical elements of activity) of analysis by using system MTM-1.

Let us consider the second version of the task. In this version, ten pins have a clearly visible flute. Therefore, the pins are put in the holes according to specific rules:

1. If the pins are regular (without a flute), they can be installed in any position.
2. If a fluted pin is picked up by a subject's left hand, it must be so placed that the flute is below the hole.
3. If a fluted pin is picked up by a subject's right hand, it must be so placed that the flute is above the hole.

If a subject erroneously installs a pin, he can correct it before finishing the task. This error, of course, increases performance time. A subject is permitted to make no more than one error per five installations. All tasks are divided into fifteen similar sub-tasks and are repeated fifteen times. This means that for sub-task, "taking and putting pins," the probability of error equals 1/75.

It was a sufficiently complex manual task because it included different logical rules and was performed multiple times in accordance with the specified time. Since the sequence of actions in this operation depended on logical conditions, an algorithmic description could be constructed. This description is given in the following manner: an afferent operator is designated by symbol O^{α}, complex logical conditions connected with decision making are designated by symbol L, and logical conditions connected with making decisions about correcting errors are designated as l.

Logical conditions for when fluted pins are absent in both the left and the right hands can be designated in the following way, $L_1 = (l_l = 0 \text{ and } l_r = 0)$.
Logical conditions for when fluted pins are present in the left hand and absent in the right hand can be designated as $L_2 = (l_l = 1 \text{ and } l_r = 0)$.
Logical conditions for when fluted pins are present in the right hand and absent in the left can be designated as $L_3 = (l_l = 0 \text{ and } l_r = 1)$.
Logical conditions for when both the left and the right hands possess fluted pins can be designated as $L_4 = (l_l = 1 \text{ and } l_r = 1)$.

Since the number of pins is known, including those with a flute, the possible different pin combinations during task performance can also be predicted. Based on this data, we can calculate the probabilistic features of production operations. There are 10 pins with a flute and 20 without a flute. The probability of events when pins will be with flutes is 1/3 and without flutes is 2/3. The probabilities of occurrences of the pins with and without flute being in either the right or the left hand are two independent events. The probability of both independent events is equal to the product of the probabilities of these independent events. Hence the probability of occurrences of the pins without flutes will be $2/3 \times 2/3 = 4/9$, the probability of events when both pins have a flute is $1/3 \times 1/3 = 1/9$, and the probability of the occurrence of the event when pins with flutes will be in the right (or left) hand is $1/3 \times 2/3 = 2/3$ (see Table 3.18).

TABLE 3.18
Probabilistic Features of Task Performance (Installation of Pins)

Logical conditions	Probability
$(l_l = 0 \text{ and } l_r = 0) = L_1$	4/9
$(l_l = 1 \text{ and } l_r = 1) = L_4$	1/9
$(l_l = 0 \text{ and } l_r = 1) = L_3$	2/9
$(l_l = 1 \text{ and } l_r = 0) = L_2$	2/9
$l_{\text{l.er.}} = 1$	1/75
$l_{\text{r.er.}} = 1$	1/75
$l_{\text{l.er.}} = 1 \text{ and } l_{\text{r.er.}} = 1$	1/75

An algorithm of task performance allows only one error with a probability of 1/75. Based on the obtained data, we can describe the algorithm in tabular form as in Table 3.19.

TABLE 3.19
Algorithmic Description of the Operation "Installation of Pins" (Second Version)

Members of algorithm	Description of members of algorithm
$\omega\downarrow O_1^{\varepsilon}$	Move both hands to the pin box and grasp two pins
O_2^{α} $({}^1O_2^{\alpha}\text{–}{}^4O_2^{\alpha})$	Determine the type and position of the pins: ${}^1O_2^{\alpha}$ — both pins without flutes; ${}^2O_2^{\alpha}$ — in the left hand pin with flute; ${}^3O_2^{\alpha}$ — in the right hand pin with flute; ${}^4O_2^{\alpha}$ — both pins with flutes
$L_1 \overset{1(1\text{–}4)}{\uparrow}$	If a flute, in either the right or left pin is absent ($l_l = 0$ and $l_r = 0$), perform O_3^{ε}. If this condition is not observed ($L_1 = 0$), transfer to L_2. If L_2 is also not noticed ($L_2 = 0$), transfer to L_3. If L_3 is not observed ($L_3 = 0$), transfer to L_4
$\overset{1(2)}{\downarrow} L_2 \overset{2(1\text{–}3)}{\uparrow}$	If the left pin is fluted and the right pin is not ($l_l = 1$ and $l_r = 0$), a decision must be made to install the right pin in any position and to turn the *left pin* so the fluted side would be placed *inside a hole*, and perform O_4^{ε}. If the flute exists on the right pin instead of the left ($L_2 = 0$), transfer to L_3. If L_3 also equals 0, then transfer to L_4
$\overset{1(3)}{\downarrow}\overset{2(2)}{\downarrow} L_3 \overset{3}{\uparrow}$	If the left pin is not fluted but the right pin is ($l_l = 0$ and $l_r = 1$), a decision must be made to put the left pin in any position and to turn the *right pin* so that the fluted side would be placed *outside a hole*, then perform O_5^{ε}. If L_3 is also not observed ($L_3 = 0$), transfer to L_4
$\overset{1(4)}{\downarrow}\overset{2(3)}{\downarrow} L_4 \overset{4}{\uparrow}$	If both pins have a flute, ($l_l = 1$ and $l_r = 1$), then a decision must be made to install the left pin according to L_2 and to install the right pin according to L_3 (perform O_6^{ε}), (if $L_2 = 1$ and $L_3 = 1$ then O_6^{ε})
$\overset{1(1)}{\downarrow} O_3^{\varepsilon}$	Put pins into a hole in any position
$\omega_1\uparrow_{\omega^1}$	Always false logical condition (return to O_1^{ε} and repeat until task is finished)
$\overset{2(1)}{\downarrow} O_4^{\varepsilon}$	A pin without flute should be put in a hole in any position. A fluted pin in the left hand must be installed according to L_2. (*left pin* with fluted side is placed *inside* a hole)
$\omega_2\uparrow_{\omega^2}$	Always false logical condition (go to O_8^{α})
$\overset{3}{\downarrow} O_5^{\varepsilon}$	A pin without flute should be put in a hole in any position. A fluted pin in the right hand must be installed according to L_3 (*right pin* with fluted side is placed *outside* a hole)
$\omega_2\uparrow_{\omega^2}$	Always false logical condition (go to O_8^{α})

(Continued)

TABLE 3.19
Continued

Members of algorithm	Description of members of algorithm
$\overset{4}{\downarrow} O_6^{\varepsilon}$	Install a right pin according to L_3 (*right pin* with fluted side is placed *outside* a hole), and install the left according to L_2 (*left pin* with fluted side is placed *inside* a hole)
$\omega 2 \downarrow O_7^{\alpha}$	Check to see if a pin was put in wrong
$l_{5er} \overset{5(1-3)}{\uparrow}$	If both pins are in correct positions ($l_{5er} \uparrow^{5(1)}$) go to ω_1; If one pin is in an incorrect position ($l_{5er} \uparrow^{5(2)}$), transfer to O_8^{ε}. If both pins are installed incorrectly ($l_{5er} \uparrow^{5(3)}$), transfer to O_9^{ε}
$\overset{5(2)}{\downarrow} O_8^{\varepsilon}$	Remove one incorrectly installed pin from its hole, and install it again according to the proper instructions
$\overset{5(1)}{\downarrow} \omega_1 \uparrow \omega 1$	Always false logical condition (return to O_1^{ε} and repeat until task is finished)
$\overset{5(3)}{\downarrow} O_9^{\varepsilon}$	Remove both incorrect pins from their holes and install again according to the proper instructions
$\omega_1 \uparrow \omega 1$	Always false logical condition (return to O_1^{ε} and repeat until task is finished)

The algorithm requires some explanation. The symbolic formula is presented on the left side of the table (vertical left column). The algorithm works from top to bottom. The always-false logical conditions ω_1 or ω_2 do not demonstrate real performance. They are only used to demonstrate the logic of the algorithm. The numbers in parentheses on top of the arrows demonstrate possible outputs from logical conditions. For example, L_1 has an arrow and above it the numbers 1–4 in parentheses. These numbers demonstrate possible outputs from logical conditions. For example, $\uparrow^{(1)}$ corresponds to situations when $l_1 = 0$ and $l_r = 0$.

If $L_1 = 0$, the algorithm starts to work according to L_2. If $L_2 = 0$, the algorithm can work according to $\uparrow^{(2)}$ (move to L_3). If this condition does not exist, the algorithm can work according to $\uparrow^{(3)}$. If $L_3 = 0$ one must transfer to L_4. Regular logical conditions possess only two meanings or output, 0 and 1. In this example, logical conditions possess more than two outputs with different probabilities. In this example, the algorithm, accordingly, has a probabilistic character. In the algorithm, two different false logical conditions always exist and are designated by the symbol ω. One is designated by the symbol ω_1 and the other by the symbol ω_2. A false logical condition ω_1 is repeated three times and there is only one false logical condition ω_2. For example, after the performance of O_5^{ε} according to false logical condition ω_2, the subject transfers to O_7^{α}. According to algorithmic rules, O_7^{α} and l_{5er} should be performed only after the pins are incorrectly put in holes.

The task was manual with "yes–no" rules; therefore we can use system MTM-1. Knowing an algorithm of task performance and the probabilistic features of task performance, one can describe the temporal structure of activity. As the first step, we describe the time structure of task in tabular form (Table 3.20).

TABLE 3.20
Time Structure of the Task "Installation of Pins" (Second Version)

Members of algorithm	Executive activity		Mental processes		Time 0.01 min
	Left hand	Right hand	During right-hand work	During left-hand work	
O_1^{ε}	R32C+G1C1	R32C+G1C1	—	—	1.31
O_2^{α}	(if ${}^1O_2^{\alpha}$–1/2EF; if ${}^2O_2^{\alpha}$–${}^4O_2^{\alpha}$–1/2EF + 1/2EF)		1/2 EF	1/2 EF	—
L_1	—	—	1/2 EF	—	
L_2	—	—	1/2 EF	1/2 EF	
L_3	—	—	1/2 EF	1/2 EF	—
L_4	—	—	1/2 EF	1/2 EF	—
	1/2EF	1/2EF			
O_3^{ε}	[M22B+mM10C ~~T90S~~ +P2SE+RL1] +	[M22B+mM10C ~~T90S~~ +P2SE+RL1] +	—	—	2.08
O_4^{ε}	[M22B+0.44EF+ +m M10C ~~T90S~~] + +P2SE+RL1	EF [M22B+mM10C ~~T90S~~] + +P2SE+RL1	—	—	2.26

O_5^{ε}			The same			2.26
O_6^{ε}	M22B+0.25T90S+ +EF +mM10C ~~T90S~~ +P2SE+RL1		EF M22B+mM10C ~~T90S~~ + +P2SE+RL1	—	—	2.59
O_7^{ε}			when	$l_{5er}^{5(1)}\uparrow$		0
			when	$l_{5er}^{5(2)}\uparrow$	1/2 EF	0.22
			when	$l_{.5er}^{5(3)}\uparrow$	EF	0.43
$l_{.5\,er}$			when	$l_{.5er}^{5(1)}\uparrow$		0
			when	$l_{.5er}^{5(2)}\uparrow$	1/2 EF	0.22
			when	$l_{.5er}^{5(3)}\uparrow$	EF	0.43
O_8^{ε}	G1A+M8A+ +T180S+P2SE+ +RL1	or	G1A+M8A+ +T180s+P2SE+ +RL1			2.08
O_9^{ε}	G1A+M8A+ +T180S+P2SE+ +RL1	and	G1A+M8A+ +T180s+P2SE+ +RL1	(Corrective actions are performed simultaneously)		2.08

It has to be taken into account that when the time structure of activity is complicated, the tabular method of time structure development is not sufficient because it is very difficult to describe complicated time structure using only the tabular method. Hence, after the table with some approximation is developed, one can switch to a graphic method of time structure description. This helps to interpret time structure and analyze possible strategies of activity. A comparison of tabular and graphical time structure models allows correcting and making more precise both methods of time structure descriptions. This corresponds to systemic–structural principles of design when a specialist uses different models of the same object within time structure analysis. The above presented model is the final description of the time structure of activity. Several intermediate results of this description that are corrected through comparison of graphical and tabular models are not shown.

A graphical model of the time structure of activity during task performance "installation of pins" (second version) is depicted in Figure 3.8.

As in the previous section the horizontal segments index the duration of the individual elements of activity. These elements are specified by symbols above the segments. At the left of each line are designated which physical or mental elements are involved — Right Hand (RH), Left Hand (LH), and Mental Process (MP). These symbols are used when cognitive and motor members of the algorithm can be performed at the same time.

In our example, cognitive components of activity significantly overlapped with motor components. For example, O_2^α has a duration of 1/2 EF and is designated by a solid line but is performed together with M22B (motions), which belong to O_4^ε or O_5^ε or O_6^ε. Therefore, when we described O_2^α, the motion M22B is designated by a dashed line (M22B does not belong to O_2^α). The probability of occurrence of a particular element of activity in a time structure is designated by the letter "P." The operators O_4^ε and O_5^ε are identical. Symbols RH and LH should be switched in their position during the performance of one or another operator. Therefore we put them together in the left column of the graph as "O_4^ε or O_5^ε." The probability of both their performance is $P = 4/9$.

The symbols used in Figure 3.8 to describe activity elements require some explanation. R32C means "move the hand to a distance of 32 cm to a precise position" (which requires a high level of concentration of attention): G1C1 — "grasp the nearly cylindrical object with a diameter of more than 12 mm"; M22B — "move the object 22 cm to the approximate location" (which requires an average level of concentration of attention); mM10C — "move the object 10 cm to an exact location" (requires a high level of concentration of attention).

Symbol m before M — "the arm continues movement after the first movement" (M22B) without interruption; T90S — "turn the object with a small weight 90°" (from 0 to 2 lb); P2SE "position symmetrically the object" (align, orient, and engage one object with another object), requires light pressure and easy to handle; and RL1 — "normal release performed by opening fingers." In the same way, according to the rules of the MTM-1 system, one can designate other elements. There are three situations: picking up two pins (one in each hand) without flutes; picking up two pins, one with a flute; and picking up two pins, both with flutes. Let us consider O_3^ε, O_4^ε, or O_5^ε and O_6^ε.

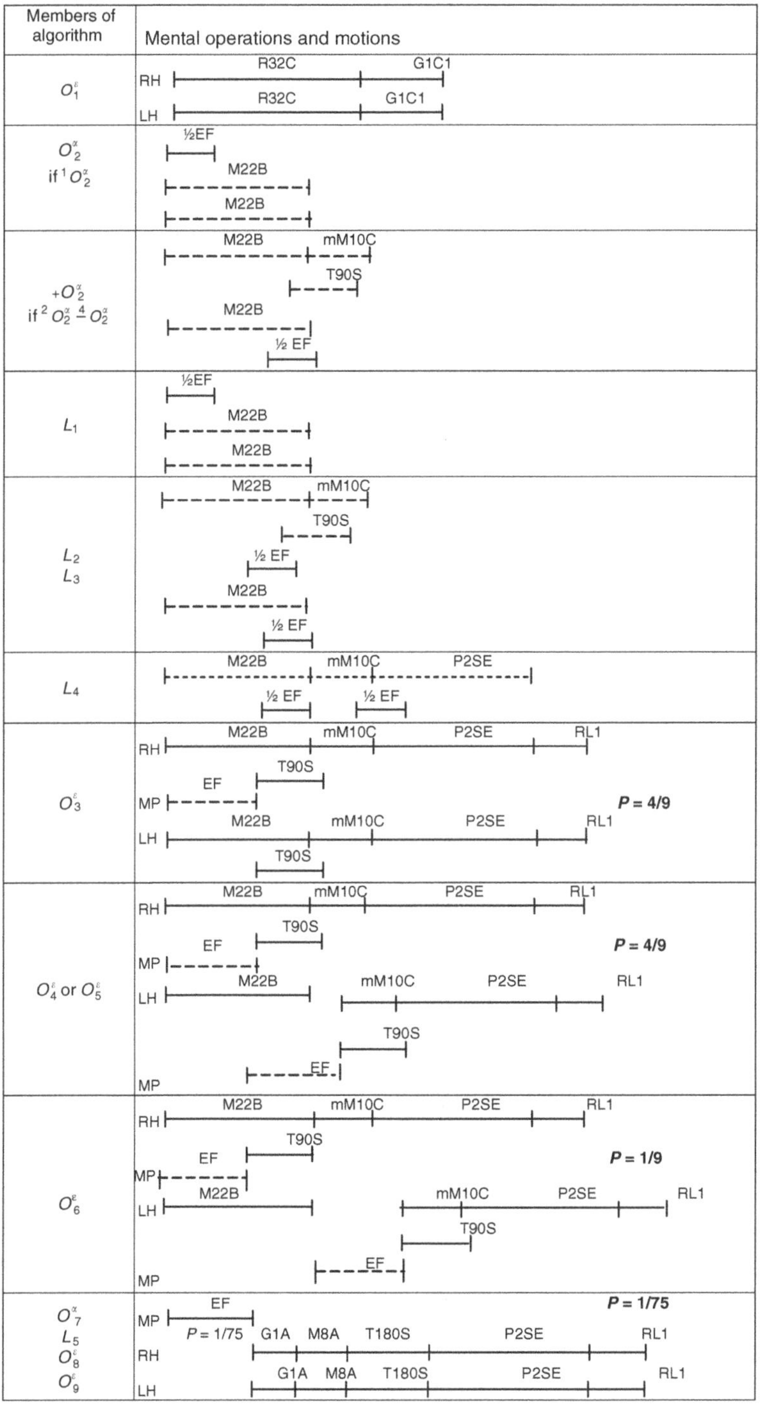

FIGURE 3.8 Graphical model of activity time structure during installation of pins (second version).

O_3^{ε} describes the situation in which two pins are picked up without flutes, with both arms moving simultaneously. In the beginning these movements were performed by M22B and at the end the movements were performed by mM10C. This is because the final step of moving the pins to the hole requires a high level of attention. At the start of a movement of the hands, the subject recognizes pins and decides that he can orient them in any position. Recognition and decision making are noted as microelement EF, designated by a dashed line (which belong to other members of the algorithm). Further, without any stopping motions of the arm, a subject turns both pins 90° to put them in the vertical position (T90S). The simultaneous installation of pins into a hole with the method P2SE is then followed. After that the subject releases the pins (RL1).

The strategy of performance changes when one pin is fluted (the subject performs O_4^{ε} or O_5^{ε}). At the beginning of the movement the subject recognizes that one pin is fluted and makes the decision (EF) to turn the other pin (without a flute) to a vertical position. Immediately after that motion, "Turn" (T90S) is executed and the pin turns 90° in any position. Then the subjects installs the pins and releases P2SE and RL1. The decision on how to orient the fluted pin in the left hand is performed immediately after deciding that the first pin can be installed in any position, which is why the second EF usually follows after the performance of the first EF.

The strategy of performance becomes even more complicated when two pins are fluted (the subject performs O_6^{ε}). At the beginning of the movement the subject recognizes that both pins are fluted and makes the decision (EF) to turn one pin in a particular position. He then turns that pin in the required position (performs T90S). After the rotation of one pin is finished, the subject concentrates on the other pin, makes the decision, and rotates the second pin into the required position. As a result, interruption between rotations of both pins increases.

Let us now consider major units of analysis used for the description of time structure of activity. For this purpose, we will use a graphical model of activity time structure (Figure 3.8). A hierarchically organized system of units of analysis, which have been covered in the previous chapters, is used in the development of activity time structure. This system consists of cognitive and motor actions and operations and members of the algorithm which are basic units of analysis for the description of time structure. Typical elements of task (technological units) are transferred into typical elements of activity at this stage of analysis. For instance in the algorithmic description of task (Table 3.19) the operator O_1^{ε} consists of two simultaneously performed similar actions "take pin from the box by left hand" and "take pin from the box by the right hand." These are technological units of analysis and do not provide a precise description of the actions. We cannot precisely imagine what kind of actions the two hands perform. This method of description does not tell us about distance and time of performance. We do not know exactly how a subject moves his hand or grasps the pin. At the next stage of time structure description, we transfer technological units of analysis into typical elements of activity or psychological units. For this purpose, we can use our example of MTM-1 where each action consists of two motor operations R32C and G1C1. These motor operations (motions) are an example of the psychological units of analysis. It is very interesting that professionals employing MTM-1 do not use concepts of technological (typical elements of tasks) or psychological units of analysis. As a result, they ignore the concept of actions. At the first step of analysis (algorithmic description), we describe actions in terms of technological units. Then these units are transferred into psychological units (at the second stage of time structure analysis). Units of analysis have a hierarchical organization. A member of the algorithm O_1^{ε} consists of two actions that are integrated by high ordered goals (take two pins simultaneously). Each action has a goal (taking a pin by the right or left arm). Actions, in turn, consists of several motions (motor operations).

Let us consider the other example. O_2^{α} includes only the simplest psychological act. We were able to extract this act by dividing element EF that includes simple recognition mental operations and decision-making operations at the sensory–perceptual level. This is performed according to the rule discussed in the Section 3.1.2.

Finally, we consider a more complicated member of algorithm O_5^{ε}. It consists of two motor actions. However, they are combined with two cognitive actions that belong to other members of the algorithm (two EF designated by dashed lines). In optimal conditions, a subject usually can combine no more than three actions at the same time (capacity, working memory, and no more than three or four elements). In this example, we can see that during the performance of O_5^{ε} the subject should interrupt movements performed by the left hand. However, this strategy of performance is sufficiently complex for the subject even if it includes interruption of movement. Only because the two cognitive and two motor actions are similar and these actions are performed multiple times can they be combined in time. This example demonstrates that not only actions that belong to the same member of algorithm can be combined in time but also actions that belong to different members of an algorithm can be combined in similar ways.

If one considers the time structure of an activity and the time-line chart presented in multiple sources one can see that they are totally different methods of describing the temporal aspects of human work. A time-line chart uses the duration of a task or subtask as a unit of analysis (technological units of analysis). The time structure of an activity is not described in this instance. Hence, time-line analysis gives only a rough evaluation of the operator's performance.

Finally, we describe the third version of the task. This version of the task is more complicated than the second one. In this version, additional rules were introduced. According to these rules, the subjects should install pins into the first and second central columns (column numbers 1 and 2) and the two edge columns (column numbers 5 and 6) using rules for the second version of the task. If subjects should install the pins in column numbers 3 and 4 that have a middle position, then they should turn the pins in the opposite direction (the pin in the right hand should be installed so that the flute will be inside the hole and the flute in the left hand will be above the hole). This requires memorization of additional rules and retrieving information from long-term memory.

Therefore, for the algorithmic description of the third version of the task we need to introduce additional operators associated with recalling the required rules from memory O_0^{μ}. We also need to introduce new logical conditions. These logical conditions are L_0 and 'L_1–'L_4. According to L_0 the subjects decide which rules (logical conditions) they should use. Logical conditions L_1–L_4 are related to the rules already described in task version two; 'L_1–'L_4 are the new logical conditions applied only in the third version. They present new rules. Operator O_0^{μ} and logical conditions 'L_1–'L_4 are used only twice when subjects complete columns 1–2, start columns 3–4, and when subjects start to work with columns 5–6. The algorithmic description of the third version of the task is presented in Table 3.21. In order to keep the same numerical symbols as in Table 3.21, we introduce a superscript as ' and subscript zero.

3.3.3 Illustration of Principles of Qualitative Analysis of Time Structure of Manual Tasks

As can be seen, algorithmic description of a task in combination with description of time structure of activity is a powerful tool for task analysis. These stages of analysis are performed in combination with qualitative analysis. In our study (installation of pins), we used such units of analysis as member of algorithm, cognitive and motor actions, and operations. We also described their logical organization and the probabilistic features of activity structure. Such analysis allowed us to describe the possible strategies of task performance. The first version of the task is the simplest one and is similar to production operations in industry. The complexity of these tasks usually increases along with an increase in the variety of motor actions. These tasks are skill based or deterministic tasks. They have a strictly determined sequence of actions and do not require logical analysis and changes in the order of their performance. The MTM-1 system is usually applied to these kinds of production operations. However, our study demonstrates that a combination of algorithmic description of task and developed procedures of time structure description significantly broaden the application of the MTM-1 system.

The main purpose of this study was to determine how informative developed models are and whether they can reflect the real strategies of manual task performance. We present below a brief analysis of these models and their application to the study of manual tasks of different complexity. We start with the simplest version of the task. Algorithmic and time structure description of activity during task performance demonstrate that this version of task does not include independent cognitive components of activity. This task is deterministic or a skill based task. Attention during task performance is divided between two motor actions that are performed simultaneously. This task does not require high levels of concentration of attention, excluding the short period of time when subjects install pins into holes (microelement mM10C). These two elements of activity that are performed by the left and right hands can be performed simultaneously because they are identical and performed in most cases in optimal visual field, and the subjects perform the same task multiple times. The pace of performance is high and can be maintained only after introducing breaks. It can also be recommended that one hand interrupts movement when the other performs mM10C. This can insignificantly increase the time of task performance and also possibly delay the corresponding duration of mM10C. It is also recommended that micropauses between production operations be introduced.

Let us consider the second version of the task. This version is not typical of regular production operations in industry. The task is more complicated because it includes four basic logical conditions. Each logical condition is associated with decision-making actions. There are also three supplementary logical conditions that

TABLE 3.21
Algorithmic Description of the Operation "Installation of Pins" (Third Version)

Members of algorithm	Description of members of algorithm
$\omega 1 \downarrow O_1^{\varepsilon}$	Move both hands to the pin box and grasp two pins
O_2^{α} $({}^1O_2^{\alpha}\text{–}{}^4O_2^{\alpha})$	Determine the type and position of the pins: ${}^1O_2^{\alpha}$ — both pins without flutes; ${}^2O_2^{\alpha}$ — in the left hand pin with flute; ${}^3O_2^{\alpha}$ — in the right hand pin with flute; ${}^4O_2^{\alpha}$ — both pins with flutes
${}^{a}O_0^{\mu}$	Recall rules associated with $L_1 - L_4$ or $`L_1 - `L_4$
${}^{a}L_0 \overset{0}{\uparrow}$	If pins should be installed in rows 1–2 or 5–6 then $L_1 - L_4$ should be used (when $L_0 = 0$ to perform L_1). If pins should be installed in rows 3–4 ($L_0 = 1$) then $`L_1 - `L_4$ should be used
$L_1 \overset{1(1\text{–}4)}{\uparrow}$	If a flute, in either the right or left pin is absent ($l_l = 0$ and $l_r = 0$), perform O_3^{ε}. If this condition is not observed ($L_1 = 0$), transfer to L_2. If L_2 is also not noticed ($L_2 = 0$), transfer to L_3. If L_3 is not observed ($L_3 = 0$), transfer to L_4
$\overset{1(2)}{\downarrow} L_2 \uparrow^{2(1\text{–}3)}$	If the left pin is fluted and the right pin is not ($l_l = 1$ and $l_r = 0$), a decision must be made to install the right pin in any position and to turn the *left pin* so the fluted side would be placed *inside a hole*, then perform O_4^{ε}. If the flute exists on the right pin instead of the left ($L_2 = 0$), transfer to L_3. If L_3 also equals 0, then transfer to L_4
$\overset{1(3)}{\downarrow} \overset{2(2)}{\downarrow} L_3 \overset{3}{\uparrow}$	If the left pin is not fluted but the right pin is ($l_l = 0$ and $l_r = 1$), a decision must be made to put the left pin in any position and to turn the *right pin* so that the fluted side would be placed *outside a hole*, then perform O_4^{ε}. If L_3 is also not observed ($L_3 = 0$), transfer to L_4
$\overset{1(4)}{\downarrow} \overset{2(3)}{\downarrow} L_4 \overset{4}{\uparrow}$	If both pins have a flute ($l_l = 1$ and $l_r = 1$), then a decision must be made to install the left pin according to L_2 and to install the right pin according to L_3 (perform O_5^{ε}) (If $L_2 = 1$ and $L_3 = 1$ then O_5^{ε})
$\overset{0}{\downarrow} `L_1 \overset{1(1\text{–}4)}{`\uparrow}$	If a flute, in either the right or left pin is absent ($l_l = 0$ and $l_r = 0$), perform O_3^{ε}. If this condition is not observed ($`L_1 = 0$), transfer to $`L_2$. If $`L_2$ is also not noticed ($`L_2 = 0$), transfer to $`L_3$. If $L_3$ is not observed ($`L_3 = 0$), transfer to $`L_4$
$\overset{1(2)}{`\downarrow} `L_2 \overset{2(1\text{–}3)}{`\uparrow}$	If the left pin is fluted and the right pin is not ($l_l = 1$ and $l_r = 0$), a decision must be made to install the right pin in any position and to turn the *left pin* so the fluted side would be placed *outside a hole*, and perform O_4^{ε}. If the flute exists on the right pin instead of the left ($L_2 = 0$), transfer to L_3. If L_3 also equals 0, then transfer to L_4
$\overset{1(3)}{`\downarrow} \overset{2(2)}{`\downarrow} `L_3 \overset{3}{`\uparrow}$	If the left pin is not fluted but the right pin is ($l_l = 0$ and $l_r = 1$), a decision must be made to put the left pin in any position and to turn the *right pin* so that the fluted side would be placed *inside a hole*, then perform O_4^{ε}. If L_3 is also not observed ($L_3 = 0$), transfer to L_4
$\overset{1(4)}{`\downarrow} \overset{2(3)}{`\downarrow} `L_4 \overset{4}{`\uparrow}$	If both pins have a flute, ($l_l = 1$ and $l_r = 1$), then a decision must be made to install the left pin according to L_2 and to install the right pin according to L_3 (perform O_5^{ε}) (If $L_2 = 1$ and $L_3 = 1$ than O_5^{ε})

(Continued)

TABLE 3.21
Continued

Members of algorithm	Description of members of algorithm
$\overset{1(1)}{`\downarrow}\ \overset{1(1)}{\downarrow}\ O_3^{\varepsilon}$	Put pins into a hole in any position
$\omega_1 \uparrow_{\omega^1}$	Always false logical condition (go to O_7^{ε} and repeat until task is finished)
$\overset{2(1)}{\downarrow}\ O_4^{\varepsilon}$	A pin without flute should be put in a hole in any position. A fluted pin in the left hand must be installed according to L_2 (*left pin* with fluted side is placed *inside a hole*)
$\omega_2 \uparrow_{\omega^2}$	Always false logical condition (go to O_7^{α})
$\overset{3}{\downarrow} O_5^{\varepsilon}$	A pin without flute should be put in a hole in any position. A fluted pin in the right hand must be installed according to L_3 (*right pin* with fluted side is placed *outside a hole*)
$\omega_2 \uparrow_{\omega^2}$	Always false logical condition (go to O_7^{α})
$\overset{4}{\downarrow} O_6^{\varepsilon}$	Install a right pin according to L_3 (*right pin* with fluted side is placed *outside a hole*), and install the left according to L_2 (left pin with fluted side is placed inside a hole)
$\omega_2 \uparrow_{\omega^2}$	Always false logical condition (go to O_7^{α})
$\overset{2(1)}{`\downarrow}\ `O_4^{\varepsilon}$	A pin without flute should be put in a hole in any position. A fluted pin in the left hand must be installed according to $`L_2$ (*left pin* with fluted side is placed *outside a hole*)
$\omega_2 \uparrow_{\omega^2}$	Always false logical condition (go to O_7^{α})
$\overset{3}{`\downarrow}\ `O_5^{\varepsilon}$	A pin without flute should be put in a hole in any position. A fluted pin in the right hand must be installed according to $`L_3$ (*right pin* with fluted side is placed *inside a hole*).
$\overset{4}{`\downarrow}\ `O_6^{\varepsilon}$	Install a right pin according to $`L_3$(*right pin* with fluted side is placed *inside a hole*), and install the left according to $`L_2$ (*left pin* with fluted side is placed *outside a hole*)
$_{\omega^2}\downarrow O_7^{\alpha}$	Check to see if a pin was put in wrong
$l_{5er}\ \overset{5(1\text{–}3)}{\downarrow}$	If both pins are in correct positions ($l_{5er} \uparrow^{5(1)}$) go to ω_1; If one pin is in an incorrect position ($l_{5er} \uparrow^{5(2)}$), transfer to O_8^{ε}. If both pins are installed incorrectly ($l_{5er} \uparrow^{5(3)}$), transfer to O_8^{ε}
$\overset{5(2)}{\downarrow}\ O_9^{\varepsilon}$	Remove one incorrectly installed pin from its hole, and install it again according to the proper instructions
$\overset{5(1)}{\downarrow}\ \omega_1 \uparrow_{\omega^1}$	Always false logical condition (return to O_1^{ε} and repeat until task is finished)
$\overset{5(3)}{\downarrow}\ O_9^{\varepsilon}$	Remove both incorrect pins from their holes and install again according to the proper instructions
$\omega_1 \uparrow_{\omega^1}$	Always false logical condition (return to O_1^{ε} and repeat until task is finished)

[a] O_0^{μ} and L_0 is used only twice during task performance, the first time when column numbers 1 and 2 are completed, and the second time when column numbers 3 and 4 are finished.

describe decision-making actions required for the correction of errors. The last happens with the least probability (see Table 3.18). This task is a rule-based task. According to activity theory, this is a probabilistic algorithm task. The existence of four basic logical conditions leads to an increase in the logical components of activity associated with decision-making actions and also results in overloading of working memory. The capacity of working memory usually does not exceed four units of memory. The process of keeping four logical rules in working memory can cause mental fatigue. During the performance of O_3^{ε} subjects can divide their attention between two actions that are performed by the left and right hands. During the performance of O_4^{ε}, when one fluted pin is in hand, it is also possible to divide attention between two hands. However, it is a more complex member of the algorithm. As a result, subjects can periodically divide attention strategy to switching strategy of attention.

Based on a comparative analysis of the time structure of O_3^{ε} and O_4^{ε}, one can conclude that in the first case subjects prefer simultaneous strategies of performance and in the second case use sequential strategies. This can be explained by the fact that instead of one decision-making action in the case of O_3^{ε} subjects perform two decision-making actions in the case of O_4^{ε} or O_5^{ε}. When one considers the time structure of activity during the performance of O_4^{ε} or O_5^{ε} it can be seen that that either the left hand or the right hand can interrupt movement. As a result, the duration of O_4^{ε} or O_5^{ε} is slightly increased in time in comparison with that of O_3^{ε}. All logical conditions are considered for tasks related to simple decision making at the sensory–perceptual level. However, due to the existence of different logical conditions, their quantity increases the workload of working memory, concentration of attention, and the variety of performed actions. This, in turn, can lead to an increase in mental fatigue. Hence, it is difficult for subjects to keep up the pace of performance during shifts that corresponds to the pace suggested by the MTM-1 system. If this task is performed multiple times during shift, the pace of performance of this version of the task should be reduced.

Next, we consider the member of algorithm O_6^{ε}. Activity becomes even more complicated during the performance of this member of the algorithm. When O_6^{ε} is performed both pins are fluted and the subject has to make two separate decision-making choices in sequence. The strategy of attention and activity in general is changed. Attention is no longer divided between two actions. In this situation, attention is switched from one action to another during decision making. This is the reason why the duration of the task is increased with the interruption of the movement of one hand.

Let us briefly consider the third version of the task. The increasing complexity of this version of the task, in comparison with the second one, can be explained as follows. New logical conditions should be introduced. The first one is L_0. This logical condition describes decision making associated with the position of rows that must be filled. It can be combined with other elements of activity in different ways. In most cases, it is not overlapped with motor components of activity. Overlapping is possible when two pins without flutes are in hand. In all other cases, it is more probable that these decision-making actions are not overlapped by motor components of activity. This strategy can significantly reduce possible errors. The decision-making action that is described by L_0 is also related to simple decision making at the sensory–perceptual level and according to (Lomov, 1982) requires 0.3 sec. Other complications are associated with logical conditions 'L_1 – 'L_4. They require turning pins in opposite directions when the subject puts them into holes. The addition of logical conditions not only increases the complexity of the logical aspects of a task, but also significantly increases the working memory workload. In such a case it becomes more difficult to maintain the pace recommended by the MTM-1 system. It is more suitable for the operator to work in time-restricted conditions when one can describe task performance with maximum speed. In general, the third version of the task is sufficiently complicated and cannot be performed multiple times without special organization of rest periods.

3.3.4 Pace of Performance in Semiautomatic and Automatic Systems

In the MTM-1 system, the pace of performance is equivalent to the walking speed of 5.7 km/h (Smidtke and Stier, 1961). However, according to physiological data the standard pace of a physical job should be 4.8 km/h. This pace guarantees that energy expenditure does not exceed 4.17 kcal/min or a workload that is equivalent to a heart rate of 100 beats/min. These are physiological criteria that are considered as the border between acceptable and unacceptable workloads during the performance of physical work. Attempts to use a higher pace of performance, according to the MTM-1 system, can be explained by the fact that this system was not developed for the study of physical work. The assumption is that this system was developed for mass production or assembly line work as in the electronics industry, etc., where one cannot observe substantial physical efforts. In such production conditions, workers perform the same production operations for an extended period of time. For such work, the pace of MTM-1 is considered optimal. However, according to the same experimental data, the pace in the MTM-1 system is too high, even for mass production, when workers produce easier work (Smidtke and Stier, 1961; Gal'sev, 1973). According to these data, it is necessary to use coefficient 1.2 to reduce the pace of performance; that is determined according to the MTM-1 system. Only after this correction, can the physiological costs of performed work approach standard physiological levels.

While studying an operator's work in semiautomatic and automatic systems, professionals encounter different situations. An operator does not perform the same production operation for a significant period of time. In contrast, one performs multiple different tasks. Moreover, tasks usually do not have the same sequence of actions as in a production line. Usually, in semiautomatic and automatic systems, they are algorithmic or rule-based and nonalgorithmic or knowledge-based tasks (Rasmussen, 1986; Bedny et al., 2005). These tasks can be classified as deterministic algorithmic tasks, probabilistic algorithmic tasks, semialgorithmic tasks, and nonalgorithmic tasks according to the activity approach. Moreover, because an operator encounters multiple tasks they can subjectively perceive algorithmic tasks as nonalgorithmic. As a result, the operator's skills do not receive as high a level of automatism as do a worker's skills on a production line. The system of expectations also influences the pace of performance. The more workers anticipate and predict what can happen in different situations, the higher can be the pace of performance. On the other hand, when there are numerous varieties of signals and possible responses in semiautomatic and automatic systems in which an operator performs a diversity of tasks, the readiness to react in different situations is reduced. The recognition of a situation and the ability to act adequately in it are complicated by the fact that it is multivariate and that the same signals may have multiple meanings. The MTM-1 system is designed for more predictable situations when a worker performs the same sequence of actions multiple times. In highly skilled repetitive tasks performed on a production line, the options are essentially nil, while in a complex task that combines cognitive and motor components, the number of options may be very large. Based on this analysis, one can conclude that the pace in the MTM-1 system is not optimal for those who work as operators in semiautomatic and automatic systems. Our chronometrical studies and subjective assessment of those who take part in experiments demonstrate that the pace of the MTM-1 system was evaluated as faster than optimal. This pace of performance is adequate for situations when operators should work in time-restricted conditions and limited emergency situations. However, this pace is much lower than the speed of separate reactions. Experiments also demonstrate that the pace that exceeds that of the MTM-1 is considered as excessive and cannot provide reliable performance in multitask conditions. In the sections discussed above, it was demonstrated that the Hick–Hyman Law, in the same way as the Fitts Law, ignores the concept of pace of performance. These laws were developed based on the study of different reactions. However, the speed of reactions significantly exceeds both the pace of the MTM-1 system and the speed of the operator's performance in time-restricted conditions. The informational approach can be used only in situations when an operator should react by single independent actions to a dependent number of stimuli. This pace is inadequate for the solution of design problems in real life situations.

In cognitive psychology, speed of reaction (particularly simple reaction) is usually considered relatively stable for each individual. However, under different conditions a subject can consciously or unconsciously change strategies of performance and thus the speed of reactions for the same individual can significantly change, even with respect to reaction time tasks. Nojivin (1974) studied the speed of simple sensory–motor reactions from the self-regulative aspect. Consequently, speed of performance significantly changes depending on the context of the operator's work.

A series of experiments was designed to introduce into instructions new requirements that could change the subjects' strategies based on the mechanisms of self-regulation. This is a functional analysis of activity according to the systemic–structural approach. There were seven experiments in this series.

In the first experiment, the subjects pressed a button after receiving a visual signal. They were asked to respond with any suitable speed; the requirements of the self-regulation process were, therefore, minimal. In the second experiment, the new requirement employed was a warning signal two seconds before the visual stimulus requiring the reaction. In the third experiment, the subjects were asked to react with maximum speed. In the fourth experiment, the subjects were given information about the speed of their reactions. In the fifth experiment, each subject was required to match his or her fastest reaction time in the previous experiment. The sixth and seventh experiments repeated the requirements of the first and second experiments to

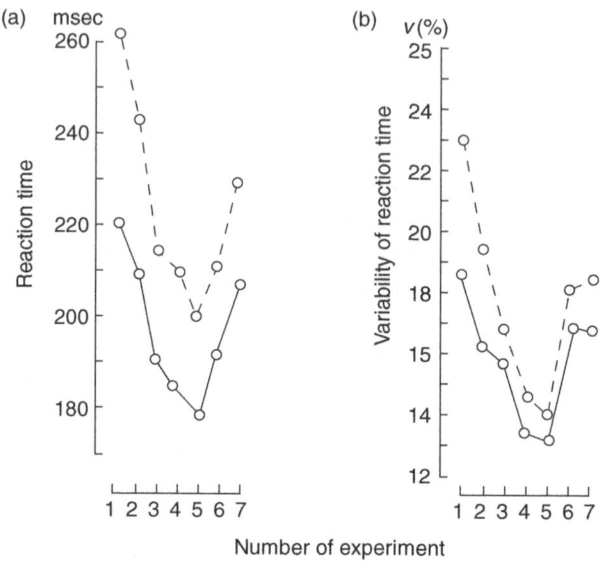

FIGURE 3.9 The effect of self-regulation on speed and variability of simple sensory-motor reaction. (a) Reaction time; (b) variability of reaction time. Dashed line — women; solid line — men.

see how the previous experiments (three to five) had influenced reaction time when all speed requirements were removed.

Figure 3.9 shows that the reaction speed increased from experiments one to five. Reaction time in experiments six and seven was less than that in experiments one and two; the differences were statistically significant. This means that preliminary trials changed the subjective perception of "suitable" speed. The high speed performance in the preliminary trials (three to five) had a positive effect on the speed of performance of the subsequent trials six and seven. As a result, the "suitable" pace changed but the subjects did not realize it. An analysis of the variability of reaction time demonstrates that increasing speed of performance and introducing information about reaction time reduces the variability of reaction time.

It is interesting to describe how the subjects learned their "suitable" pace in the first series of experiments. In the beginning, the subjects increased reaction time to a certain level during the first 10 to 15 trials, and then decreased it. Hence, the strategy of the subject's activity was as follows: they started working slowly, then increased the speed, reached the "suitable" pace, continued to speed up, and then went back to working at the "suitable" pace. That is because a verbal criterion such as "suitable" is not sufficiently precise for self-regulation of activity. That is why subjects tested various speeds of performance to try and find a "standard" for suitable speed based on their subjective feeling. This means that using an ambiguous criterion, such as a verbal requirement should be transferred into more subjective feelings about speed. This can be achieved based on a subjective comparison of different speeds.

The results of these experiments reveal that even in strictly predetermined stimulus–response situations the subjects develop different strategies that influence their response time. This suggests that complicated self-regulative processes take place even in the simplest reactions. The experiment discussed also demonstrates that the speed of a simple reaction can be changed significantly and depends on those variables not discussed by information theory. It is suggested that a subject organizes his behavior according to mechanisms of self-regulation and that pace depends on these mechanisms (Bedny, 1987).

An operator's activity in multitask performance has a complicated logically organized system of cognitive and behavioral actions that cannot be regarded as independent reactions. The readiness to perform voluntary acts is, in this case, significantly low in comparison with different reactions. The pace is regulated in the same acceptable range. In time-restricted conditions, when an operator must work quickly and precisely, the speed of performance corresponds to the pace used in the MTM-1 system (Bedny, 1979). This pace provides not only speed but also reliability of performance and is subjectively evaluated as expected. In some situations it is possible to introduce special coefficients that can correct pace, depending on the specificity of the particular situation. The application of these coefficients can be particularly useful in training conditions. It should be taken into consideration that the time of task performance depends not only on speed but also on the strategies of performance.

We present below an example that demonstrates how time structure is changed during the skill acquisition stages. Figure 3.10 shows the four strategies of performance required to install pins into the holes of a pin board during skill acquisition.

The pin's position in the hole depends on the following: whether the pins are fluted, and whether they are grasped by the right or left hand. Any line segments designate the duration of motor or mental components of activity (actions or operations). In Figure 3.10 the elements designate the movement of pins by method M22B

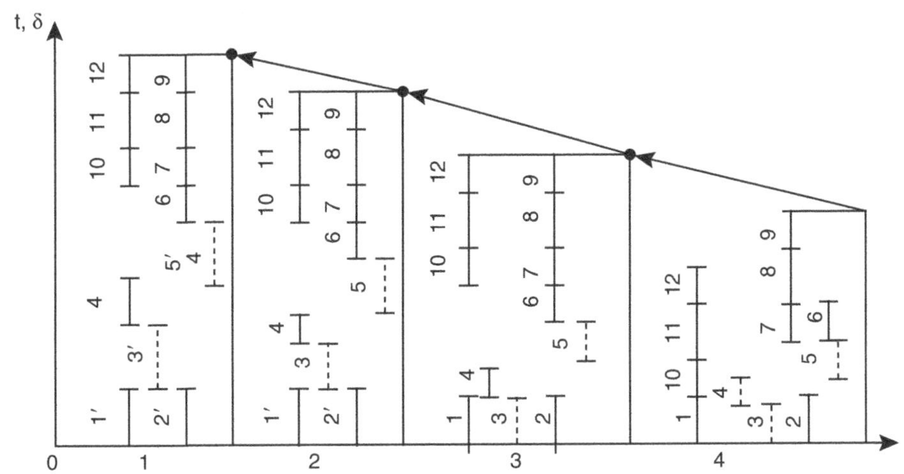

FIGURE 3.10 Time structure during skill acquisition stages.

(at a distance of 22 cm with average concentration); 3 and 5 designate the decision-making process when the skill has been acquired (in this situation, microelement EF is required); 4 and 6 designate turning pins at a 90° angle (microelement T90S); 7 and 10 designate moving pins by method M10C (at a distance of 10 cm with maximum concentration); 8 and 11 designate putting pins in a hole (microelement P2SE); and 9 and 12 designate release (microelement RL1).

The first strategy is characterized by the existence of motor pauses, the sequential recognition of signals and decision-making processes, the longer duration of the cognitive components of activity, a sequential turning of pins, and a high level of visual control of any movements. The second strategy has a similar allocation of different elements in time. The distinguishing factor is a reduction in the duration of cognitive components (duration of elements 3 and 5 is less than that of 3′ and 5′). As a result, the duration of motor pauses is also reduced. The third strategy is characterized by a reduction in visual control when moving pins with both hands from a container to the holes. As a result, elements 1 and 2 require less concentration during their performance and become shorter in duration (elements of activity 1′ and 2′ are performed by method MC substituted by elements 1 and 2, performed by method MB). The cognitive components of activity (element 3) are combined with motor elements 1 and 2, and element 4 (turning one pin) is executed immediately after elements 1 and 2. As result, the general duration of a task is further reduced. The last strategy involves a further reduction of motor pauses because the elements of activity 10, 11, and 12 are performed immediately after completing element 1. Element 6 (turning the pin) is combined with element 7 (moving pins to the hole). From Figure 3.10, one can see that the trainee transfers to the last strategy not directly but through intermediate strategies.

It was discovered that the simpler the task (e.g., when no pins were fluted), the greater the reduction in the number of intermediate strategies. This means that the more complicated the task, the more intermediate strategies needed. Therefore, learning can be considered to be changes in strategies of performance. We will consider this problem in a more detailed manner in Section 5.4. An analysis of the figure shows that time structure demonstrates how deferent components of an activity are allocated in time and how they are changed during skill acquisition. The transformation from one strategy to another is based on the trainee's past experience and evaluation of intermediate results, either consciously or unconsciously. This means that the skill acquisition process has a self-regulative nature. During skill acquisition the relationship between an automated level and a conscious level of self-regulation is changed (Bedny and Karwowski, 2004).

This example demonstrates that reducing the time of task performance involves not only increasing pace but also reconstruction of skills. Sometimes it is useful to use a more rapid pace of performance during training (above real-time training).

Experimental data presented in Table 3.13 demonstrate that even simple activity such as turning a switch and then pushing a button can be performed at different speeds when performed in isolation or in combination with other elements of a task. We have a number of other data which demonstrates that the same actions that are performed independently and in the structure of a holistic task are performed at different speeds (Bedny, 1987). According to Zarakovsky (2004), the quantitative characteristics of actions include time performance, precision, and reliability of actions. For evaluation of these characteristics, a specialist can determine tabular quantitative characteristics of standard actions and then determine the specificity of performing an action in an actual situation. Only after that can we introduce corrective coefficients for the estimation of quantitative characteristics of actions in specific conditions. Cognitive psychology, which does not use the concept of action, nor relate this concept of structure of activity, ignores critical aspects of time measurement.

In general, our studies demonstrate that we distinguish three levels of pace: very high, high, and average. A very high pace of activity is slightly slower than the operator's reaction time for various stimuli. This pace is possible only in those cases when an operator reacts to isolated signals, using discrete actions in highly predictable situations. For example, an operator can have a high level of readiness to push a button or throw a switch when a particular signal appears.

A high pace is that in which an operator performs a sequence of logically organized mental and physical actions in response to the appearance of different signals. It is essentially the same as that reported in MTM-1for motor activity. The pace of performance of mental actions should be determined based on an analysis of the strategies of their performance in a particular situation. This refers us to the functional analysis of activity which we will discuss later. The condition when a subject performs actions in a logically organized sequence lowers the degree of an operator's readiness to perform particular actions.

An average pace is that in which an operator performs tasks at his own subject time scale (there are no time constraints).

3.3.5 Algorithmic Description of Activity and Design of Activity Time Structure in HCI Tasks

Activity is a multidimensional system which requires different methods of study. Therefore, we cannot demonstrate one method of study in complete isolation from the other methods. In this section, we attempt to demonstrate the combination of algorithmic description with time structure design. During motor activity the user's major tools are the mouse and keyboard. There is a problem with segmentation of motor activity while the user works with a keyboard. The major criteria for segmentation of this kind of motor activity are the existence of the goal of motor action and the principle of rhythmic organization of repetitive motor motions. These criteria are interdependent and the potential goal of motor action demonstrates what should be achieved during this action. A goal reached informs the user that purposeful action is completed. A single keystroke or a mouse click without the awareness of a goal should be regarded as a motor operation which is a component of action. During repetitive performance of a sequence of motions they are segmented into rhythmically organized groups. The time intervals between motions inside a group are shorter than those between separate groups of motions (Gorbynova and Rokotova, 1970). According to Gordeeva and Zinchenko (1982) the intervals between groups include time for creation of the program which determines the sequence of motions. Therefore, the rhythmic organization of repetitive motions can also be used in the segmentation of sequential motor activity. At the same time the extraction of actions depends on the strategies utilized by the user. For example, if a single stroke becomes very significant it can be performed under conscious control and careful evaluation of the result. This stroke should be regarded as a separate motor action. Hence, any segmentation into separate actions depends on the strategies of performance. From this follows the dividing of activity into actions always associated with particular strategies of performance. Usually for design purposes a specialist utilizes a standardized strategy of performance. However, some other strategies can often be considered. At the same time, one should understand that any description of activity is only an approximation of real performance. Let us consider an example: suppose the first member of algorithm describes actions that include typing the user name and password. This is a description of human activity in technological terms. It is a highly automated sequence of motions the goal of which is to introduce into the computer the user name and password. It can be considered a separate motor action. The typing of a single and simple statement also should be considered a separate action.

As discussed before, there are macro- and microstructural levels of analysis. In a later section, we will use a macrostructural level of analysis during an algorithmic description of a computer based task because our task includes not only work with a computer but also the physical components of the equipment. However, the user often works with a computer only. In such a situation the physical components of work are eliminated. In these tasks, short cognitive and motor components prevail. Hence, the members of the algorithm include smaller actions and the algorithm consists of detailed units of analysis. In this case, the algorithmic analysis of HCI tasks are performed at the microstructural level of analysis. Time structure design is always performed at a microstructural level. Therefore, in this section we demonstrate microstructural analysis during an algorithmic description of the task and time structure design.

Experimental data presented here were prepared by Sengupta and Jeng (2003) under our supervision. As an example, they selected a drawing task. Subjects who had experience in performing simple drawing tasks were involved in the experiment. The subjects were required to recreate a figure in a particular area of the screen. The standard figure from which they were required to recreate their picture was located to the lower left corner of the screen. The study procedures included qualitative cognitive analysis, eye movements and mouse movement registration, video registration, an algorithmic description of the task and time structure analysis. Therefore, in this study, according to the principle of unity of cognition and behavior, cognitive and motor activities were studied together. All methods can be presented as three stages of analysis as suggested by systemic–structural theory of activity (qualitative stage, algorithmic analysis, and time structure analysis). The last stage of analysis associated with the evaluation of task complexity was not used in this study. We consider below only the algorithmic description of task and time structure analysis for a computer based drawing task. We also present a fragment of the task performed by the subjects (Table 3.22).

In this table, in the last column on the right, the time performances of algorithm members were also presented in milliseconds. This data was extracted from the third stage of analysis. The algorithm was derived from qualitative analysis which included retrospective protocol analysis, which consists of observation and expert analysis. It

TABLE 3.22
Algorithmic Description of Drawing Task Performance (Fragment)

Symbol	Member of algorithm	Classification of actions	Time (in msec)
O_1^{α}	Visual perception of the given figure	Simultaneous perceptual action	300
$l_1^{\mu}\overset{1}{\uparrow}$	Mental selection of required figure	Decision making at perceptual level (include following operations): 1. Actualization of information from memory 2. Decision making	250 370
$\overset{1\mu}{O_2}$	Sustain actualized information about selected circle in working memory (abandoned option)	See description below	—
$\underline{\underline{\downarrow \overset{1\ 2\mu}{O_2}}}$	Sustain actualized information about selected square in working memory during performance of O_3^{ε} (correct action)	Mnemonic action of maintaining information in working memory while performing O_3^{ε} (direct connection action)	300
$\underline{\underline{O_3^{\varepsilon}\ *}}$	Move pointer from the starting position to the drawing toolbox	Motor positioning under visual control (required coordination of eye and arm movement)	300
O_4^{α}	Examine area with related tools	Perceptual actions	200
$l_2\overset{2}{\uparrow}$	Selection of required tool	Decision making at perceptual level	200
$\overset{1\varepsilon}{O_5}$	Move to circle drawing icon and click (abandoned option)	See below	—
$\downarrow \overset{2\ 2\varepsilon}{O_5}$	Move to rectangle drawing icon and click (Correct action)	Motor action	380
O_6^{ε}	Move mouse to work (dotted) area (include coordination of eye and mouse movement operations)	Complex motor actions (involve visual motor coordination. Include following operations): 1. Quick eye movement to dotted area 2. Quick mouse movement to the dotted area without visual control 3. Slow mouse movement under visual control for eye and mouse position coordination	200 250 170

(Continued)

TABLE 3.22
Continued

Symbol	Member of algorithm	Classification of actions	Time (in msec)
$\underline{\underline{O_7^{\alpha}}}$	Visual evaluation of pointer position. Dynamical perceptual evaluation of mouse position during performance of O_8^{ε}	Perceptual action	780
$\underline{\underline{O_8^{\varepsilon}}}$	Motor movement of mouse pointer to the required position	Motor action	780
O_9^{ε}	Initial drawing of rectangular shape	Motor movement, which involved visual motor coordination	670
$\underline{\underline{O_{10}^{\varepsilon}}}$	Visual evaluation of shape and size of the rectangle	Perceptual action	400
$\underline{\underline{l_3\overset{3}{\uparrow}}}$	Decision about correction during performance of O_{10}^{ε}		400
$\overset{1\,\varepsilon}{O}_{11}$	Not to increase the size of the rectangle (abandoned action)		—
$\overset{3}{\downarrow}\overset{2\,\varepsilon}{O}_{11}$	Increase size of the rectangle	Motor movement, which involved visual–motor coordination	340

is also derived from instrumental analysis as eye movement registration and analysis of video.

In Table 3.22, the algorithmic description of the HCI task has certain differences from the algorithmic description given in the previous chapters. All members of the presented algorithm have a short duration. Therefore, this is a microlevel of the algorithmic description of the task. Some members of the algorithm are prone to abandonment, because in HCI tasks a user often does not know in advance the sequence of actions he has to perform. As a result, the user explores the possibility of different actions. This exploration can be performed in external or internal mental form. According to activity theory terminology, the external explorative activity with a database are called externalized. A user can observe the result of externalized explorative actions on the screen and evaluate them either positively or negatively. The members of an algorithm that contain negatively evaluated actions are called "abandoned options." The users correct the result of abandoned actions and then choose the next one. Finally, they select those actions that they evaluate as positive. This means that an HCI task includes in itself explorative externalized activity. This kind of activity is an important component of orienting activity that will be considered later during the description of functional analysis of activity. In the production environment, some externalized explorative components of activity can lead to corruption of the database. The possibility of such kind of actions should be eliminated during the design of HCI tasks. The less the possibility of any kind of abandoned options, the more efficient HCI task performance. Therefore, the most common abandoned options of a particular task

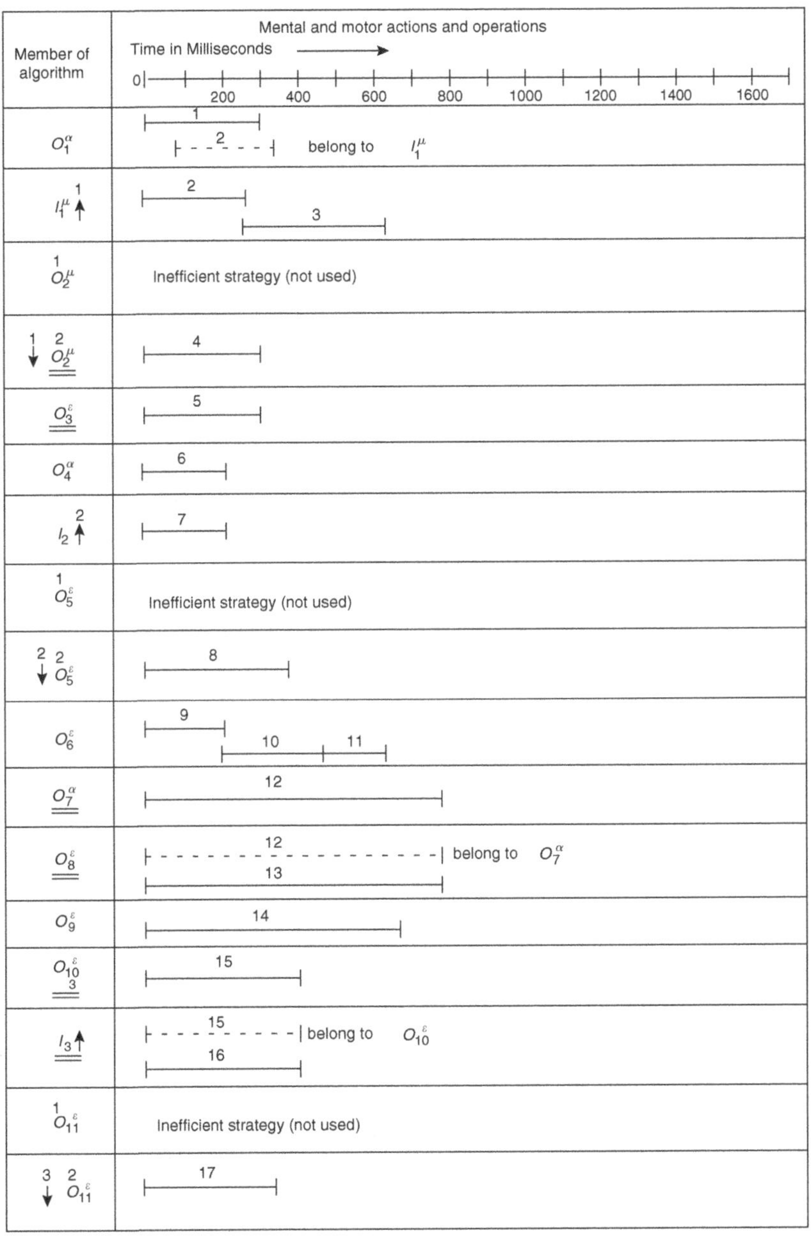

FIGURE 3.11 Time structure of activity during performance of HCI task (fragment).

should be presented in algorithmic analysis. Depending on the purpose of the study of the time performance of abandoned options, actions can be considered or ignored. There are members of the algorithm in Table 3.22 that are underlined by two lines. This means that such members of the algorithm are performed simultaneously with other members of the algorithm. For example, O_7^α and O_8^ε are "performed simultaneously." It is interesting to analyze the member of the algorithm $\overset{1}{O}{}_2^\mu$. Its performance involves sustaining information in memory about selected figures. This information is necessary in decision-making associated with logical condition l_2. However, the information selected for decision making is not immediately used. The user performs O_3^ε and O_4^α first and only after that uses information from memory for decision making (performance of logical condition l_2). Hence, when designing an HCI task one should strive to eliminate or minimize the time between getting information and using it for decision making. In general, the presented fragment of an algorithmic description of an HCI task allowed us to analyze the logical organization of cognitive and behavior actions and perform their qualitative analysis more efficiently.

The next stage of analysis involved time structure design. The time structure of the same fragment of the task is demonstrated in the Figure 3.11. The time structure of activity performance should be regarded as microanalysis of a computer based task. It gives us a very detailed description of the strategies of activity performance and information about their efficiency. Knowledge about the time structure of a human–computer interaction task can provide insight into the complexity of the various tasks that a subject has to perform. The time structure of task performance can be used in general for usability evaluation, such as user interface design and other related purposes.

4 Complexity of Task Performance

4.1 TASK COMPLEXITY EVALUATION

4.1.1 Complexity and Difficulty

There are different concepts of task complexity. For example, one can distinguish complexity of design process (Braha and Maimon, 1998), complexity of socio-technical systems (Vicente, 1999), task complexity, or complexity of task from a behavioral standpoint (Meister, 1999). We will consider task complexity which describes the relationship between the behavioral and physical domains during task performance. Task complexity is a major factor creating a challenge for operator performance. According to Lee et al. (1989) a complex task apparently requires greater use of cognitive and other abilities than simple tasks and, therefore, also influences the relationship between goal and task performance. It is one of the most important integrative characteristics of a task. This characteristic is also critical in determining cognitive efforts needed during task performance.

The term complexity may be said to be the converse of simplicity. Thus, the lower the complexity, the greater the simplicity. Therefore, the concept of complexity is critically important for usability evaluation. Different authors have considered different aspects of task complexity. For example, in multiple-choice tasks the number of alternatives is used as a measure of task complexity (Payne, 1976). In a case discussed by Kieras and Polson (1985), the number of production rules or condition–action pairs, in the form "if" as condition and "then" as action, that the user has to learn serves as the major criterion of complexity. For an inspection task, the number of different fault types is suggested as a measure of task complexity (Gallwey and Drury, 1986). The nature of the cause and effect relationship, in terms of whether it is obscure and unpredictable or relatively straightforward, influences task complexity (Meister, 1999). Task complexity may depend on the quantity of task elements and the specification of the interactions of different task elements and can also be defined in terms of the number of static and dynamic components of the task and interaction among these components.

The degree to which a task is unpredictable, and the uncertainty associated with that unpredictability, are other important factors influencing task complexity. Uncertainty in a task depends on both objective and subjective task characteristics. While the number of possible alternatives available in any task is one example of an objective characteristic of uncertainty, it is clear that an objective analysis of the number of alternatives presented in any given situation will not always coincide with subjective perceptions; an individual's lack of knowledge about the external world may result in his being unable to accurately predict environmental events or the outcomes of his decisions. Another aspect of task uncertainty is temporal uncertainty, which determines the degree of predictability of events in time. The dynamics of events in time, that is, how quickly they change, or can be changed, is also a source of uncertainty. The relationship between time restrictions and the risks consequent to making incorrect decisions also contributes to task complexity. In general, the wide range of views in the literature suggests that there are multiple factors determining the complexity of any task.

Having described some of the general factors of complexity, let us now consider more specific, cognitive factors of complexity. The specificity of memory workload is one important source of task complexity. According to Zarakovsky and Magazannik (1981) having more than three intermediate data on dynamic objects in memory is subjectively evaluated as a very difficult situation and produces errors. The duration for which information is kept in memory is also a critical factor.

The specificity of the extraction of information from long-term memory also influences any evaluation of task complexity. In relation to this the "degree of familiarity" of retrieving information is a factor. If the retrieved information has similarities with task-irrelevant information, then the complexity of the task is increased. Moreover, if relevant and irrelevant units of information are similarly familiar, the complexity of recalling is increased. Another factor of task complexity is the number of steps involved in the processing of information, a number that can change during training. Each step in information processing can have its own specific characteristics, and the conceptual, internal quantity of steps may differ from the quantity of formally described steps.

The sensory–perceptual characteristics of signals also influence the complexity of task performance. If perceived signals are in the threshold area, this results in an increase in complexity. Such threshold data represent the extreme performance which the individual is capable of, meaning that he must exert the maximum effort to perform successfully. The relationship between stimulus differences in the discrimination process is another sensory–perceptual factor that contributes to complexity. These sensory–perceptual factors are closely interconnected with decision making at the sensory–perceptual level; therefore, sensory and nonsensory factors are interdependent. Increasing the level of complexity in a thinking problem can decrease sensitivity. The factor of similarity between different signals influences the complexity of the perceptual components of the task. The more similar signals in the process of recognition, the more complicated the perceptual components of the task.

The various characteristics of the decision-making process at the verbal–thinking level are also important factors of task complexity. For example, the number of contradicting solutions influences the complexity of a task. The decision-making process is more complicated when it is determined by information extracted from memory. On the other hand, decision-making processes become easier to perform when they are predominantly determined by external stimuli or by information provided from external sources. Complexity also increases when the subject is required to alter stereotypical actions.

Finally, the level of concentration of attention is also a critically important characteristic of complexity. The more a subject concentrates, the more complex the task. The foregoing considerations make it clear that task complexity is the basic characteristic determining demands on the cognitive components of activity during task performance. Consideration of task complexity is critical for the successful evaluation of reliability and efficiency of performance, the dynamics of skill acquisition, and for the prediction of time performance.

Complexity can also be used to evaluate motor components of activity. For example, the more precise a motor action is, the more concentration of attention it requires, and the more complex the action becomes. If when studying motor actions one pays attention to movement coordination and regulation, this can be regarded as evaluating the complexity of the motor components of a task. However, when one considers the amount of physical effort expended by an individual during the performance of motor actions, this is an evaluation of the physical characteristics of a task. Therefore, the concept of complexity can be applied to the motor aspect of task or — more specifically — to the mental regulation of movements. In Section 4.3 it is demonstrated that motor actions include cognitive mechanisms. In reality, one cannot perform a complex motor task without significant mental effort and concentration. The relationship between these different components of a task (cognitive, motor) is also critical in evaluating the complexity of task performance.

The concept of complexity is associated not only with the cognitive and motor aspects of activity but also with its emotional–motivational components. Emotional tension and motivational forces increase as task complexity increases. For example, in time-restricted conditions increasing task complexity is conveyed by an increase in emotional–motivational tension. Therefore, it is important to distinguish between those cognitive aspects of complexity that depend on the specificity of information processing and those emotional–motivational aspects of complexity that reflect the energetic aspects of activity. These two aspects of complexity are interdependent and influence each other.

With the concept of task complexity, the concept of task difficulty also arises. Although complexity and difficulty are often considered synonymous, they must in fact be carefully differentiated. If complexity is an objective characteristic of the task, then difficulty is the performer's subjective evaluation of the effects of task complexity. Depending on the skills and the individual features of the subject, the same complex task will be evaluated by him as relatively more or less difficult. An increase in the complexity of a task will increase the probability of the performer being required to exert more cognitive effort, and hence motivational mobilization will increase. Yet complexity itself does not have a subjective component; performers cannot directly experience complexity in itself but rather only perceive a subjective difficulty.

There are some aspects of design complexity that have nothing to do with psychology and ergonomics. For example, one can identify complex technical components of a production or control system that influence the manufacturing process but do not influence human performance. Ergonomics is concerned only with those aspects of man–machine system that affect human performance. From the standpoint of activity, the measurement of complexity involves translating characteristics of the physical configuration of equipment into the specificity of human actions. There is a probabilistic relationship between the physical characteristics of work equipment and methods of task performance; different individuals will employ different methods of performance when working with identical equipment. The concept of complexity can be used when evaluating the efficiency of a performance: those methods of performance which are more complex are generally less efficient. Likewise, the concept of complexity is also important when evaluating the efficiency of training processes.

The notion of "physical exertion of task," that is, the amount of physical effort expended by an individual on motor actions, can be used when evaluating the physical efforts an operator must exert during task performance. This physical effort also has a subjective component, the feeling of stress or physical tension. Physical exertion of task is affected not only by objective task characteristics but also by the individual's physical conditions. An increase in physical effort can increase the complexity of regulation of motor actions; the more physical effort required for task performance, the harder the task becomes for the operator. Conversely, the more mental effort the task requires, the higher the task complexity becomes.

The materials presented in this section demonstrate that there is no unitary integral measure of complexity. Rather, there are a number of different criteria and possible measures for complexity evaluation. Complexity can be evaluated both experimentally and theoretically. Experimental evaluation is based on criteria such as evaluation of probability of errors, measurement of time performance, evaluation of duration of skill acquisition, and measurement of mental fatigue. Expert judgments — such as the use of a five-point scale for complexity evaluation — and the subjective opinion of the performer can also be taken into consideration. However, for design purposes it is critically important to develop analytical procedures for the objective evaluation of task complexity. This problem is considered in the following sections.

4.1.2 Theoretical Foundation of the Evaluation of Task Complexity

Task complexity evaluation is a fundamental problem for ergonomics and engineering psychology. Various practitioners have attempted to develop suitable methods for task complexity evaluation, including the use of various units of measure such as the number of controls and indicators, or the number of actions (Mirabella and Wheaton, 1974; Venda, 1975). Another approach has been to generate an algorithmic description of activity, calculate the number of different members of the algorithm, and then study their interrelationships (Zarakovsky, 1966). However, task complexity cannot be successfully evaluated by such methods, principally because they employ incommensurable units of measure. The quantitative method of task complexity evaluation suggests a requirement for units of measurement, and measurement procedures, that permit the comparison of different elements of activity. However, this important issue has not yet been resolved.

As a result, when specialists tried to evaluate task complexity by calculating, for example, the number of controls, they may have neglected the fact that sometimes manipulation with one control can be a more complex procedure than manipulation of several controls. Similarly, an operator can perform three simple decision-making actions in one task and one complicated decision-making action in another; in this case, the latter may be the more complicated. Hence the evaluation of task complexity cannot be reduced to simply counting the number of controls, indicators, actions, and the like, particularly when these variables use noncommensurable units of measure.

Failure to develop objective procedures for the evaluation of complexity has encouraged attempts to reduce this problem to a subjective judgment using rating methods; usually the use of scales on the order of 5 to 10 is suggested (Meister, 1999). However, such methods are highly subjective and therefore cannot be used alone. More generally, complexity is a multidimensional phenomenon that cannot be effectively evaluated by one measure; single measures are insufficiently informative to guide designers in reducing complexity. As the purpose of this work is to demonstrate the possibility of using formalized methods of task complexity evaluation, there will be no further discussion of subjective rating methods.

An analysis of the data presented above allows the formulation of several basic theoretical statements that should be considered during any evaluation of task complexity:

1. Evaluating task complexity is only possible when one analyzes the structure of activity during task performance. Therefore, typical activity elements should be used as units of analysis rather than typical task elements.
2. Since activity is a multidimensional system, several measures, instead of just a single one should be used to evaluate task complexity. If only one measure is used, significant information on complexity will be lost.
3. Since activity is a process, a time structure of activity must be built before complexity is assessed. Intervals of time devoted to different elements of activity as units of measurement should be used to evaluate task complexity.
4. Any quantitative measure of complexity should reflect the possibility of simultaneous and sequential performances of activity elements and their probabilities of occurrence.
5. Quantitative measures should reflect the relationship between different components of activity and demonstrate how this relationship influences the complexity of task performance.

In order to develop activity-derived measures of complexity a preliminary classification system for typical elements of activity should be created and the duration of these typical elements of activity should be determined. The following criteria may be used for classifying these units of measurement: (1) The substantial characteristics of activity during an interval of time, and (2) the complexity of these activity elements during the chosen interval.

According to the first criteria, one must initially distinguish among activity elements and then classify them by the cognitive and behavioral processes performed during the chosen time interval. Hence, one identifies typical elements devoted to the cognitive and motor components of the activity under study and the time required for their performance (later simply "interval of time"). The cognitive activity elements, with the intervals of time required for their performance, are classified based on their dominant psychic processes; for example, intervals of time for perceiving information, memorization, and decision making.

In task complexity evaluation, intervals of time during which actual performance occurs must be distinguished from periods in which, for example, the operator actively waits for signals that will inform him of the state of the work and what to do next. According to the second criteria set out above, such active waiting periods should be evaluated according to their complexity, which depends on the level of concentration of attention required, the character of the combined elements of activity, and existing emotional stress.

In the active waiting period associated with securing information as a basis for continuing operations, the operator attempts to predict what will occur. Konopkin (1980) labeled this the "intellectual" waiting period. He also distinguished a "physiological" waiting period, which is concerned with the general mobilization of an organism in response to an indeterminate situation. Clearly, both these waiting periods are linked and it may or may not be possible to distinguish them, as underlying any waiting period is the general state of the operator in terms of neural center activation and level of wakefulness.

Active waiting periods are an important part of orienting activity. Orienting activity is not concerned with the transformation of a situation but rather with the comprehension and interpretation of a situation and with the prediction of future events. Such activity is particularly important in vigilance tasks. Behind an externally passive operator's state can be discovered hidden and continuously performed mental actions and operations. Activity during an active waiting period includes the continuous generation of expectations and the promotion of hypotheses about the nature of ongoing events. These expectations guide strategies of attention and thinking and specify possible selections among sensory data. Anticipation and prognosis of possible events is the more significant aspect of the active waiting period in which prognosis is not simply the utilization of past experience but also involves extrapolation of regularities encountered in the past. Such prognosis has probabilistic features. Any distraction of attention during the active waiting period tends to reduce the ability to utilize past experience and anticipate future events.

Active waiting periods may differ in terms of their content, one major difference being the presence or absence of external motor activity, but all share similar features. One such important feature is the operator's readiness to react to changes in the environmental situation in the absence of external activity. Therefore, the concentration of attention during an active waiting period is a basic characteristic that can be used for evaluating a period of time associated with this kind of activity.

The levels of concentration of attention and stress during various time intervals influences task complexity, both in situations where operators actively transform the situation or when they are involved in active waiting periods. Thus, the level of concentration of attention can be considered an important criterion for evaluating task complexity. The more complex a task, the more mental effort required, and the higher the level of attention required during task performance. Actions requiring a higher level of concentration of attention require more time to perform, even when other conditions remain the same. The MTM-1 (Methods-Time-Measurement) system divides microelements into three separate levels of control (equivalent to levels of attention); low, average, and high, where the performance is easier with low and average levels of control. Task complexity correlates with the performance time.

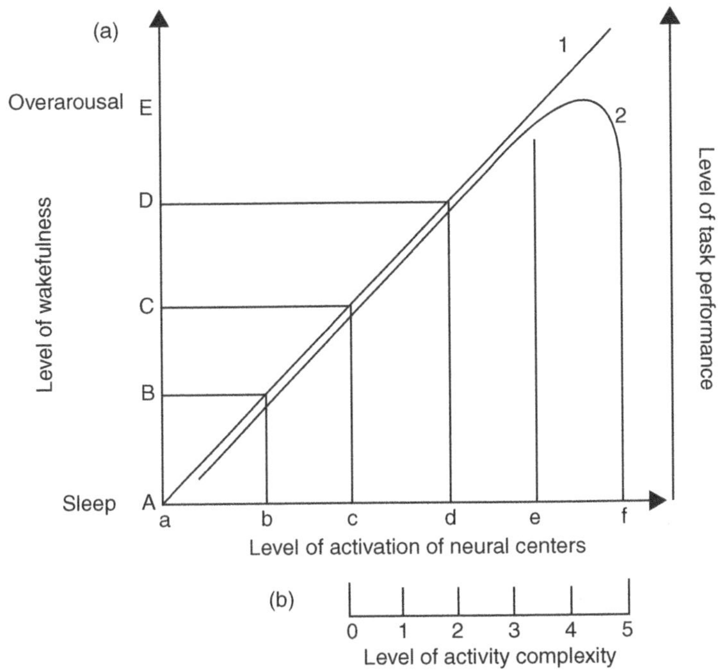

FIGURE 4.1 Relationship between wakefulness, activation, and complexity. (a) Relationship between level of wakefulness and level of activation according to Bloch (1966). (b) Scale of complexity and its relationship to activation of neural system and wakefulness.

However, sometimes a more complex task can be carried out in less time if the operator is able to mobilize his effort or combine different task elements, and as a result increase the speed of performance.

Bloch (1966) characterized attention as the intensity of neuropsychological energy, or the level of activation involved in performance. The higher the level of activation, the higher the level of wakefulness. Performance is a linear function of these two variables (Figure 4.1). In this scheme, attention is considered to be one of the levels of wakefulness. Therefore, it follows that the transition from sleep to wakefulness is, at the same time, a shift of attention, and the level of active wakefulness is, at the same time, a level of attention activation.

The next stage is connected to those levels of concentration of attention that convey emotions. From Figure 4.1a, we can see that performance deteriorates after achieving a particular emotional state.

Task complexity can be linked to the activation level of the brain (Bloch, 1966; Lazareva et al., 1979, etc.) At present two kinds of functional brain states can be described as either specific or nonspecific levels of activation. Global or general changes in the functional state of the brain are seen as simultaneous with regional changes in the different subsystems (sensory, motor and associative); this is considered to be the nonspecific form of activation, which occurs on the basis of general, nonspecific stimulation and is closely connected with the functioning of the reticular activating system. Specific forms of activation are connected to different parts of the brain and produce different types of activation in relation to their content. It has been demonstrated that nonspecific forms of activation are connected with the difficulty of performing particular activities but are not directly connected to the content of activity (Aladjanova et al., 1979; Lazareva et al., 1979). This being the case, specific activation need not be considered during complexity evaluation.

In this work, only those particular ranges of wakefulness and nonspecific activation related to work activity are considered. Point "c" on the scale of activation (Figure 4.1a) corresponds to a minimal level of excitement for task performances. Point "f" corresponds to a maximum level of activation for task performance. After point "e," the efficiency of activity decreases, and after point "f" it sharply deteriorates. Points "c" and "f" on the scale of activation, correspond to points "C" and "E" on the scale of wakefulness. Point "C" corresponds to the level of wakefulness or attention present when a person performs the simplest automatic action. Point "D" corresponds to a situation where possible performance actions do not require a high level of concentration of attention and do not convey emotional attention. Starting from point "D," the level of attention sharply increases. This is a result of emotional tension. Point "E" is characterized by a high level of concentration of attention and emotional stress. After this point, efficient task performance is impossible and activity deteriorates.

The complexity of activity can also be viewed as a continuum. Actions that require a minimum level of wakefulness (attention) are the simplest, whereas actions requiring a maximum level of wakefulness are more complex. This continuum is illustrated as a straight line in Figure 4.1b. Point 0 corresponds to the simplest level of activity and Point 5 to the more complex. At Point 5, activity cannot be performed with a high level of efficiency. If this five-point scale is compared with the scales in Figure 4.1a, it appears that the greater the concentration of attention needed, the higher the level of activation required. This implies that the complexity of a task or the difficulty of its performance can be compared with the level of wakefulness and activation. Zarakovsky and Magazannik (1981) also suggested that the increasing complexity of the decision-making process is correlated with increasing emotional–motivational levels. These authors described two criteria for evaluating the complexity of the decision-making process. The first is intentional, being associated with the probability of failure and strengths of the motives underlying individual activities. This criterion is connected with the significance of the decision-making process. The second criterion is more operational in nature, being associated with the specificity of information processing during decision-making operations.

The analysis of attention, wakefulness, and activation presented in this section corresponds to the model of attention developed by Bedny (Bedny and Meister, 1997), which suggests that an increase in the complexity of a task requires more resources of attention and, as a result, higher levels of motivation and activation of the neural center. The model of self-regulation in Section 5.2 is also consonant with this hypothesis. The blocks, "significance" and "difficulty," in the model of self-regulation are intimately connected with the function block, "motivation." It follows that the significance of the goal and the difficulty of its attainment are the basic factors influencing the motivation for an activity. The data presented in the literature (Bloch, 1966;

Aladjanova et al., 1979; Bedny, 1987, etc.) demonstrate that increasing task complexity incrementally increases the levels of concentration of attention and motivation. From the self-regulation point of view, the relationship between complexity and difficulty of the task and the interaction between the difficulty and significance of the task are also important. For example, increasing the difficulty of task performance and decreasing the significance of the task can decrease the level of motivation.

Usually, the motivational aspects of an activity are ignored in the design process. However, these function blocks of self-regulation connected with evaluating the difficulty and significance of the task play a central role in integrating the cognitive and motivational aspects of activity. The fundamental notions of the complexity, difficulty, and significance of the task, and concentration of attention on the performed activity permit the designer to take not only cognitive and behavioral but also the motivational aspects of the activity into consideration.

The theoretical data outlined above allowed the development of an ordered scale with different categories of complexity in which any category of complexity can be considered an interval. Within an interval, there is a continuum of complexity (for different elements of activity) related to a continuum of wakefulness or attention. Clearly, any category of complexity will include both complex and more simple elements of activity. However, these differences inside an interval can be safely neglected, employing the technically scientific principle of "interchangeability," which is applied to mass manufacturing processes. For example, size differences in particular equipment parts can be neglected when these differences fall inside a particular range.

This analysis of attention, wakefulness, activation and self-regulation, taken together with the MTM-1 system, enables the precise definition of the five categories of complexity, as shown in Figure 4.1b. In MTM-1 elements of activity are clustered into three groups according to their complexity. The simplest group requires a low level of attention. For example, element of activity "RA" (reach to object in fixed location) requires a minimal concentration of attention. More complicated is element "RB" (reach to single object in location which may vary slightly from cycle to cycle), which according to MTM-1 requires an average level of concentration of attention, comparable to the second category of complexity. More complicated again is element "RC" (reach to an object jumbled with other objects in a group so that search and select occur), which requires a higher level of attention and thus is scaled as belonging to the third category of complexity.

In the MTM-1 system there exists an element of activity that is used when the operator is required to recognize an object and make a decision of the kind "If-Then," "Yes-No," etc. According to MTM-1, this simplest cognitive element of work activity requires a high level of concentration of attention and thus can also be related to the third category of complexity.

However, there exist more complicated components of activity. For example, decision making when the required responses are unknown in advance is more complicated than that when the required responses are already known. Therefore, this type of decision making should be related to the fourth category of complexity shown in Figure 4.1b. Very often, the operator must perform actions or take decisions in an ambiguous situation, for example, when a signal on a screen moves forward, but the operator is required to move a lever backwards. This kind of situation requires

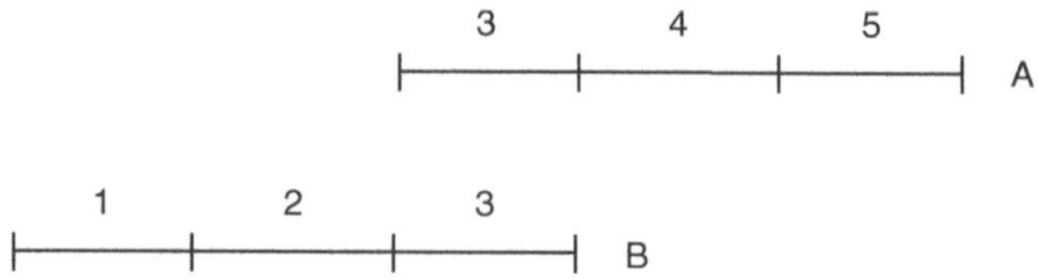

FIGURE 4.2 Five-category ordered scale for evaluation of complexity. A — Order scale for cognitive activity; B — order scale for motor activity.

remembering instructions, a greater level of concentration of attention, etc. Because of this, mental actions connected to overloading attention, recognition actions involved with perceiving unclear signals, and decision making and performing actions in a contradictory situation can be placed in the fourth category of complexity shown in Figure 4.1b. These actions require a greater concentration of attention than the simplest cognitive ones that were referred to in the third category.

In some cases the operator must perform a task in a stressful situation, such as during emergency conditions. Actions performed in such stressful situations can be transferred to an even higher level of complexity. For this reason it is necessary to develop a five-point order scale for motor and cognitive activity (see Figure 4.1).

According to the above analysis motor activity has three categories of complexity. The simplest motor activity element corresponds to category 1, and the most complicated is related to category 3. Cognitive components of activity also have three categories of complexity. However the simplest one is related to the third category and the most complicated corresponds to the fifth one (see Figure 4.2)

From this figure, one can see that the motor and cognitive scales partly overlap each other, with the complexity of the more complicated motor components of activity corresponding to that of the simpler cognitive components.

Based on the materials presented above, it is possible to develop a formalized system of rules and procedures, which permits the translation of qualitative concepts into quantitative indices and that enables the identification of a strictly mono-semantically determined category of complexity of activity elements associated with time intervals. In this system the concept of "complexity of time intervals" is used to describe the complexity of different elements of activity performed in a given time interval. This system of formalized rules is described below and a graphical interpretation of the system is shown in Figure 4.3.

Rule 1. Time intervals for motions requiring a lower (A), average (B), or higher (C) level of concentration of attention can be related to the first, second, and third categories of complexity.

In Figure 4.3a three actions are designated by intervals A, B, and C (low, average, and high levels of concentration). The complexity of these intervals of activity can be designated by the numbers 1, 2, and 3.

Rule 2. If two simultaneous actions requiring a high level of concentration of attention (third category of complexity) are combined, then this period of time is related to the fourth category of complexity.

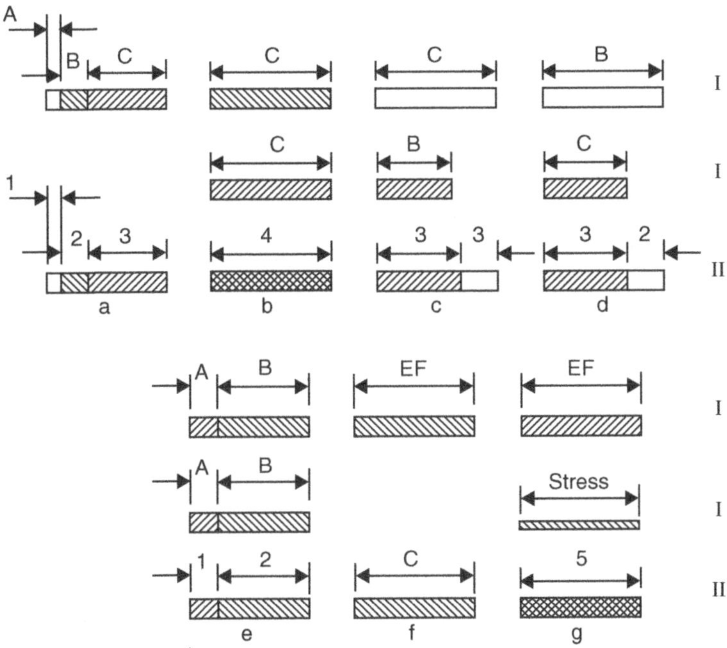

FIGURE 4.3 Graphical interpretation of formalized system of rules for evaluation of complexity of time intervals for different elements of activity. I — Duration of activity elements; II — complexity of activity elements. A, B, and C — low, average and higher levels of concentration.

This rule was concluded from the experimental evidence demonstrating that the combination of these actions at the same time requires the highest level of resource mobilization and significantly increases their performance time (Hassan and Block, 1968). This rule is depicted graphically in Figure 4.3b. Suppose two arms perform two motions that require a high level of concentration of attention (the third category). Both actions are performed in the normal visual field. The motions for the right (r) and left (l) hands begin and end at the same time. According to rule 2, the complexity of this time interval of activity equals four.

Rule 3. If two motions (or motions and cognitive elements) with different levels of complexity are performed concurrently and the complexity for one element corresponds to the third category or higher, the complexity category of the time interval for these simultaneous elements of activity is determined by the complexity of the more difficult element. The graphical interpretation of this rule is shown in Figure 4.3c.

When a simpler action involves moving a greater distance and thus requires more time than a more complex action, the complexity evaluation can be presented as in Figure 4.3d.

Rule 4. If two motions with the first or second categories of complexity are performed simultaneously, any overlapping of the motions also assumes the value of the first or second categories of complexity. That is to say, the complexity of the overlapping time remains unchanged. See Figure 4.3e.

This rule was concluded from experiments that established that the performance of the simultaneous motions that require low or average levels of attention (that is, of low and average difficulty) does not slow them down (Hassan and Block, 1968).

Rule 5. Time intervals related to simple decision making (where an operator knows in advance how to react to a particular situation) can be related to the third category of complexity (Figure 4.3f). This rule follows from the data embodied in the MTM-1 system. According to MTM-1, the simplest decision-making processes require a higher level of concentration of attention. If an operator must make a more complex decision, that is, one where he does not know in advance how to react to varying situations, then this interval of time is related to the fourth category of complexity.

Decision making in a contradictory situation, or in a condition where stereotypical decision-making processes are destroyed, is more complicated than choosing from known alternatives. Therefore, this kind of decision making is also related to the fourth category of complexity.

Rule 6. Intervals of time devoted to an activity where the information presented to an operator interferes with other data, hindrances exist, feedback is destroyed, or the performed actions contradict existing stereotypical behavior should be transferred to a higher level of complexity. For example, if an arm moves a lever ahead, requiring a low level of concentration of attention (nominally the first category of complexity), but this results in moving the control object back, the activity must be evaluated as belonging to the second category of complexity.

Rule 7. If motor activity coincides with mental–perceptual and decision-making actions, the same level of defined complexity should be assigned to cognitive and motor activity. At the later stage of analysis, the determined level of complexity should be taken into consideration separately for cognitive and motor activity. Complexity of an interval, when cognitive and motor elements of activity are performed at the same time, depends on specificity of how these elements are combined. For example, if a motor element of activity has the second level of complexity and the cognitive element has the third level of complexity, according to rule 3 a cognitive element retains the same level of complexity. On the other hand, the motor activity complexity is determined by complexity of the most difficult element. Hence, a motor level of complexity has the third category in this case.

Rule 8. If an activity is performed in a stressful situation, then time intervals related to the third and fourth category of complexity should be transferred to the fifth category of complexity. Time intervals related to the first and second categories of complexity should be transferred to the third category.

For instance, if the operator performs a simple decision-making action (third level of complexity) but this action takes place under stressful conditions, the action should be considered as being of the fifth level of complexity (Figure 4.3g). We want to mention here that rules of combining different elements of activity have been described in Section 3.2.4.

There are also additional rules for applying MTM-1 element EF (eye focus time). This element of activity is associated with the time required to focus the eye on an object, and then visually determine certain clearly distinguishable characteristics on the basis of which the worker must perform the simplest decision making ("yes" or "no"; "if-then"). Therefore, activity element EF consists of two inseparable mental operations: one strictly sensory–perceptual, the other involving decision making at the sensory–perceptual level. This corresponds to the data from psychophysics (Green and Swets, 1966), where it has been established that even simple signal detection involves not only sensory processes but also nonsensory factors (e.g., decision making). Bardin and Zabrodin (1988) demonstrated that in more complex psychophysical tasks signal detection is closely linked with cognitive processes and that cognition can influence sensory sensitivity. From these findings it follows that sensory–perceptual and thinking (decision making in particular) processes are closely interconnected.

Real-world situations are, of course, even more complex than simple laboratory studies. Hence, when applying activity element EF during the design of time structures it is important to pay attention to the relationship between the detection or recognition stage and the decision-making stage. Based on this theoretical analysis we derive a conventional rule, which states that during the design of a time structure, a value of 1/2 EF should be related to afferent operators (O^{α}), and that 1/2 EF should also be related to the simplest decision on "what should be done based on obtained information?" (logical condition, l). When the characteristics of the object to be perceived are not easily distinguishable (i.e., when the characteristics of the object are in the threshold area) it is necessary to introduce two elements of EF during the design of a time structure: the first element associated with recognition of the object (operator O^{α}) and the second with the decision (logical condition, l) as to what should be done based on the data obtained.

The formalized system of rules and procedures described above, which evaluates the complexity of the different components of activity according to an order on the scale of five categories, permits the evaluation of task complexity with an acceptable level of precision.

4.1.3 Measures of Complexity

The quantitative approach to task complexity evaluation outlined here involves the development of methods for describing the time structure of activity during task performance, using the intervals of time taken for different elements of activity as units of measure with which to conduct measurement procedures and compare activity elements. The following criteria are used to classify units of measure of task complexity: the substantial characteristics of activity during a particular interval of time, and the complexity of these activity elements, which is dependent on the subject's concentration of attention on particular elements of activity, the possibility of their performance simultaneously or sequentially, and the probability of the appearance of particular elements of activity during task performance. For example, according to the first criterion, it is possible to distinguish time intervals devoted to either mental or motor activity. Time intervals devoted to mental activity are classified on the basis of their dominant psychic processes: for example, intervals of time for the perception of information, memorization, and decision making. The complexity of time intervals depends on the level of concentration of attention during those intervals, the way elements of activity are combined, and the emotional stress on the operator during those intervals.

The quantitative evaluation of task complexity is made possible using the principles of systemic–structural activity analysis. The development of measures of task complexity involves several steps (Bedny et al., 2001). First, specialists must develop an algorithm of task performance and then a time structure of activity during task performance. They are then required to determine whether an element of activity is performed simultaneously or sequentially. The next step is to describe the probabilistic characteristics of activity during task performance. Following this, quantitative measures of task complexity are calculated. These measures are the mathematical means used to evaluate fractions of different components of activity in the holistic structure of task performance. Measures of complexity should reflect the workload on separate cognitive processes and demonstrate the relationships between different components of activity. We now consider in detail a range of quantitative measures of task complexity.

The first step in the evaluation of task complexity is to determine the overall duration of the task performance and individual durations of those different elements of activity involved in task performance.

The duration of task performance can be determined according to the following formula:

$$T = \sum P_i t_i, \tag{4.1}$$

where P_i — probability of the ith member of algorithm; and t_i — time of performing ith member of algorithm.

The time taken for logical conditions, afferent operators, and the executive (response) components of activity can be determined similarly:

$$L_g = \sum P_i^l t_i^l; \quad T_\alpha = \sum P_r^\alpha t_r^\alpha; \quad T_{ex} = \sum P_j^o t_j^o, \tag{4.2}$$

where P_i^l, P_r^α, P_j^o — the probability of ith logical conditions, rth afferent, and jth efferent operators; t_i^l, t_r^α, t_j^o — time performance of ith logical conditions, rth afferent, and jth efferent operators.

Time related to recognizing and identifying weak (i.e., approaching to threshold range) task signals can be determined from the following formula:

$$T'_\alpha = \sum P_{rep'}^{\alpha'} t_{rep'}^{\alpha'}, \tag{4.3}$$

where $P_{rep'}^{\alpha}$ — the probability of emerging r'th afferent operator, characteristics of which approach the threshold area; $t_{rep'}^{\alpha}$ — time for the recognition or identification of weak signals with characteristics approaching the threshold range.

In the next step of task complexity evaluation, one determines the relationship between the time spent on logical conditions to the time used for the executive components of activity (i.e., time for efferent operators), or the time for overall task performance.

$$N_\alpha = L_g/T_{ex}; \quad N_1 = L_g/T, \tag{4.4}$$

where L_g — time for logical conditions, T_{ex} — time for response (executive) components of activity and T — time of task performance. Measure (4.4) demonstrates the relationship between the logical (decision making) and executive components of activity, giving the fraction of the logical components in the task performance.

Activity may be either stereotyped (repetitive) or changeable (variable). The performance of a stereotyped activity is normally easier; if procedures always take place in a set order or a given procedure always follows some particular member of an algorithm, these logical components of activity are stereotyped. When procedures and the transition from one action to another have probabilistic features, these procedures are considered variable. In activity analysis, those members of algorithm that always follow in the same sequence can be considered stereotyped components of activity. Their sequence is subjectively perceived by the operator as the habitual performance of the same order of actions. If the habitual performance of a stereotyped efferent operator is always followed by the same afferent operator and its associated logical condition, then the afferent operator and its logical condition are also related to the stereotyped activity. It can be hypothesized that more the time in a process is devoted to variable procedures, the more complex this process is. It is possible to calculate measures of stereotyped and variable (changeable) components for executed activity and logical conditions. The time devoted to stereotyped and variable operators and logical conditions during activity performance can be determined according to the following formulae:

$$t_{st} = \sum P^o_{jst} t^o_{jst}; \quad t_{ch} = \sum P^o_{jch} t^o_{jch}, \tag{4.5}$$

$$l_{st} = \sum P^l_{ist} t^l_{ist}; \quad l_{ch} = \sum P^l_{ich} t^l_{ich}, \tag{4.6}$$

where P^o_{jst}, P^o_{jch} — probability of the appearance of the *j*th stereotyped and variable operators; t^o_{jst}, t^o_{jch} — performance time of the *j*th stereotyped and variable operators; P^l_{ist}, P^l_{ich} — probability of the appearance of the *j*th stereotyped and variable logical conditions t^l_{ist}; t^l_{ich} — performance time of the *i*th stereotyped and variable logical conditions.

Accordingly, the measure of stereotyped and variable logical components of activity can be determined from the following expressions:

$$L_{st} = l_{st}/L_g, \tag{4.7}$$

$$L_{ch} = l_{ch}/L_g, \tag{4.8}$$

where l_{st} and l_{ch} — mathematical mean of performance time of stereotyped and variable logical activity.

In the same manner, we can determine the stereotyped and variable executive components of activity:

$$T_{st} = t_{st}/T_{ex}, \tag{4.9}$$

$$T_{ch} = t_{ch}/T_{ex}, \tag{4.10}$$

where t_{st} and t_{ch} — mathematical mean of performance time of the stereotyped and variable afferent and efferent operators. It should be noted that the measures of changeable and stereotyped logical and executive components of activity are interconnected.

The increase in number of logical conditions and their outputs makes the activity more variable. However, in some cases, different logical conditions may have an output that leads to the same operator of an algorithm. This makes it necessary to introduce differing measures of the "stereotypy" and "changeability" of logical and executive components of activity.

There are two main aspects impinging on the complexity of work activity. The first is where the sequence of actions or character of the decision-making process is predominantly determined by stimuli or information external to the individual. The second aspect is where the performance of actions and decision-making processes are determined largely by information retrieved from long-term memory. The latter situation is more complex for the performer (Bedny, 1979; Konopkin, 1980). When attempting to evaluate the complexity of logical conditions it is therefore necessary to distinguish between these two aspects. The following measure of complexity applies to logical conditions performed on the basis of information extracted from long-term memory:

$$L_{\mathrm{ltm}} = l_{\mathrm{ltm}}/L_{\mathrm{g}}, \tag{4.11}$$

where L_{ltm} — proportion of time for logical components of work activity depending mostly on memory; and l_{ltm} — mean performance time for logical conditions predominantly governed by memory.

For example, imagine a single digital display which presents either the number 1 or 2 to a subject. When number 1 is displayed the operator should press a button on the left and when number 2 is displayed, a button on the right. This decision-making action may not be difficult for an operator working under laboratory conditions where he acts only in response to the display information. However, in a real-world situation where the operator is required to respond to multiple displays it may become difficult for him to remember the appropriate response to a particular number; before responding he must retrieve information from long-term memory. In another situation an operator may be required to act in response to indicator bulbs; if a red bulb is lit he should press a red button, if a green bulb is lit a green button. In this situation, even though the operator must attend to multiple displays he is able to react correctly because the externally presented information reminds him of the correct action. In the first situation the operator acts based on internal information extracted from memory; in the second case he performs decision making based on externally presented information. This significant difference in the complexity of decision-making actions can be evaluated by Formula 4.11.

Another important measure of task complexity is the workload imposed on working memory. This is the proportion of time spent retaining current information in working memory relative to the duration of task performance. The greater this proportion, the greater the workload imposed on working memory. Working memory workload also depends on the specificity of the information in working memory and the specificity of those actions performed during the time the information is retained. Evaluation of these aspects of task performance can be evaluated using the following rules:

1. If during the decision-making process the operator performs more than three logical conditions sequentially, this time interval should be related

to the fourth category of complexity. This rule is based on Zarakovsky (1966), who experimentally determined that the presence of three or more logical conditions in a sequence sharply increases the mental workload of the operator and the probability of error.

2. If the operator retains information in working memory which approaches its maximum capacity (i.e., 3–4 elements), this interval of time should also be related to the fourth category of complexity.
3. If, during the period of time in which an operator keeps information in working memory he performs complicated actions that require a high level of concentration of attention, or if the operator performs actions in a stressful situation, this interval of time should be related to the fourth category of complexity.

These rules permit us to determine not only the duration of the workload of the working memory but also the level of complexity of time which is connected to the loading of the memory.

The measure that characterizes the workload of the working memory can be determined according to the following formula:

$$N_{\mathrm{wm}} = t_{\mathrm{wm}}/T, \tag{4.12}$$

where N_{wm} — proportion of time in which current information is retained in working memory; t_{wm} — mean time for activity related to the storage in working memory of current information concerning task performance.

The measure of complexity related to the retrieval of information from long-term memory should be considered an independent measure when it takes a notable fraction of the time for task performance. It can be determined as follows:

$$N_{\mathrm{ret}} = t_{\mathrm{ret}}/T. \tag{4.13}$$

In those cases where the retrieval of information from long-term memory does not use a notable fraction of the time for overall task performance, this period of time is considered to be included as a component of related measures such as N_{wm}.

The complexity of the afferent components of activity (those involving sensory–perceptual processes) can be evaluated in relation to the detectability of perceived stimuli. In those cases where the operator receives information from stimuli with low detectability, the task should be considered more complex. The measure of that complexity is the proportion of time required for the discrimination and recognition of stimuli with low detectability features (i.e., working in threshold area). The relevant formula is given as follows:

$$Q = T'_{\alpha}/T_{\alpha}, \tag{4.14}$$

where T'_{α} — the time for discrimination and recognition of different features of the task in conditions approaching the threshold characteristics of sense receptors (low level of detectability).

Another factor affecting task complexity is the presence of repetitive components in work activity. The more often the same elements of activity are performed during task execution, the easier the task is for the performer. For measuring the repetitive characteristics of logical conditions and afferent and efferent operators we introduce the following formulae:

$$Z^l = t^l_{\text{rep}}/L_g; \quad Z^\alpha = t^\alpha_{\text{rep}}/T_\alpha; \quad Z^{\text{ef}} = t^{\text{ex}}_{\text{rep}}/T_{\text{ef}}, \tag{4.15}$$

where Z^l, Z^α, Z^{ef} — are the proportion of time for repetitive logical conditions, afferent and efferent components of work activity; and t^l_{rep}, t^α_{rep}, $t^{\text{ex}}_{\text{rep}}$ — mean time needed to perform repetitive logical conditions, afferent and efferent components of activity.

The category of complexity (level of complexity) of any member of an algorithm can be evaluated according to the five-point scale described before. If any one category of complexity is predominant, that is it exceeds 70% of the time taken for a particular task element, the general complexity of this component of activity belongs to this category of activity. For example, if 70% of the time is associated with the third category of complexity and the rest with the first and second categories, the total category for that element of activity is the third level of complexity. If an activity element consists of 30% of the first category, 10% of the second, and 60% of the third, the total category of complexity will be of the second order.

In more complex tasks there can be even higher levels of complexity. This depends on the specificity of the combination of activity elements that belongs to the third category or on the independent complexity of activity elements. Although according to the formalized rules of the MTM-1 system a fourth category of complexity does not exist, from the point of view of the activity approach the simultaneous performance of two microelements related to the third category of complexity (if it is possible) should be evaluated as a fourth category of complexity. Often this fourth category of the complexity of separate activity elements can be found in semiautomatic and automatic systems. This is an extreme and undesirable situation. It is recommended that when dealing with this fourth category of complexity a coefficient of 2 should be applied in order to evaluate the fraction of category 4 elements in an activity. If multiplying the performance time allocated to fourth category complexity elements by 2 produces more than 70% of the total time for task performance then the total complexity of the task will be in the fourth category. If the product is less than 70%, it should be added to the time for elements that belong to the third category of complexity. If this produces a time for the third category elements that exceed 70% of total performance time, then the task belongs to the third category of complexity.

Let us now consider methods of evaluating the complexity of a time interval connected with an active waiting period. Sometimes an operator may perform tasks that periodically do not require any active involvement in performance; such intervals of waiting time may be encountered both within and between tasks. In spite of the absence of externally observable behavior, such waiting times require concentration of attention; the operator must be ready to become immediately involved in the performance as the situation requires, for example, if an emergency arises. The complexity of an active waiting period can be evaluated with respect to the level of concentration of attention required during the waiting period, in accordance with the following rules:

1. If waiting periods require a low, average, or high level of concentration of attention, they are described by the first, second, and third categories of complexity, respectively.
2. When waiting periods convey emotional stress (i.e., there is a danger of trauma or accident) they are described by the fourth category of complexity.
3. When waiting periods of any level of complexity require that information be retained continuously in working memory, their complexity category should be increased by one.

The existence of an active waiting period in a task requires the introduction of additional measures of task complexity. One such measure is the "proportion of active waiting period," which is calculated according to the following formula:

$$\Delta T_w = t_w / T, \qquad (4.16)$$

where t_w — mean time for an active waiting period; and T — total duration of task performance.

If an active waiting period consistently occurs following a particular element of the activity or task, it is considered to be a repetitive active period, measured by the following formula:

$$W^{\mathrm{st}} = t_{\mathrm{wst}} / t_w, \qquad (4.17)$$

where t_{wst} — time for repetitive waiting components; and t_w — total duration of the waiting period. If the internal psychological content of waiting periods of time is identical, they are repetitive. The proportion of repetitive active waiting period can be calculated in a similar fashion. It is possible to evaluate stereotyped and variable components of the waiting period the same way we evaluate T_{st} and T_{ch}. However we will not discuss this in more details. Table 4.1 summarizes the various complexity measures.

Let us now consider some hypothetical examples that demonstrate the evaluation of performance complexity, first, in a situation where all elements of an activity are performed sequentially and, second, in a situation where some elements are performed simultaneously. Suppose that a subject performs three elements of activity, two of which are related to motor activity, and one to cognitive activity, and where all the elements can be described using the MTM-1 system. In the case "a," the subject performs a careful arm movement to an exact position (element RC), then a simple decision-making action at the sensory–perceptual level (element EF), and finally moves his arm to a fixed location (element RA). All activity elements are performed sequentially one after another. This is shown in Figure 4.4a, where horizontal lines depict elements of activity (*1-duration of activity elements*), while line II below it indicates the complexity of a different period of time. Element RC requires a high level of concentration of attention and is thus related to the third category of complexity. According to the rules developed above, EF is also related to the third category of

TABLE 4.1
Measures of the Complexity of Task Performance and Their Psychological Interpretations

Name of measure	Formula for calculation	Variables	Psychological meaning
Time for algorithm execution	$T = \Sigma P_i t_i$	P_i — occurrence probability, t_i — occurrence time of *i*th member of algorithm	Duration of task performance
Time for performance of logical conditions	$L_g = \Sigma P_i t_i$	P_i — occurrence probability, t_i — occurrence time of *i*th logical conditions	Duration of decision-making process
Time for performance of afferent operators	$T_\alpha = \Sigma P^\alpha t^\alpha$	P^α — occurrence probability, t^α — occurrence time of *r*th afferent operators	Duration of perceptual workload
Time for performance of efferent operators	$T_{ex} = \Sigma P_j t_j$	P_j — occurrence probability, t_j — occurrence time of *j*th efferent operators	Duration of executive components of activity
Time for discrimination and recognition of distinctive features of task approaching threshold characteristics of sense receptors	$'T_\alpha = \Sigma P_{r'} t_{r'}$	$P_{r'}$ — occurrence probability, $t_{r'}$ — occurrence time of r'th afferent operators, characteristics of which approach threshold value (required additional EF)	Duration of perceptual process connected with weak stimuli (closely with threshold characteristics)
Proportion of time for logical conditions to time for executive activity	$N_l = L_g / T_{ex}$	L_g — time for performance of logical conditions, T_{ex} — time for executive components of activity	Relationship between decision-making process and executive components of activity

Measure of stereotyped logical components of activity	$L_{st} = l_{st}/L_g$	l_{st} — time for stereotyped logical component of activity	Characteristics of inflexibility or rigidity of decision-making process
Measure of variable logical components of activity	$L_{ch} = l_{ch}/L_g$	l_{ch} — time for variable logical component of activity	Characteristics of irregularity of decision-making process
Measure of stereotyped executive components of activity	$N_{st} = t_{st}/T_{ex}$	t_{st} — time for stereotyped executive components of activity, T_{ex} — time for executive components of activity	Characteristics of inflexibility or rigidity of executive components of activity
Measure of variable executive components of activity	$N_{ch} = t_{ch}/T_{ex}$	t_{ch} — time for variable executive components of activity	Characteristics of irregularity or flexibility of executive components of activity
Scale of complexity (a) algorithm (b) member of algorithm	X_r — level of complexity (1,2,...,5)	Based on level of concentration of attention during task performance (1 — minimum concentration, 5 — maximum)	Level of mental effort during task performance and performance of different elements. Unevenness of mental effort and critical points of task performance
Proportion of time for repetitive logical components of work activity	$Z^l = t_{rep}/L_g$	t_{rep} — time for performance of identical logical conditions	Characteristics of habitualness of information processing

(Continued)

TABLE 4.1
Continued

Name of measure	Formula for calculation	Variables	Psychological meaning
Proportion of time for repetitive afferent components of work activity	$Z^{\alpha} = t_{rep}/T_{\alpha}$	t_{rep} — time for performance of identical afferent components	Characteristics of habitualness of perceiving process
Proportion of time for repetitive efferent components of work activity	$Z^{ef} = t_{rep}/T_{ex}$	t_{rep} — time for performance of identical efferent components	Characteristics of habitualness of executive components of activity
Proportion of time for logical components of work activity depending largely on information selected from long-term memory rather than exteroceptive information	$L_{ltm} = l_{ltm}/L_{g}$	l_{ltm} — time for logical components of activity whose operational nature is predominantly governed by information retrieved from long-term memory	Level of memory workload and complexity of decision-making process
Proportion of time for retaining current information in working memory	$N_{wm} = t_{wm}/T$	t_{wm} — time for activity related to storage in working memory of current information concerning task performance	Level of workload of working memory and complexity of decision-making process

Proportion of time for discrimination and recognition of distinct features of task approaching threshold characteristics of sense receptors	$Q = T\alpha/T\alpha$	$T\alpha$ — time for discrimination and recognition of different features of task approaching threshold characteristics of sense receptors	Characteristics of complexity, sensory, and perceptual processes
Proportion of active waiting period in total work process	$\Delta T_w = t_w/T$	t_w — time for active waiting period in work process	Relationship between active waiting period and performance
Category of complexity of active waiting periods	$X_w = 1 \ldots 4$	Concentration of attention during waiting period (1 — minimum, 4 — maximum)	Level of mental effort during active waiting period
Proportion of time for repetitive waiting periods of work activity	$Z^w = t_{wrep}/t_w$	t_{wrep} — time for repetitive waiting periods	Characteristics of habitualness of waiting periods
Measure of changeability of waiting periods in work process	$W^{ch} = t_{wch}/t_w$	t_{wch} — time for changeable waiting components in work process	Characteristics of irregularity of waiting periods
Measure of stereotypy of waiting periods in work process	$W^{st} = t_{wst}/t_w$	t_{wst} — time for stereotypy wait component in work process	Irregularity of waiting periods

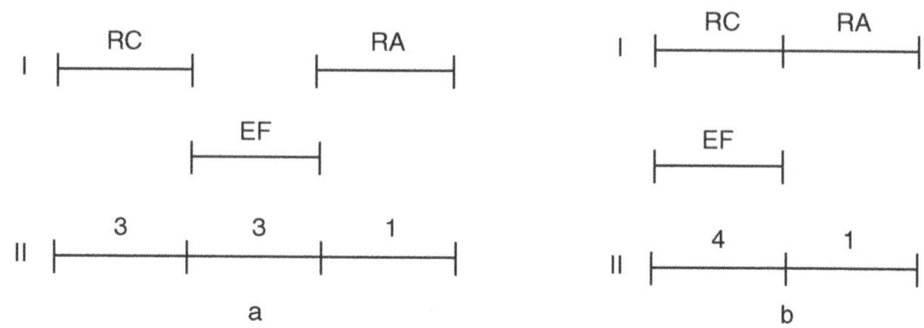

FIGURE 4.4 Task complexity during sequential (a) and simultaneous (b) performance of activity elements. I — Elements of activity; II — complexity of these elements (complexity of different period of time).

complexity. Finally, element RA is related to the first category of complexity as it requires a low level of concentration of attention.

In the second example (see Figure 4.4b), the subject performs the same elements of activity. However, in this situation elements RC and EF are performed simultaneously. Horizontal line II again depicts the complexity of different periods of time performance; in this case, because elements RC and EF belong to the third category of complexity and are performed simultaneously, the period of time during which they are performed is related to the fourth category of complexity.

The complexity of the period of time during which element RA is performed (category 1) remains the same as in the previous example. In the first example the fraction of time expended on cognitive processes is 0.33, in the second example it is 0.5. In the first example 0.66 of the period of time is devoted to the third category of complexity and 0.33 to the first category. In the second example this becomes 0.5 of the period of time related to the fourth category and the same fraction related to the first category. These simple examples demonstrate how the procedures and units of measure developed and described above provide mathematical methods of task complexity evaluation.

4.2 TESTING AND EVALUATION OF TASK COMPLEXITY UNDER LABORATORY CONDITIONS

4.2.1 Evaluation of Manual Task Complexity

4.2.1.1 Evaluation of Manual Task Complexity of the task "Installation of Pins" (Second Version)

In Section 3.3.2 a physical model of a production operation was used as an example of the study of a manual task under laboratory conditions. The model consisted of a pin board containing 30 holes for metal pins. In the first version of this task all pins were unfluted, while the second and third versions involved fluted pins. This section considers how one can evaluate the varying complexity of these different versions of the task. Although manual tasks with such a complex logical structure as the second and the third versions are rare in the real production environment, these versions will be considered in detail. This will demonstrate a new approach to the quantitative evaluation of task complexity and introduce the skills required for such evaluation. Practical examples involving real production processes will be considered in later sections. Another purpose of this study is to demonstrate that the suggested approach is sufficiently sensitive for the evaluation of objectively existing complexity; for this purpose, experimental methods are also used. Time of task performance, duration of skill acquisition, subjective evaluation of complexity by performers, and expert analyses provide some experimental criteria of task complexity.

We begin with an evaluation of the second version of task as in the first, simpler version (where the subject uses unfluted pins) the various complexity measures often return a value of zero. In the second, more complex version of the task in which ten of the pins were fluted, the pins were to be put into the holes according to the specific rules. Regular pins may be installed in any position. If a fluted pin is picked up by the left hand, it should be installed so the flute is placed above the hole. If a fluted pin is picked up by the right hand, it must be placed so that the flute is placed below the hole. An algorithmic description of this task was presented in Table 3.19. In order to evaluate its complexity one needs to know the duration of different members of algorithm and the total time of algorithm (i.e., task) performance. In order to achieve this, it is necessary to determine the probability and time of performance of different members of the algorithm. In the example, cognitive and motor components of activity overlap. This implies that first, only those members of algorithm upon which the total time of task performance depends should be selected and once identified their probability of occurrence and duration should be determined. These data can be obtained from the description in Table 3.20. They can be presented in a tabular form as shown in Table 4.2.

Here we recall that operator O_7^{α} is associated with the recognition of the incorrect positions of two pins and operator O_9^{ε} is associated with the correction of their positions (see Table 3.19).

According to Formula 4.1 the time of task performance T can be determined in the following manner:

$$T = 1 \times 1.31 + 4/9 \times 2.08 + 2/9 \times 2.26 + 2/9 \times 2.26 + 1/9 \times 2.59 \\ + 2/75 \times 0.42 + 2/75 \times 4.16 = 3.64$$

TABLE 4.2
Probability and Time Performance of Different Members of Algorithm for the Second Version of Task

Members of algorithm	O_1^{ε}	O_3^{ε}	O_4^{ε}	O_5^{ε}	O_6^{ε}	O_7^{α}	O_9^{ε}
Probability of occurrence	1	4/9	2/9	2/9	1/9	$1/75 + 1/75$	$1/75 + 1/75$
Time (0.01 min)	1.31	2.08	2.26	2.26	2.59	0.42	$2.08 + 2.08$

In order to transform this result into 0.01 sec units of measure we must multiply it by 60. Therefore $3.64 \times 60 = 218.4$ or 2.18 sec. This is the mean time for taking two pins and putting them into two holes. In order to calculate the overall time of task performance we must again multiply the obtained result by 15, as this model production operation uses 30 pins. Therefore the overall time for performance of the task will be as follows:

$$T = 2.18 \times 15 = 32.7 \text{ sec.}$$

In order to calculate T_{ex} (time for performance of efferent operators) we must eliminate the time required to perform afferent operator O_7^α. However, as in this case O_7^α only requires 0.01 sec, we can safely neglect this difference. T and T_{ex} are thus approximately the same. This frequently happens when the motor and cognitive components of activity overlap.

It should be noted that in these and the following calculations we employ very small units of measure, as is common in engineering psychology and ergonomics. This is because behavioral (such as motor operations) and cognitive (cognitive actions) units often have very short durations. Both reaction-time measurements such as Hick–Hyman (Hick, 1952; Hyman, 1953) and Fitts' Law (Fitts, 1954) and subtractive and additive methods such as those of Donders (1862 translated 1969) and Sternberg (1969) use milliseconds. However, in order to successfully apply such small units of measure in practice it is important to use systemic-structural methods of analysis, where holistic activity or behavior is divided into smaller hierarchically organized units and then built up into a holistic activity system.

The next step of the complexity analysis involves determining the duration of the sensory–perceptual components of the work process (i.e., the afferent operators) using Equation 4.2. The calculations are performed in a similar way: all afferent operators are selected, and then the probability of their occurrence and time for their performance are determined. The simple decision-making actions involved in the task include those cognitive operations associated with receiving information and decision making. Hence, half the time is assigned to afferent operators (1/2 EF) and the other half (1/2 EF) to logical conditions (see the description of formalized rules for complexity evaluation above). There are four versions of operator O_2^α (see Table 4.3); for each we must determine the probability of occurrence and time of performance. Version ${}^1O_2^\alpha$ (pins without flute) requires only 1/2 EF. Versions ${}^2O_2^\alpha - {}^4O_2^\alpha$ require $1/2\text{EF} + 1/2\text{EF} = \text{EF}(\text{EF} = 0.43$ units). This produces the following data (see Table 4.3).

TABLE 4.3
Time and Probability of Occurrence of Different Versions of O_2^α

Versions of operator O_2^α	${}^1O_2^\alpha$	${}^2O_2^\alpha$	${}^3O_2^\alpha$	${}^4O_2^\alpha$
Probability of occurrence	4/9	2/9	2/9	1/9
Time (0.01 min)	$1/2 \times 0.43$	$1/2 \times 0.43 + 1/2 \times 0.43$	$1/2 \times 0.43 + 1/2 \times 0.43$	$1/2 \times 0.43 + 1/2 \times 0.43$

As the calculations are performed in the same way as described above they will not be considered in detail here. Using Equation 4.2 we can determine the mathematical means for the afferent operators (T_α).

$$T_\alpha = 4/9 \times 0.21 + 2/9 \times 0.43 + 2/9 \times 0.43 + 1/9 \times 0.43 = 0.33 \text{ units.}$$

By multiplying 0.33 by 15 (to produce the time for whole task) we obtain 4.95 units. In order to transfer this result into seconds we multiply by 60. The result is a time of 2.97 sec (approximately 3 sec).

In the same manner we determine the duration and probability of occurrence of all logical conditions in the task. Logical conditions (L_g) will be the same, that is 3 sec.

In this task all received signals are above the threshold level. Therefore the time for discrimination and recognition of distinctive features of tasks at threshold level $T'_\alpha = 0$. The ratio of time for logical conditions to time for executive actions (efferent operators) is determined using Formula 4.4.

$$N_l = L_g / T_{ex} = 3/32.7 = 0.09.$$

This formula demonstrates the relationship between decision-making actions and motor actions (executive activity). The greater the value of N_l, the heavier the cognitive workload during task performance.

The next complexity measures to be considered are those associated with the evaluation of stereotyped (inflexible, rigid) or, conversely, variable (flexible) components of activity. The more stereotyped the components of a task are, the more that task approaches a skill-based performance. The more changeable or flexible the task components are, the more the task approaches a rule- or knowledge-based task.

Measures of stereotyped components of activity can be calculated separately for logical conditions, executive components, etc. The stereotyped logical components of activity, L_{st} is determined according to 4.7. At the first step one needs to know how much time is required for stereotyped kinds of decision making (l_{st}). If the task is such that a logical condition (decision-making action) can emerge during task performance at the same position in the algorithm, and thus the subject expects that he will need to perform this decision-making action, then this logical condition can be regarded as stereotyped. In the example we have been considering, logical conditions can change their position during task performance (as their position depends on the kind of pins grasped by subjects). Therefore all the logical conditions in the example are related to variable activity. From this it follows that $l_{st} = 0$ and $L_{st} = 0$. In contrast, the measure of variable logical condition (L_{ch}) derived according to 4.8 will be 1 (as $l_{ch} = L_g$).

In a similar fashion it is possible to determine the stereotyped or variable executive components of activity (T_{st} or T_{ch}) according to Formulas 4.9 and 4.10. In our example, only one efferent operator, O_1^ε, can be related to stereotyped components of activity (the task always starts by moving the hands to the pin box). All the other efferent members depend on chance events. This means that the relationship between the time required to perform O_1^ε (where the probability of O_1^ε is equal to 1) and the mathematical mean of the time required for the performance of the other efferent operators (where the probability of $O_1^\varepsilon - O_6^\varepsilon$ and O_9^ε is less than 1) is that fraction of the executive components of the task related to the stereotyped components of activity. Therefore, the measure of stereotyped executive components of activity $T_{st} = 0.37$ and the measure of variablity of the executive components of activity is $T_{ch} = 0.63$.

Another important factor in task performance is the specificity of memory function. It is possible to evaluate various aspects of memory functioning. Memory is critical for decision-making actions. In order to evaluate this aspect of work one needs to determine l_{ltm} — the time for logical conditions performed on the basis of information extracted from long-term memory. Then we determine the proportion of time spent on those logical components of work activity that depend largely on information selected from memory rather than exteroceptive information. This is done according to the formula $L_{ltm} = l_{ltm}/L_g$ (see Formula 4.11). This kind of decision making is particularly difficult for the performer. In the example, a pin's flute does not remind subjects whether to turn the pin to the left or right; subjects must remember the task instructions. However not all pins have a flute. Those logical conditions which must be performed according to information extracted from memory apply only when the pins are fluted. The probability of nonfluted pins appearing is 4/9. It follows that the time for logical components of work activity depending largely on information selected from memory l_{ltm} is 1.68 sec, and therefore $L_{ltm} = 0.56$.

Another measure of memory workload, N_{wm} is associated with the evaluation of working memory. In the example, the interval of time during task performance for which the subject is required to retain current information (i.e., the intermediate result of performance) in working memory is equal to zero, that is $N_{wm} = 0$. The measure, Q, which determines the proportion of time allotted to the discrimination and recognition of those features of the task approaching the threshold characteristics of sense receptors is also 0.

Sometimes the same actions or elements of a task may be repeated multiple times during task performance. An increase in the diversity of performed actions influence the complexity of a task. One can use those measures associated with the proportion of time given over to repetitive task elements in order to evaluate these characteristics of task complexity. These measures reflect the habitual nature of task performance. Let us consider how one can define the proportion of time used for repetitive logical conditions according to the formula $Z^l = t^l_{rep}/L_g$, where t_{rep} is the time for performing those identical logical conditions that are repeated more than once, and L_g is the total time for logical conditions. In the example, when the subject performs $L_1 - L_4$ for the first time during task execution they are not repetitive. When any of these logical conditions is repeated it becomes a repetitive element. Therefore, the mean time for nonrepetitive logical conditions is equal to the total time associated with the performance of $L_1 - L_4$ when they are performed only once. However, as the probability of occurrence of logical conditions can be less than one, it is necessary to check if logical conditions occur more or less than once during task performance. Let us determine how many times each logical condition occurs during the performance of the example task. For this purpose we multiply the probability of each logical condition by 15 (as the subject takes two pins with two hands fifteen times). $L_1 = 4/9 \times 15 = 6.6$ times, $L_2 = 2/9 \times 15 = 3.3$ times, $L_3 = 2/9 \times 15 = 3.3$ times, and $L_4 = 1/9 \times 15 = 1.6$ times. It is clear that each logical condition occurs more than once during task performance, and therefore the probability of each logical condition occurring once is one. Thus, the total time for nonrepetitive logical conditions equals $L_1 + L_2 + L_3 + L_4$. The first logical condition requires only 1/2 EF. All the other logical conditions require $1/2\text{EF} + 1/2\text{EF} = \text{EF}$ (see Figure 4.6 which shows the time structure for the installation of pins in the second version of the task). From this it follows that the time for nonrepetitive logical conditions ($t^l_{\text{non rep}}$) will equal $0.13 + 0.26 + 0.26 + 0.26$, a total of 0.91 sec. As the total time for the performance of logical conditions is 3 sec, the total time for repetitive logical conditions is $3 - 0.91 = 2.09$ sec. From this we can infer that the fraction of time used for repetitive logical conditions (Z^l) will be 0.7. In the same way we can determine the proportion of time given over to repetitive efferent components of work (Z^{ef}) and to repetitive afferent components of work Z^α (see Table 4.4). The evaluation of Z^α is exactly the same as for Z^l. When analyzing Z^{ef} we can see that the subject performs the same motor actions multiple times. However, this involves two different 90° rotations of the pins, one clockwise, and the other counterclockwise. When these two rotations are performed for the first time during task execution they are nonrepetitive elements, and their performance takes 0.17 sec. All subsequent rotations are repetitive components of task performance. Similarly, reaching and grasping and then moving and releasing pins, when performed for the first time with unfluted pins, are also nonrepetitive elements of work which take 1.96 sec. Therefore, the time for nonrepetitive work in the example task is $0.17 + 1.96 = 2.13$. The time for repetitive efferent components of the task will be $32.7 - 2.13 = 30.57$, making $Z^{\text{ef}} = 0.93$.

The next measures to be discussed are associated with a five-point order scale of complexity of algorithm (task) and members of algorithm. In order to evaluate complexity according to this scale it is necessary to determine the fraction of task time devoted to the microelements of different complexity levels. All microelements of the task, along with their complexity levels and performance time, are listed below. According to the rules established earlier, the level of complexity of different task microelements depends on the level of concentration of attention during their performance. In the following code sequences, the first alphanumeric element designates the microelement, the second numeral the level of complexity, and the final numbers the time of performance in conventional units: R32C − 3 − 0.87; RL2 − 1 − 0; G1C1−2−0.44; M22B−2−0.67; mM10C−3−32; P2SE−2−0.97; T90S−1−0.32; RL1 − 1 − 0.12; M8A − 1 − 0.31; T180S − 1 − 0.56; P2SE − 2 − 0.97; RL1 − 1 − 0.12; G1A − 1 − 0.12; EF − 3 − 0.43.

Let us analyze the complexity of each member of the algorithm taking into consideration the specificity of combination and simultaneity of the performed actions. 1.31 units of time (0.01 min) are assigned to O_1^ε (see Table 3.20). This time consists of R32C–3–0.87 units and G1C1–2–0.44 units. Hence, 66% of the time belongs to the third category of complexity, and 34% of the time to the second category of complexity. Therefore, according to the rule, the second category is the combined category of complexity for this member of the algorithm.

The time of performance for O_2^α and for logical conditions $L_1 - L_4$ is described by microelements EF related to the third category of complexity (i.e., they require a high level of concentration of attention). However, these microelements are combined with either simultaneous hand movements belonging to the third category of complexity or require decision-making actions in the fourth category of complexity; therefore, the time for their performance is related to the fourth category of complexity.

Combinations of microelements where cognitive processes of perception and decision making are combined with hand movements in the third category of complexity are possible in this task. Therefore, the time for $O_3^{\varepsilon} - O_5^{\varepsilon}$, excluding P2SE and RL1, must be assigned to the fourth category of complexity. P2SE is related to the second category and RL1 to the first. After determining the fraction of time for each category of complexity and taking into consideration that there are elements related to the fourth category of complexity, one can conclude that in general O_5^{ε} should be regarded as related to the third category of complexity. We do not independently evaluate the complexity of logical conditions because they are combined with executive components of activity. Quantitative measures of complexity for the second version of the task are presented in Table 4.4.

4.2.1.2 Evaluation of Manual Task Complexity of the Third and the First Versions of Task in Laboratory Conditions

Let us consider the third, more complicated version of the task in which additional rules were introduced. According to these rules, if the subject installs pins into the first and second central rows (rows number 1 and 2) and two last rows on the right and the left-hand sides (rows number 5 and 6) then the rules from the second version of task are used. If a subject installs the pins into rows 3 and 4, those in the middle position, then the pins should be turned in the opposite direction, that is, the pin on the right-hand side should be installed so that the flute will be inside the hole, and that on the left-hand side so that the flute will be above the hole. This requires the memorization of additional rules.

Therefore, additional logical conditions, L_0 and $'L_1 - 'L_4$, were introduced for the algorithmic description of the third version of the task. According to L_0, the subjects decide which rules (logical conditions $L_1 - L_4$ or $'L_1 - 'L_4$) they should use, where logical conditions $L_1 - L_4$ relate to the rules already described in task version two, and operator O_0^{μ} and logical conditions $'L_1 - 'L_4$ are used only in the third version of the task (see Table 3.21). As can be seen from Figure 3.7, the panel has 6 columns of holes. The subject performs three cycles, simultaneously using two hands, in order to install pins into all six columns, where each cycle involves filling up two columns. During these three cycles, the subject must extract the required information and switch rules twice (between cycle 1 and cycle 2 and between cycle 2 and cycle 3). The probability of switching rules in the transition from cycle 1 to cycle 2 is 1, as it is for the transition from cycle 2 to cycle 3; both events are certain and independent. These switching rules are associated with O_0^{μ} — the retrieval of information from long-term memory (a simple mnemonic action) and L_0 — making a decision about which rules are required (see Table 3.21). According to Zarakovsky (2004), retrieving well-known information from long-term memory requires, on average, 0.3 sec. The simple decision-making action includes two mental operations: a receiving information operation (1/2 EF) and a decision-making operation (1/2 EF). Referring to MTM-1, these two operations can be assigned a duration of 0.26 sec. However, the decision-making action associated

TABLE 4.4
Quantitative Measures of Evaluation of Task Complexity for Three Versions of Task "Installation of Pins" (Time in Table is Presented in Seconds)

Measures of complexity	The first version	The second version	The third version
Time for algorithm (task) execution (T)	29.8	32.7	33.9
Time for performance of logical conditions (L_g)	0	3	3.76
Time for performance of afferent operators (T_α)	0	3	3.16
Time for performance of efferent operators (T_{ex})	29.8	32.7	33.3
Time for discrimination and recognition of distinctive features of task at threshold level (${}^{\prime}T_\alpha$)	0	0	0
Proportion of time for logical conditions to time for executive activity (N_l)	0	0.09	0.11
Measure of stereotyped logical components of activity (L_{st})	0	0	0
Measure of variable logical components of activity (L_{ch})	0	1	1
Measure of stereotyped executive components of activity (N_{st})	1	0.37	0.35
Measure of variable executive components of activity (N_{ch})	0	0.63	0.65
Scale of complexity (1 to 5 scale) (a) algorithm (b) members of algorithm	(a) 2 (b) ($O_1^\alpha - O_2^\varepsilon$): 2; 2	(a) 3 (b) ($O_1^\alpha - O_9^\varepsilon$): 2; 4; 4; 4; 3; 4; 3; 3; 2.	(a) 4 (b) ($O_1^\alpha - O_9^\varepsilon$): 4; 4; 3; 4; 4; 4; 4; 4; 4; 3; 4; 4; 2
Proportion of time for repetitive logical components of work activity (Z^l)	0	0.7	0.4
Proportion of time for repetitive afferent components of work activity (Z^α)	0	0.7	0.7
Proportion of time for repetitive efferent components of work activity (Z^{ef})	0.94	0.93	0.93

(Continued)

TABLE 4.4
Continued

Measures of complexity	The first version	The second version	The third version
Proportion of time for logical components of work activity depending on information selected from long-term memory rather than exteroceptive information (L_{ltm})	0	0.56	0.56
Proportion of time for retaining of current information in working memory (N_{wm})	0	0	1
Proportion of time for discrimination and recognition of distinct features of task approaching threshold characteristics of sense receptors (Q)	0	0	0

with L_0 is more complicated, as the subject must keep in memory and manipulate three units of information (about which column is involved, and the rules for turning the pins on the left and right hand sides, respectively). This is a more complex decision-making action in comparison with those described in the MTM-1 system and therefore should be assigned a duration of 0.3 sec (for comparison see Zarakovsky, 2004). The mnemonic action "recalling the required rule" overlaps motor component G1C1 which belongs to O_1^ε. In fact, the only logical operator $L_0$is not overlapped by motor components of activity. Therefore, O_1^ε has the same duration as in the second version of the task.

Operator O_3^ε also has the same duration as in the second version of the task as the pins are not fluted (see Figure 4.5).

In Figure 4.5 it can be seen that the mental action MP includes two mental operations (line segment EF is divided by dot). The first mental operation is associated with receiving the information that both pins are unfluted. The second includes the decision that the pins may be turned in any direction. LH and RH indicate the microstructure of motor actions performed simultaneously by the left and right hands. The bottom line, B, indicates the complexity of time intervals during the performance of operator O_3^ε. This is discussed further below.

In the second version of the task the efferent operators associated with the correction of errors are the same as in the first version. However, the decision making associated with error correction (l_{er}) becomes more complicated, as does L_0. This alters the performance time for these efferent operators. The additional time needed

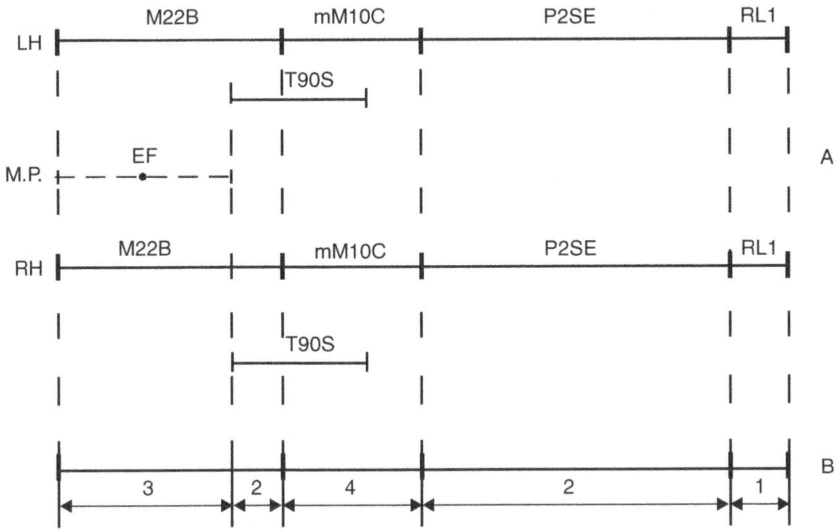

FIGURE 4.5 Time structure and category of complexity of activity during filing the pinboard when both pins are not fluted in the third version of task (performance of O_3^ε). A — Time structure; B — category of complexity.

to perform L_0 twice can be calculated in the following manner:

$$\text{“}T = 0.3 + 0.3 = 0.6 \text{ sec.}$$

In calculating the complexity of task performance we also need to know how much time is required to perform the mnemonic actions associated with O_0^μ. These mnemonic actions are also performed twice and hence can be calculated in the same way:

$$\text{“}T^\mu = 0.3 + 0.3 = 0.6 \text{ sec.}$$

The combined time structure for O_1^ε and O_0^μ is shown below:

R32C+G1C1

“t^μ

where “t^μ is the time required by the mnemonic action “retrieving well-known information from long-term memory.” This means that when grasping the pins (i.e., performing G1C1) the subject must recall the appropriate rule (perform O_0^μ with time t^μ). If the subject recalls this rule before grasping he must retain this information during the final stages of performing R32C and G1C1. This puts an additional load on working memory — a less efficient strategy in comparison with that where G1C1 is combined with O_0^μ.

The decision making associated with logical condition L_0 is performed on the basis of information extracted from long-term memory, so in this case no time is required for an afferent operation. When the subject retrieves information from long-term memory (O_0^μ) he must take into consideration both the column number in the pinboard and the rules for turning two pins. This implies that three units of information must be retained in working memory. Therefore, during the performance of O_0^{μ} the information stored approaches the maximum capacity of working memory. This means that this mnemonic action is related to the fourth category of complexity.

Finally, the time required to perform O_0^{μ} is included as a component of the measure N_{wm}. The general amount of logical conditions in this second version of the task is 9. The decision making associated with $L_1 - L_4$ and $'L_1 - 'L_4$ requires the manipulation of information approaching the maximum capacity of working memory. The continual switching from one kind of rule to another complicates the decision-making process. Hence, all the logical conditions associated with fluted pins in this version of the task are related to the fourth category of complexity. When the subject recognizes that a pin does not have a flute, he can decide to install the pin in any position; the decision making involved here is easier than that associated with fluted pins. We can assign microelement EF (0.26 sec) to this easier decision making, which is associated with the third category of complexity (see Figure 4.5).

In order to calculate performance time for the third version of the task and evaluate its complexity it is also necessary to determine the duration of $O_4^{\varepsilon} - O_6^{\varepsilon}$ and $'O_4^{\varepsilon} - 'O_6^{\varepsilon}$, and of those elements of activity related to them. This analysis can be facilitated by a graphical representation of their complicated time structure, shown in Figure 4.6. It should be noted that members of algorithm $O_4^{\varepsilon} - 'O_4^{\varepsilon}$ and $O_5^{\varepsilon} - 'O_5^{\varepsilon}$ have an identical time structure, inasmuch as one fluted pin can be in either the right or left hand.

Figure 4.7 presents the time structure of $O_6^{\varepsilon} - 'O_6^{\varepsilon}$ when both pins are fluted. In this case the two operators also have an identical time structure.

In Figure 4.5 through Figure 4.7 part A demonstrates the time structure of activity during the performance of the corresponding members of the algorithm. Line B

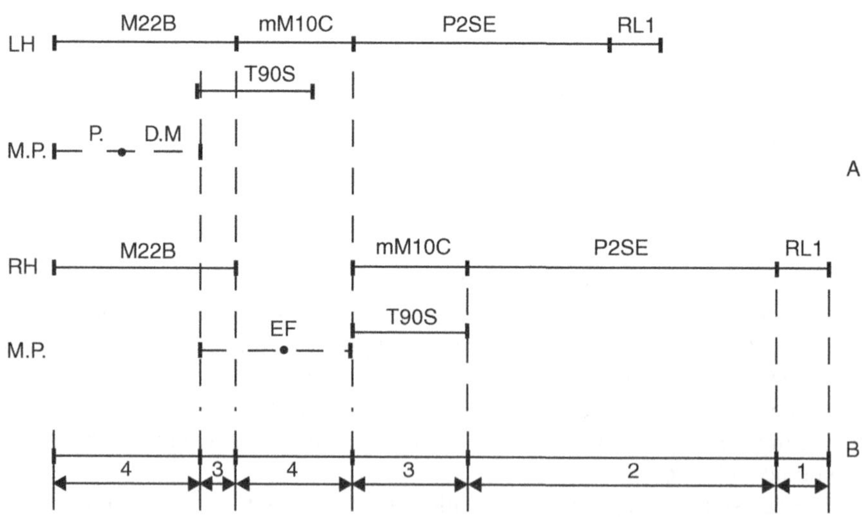

FIGURE 4.6 Time Structure and category of complexity of activity during filing pin-board when one pin is fluted in the third version of task. A — Time structure; B — category of complexity; performance of one of the following members of the algorithm (O_4^{ε}; $'O_4^{\varepsilon}$; O_5^{ε}; $'4O_5^{\varepsilon}$).

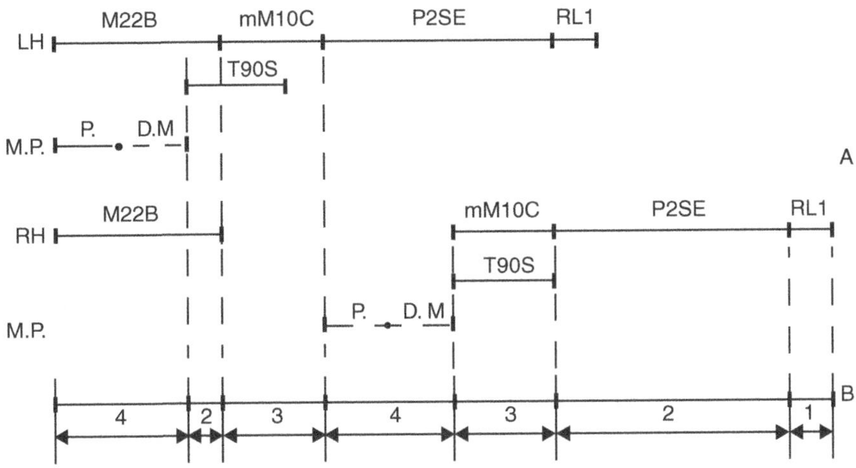

FIGURE 4.7 Time structure and category of complexity of activity during filling the board when both pins are fluted in the third version of task. A — Time structure; B — category of complexity; performance of one of the following members of the algorithm (O_6^ε; 'O_6^ε).

presents the complexity of the time intervals associated with different elements of activity. These figures require some additional explanation.

During the performance of O_3^ε (Figure 4.5) the time structure is the same in both the second and the third versions of the task. Only one element, EF, is related to cognitive activity (i.e., receiving information and decision making at the sensory–perceptual level), while all the other elements are related to motor activity. Element EF is in the third category of complexity and is combined with two elements (M22B) that have the second level of complexity. According to our rules for the evaluation of complexity (see Figure 4.3c) this combination of elements sums to the third level of complexity (see left-most interval, line B). According to the rule demonstrated in Figure 4.3e (the combination of two elements with the second level of complexity does not increase the complexity of the interval) the second interval in line B (Figure 4.5) can be assigned to the second level of complexity. The third interval on line B involves the combination of the two motor elements of activity mM10C, which are in the third category of complexity; thus, according to the rule shown in Figure 4.3b this interval exhibits the fourth level of complexity. In the fourth interval on line B the two second-level elements P2SE combine to give the second level of complexity overall. The right-most interval in the figure is assigned to complexity level one as both its constituents (RL1) are of the first category of complexity. The overall, general level of complexity for this member of the algorithm is in the third category and its time for performance is 2.08 units.

Figure 4.6 shows that during the performance of O_4^ε or O_5^ε and 'O_4^ε or 'O_5^ε the time structure is identical for all these members of the algorithm. In comparison with O_3^ε, the rules for turning the pins using the left or right hand are changed. The two cognitive actions P.D.M and EF are designated by dashed lines because they belong to other members of the algorithm that are performed at the same time as one of the members illustrated here. In the figure, cognitive action can consist of P.D.M. (associated with the left hand) or EF, associated with the right hand. Depending on the situation (i.e., whether the fluted pin is in the right or left hand) the combination of these cognitive actions, which differ in content depending on whether they are performed with the left or right hand) can change. This time structure is similar to O_5^ε in the second version of the task, but is not identical (cf. Table 3.20). The third version of the task involves a longer interruption in the movement of the second (right) hand. Element M.P., associated with the left hand, requires slightly more time than element EF, associated with the right hand. For this purpose one should compare O_5^ε in the second version of the task and time structure for the same member of the algorithm in the third version of the task (see Figure 4.6). In the second version of the task during calculation of time performance O_5^ε we utilize only microelement EF. In the third version of the task we use EF and P.D.M. The last cognitive element has a longer duration. As a result, the time required to perform this member of the algorithm slightly increases to 2.34 units as opposed to 2.26 units in the second version of the task.

Line B in Figure 4.6 demonstrates the complexity of different time intervals associated with these members of the algorithm. Reading from left to right, the first interval of time is related to the 4th category of complexity, as at this time the decision-making process, P.D.M (which belongs to the fourth category of complexity) is being performed (according to the rule 5(page 202). The second, short time interval is when element EF is performed , and thus is related to the 3rd category of complexity (according to rule shown in Figure 4.3d). The third time interval is related to the 4th category of complexity as two third category elements (mM10C and EF) are performed simultaneously (rule 4.3b). The fourth interval is assigned to the 3rd category of complexity because at this time one element (mM10C) of the 3rd category is combined with elements from the first or the second categories of complexity (see rule 4.3d). The fifth interval, that associated with microelement P2SE, requires only an average level of concentration of attention and thus belongs to the second category of complexity. According to rule 4.3e, if the two elements P2SE (second category) are performed simultaneously, it does not increase the complexity of the interval. The last interval of time in Figure 4.6, that associated with microelement RL1, requires only a low level of concentration of attention and is thus related to the first category of complexity (shown on the Figure 4.3a).

Let us now consider how we can evaluate the total category of complexity for these members of the algorithm (O_4^ε or O_5^ε and 'O_4^ε or 'O_5^ε) based on the results obtained so far and in accordance with the five-point order scale shown in Figure 4.2. The overall duration of these members of the algorithm is 2.34 units (where one unit is 0.01 min). The decision-making action (element P.D.M.) is in the fourth category of complexity, and its duration is 0.5 units. According to the rules described earlier, when we consider the fourth category of complexity we introduce a coefficient of 2 in order to evaluate the fraction of activity during task performance that falls into this category. This means that the duration of element L as a fraction of the total task time will be $0.5 \times 2 = 1$ unit. Element EF begins immediately upon element P.D.M's completion. Element mM10C starts when M22B (the first hand) finishes. Element M22B has a duration of 0.67 units, that is, it takes longer than element P.D.M. Therefore, element EF begins before mM10C. The period during which the performance of EF overlaps that of M22B is equivalent to the time for M22B minus the time for P.D.M (see Figure 4.6). This difference is $0.67 - 0.5 = 0.17$. From this it follows that element EF (0.43 units of time) overlaps mM10C only during 0.26 units of time (as $0.43 - 0.17 = 0.26$). This overlap period is related to the fourth category of complexity as during this time two third category elements (mM10C and EF) are performed simultaneously. Therefore, in calculating this time as a fraction of the overall performance we must again apply the coefficient, so that $0.26 \times 2 = 0.52$. Thus, the total running time during performance of this series of operators allotted to the fourth category complexity at this point in the evaluation is $1 + 0.52 = 1.52$ units. Furthermore, we must also consider operation mM10C, which is carried out with the second hand; this gives a total time for the third category and higher complexity of $1.52 + 0.32 = 1.84$. Hence, we see that the fraction of time for the third and fourth category complexities during the total performance time for O_4^ε or O_5^ε ('O_4^ε or 'O_5^ε) is $1.84/2.34 = 0.78$. As this value exceeds 0.7 of their total performance time all these members of the algorithm must be regarded as being related to the third category of complexity.

The time structure of operators O_6^ε and 'O_6^ε is presented in Figure 4.7. This time structure is also the same as in the third version of the task, but in this case two of the pins are fluted, and as described earlier, in the two associated decision-making actions (those elements with symbol P.D.M.) belong to the 4th category of complexity. This changes the strategy of performance and the distribution of elements of activity during performance, as a 4th category element cannot be performed simultaneously with 3rd or 4th category elements. Therefore, it is only after mM10C (the left hand, LH) has been completed that the subject can perform the second mental action which includes P.D.M, associated with the decision on how to turn the right hand (RH). Similarly, it is only when the decision-making action P.D.M. associated with the second hand (which is in the fourth category of complexity) is completed that the subject can perform element mM10C, a third level of complexity element associated with the right hand. The interruption to the movement of the right hand is increased significantly, and as a result, the duration of O_6^ε or 'O_6^ε is also increased. In this case, the time taken to perform these members of algorithm becomes 2.9 units.

We can calculate the complexity of different intervals of time (see line B, Figure 4.7) in a similar fashion. In this example we have two elements P.D.M, which belong to the fourth category of complexity and have a duration of 0.5 units. Here we once again use a coefficient of 2. Thus, the time for the fourth category of complexity action as a fraction of the total time for performing operators O_6^ε or 'O_6^ε will be $0.5 \times 2 + 0.5 \times 2 + 2$. Therefore, the fraction of time for the fourth category complexity is $2/2.9 = 0.69$ or almost 0.7 of the total. We also have the two elements mM10C, which have a duration of $0.32 + 0.32 = 0.64$. Based on our data we can assign these members of the algorithm to level four, as these members of the algorithm are very difficult for the subject to perform. The two examples described above demonstrate that graphical representations of time structure are useful when the time structure of the activity under study is highly complicated.

Let us consider how we can calculate the complexity of O_1^ε in accordance with the five-point order scale when the subject combines this member of the algorithm with operator O_0^μ, a combination which occurs only twice during task performance. The duration of O_1^ε is 1.31 units. The motor operation (motion) of activity G1C1 that is a component of member of the algorithm O_1^ε takes almost the same time as O_0^μ. Element G1C1 belongs to the second level (category) of complexity, mnemonic action O_0^μ to the fourth. Hence, when these two elements have to be carried out simultaneously the period of time for their performance belongs to the fourth category of complexity. The duration of O_0^μ is 0.3 sec. In order to transfer seconds into our conditional units 0.3 sec must be divided by 60 and multiplied by 100, giving a result of 0.5 units of time for O_0^μ. According to the rules described above, when we consider the fourth category of complexity we should introduce a coefficient of 2 when evaluating the fraction of this category in activity during performance. Thus, the time for O_0^μ as a fraction of the performance of O_1^ε will be $0.5 \times 2 = 1$ unit. As the total time for O_1^ε is 1.31 units, the fraction of time in the fourth category of complexity is 0.76. Again this value exceeds 0.7 and so this member of algorithm (O_1^ε) must be related to the fourth category—even if we ignore the duration of the two motions R32C. If the same member of algorithm (O_1^ε) is performed in a regular way (without combination with O_0^μ) its category of complexity will remain the same, as element R32C is related to the third category of complexity, its duration is 0.87 and its fraction of total performance time for O_1^ε is 0.66. Element R32C is performed simultaneously by two hands, and according to the rules when two third-category elements are performed simultaneously the period of time involved should be regarded as belonging to the fourth category of complexity. Again, when we use the five-point order scale we should apply a coefficient of 2 to the fourth category, and from this it follows that even when O_1^ε is performed in a regular way it will also have a fourth category complexity, as its fractional component having this complexity again exceeds 0.7.

We will now go on to consider other complexity measures. In order to evaluate task complexity we need to determine the probability of appearance of the efferent operators, because the cognitive components of activity are largely performed together with the executive (motor) components of activity. However, decisions such as whether to turn the pins left or right do not change the complexity of efferent operators. Therefore, we can calculate the probability of O_4^ε and 'O_4^ε together, as we can the probability of O_5^ε and 'O_5^ε, and of O_6^ε and 'O_6^ε. The probability of these events is the same as that of the probability of O_5^ε, O_6^ε, and O_6^ε. The probability of O_3^ε does not change as we have the same quantity of unfluted pins. The probabilities and time of performance for those members of the algorithm which determine the overall duration of task performance are presented in Table 4.5.

According to Formula 4.1, the time t for subtask "take and install pins" can be calculated as follows:

$$t = 1 \times 1.31 + 4/9 \times 2.08 + 2/9 \times 2.34 + 2/9 \times 2.34 + 1/9 \times 2.9 \\ + 2/75 \times 0.42 + 2/75 \times 4.16 = 3.71.$$

In order to transform this result into 0.01 sec units of measure, it must be multiplied by 60 : $3.71 \times 60 = 222.6$ or 2.22 sec. In order to calculate the total time for task performance we need to multiply the obtained result by 15, giving $2.22 \times 15 = 33.3$ sec. However, in the third version of the task we have an additional cognitive member of the algorithm L_0, which does not overlap with the members of

TABLE 4.5
Probabilities and Time Performance of Different Members of the Algorithm for the Third Version of the Task

Members of algorithm	O_1^{ε}	O_3^{ε}	O_4^{ε} or ‘O_4^{ε}	O_5^{ε} or ‘O_5^{ε}	O_6^{ε} or ‘O_6^{ε}	O_7^{α}	O_9^{ε}
Probability of occurrence	1	4/9	2/9	2/9	1/9	$1/75 + 1/75$	$1/75 + 1/75$
Time (0.01 min)	1.31	2.08	2.34	2.34	2.9	0.42	$2.08 + 2.08$

the algorithm considered so far. In fact, L_0 is repeated twice during task performance. From this, the time required to perform the third version of the task is slightly longer, as $T = 33.3 + 0.6 = 33.9$ sec.

If we wish to determine the time spent only on the executive components of the task, T_{ex}, we must eliminate the time for performing O_7^{α} from our overall time of 33.3 sec. While we cannot ignore this time, in our example the probability of O_7^{α} occurring is so small that the probable time taken by O_7^{α} is $0.02 \times 0.42 = 0.008$. We can safely ignore this negligible time period, and hence we preserve our value for T_{ex} of 3.33 sec.

The next step in the quantitative analysis is to determine the duration and probability of occurrence of the afferent operators and the logical conditions (for an algorithmic description of the third version of task and the symbolic designation of different members of the algorithm see Table 3.21). This is achieved by calculating the mathematical means for the afferent operators (T_{α}) and logical conditions (L_g) according to 4.2. While the mean time for O_2^{α} can be defined using the same procedures as for the second version of the task, in the third version the perceptual operations involved have differing durations. Hence, microelement EF (with a value of 0.43 unit) is assigned to one kind of perceptual operation and L (0.50) is assigned to other kinds of afferent operations (see Table 4.6).

From this it can be seen that the total time for O_2^{α} is 3.15 sec. Afferent operator O_7^{α} has an occurrence probability of 2/75, and thus requires $2/75 \times (1/2 \times \text{EF} + 1/2 \times \text{EF})$, where EF $= 0.43$ units. Therefore, the performance of O_7^{α} requires 0.01 units, and the total time for the overall task will be $0.01 \times 15 = 0.15$ units or 0.09 sec. Thus, the total time taken by afferent operators is the time for O_2^{α} plus the time for O_7^{α}, that is 3.16 sec. Of course, in a practical situation we may wish to simplify the calculation and, for example, assign the same time 0.5 units (L) to all versions of afferent operator O_2^{α}.

According to our formalized rules, the same time is required for logical conditions, at least in the case of decision-making actions at the sensory–perceptual level. However, in our example the time for logical conditions (L_g) is changed by the presence of an additional logical condition, L_0, which requires 0.3 sec to perform. The subject is required to install 30 pins using two hands; the taking and putting of pins is repeated fifteen times, yet the subject uses different rules only twice (2/15). Therefore, the performance time for L_0 is $2/15 \times 0.3 = 0.04$ sec. This time must

TABLE 4.6
Probabilities and Time Performance of Different Versions of Afferent Operator O_2^α

Versions of operator O_2^α	${}^1O_2^\alpha$ (1/2 EF)	${}^2O_2^\alpha$ $(1/2L + 1/2\text{EF})$	${}^3O_2^\alpha$ $(1/2L + 1/2\text{EF})$	${}^4O_2^\alpha$ $(1/2L + 1/2L)$
Probability of occurrence	4/9	2/9	2/9	1/9
Time (0.01min)	$1/2 \times 0.43$	$1/2 \times 0.50 + 1/2 \times 0.43$	$1/2 \times 0.50 + 1/2 \times 0.43$	$1/2 \times 0.50 + 1/2 \times 0.50$

then be multiplied by 15, giving a total time for L_0 of $0.04 \times 15 = 0.6$ sec, as was demonstrated earlier. When calculating the time for logical condition L_0 the time for the afferent operator associated with receiving information was not allocated, as this decision is made on the basis of information extracted from long-term memory rather than exteroceptive information). Therefore, the total time for logical conditions L_g sums to $3.16 + 0.6 = 3.76$ sec.

In both the second and the third versions of the task all signals are above the threshold level so that the time for discrimination and recognition of distinctive features of the task at threshold level sums to zero, that is '$T_\alpha = 0$.

The proportion of performance time involved with logical conditions in relation to that for executive (efferent) operators (N_l) can be determined from the formula:

$$N_l = L_g/T_{ex} = 3.7/33.3 = 0.11.$$

The next measure evaluates the stereotypical, or conversely the changeable (variable) components of activity. This "stereotyped logical component of activity (L_{st}) is determined according to Formula 4.7. Initially we determine how much time is required for stereotyped kinds of decision making (l_{st}). In the third version of the task, logical conditions can change their position during task performance depending upon chance events, that is what kind of pins (fluted or unfluted) are grasped by the subject. From this it follows that $l_{st} = 0$ and thus L_{st} is zero. Correspondingly, the measure of the variability of logical conditions, L_{ch}, is 1.

The stereotyped and variable executive components of activity is evaluated according to Formula 4.9 and Formula 4.10. Here, we recall that only operator O_1^ε is related to stereotypical components of activity. The total time taken up with performing O_1^ε will be $1.31 \times 15 = 19.65$ units or 11.7 sec. The overall time spent on the executive components of activity during the task is 33.3 sec. Therefore, the fraction of time dedicated to stereotypical executive activity, N_{st}, will be $t_{st}/T_{ex} = 11.7/33.3 = 0.35$. Correspondingly, the measure of the variable executive components of activity N_{ch} is evaluated at 0.65.

The other important complexity measures are associated with memory functions; these are determined using Formula 4.11. The first of these is l_{ltm} — the time taken up by logical conditions that are performed on the basis of information extracted from long-term memory. Establishing a value for l_{ltm} allows us to then determine L_{ltm}, the proportion of overall task time spent on logical conditions depending significantly on information selected from memory rather than from exteroceptive information. This is of interest as such logical conditions indicate more difficult decision-making processes.

In the example task, the probability of unfluted pins appearing is 4/9, while the overall time taken by the logical components of activity is 3.76 sec. Thus, the time taken up by those logical components of work activity that do not require any extraction of information from memory is $0.44 \times 3.76 = 1.65$ sec. This implies that the time spent on those logical components of work activity that do depend largely on information selected from memory (l_{ltm}) is $3.76 - 1.65 = 2.11$ sec. Therefore $L_{\text{ltm}} = 0.56$.

The measure associated with the extraction (i.e., retrieval) of information from long-term memory (N_{ret}) is calculated based on Formula 4.13. This measure should be used when the algorithmic description of the task includes operators that describe processes involving the retrieval of information from long-term memory. Hence, in the example T_{ex} is 33.3 sec and T^{μ} (the time for performing O_0^{μ} twice) is 0.6 sec. Thus, $N_{\text{ret}} = 0.02$. This is a very small fraction of overall task time which can safely be discounted. However, it should be noted that O_0^{μ} makes the performance of L_0 more complicated for the subject.

Another measure that evaluates memory is N_{wm}, which determines the proportion of performance time during which the subject must retain current information in working memory. In the third version of the task, the subject must constantly retain in working memory information about the columns and the rules (on how to turn pins) associated with them. Hence, $N_{\text{wm}} = 1$. The values obtained for mnemonic measures, L_{ltm} and N_{wm}, demonstrate that memory is a critical factor affecting the performance of this task.

The measure Q is used to determine the proportion of performance time spent on the discrimination and recognition of distinct features of the task approaching the threshold characteristics of sense receptors. The value of Q is the same here as in the second version of the task, that is, zero.

The next measure to be considered is Z^l — the proportion of time spent on the repetitive logical components of work activity. In the third version of the example task, subjects installed pins into four columns (columns 1–2 and 5–6) of holes using both hands, according to logical conditions $L_1 - L_4$. In total, these four columns comprise 20 holes, making it necessary to perform the subtask "grasp and put pins into hole" 10 times with each hand in order to complete the task. The probability of a fluted pin being in either the right or the left hand is 2/9. Therefore, when a subject carries out logical conditions $L_2 - L_3$, fluted pins will be in either the right or the left hand $2/9 \times 10$, that is, 2.2 times. However, when the subject carries out logical conditions $`L_2 - `L_3$ he should only install pins into two of the four columns (columns 3–4). In this case there are only 10 holes to be filled, requiring that the subtask "grasp and put pins into hole" is performed five times with each hand. Hence when the subject uses $`L_2 - `L_3$ fluted pins will be in either the right or the left hand $2/9 \times 5 = 1.1$ times. We can conclude that one occurrence of logical conditions $L_2 - L_3$ or $`L_2 - `L_3$ (when a fluted pin will be in one hand) is certain, that is these events must happen at least once.

The probability that the pins in both hands will be fluted is 1/9 (see Table 3.18). This means that when the subject makes a decision according to condition L_4, fluted pins will be in both hands $1/9 \times 10 = 1.1$ times. This is also a certain event, that is, one that must happen at least once. The subject makes a decision associated with logical operator $L_4 1/9 \times 5 = 0.55$ times. Our calculation shows that this event can be expected to occur 0.55 times when installing pins in columns 3 and 4 (i.e., this event will not always happen at least once).

Clearly, the logical conditions L_1 or $'L_1$, carried out when unfluted pins are in both hands will occur more than once. Due to the fact that $L_1 - L_4$ occurs at least once the measure of nonrepetitive time $t^l_{\text{non rep}}$ will be $L_1 + L_2 + L_3 + L_4$. The logical condition L_1, carried out when unfluted pins are in both hands requires 1/2 EF or 0.13 sec. Logical conditions L_2, L_3, and L_4 take more time, 0.3 sec each. Hence $t^l_{\text{non rep}} = 0.13 + 0.3 + 0.3 + 0.3 = 1.03$ sec.

We can use the same method to determine the time spent on nonrepetitive logical conditions $'L_1 - {}'L_4$. This will be $'t^l_{\text{non rep}} = {}'L_1 + {}'L_2 + {}'L_3 + 0.55 \times {}'L_4$ or $0.13 + 0.3 + 0.3 + 0.55 \times 0.3 = 0.89$ sec. We must also consider logical condition L_0, which determines the transition from one set of rules to the other. This condition, which takes 0.3 sec, appears twice; $0.3 + 0.3 = 0.6$ sec. The first appearance taking 0.3 sec is clearly related to a nonrepetitive logical condition. Therefore the total time for nonrepetitive logical conditions will be $t^l_{\text{non rep. total}} = t^l_{\text{non rep}} + {}'t^l_{\text{non rep}} + L_0 + = 1.03 + 0.89 + 0.3 = 2.22$ sec. We have already calculated the total time spent on logical conditions in the third version of the task as 3.76 sec. Therefore, the time spent on repetitive logical conditions $t^l_{\text{rep}} = 3.76 - 2.22 = 1.54$ sec. Thus, the proportion of time taken up by repetitive logical conditions will be $Z^l = 1.54/3.76 = 0.4$. The greater this coefficient, the more habitual the decision-making actions. The remaining measures such as Z^α , Z^{ef}, and L_{ltm} have the same values in both the second and the third versions of the tasks. A summary of all quantitative measures for the third version of the task are presented in Table 4.4.

Finally we consider the first, simplest version of the task, in which all the pins are unfluted. We will not discuss all the relevant calculations in detail, but rather will just present the results obtained in ready forms. There are neither afferent operators nor associated logical conditions in the first version of the task. This means that independent cognitive components of activity are not present in this version of the task; rather, the cognitive aspects of task performance are regarded as those components of motor actions responsible for their regulation. In considering this case, we would like to stress the importance of the principle of unity of cognition and behavior (Bedny et al., 2001). According to this principle, motor activity is not regarded as a series of independent reactions to different stimuli but rather as a system of object-oriented actions where each motor action is organized according to the mechanisms of self-regulation that integrate different cognitive processes into a holistic system directed toward achieving the goal of action. Hence, in spite of the fact that in this version of the task there are no independent cognitive actions associated with afferent operators and logical conditions, the cognitive elements of activity are not ignored but rather are regarded as constituent parts of motor activity. This specific aspect of the cognitive regulation of motor actions is reflected in the complexity measures of the first version of the task.

An analysis of the quantitative complexity measures of the first version of the task demonstrates that the time involved in performing afferent operators and logical conditions equals zero. As a result, all other measures that describe independent cognitive components of activity also have zero value. Therefore, in this case all the measures presented describe the complexity of motor components of activity associated with efferent operators. The measure of stereotypy of the executive components of activity, N_{st}, is one, as the subject repeats the same elements in the same sequence. According to the five-point ordered scale the category of complexity for the motor components of activity is two. The other members of this algorithm are also of the second level of complexity, as the performance of this kind of motor activity requires an average level of concentration of attention or an average level of mental effort. The general time of performance, T, is approximately 30 sec. The proportion of time spent on the repetitive efferent components of work activity, $Z^{ef} = 0.94$. This means that a significant part of this version of the task involves the repetitive performance of the same elements of motor activity. This task is extremely repetitive, being performed on the basis of what is called skill-based behavior in Rasmussen's (1983) taxonomy. All complexity measures for the first version of the task are presented in Table 4.4.

4.2.1.3 Comparative and Experimental Analysis of Manual Task Complexity in Different Versions of the Task

We will now perform a comparative analysis of different versions of the task based on the data presented in Table 4.4. In the second version of the task 10 of the pins were fluted, and pin installation was conducted according to particular rules. While the motor components of activity described by efferent operators practically did not change even a little from the first, simplest version of the task, this second version also contained afferent operators and logical conditions used to describe the cognitive components of activity. Task performance time T increased by 2.9 sec, while the performance time for efferent operators (total executive activity T_{ex}) remained identical to the total task performance time T. Hence, we can conclude that the cognitive components of activity were overlapped by (i.e., performed simultaneously with) motor components. The time for performing afferent operators, T_a, was 3 sec, which was the same as that for performing all logical conditions. The ratio of time for logical conditions to that for executive activity (T_l) was 0.09. When this measure exceeds 0.1 for a repetitive manual task this implies that decision-making actions play a noticeable role in task performance.

The measure of variability of the logical component (L_{ch}) was 1. This shows that during task performance the subject was required to perform decision-making actions, that appeared at random intervals and in an unpredictable sequence. In the first version of the task the executive activity was wholly stereotypical ($N_{st} = 1$). In the second version, executive activity became highly changeable ($N_{ch} = 0.63$).

According to the five-point ordered scale, task complexity increased from the second level in the first version of the task to the third level in the second version. Indeed, in the second version some members of the algorithm were even related to the fourth category of complexity.

A significant proportion of the decision-making actions involved in the more complex tasks depend on information extracted from memory rather than external (exteroceptive) information (see L_{ltm}). This not only complicates these actions but also produces a memory workload during the decision-making process. However, neither the first nor the second versions of the task requires that current information be retained in working memory. Finally, neither the first nor the second versions of the task requires that subjects receive information that is difficult to recognize ($Q = 0$).

In the third version, we saw that the time required to perform the task (T) slightly increased. In this version, however, the total time spent performing afferent operators (T_{ex}) is less than the total task time, T. This means that in this version there are some cognitive elements of the task that are not overlapped by motor activity. The time spent performing afferent operators and logical conditions (L_{g} and T_{a}) is also increased in comparison with the second version of the task. Moreover, the time spent on logical conditions L_{g} is greater than that for the afferent components T_{a}. This signifies that some of the decision-making actions being performed are totally dependent upon information extracted from memory. This finding is also confirmed by complexity measure L_{lfm}. The ratio of time spent on logical conditions to time spent on executive activity (efferent operators) is also slightly increased (see N_l).

Thus, task complexity is significantly increased in the third version, which is assigned to the fourth category of complexity according to the five-point ordered scale. This implies that the third version of the task requires the mobilization of the maximum mental effort. It is highly unlikely that this kind of manual task will be performed successfully many times during one shift. Whereas in the second version of the task a significant portion of the logical components of activity can be considered repetitive, in the third version the diversity of decision-making actions is significantly increased by comparison (see Z^l for the second and the third versions of the task). However, at the same time the proportion of time spent on the repetitive afferent components of work activity (Z^{a}) remains unchanged. This fact can be explained by noting that decision-making actions depend on information extracted from memory rather than from external sources. By way of contrast, the motor components of activity are highly repetitive (Z^{ef} is 0.93), which reduces the diversity of the required motor actions and simplifies performance.

The next significant measure is N_{wm}, which in the second version of the task was 0, but in this case becomes 1. During the performance of the third version of the task the subject is required to keep current information in working memory, a very difficult requirement that can give rise to mental fatigue and errors.

In general, the third version of the task differs from the second by requiring more complicated decision-making actions, increasing the subject's memory workload and general level of mental effort. In reality, manual tasks of this level of complexity are very seldom found in actual working environments. We selected this task as an object of study mainly to demonstrate the strength and range of the suggested approach. This example is also useful for training purposes and for the development of task complexity evaluation skills.

The next step in evaluating the complexity of various versions of the manual task was to conduct experimental studies. The purpose of these experiments was to determine the time needed for subjects to perform the different versions of the task and to investigate how many repetitions of the task were required for the subjects to become skilled in its execution.

In the experiments, the overall time of task performance, T, was used as a time standard. Skill was considered to have been acquired when the subject could perform the task five times in a row, making no more than one error and not exceeding the time standard. The methods used during the studies involved observation, interviews, discussion, recording the number of trials, measurement of time performance, and scaling procedures. The task was performed by 15 subjects, divided into 3 groups of 5 subjects each. The experimental task consisted of a single subtask "take pins and put them into the holes," repeated 15 times. The first group of subjects performed versions 1, 2, and 3 of the task, in that order. The second group performed versions 3, 2, and then 1. The third group began with version 2, then performed versions 1 and 3.

The results demonstrated that an average of 8 trials were needed for the subjects to acquire the skill required to successfully perform the first version of the task, 33 trials for the second, and 45 trials for the third. Therefore, on average each subtask was repeated 120 times in the first, 495 times in the second, and 675 times in the third.

After performing each version of the task, the subjects were asked to evaluate their own pace of performance using a rating method. The questions were formulated as follows: "Evaluate the pace of your performance, based on the premise that you will have to perform this task for the duration of the shift." Ratings were assigned against a 7-point scale, for example, the pace of performance is: very slow — score 1; slow — score 2; slightly below optimal — score 3; optimal — score 4; slightly above optimal — score 5; high — score 6; very high — score 7. The rating results showed that the subjects considered that the pace of performance for the first version of the task (pins without flutes) was slightly higher than optimal; for the second version of task it was high; and for the third version, the pace was closer to very high. Here we should recall that the initial evaluation of task performance time was conducted using standardized MTM-1 system data. The subjects' own evaluation indicates that the task pace suggested by the MTM-1 system exceeds the actual optimal pace, even for the simplest version of the task. These findings are in agreement with other scientists' data (Gal'sev, 1973).

Another important conclusion that can be drawn is that increases in task difficulty lead to subjective increases in the evaluation of pace; according to the subjects' own evaluation, the pace of the second and third versions of the task was high. Such a pace can be sustained only for a short period of time under emergency conditions; if the pace is subjectively evaluated as inconvenient or excessive the reliability of performance will suffer. It needs to be emphasised that the evaluation of pace should always be conducted using both objective and subjective criteria. It should also be taken into consideration that under stressful conditions it is possible for the pace of performance to increase as well as decrease, while the reliability of performance diminishes (Bedny and Zelenin, 1988). That task performance time often increases under stressful conditions is not due to a decrease in the pace of performance but rather to the use of less efficient strategies of performance. For example, it is often the case that elements that are normally performed simultaneously begin to be performed sequentially.

In the experiments, task complexity was subjectively evaluated in a similar fashion to performance pace. According to the subjects, the first version was the easiest, the second was considered to be more difficult, and the third version was felt to be the hardest. The third version of the task was described by the subjects as requiring their maximum effort.

The experimental studies demonstrated that task performance time cannot be the only criterion for the evaluation of task complexity, as, due to the possibility of simultaneously performing some activity elements, a more complicated task may take the same time as an easy one. Often, skill acquisition time is a more sensitive experimental criterion for task complexity evaluation than task performance time. The experiments also demonstrated that the quantitative measures of complexity developed earlier correlate closely with qualitative and subjective evaluations of task complexity.

4.2.2 Evaluating Task Complexity in Semiautomatic and Automatic Systems

In the previous section we considered a complicated manual task. In this section we will attempt to demonstrate the possibility of using our approach to evaluate task complexity in semiautomatic or automatic systems. In such systems, the operator's function is essentially one of monitoring and control, through the use of displays and control devices. In Section 3.3.1 we considered an example of time structure design that involved subjects working with a specially designed control panel. In this section we consider how one can evaluate the complexity of this sample task.

First, we will briefly recap this task. The operator may use all the controls or only some of them, depending on the information displayed by the five indicators. The first indicator is a digital display that can only present the numbers 1 or 2. Based on this information, the operator turns the switch to the appropriate position. He then grasps a 4-position lever which has a button on the handle. The operator is required to move the lever into one of four positions, but before he can do so he must depress the top button using his thumb. After moving the lever, the operator must wait 3 sec and then use a 10-position switch that is dependent on the position of a pointer on the display. Turning the switch to a particular position is linked to information presented on the digital display. Following this the operator is required to press a green or red button, depending on whether a green or red indicator on the control panel is illuminated. The general algorithmic description of this task and the time structure of activity during its performance are presented in Section 3.1.

The design of the experimental panel allowed 110 different versions of task performance to be carried out. Each of these versions can be described by a specific version of the task algorithm. Hence, it is theoretically possible to evaluate the complexity of each of the 110 versions of the algorithm. However, it is more practicable to utilize only those four distinctively different versions of the algorithm which clearly represent the four major possible strategies of task performance. Of these four, the first version of the algorithm deals with the situation where the control panel presents information that requires all five controls to be used. In the second version, the four-position hinged lever 7 is not used. In the third version of the algorithmic description of the task performance the multiposition switch 9 is not used. In the fourth and final version both lever 7 and switch 9 are not used.

These four versions of the task algorithm have been selected so as to include in our complexity evaluation the most representative and critical methods of performance. For example, an angle of rotation of 150° is selected for those occasions when the subject utilizes multiposition switch 9. When this is the case, the time for rotating the switch approaches the maximum possible. In the fourth version of the task algorithm the subject is required to press the most remote red button. For the purposes of our evaluation, it is essential that from the many different versions of the task we choose those which exhibit a marked degree of difference between their algorithmic descriptions. For example, from a practical point of view the difference between two versions of task performance is negligible if they differ only by one or two marks in the position of a switch. This approach to task complexity evaluation is in agreement with general constraint-based principles of design as, for example, during anthropometrical studies when analysts select only critical body size characteristics.

The use of the following heuristics is recommended when developing algorithmic descriptions of different versions of a task. A control or an indicator may be manipulated in a number of different ways. If during manipulative activity the quantitative and qualitative parameters of the task change slightly then during the algorithmic description of the task only one version of algorithmic description of the activity can be utilized. In this case, the general algorithmic description of the task and the algorithmic description of a particular version of the task are identical. However, if the qualitative and quantitative characteristics of activity vary depending on the specificity of control and indicator manipulation, then the method of using those controls and indicators should be strictly determined in a particular version of the algorithmic description. For example, the four-position hinged lever 7 has four identical directions of movements. Therefore, the direction of lever movement is irrelevant to the algorithmic description of a particular version of the task. In contrast, the rotation angle of multiposition switch 9 clearly influences the amount of time taken to manipulate it. This is why we specify a certain rotation angle when describing the method of manipulating this control in a particular version of the task.

According to the first version of the task performance algorithm, the operator utilizes all the indicators and controls. In the second version, pointer indicator 2 and the four-position hinged lever 7 associated with it are not used. In the third version, the operator does not use digital indicator 3 and its associated multiposition switch 9. In the fourth the operator uses neither pointer indicator 2 nor its associated four-position hinged lever 7, nor digital indicator 3 and its associated multiposition switch 9. Therefore, during all our calculations we need take into consideration only these four more representative versions of the task which show marked differences in performance, and consider that they have equal probability of occurrence, that is, 0.25. Almost all members of these algorithms are performed in sequence. Only the last two cognitive components of activity, those related to operators O^{α}_{10} and l_5, are overlapped by motor components of activity (receiving information from signal bulbs 4 or 5 and making a decision as to which button should be pressed and which activity can be performed while the hand is moving). Below we present the different, specific, versions of the task. Table 4.7 presents the algorithmic description of the

TABLE 4.7
Algorithmic Description of the Second Version of Task

Members of algorithm	Description of members of algorithm
O_1^α	Look at first digital indicator
1 $l_1 \uparrow$	If the number 1 is lit, perform ${}_1O_2^\varepsilon$; if the number 2 is lit, perform ${}_2O_2^\varepsilon$
${}_1O_2^\varepsilon$	Moves hand to the two-position switch 6 and turn it to the right
1 $\downarrow {}_2O_2^\varepsilon$	Moves hand to the two-position switch 6 and turn it to the left
O_3^α	Determine whether turn on the digital indicator 3 or the signal bulbs 4 or 5
[2(1−3) $L_2 \uparrow$	If digital indicator 3 presents numbers 5 ($L_2 = 1$) perform L_4
2(2) 4(1−10) $\downarrow$ $L_4 \uparrow$]	If the digital indicator 3 displays a number 5 decide to perform ${}_5O_9^\varepsilon$
4(1) $\downarrow$ ${}_5O_9^\varepsilon$	Turn multipositioning switch 9 to position 5
O_{10}^α	Determine that bulbs 5 (green) turn on
2(3) 4(10) 5 $\downarrow$ $\downarrow$ $l_5 \uparrow$	Decide to press red button 11 (perform O_{12}^ε)
5 $\downarrow O_{12}^\varepsilon$	Move the arm to the red button 11 and press it

second version of the task (the first version of the algorithmic description of this task and time performance is presented in Table 3.13 and Figure 3.6).

When considering logical conditions L_2 and L_4 in Table 4.7 it can be noted that a transition from one logical condition to the other is possible. In actual performance, the subject performs only one logical condition, L_4. In this example the logical condition L_2 simply demonstrates the transition from O_3^ε to L_4 (this is designated in Table 4.7 by the use of enclosing brackets). This means that in this version of the task a time value is not assigned to logical condition L_2. Table 4.8 demonstrates the algorithmic description of the third version of the task.

Table 4.9 presents the fourth version of the task.

In this fourth version of the task algorithm logical operator L_2 also designates the transition from O_3^α to logical condition l_5. This means that in this version of the task performance time is again not assigned for logical condition L_2.

According to the data presented above we can assign a time value of 0.3 sec for the performance of a simple decision-making action at the sensory–perceptual level (such as afferent operator O^α and logical condition L or l). Half of this time can be assigned for mental operations involving recognition (e.g., O^α) and half to logical conditions, of which there are five (l_1, L_2, L_3, L_4, and l_5).

The total time taken up by the performance of logical conditions, L_g, can be determined as the mathematical mean of the performance time for logical conditions

TABLE 4.8
Algorithmic Description of the Third Version of the Task

Members of algorithm	Description of members of algorithm
O_1^{α}	Look at first digital indicator
$l_1 \overset{1}{\uparrow}$	If the number 1 is lit, perform ${}_1O_2^{\varepsilon}$; if the number 2 is lit, perform ${}_2O_2^{\varepsilon}$
${}_1O_2^{\varepsilon}$	Moves hand to the two-position switch 6 and turn it to the right
$\overset{1}{\downarrow} {}_2O_2^{\varepsilon}$	Moves hand to the two-position switch 6 and turn it to the left
O_3^{α}	Determine whether turn on the digital indicator 3 or the signal bulbs 4 or 5
$L_2 \overset{2(1-3)}{\uparrow}$	If neither the digital indicator 3 nor the signal bulbs 4 or 5 are turned on ($L_2 = 0$) perform O_4^{ε}
$\overset{2(1)}{\downarrow} O_4^{\varepsilon}$	Move right arm to the fourth-positions hinged lever 7, grasp the handle and press button 8 with the thumb
$O_5^{\alpha w}$	Wait for 3 sec
O_6^{α}	Determine the pointer's position on the pointer indicator 2
$L_3 \overset{3(1-4)}{\uparrow}$	If the pointer position is 1 perform ${}_1O_7^{\varepsilon}$; if ...2 perform ${}_2O_7^{\varepsilon}$.... if 4 perform ${}_4O_7^{\varepsilon}$
$\overset{3(1)}{\downarrow} {}_1O_7^{\varepsilon}$	Move the four-position hinged lever 7 to the position that corresponds to the number 1
$\vdots$	
$\overset{3(4)}{\downarrow} {}_4O_7^{\varepsilon}$	Move the fourth-position hinged lever 7 to the position that corresponds to the number 4
O_8^{α}	Determine that signal bulb 5 (green) turns on
$\overset{2(3)}{\downarrow} \; \overset{4(10)}{\downarrow} \; l_5 \overset{5}{\uparrow}$	Decide to press red button 11 (perform O_{12}^{ε} turn on)
$\overset{5}{\downarrow} O_{12}^{\varepsilon}$	Move the arm to the green button 11 and press it

in each of the four versions of the algorithm, using the following formula:

$$L_g = 0.25 \times L_{g1} + 0.25 \times L_{g2} + 0.25 \times L_{g3} + 0.25 \times L_{g4}.$$

where $L_{g1} - L_{g4}$ is the time for logical conditions in the first to fourth versions of the task.

In the first version of the task (Table 3.13, or Figure 3.6) all five logical conditions are used. Therefore, the time taken up by logical conditions will be

$$L_{g1} = l_1 + L_2 + L_3 + L_4 + l_5 = 1/2 \times 0.3 + 1/2 \times 0.3 + 1/2 \times 0.3 + 1/2 \times 0.3 + 1/2 \times 0.3 = 0.75 \text{ sec.}$$

TABLE 4.9
Algorithmic Description of the Fourth Version of Task

Members of algorithm	Description of members of algorithm
O_1^{α}	Look at first digital indicator
$\overset{1}{l_1 \uparrow}$	If the number 1 is lit, perform ${}_1O_2^{\varepsilon}$; if the number 2 is lit, perform ${}_2O_2^{\varepsilon}$
${}_1O_2^{\varepsilon}$	Moves hand to the two-position switch 6 and turn it to the right
$\overset{1}{\downarrow}{}_2\, O_2^{\varepsilon}$	Moves hand to the two-position switch 6 and turn it to the left
O_3^{α}	Determine that bulb 5 (green) turns on
$\overset{2(1-3)}{L_2 \uparrow}$	If digital indicator 3 presents number 0 and bulb 5 turns on perform l_5
$\overset{2(3)}{\downarrow}\ \overset{4(10)}{\downarrow}\ \overset{5}{l_5 \uparrow}$	Decide to press red button 11 (perform O_{12}^{ε})
$\overset{5}{\downarrow}\, O_{12}^{\varepsilon}$	Move the arm to the green button 11 and press it

In the second version of the task we have only three logical conditions (see Table 3.14). This means that L_{g2} can be defined as:

$$L_{g2} = l_1 + L_4 + l_5 = 1/2 \times 0.3 + 1/2 \times 0.3 + 1/2 \times 0.3 = 0.45 \text{ sec.}$$

In the third version of the task we have four logical conditions. Their performance time is determined below as

$$L_{g3} = l_1 + L_2 + L_3 + l_5 = 1/2 \times 0.3 + 1/2 \times 0.3 + 1/2 \times 0.3 + 1/2 \times 0.3 = 0.60 \text{ sec.}$$

The time for logical conditions in the fourth version of the task is

$$L_{g4} = l_1 + l_5 = 1/2 \times 0.3 + 1/2 \times 0.3 = 0.3 \text{ sec.}$$

On the basis of this data we can now obtain a value for L_g which is

$$L_g = 0.25 \times 0.75 + 0.25 \times 0.45 + 0.25 \times 0.60 + 0.25 \times 0.3 = 0.52 \text{ sec.}$$

Logical conditions l_1, L_2, and L_5 are stereotyped, as l_1 and L_2 are always performed at the beginning of the task, and l_5 is always performed at the end; this is the case for all versions of the task. Logical conditions L_3 and L_4 appear with a probability of less than one, and are dependent upon which version of the task is being performed. Therefore, the measure of stereotypy of logical information-processing, L_{st}, can be determined as the relationship between the performance time for stereotyped logical conditions to that spent on all logical conditions:

$$L_{st} = 0.45/0.52 = 0.86.$$

In this example, the measure of variability of logical components, L_{ch}, can be determined as $1 - 0.86 = 0.14$. From this we can conclude that a significant portion of the logical information-processing in the task is stereotyped.

Similarly, we can calculate the time required for afferent operators. It should be noted that in this example it is possible that some afferent operators and logical conditions may be skipped due to the simultaneous perception of information from several displays. In this task a value of 0.15 sec has been assigned to each afferent operator and logical condition. This allows us to briefly evaluate the relationship between afferent operators and logical components of activity in a simplified manner. One method is to calculate the quantity of afferent operators and logical conditions in each version of the task. There are five afferent operators and five logical conditions in the first version, three afferent operators and three logical conditions in the second, four afferent operators and four logical conditions in the third, and two afferent operators and logical conditions in the fourth version of the task. Taking into consideration that the performance time for each logical condition and afferent operator is 0.15 sec, we can conclude that the subject spends exactly the same time (0.52 sec) for performing afferent operators as for logical conditions.

In this task all signals presented to the operator can be easily detected and recognized. Therefore, no time is allocated for the recognition or identification of weak signals approaching the threshold range ($T'_{\alpha} = 0$).

The total time of task performance T can be determined as the mathematical mean of the performance time of each of the four versions of the algorithm, using the following formula:

$$T = 0.25 \times T_1 + 0.25 \times T_2 + 0.25 \times T_3 \times 0.25 \times T_4.$$

In each version of the algorithm all members have probability 1. Therefore $T_1 - T_4$ can be determined as a sum of all the listed members of the relevant algorithm. The members of the algorithm required for each version of the task are listed in the algorithmic descriptions of these versions shown in Table 3.13 and Table 4.7 through 4.9. All members of the algorithm and their performance times in seconds for the first version of the task are listed in Table 4.10.

The total performance time for the first version of the task, T_1, can be calculated in the following way:

$$T_1 = 0.3 + 0.62 + 0.3 + 0.69 + 0.3 + 0.16 + 0.30 + 0.70 + 0.90 = 4.27 \text{ sec.}$$

TABLE 4.10
Members of the Algorithm and the Times of Their Performance of the First Version of the Task

O_1^{α} and l_1	O_2^{ε}	O_3^{α} and L_2	O_4^{ε}	O_6^{α} and L_3	O_7^{ε}	O_8^{α} and L_4	O_9^{ε}	O_{10}^{α} and l_5	O_{12}^{ε}
0.3	0.62	0.3	0.69	0.3	0.16	0.3	0.70	0.3 (overlapped by motor components)	0.90

TABLE 4.11
Members of the Algorithm and the Times of Their Performance of the Second Version of the Task

O_1^{α} and l_1	O_2^{ε}	O_3^{α} and $L_2 - L_4$	O_9^{ε}	O_{10}^{α} and l_5	O_{12}^{ε}
0.3	0.62	0.3	0.70	0.3 (overlapped by motor components)	0.90

TABLE 4.12
Members of the Algorithm and the Times of Their Performance of the Second Version of the Task

O_1^{α} and l_1	O_2^{ε}	O_3^{α} and L_2	O_4^{ε}	O_6^{α} and L_3	O_7^{ε}	O_8^{α} and l_5	O_{12}^{ε}
0.3	0.62	0.3	0.69	0.3	0.16	0.3 (overlapped by motor components)	0.9

TABLE 4.13
Members of the Algorithm and the Times of Their Performance of the Fourth Version of the Task

O_1^{α} and l_1	O_2^{ε}	O_3^{α}, L_2, and l_5	O_{12}^{ε}
0.3	0.62	0.3 (overlapped by motor components)	1.00

The members of the algorithm and their performance times for the second version of the algorithm are presented in Table 4.11.

In this second version of the task the total performance time, T_2 will be:

$$T_2 = 0.3 + 0.62 + 0.3 + 0.70 + 0.90 = 2.82 \text{ sec.}$$

Members of the algorithm and their performance times for the third version of the task are presented in Table 4.12.

Hence, the performance time for T_3 is

$$T_3 = 0.3 + 0.62 + 0.3 + 0.69 + 0.3 + 0.16 + 0.9 = 3.27 \text{ sec.}$$

Members of the algorithm and their performance times for the fourth version of the task are presented in Table 4.13.

Therefore,

$$T_4 = 0.3 + 0.62 + 1 = 1.92 \text{ sec.}$$

Thus, the mean performance time for the task will be:

$$T = 0.25 \times 4.27 + 0.25 \times 2.87 + 0.25 \times 3.27 + 0.25 \times 1.92 = 3.05 \text{ sec.}$$

We recall that the task also includes a 3 sec waiting period associated with the pointer display and the four-position hinged lever. Therefore, the actual total task performance time (T_τ) must include this period, so $T_\tau = 3.05 + 3 = 6.05$ sec. We can evaluate the complexity of this waiting period in a similar way as for that of efferent operators. In the task under consideration the pointer-indicator 2 has four possible positions. Depending on the pointer-indicator's position, the operator is required to move the hinged lever 7 to one of its four possible positions. Let us suppose that when the pointer is in position 1 or 2 it indicates a normal work regime, but if the pointer is in positions 3 or 4 this indicates a dangerous situation, and that position 4 is even more dangerous than position 3. The closer the pointer approaches positions 3 and 4, the more the operator will concentrate his attention on the task and the more he will experience emotional tension. For example, when the pointer exhibits position 1 — a normal work regime — this requires a relatively low level of concentration of attention by the operator. If the pointer shifts to the second position the operator begins to take more care and his attention rises to the second level. If the pointer moves to the third or even the fourth positions — indicating a dangerous state — the operator demonstrates extreme mobilization and maximum concentration of attention. The waiting periods associated with the state of the indicator and lever can therefore, at times, be related to the third or even the fourth category of complexity. If we can determine the probability of a pointer position appearing and the duration of time spent in the different positions we can calculate the complexity of the associated waiting period. For example, it is possible to calculate the proportion of active waiting period in the total task using Formula 4.16:

$$\Delta T_w = 3/6.35 = 0.47.$$

We will not consider the active waiting period further in what follows.

As the next step we can calculate the time spent on executive activity (the performance time for efferent operators — T_{ex}). In order to do so we must eliminate the time required for afferent operators and logical conditions from the time performance of the different versions of the algorithm. For example, from the first version of the task (where the total time is T_1) we must eliminate the time required for O_1^α and l_1; O_3^α and L_2; O_6^α and L_3; and O_8^α and L_4 (see Table 3.13). O_{10}^α and l_5 are not eliminated from T_1 as the time taken for their performance is overlapped by motor components of activity. Thus, in the first version of the task the time spent on executive activity ($T_{\text{ex}1}$) will be

$$T_{\text{ex}1} = 4.27 - (0.3 + 0.3 + 0.3 + 0.3) = 3.07 \text{ sec.}$$

We can calculate T_{ex2}, T_{ex3}, and T_{ex4} in a similar manner:

$$T_{ex2} = 2.82 - (0.3 + 0.3) = 2.22 \text{ sec.}$$

$$T_{ex3} = 3.27 - (0.3 + 0.3 + 0.3) = 2.37 \text{ sec.}$$

$$T_{ex4} = 1.92 - 0.3 = 1.62 \text{ sec.}$$

Hence, the mean performance time for executive activity is:

$$T_{ex} = 0.25 \times 3.07 + 0.25 \times 2.22 + 0.25 \times 2.37 + 0.25 \times 1.62 = 2.31 \text{ sec.}$$

The proportion of time taken for logical conditions to that spent on executive activity (N_l) can be determined by the following calculation:

$$N_l = 0.52/2.31 = 0.22.$$

Using Formula 4.9 and Formula 4.10 we can also determine the stereotyped and variability of the executive components of activity (t_{st} and t_{ch}). Only one executive operator, O_3^{ε}, can be related to stereotyped components of activity; the task always begins with moving the hand to switch 6 and turning it to the required position. All the other efferent operators depend upon chance events, that is, they are not performed in a predetermined order. The performance time for O_3^{ε} is 0.62 sec, and for executive activity overall it is 2.31 sec. So t_{st} can be computed as

$$N_{st} = 0.62/2.31 = 0.26.$$

Consequently, the measure of changeable executive components of activity t_{ch} can be more simply defined as

$$N_{ch} = 1 - 0.26 = 0.74,$$

confirming that most of the executive components of the task (efferent operators) are not performed in any habitual order.

The next measure we will consider is associated with the part played by memory in the decision-making process — L_{ltm}. In order to calculate this measure we must first determine the value of l_{ltm}. The task includes the following logical conditions: l_1, $L_2 - L_4$, and l_5. Logical condition l_1 is performed according to the rule "if the digital indicator presents the number 1 turn the switch forward, if the number 2 turn the switch backward." Clearly this logical condition is performed on the basis of a remembered rule, namely, information extracted from memory. In contrast, the logical conditions $L_2 - L_4$ are performed on the basis of information extracted from external sources. For example, L_3 is associated with the four-position hinged lever 7. When the pointer indicates number 1, the operator should move lever 7 in the direction designated by number 1; if number 2 is presented he should move the lever 7 in the direction number 2, and so on. This means that logical condition L_3 is performed on the basis of externally presented information, as is logical condition L_4. For L_4, if the indicator presents number 1 then multiposition switch 9 must be turned to position 1; if number 2 is presented than the switch should be turned to position 2, etc. — all the possible positions of the multiposition switch 9 also have corresponding numbers.

Logical condition l_5 is performed according to the following rule: "when the green bulb is lit press the red button" and "when the red bulb is lit press the green button." Therefore in this example, the color of the bulb and the color of the button cannot be used as external sources of information for the performance of l_5. From this it follows that logical condition l_5 is performed on the basis of information extracted from memory. This means that not only is this decision performed using a rule extracted from memory, it is also performed in a situation where contradictory information is being presented to the operator. From this it follows that this logical condition is related to the fourth, rather than the third category of complexity. Logical conditions l_1 and l_5 are presented in all the four versions of the task (see $L_{g1} - L_{g4}$). From this it follows that the time taken up by logical components of activity whose operational nature is predominantly governed by information retrieved from long-term memory can be calculated as

$$l_{\text{ltm1}} = 1/2 \times 0.3 + 1/2 \times 0.3 = 0.3 \text{ sec.}$$

The same times are required for l_{ltm2}, l_{ltm3}, and l_{ltm4}. Thus, the total time for l_{ltm1} will be:

$$l_{\text{ltm}} = 0.25 \times 0.3 + 0.25 \times 0.3 + 0.25 \times 0.3 + 0.25 \times 0.3 = 0.3 \text{ sec.}$$

According to Formula 4.11 we can calculate the proportion of time for those logical components of work activity which depend largely on information selected from long-term memory rather than exteroceptive information (L_{ltm}) as follows:

$$L_{\text{ltm}} = 0.3/0.52 = 0.58.$$

This measure demonstrates that much of the decision making involved in the sample task is of a more demanding nature. In a real-world situation, if an operator were required to perform many such tasks this would significantly complicate his performance.

However, the task under consideration does not require that the operator keep information in working memory during performance. Hence, the proportion of time spent retaining current information in working memory (N_{wm}) is zero.

In all versions of the task the signals presented to the operator are clear and can be easily recognized. Thus we can conclude that the proportion of time used for the discrimination and recognition of the distinct features of the task approaching the threshold characteristics of sense receptors (measure Q) is also zero. There are no repetitive components of activity during task performance. Therefore those measures associated with the measurement of repetitive components (Z^l, Z^α, and Z^{ef} also have zero value.

Finally, we can evaluate the complexity of the algorithms (and of individual members of the algorithms) for each version of the task according to the five-point ordered scale. According to the rules, simple cognitive components of activity are always associated with the third level or category of complexity. As all afferent operators in the algorithms do not require that the operator expend any extra effort in order to recognize the information presented by the control panel they can also be related to the third category of complexity. Similarly, the majority of logical conditions are simple and are thus related to the third category of complexity. Only logical condition l_5 can be regarded as related to the fourth category of complexity, as it requires the performance of decision-making actions on the basis of contradictory information.

The MTM-1 system is utilized to describe the motor components of activity (efferent operators) in terms of microelements; the level of concentration of attention required during the performance of these elements is used to evaluate their complexity. For example, efferent operator O_2^{ε} includes the following elements: RL1, R25B, G1A and M2.5A, and requires 0.62 sec for its performance. Element R25B is related to the second category of complexity as it requires an average level of concentration of attention. All the other elements require only a low level of concentration of attention and are thus related to the first category of complexity. The second category element R25B takes 65% of the overall time required to perform O_2^{ε}, with the first category elements taking the remaining 35%. Therefore, according to the rules developed earlier, a member of the algorithm O_2^{ε} is related to the second category of complexity. We can evaluate the complexity of other members of the algorithm in a similar fashion. Let us consider the complexity of the first version of the task. Table 4.14 demonstrates the complexity of different members of its algorithm.

According to the rules, when calculating the portion of the fourth category of complexity involved in the task we are required to multiply the performance time for fourth category elements by a coefficient of 2. This means that the conventional time given over to fourth category complexity in this task is $0.15 \times 2 = 0.3$ sec. The time required to perform O_{10}^{α} is 0.15 sec. Therefore, the total time given over to cognitive components is 1.65 sec of the conventional time. The total time required to perform the first version of the task (T_1) is 4.38 sec, giving a ratio of $1.65/4.35 = 0.37$ or 37%. This indicates that overall the first version of the task is related to the second category of complexity. We can similarly determine that the other three versions of the task are also related to the second category of complexity. Table 4.15 presents all the complexity measures and their values for the control panel task.

The purpose of working through this example in detail has not been to show how to improve the design of a control panel but rather to demonstrate our method for evaluating the complexity of operational-monitoring tasks involving automated and semiautomated systems. The kind of task described in this example is usually performed by an operator–manipulator. The accurate evaluation of their complexity becomes critically important in production settings where operators are required to perform a number of these tasks.

TABLE 4.14
Complexities of Different Members of the Algorithm

Members of algorithm	O_1^{α} and l_1	O_2^{ε}	O_3^{α} and L_2	O_4^{ε}	O_6^{α} and L_3	O_7^{ε}	O_8^{α} and L_4	O_9^{ε}	O_{10}^{α}	l_5	O_{12}^{ε}
Category of complexity	3	2	3	2	3	2	3	2	2	4	2

TABLE 4.15
Measures of Complexity during Performance of the Task on a Control Panel

Measures of complexity	Value
Time for algorithm (task) execution (T) or T_τ	3.05 sec or 6.05 sec
Time for performance of logical conditions (L_g)	0.52 sec
Time for performance of afferent operators (T_α)	0.52 sec
Time for performance of efferent operators (T_{ex})	2.31
Time for discrimination and recognition of distinctive features of task at threshold level ($'T_\alpha$)	0
Proportion of time for logical conditions to time for executive activity (N_l)	0.22
Proportion of time for logical components of work activity depending on information selected from long-term memory rather than exteroceptive information (L_{ltm})	0.58
Measure of stereotyped logical components of activity (L_{st})	0.86
Measure of variable logical components of activity (L_{ch})	0.14
Measure of stereotyped executive components of activity (N_{st})	0.26
Measure of variable executive components of activity (N_{ch})	0.74
Scale of complexity (1 to 5 scale)	
(a) algorithm	(a) 2
(b) members of algorithm	(b) $O_1^\alpha - O_{10}^\alpha - 3$; $l_1 - L_4 - 3$; $l_5 - 4$; $O_2^\varepsilon - O_{12}^\varepsilon - 2$
Proportion of time for repetitive logical components of work activity (Z^l)	0
Proportion of time for repetitive afferent components of work activity (Z^α)	0
Proportion of time for repetitive efferent components of work activity (Z^{ef})	0
Proportion of time for retaining of current information in working memory (N_{wm})	0
Proportion of time for discrimination and recognition of distinct features of task approaching threshold characteristics of sense receptors (Q)	0
Proportion of active waiting period in total task (ΔT_w)	0.47

In considering this example we have also demonstrated that the complexity measures used accurately reflect the subjective difficulties involved in performing these tasks. We can see that although motor components dominate in this kind of task, cognitive components, and in particular decision-making actions, are still important, as they take approximately one-fifth of the overall task performance time. Only one decision-making action is in the fourth category of complexity, while the remaining four belong to the third, lowest level, of decision-making action complexity. Nevertheless, more than half of the decision-making actions are performed based on information extracted from memory. When performing a number of tasks of this kind, the operator's memory can overload during the decision-making process, leading to mistakes. While the stereotypy of these decision-making actions is relatively high, the executive components of activity are highly changeable.

4.3 APPLYING SYSTEMIC-STRUCTURAL ACTIVITY THEORY TO DESIGN (MORPHOLOGICAL ANALYSIS)

4.3.1 An Example of the Redesign of Manual-Based Manufacturing Operations

In this section we will demonstrate the application of systemic-structural activity theory to the design of an assembly operation in a real manufacturing process. This analysis utilizes all four stages of the SSAT (Systemic Structure Activity Theory) design process: qualitative analysis, algorithmic analysis, time structure analysis and mathematical modeling. The use of symbolic models of activity allows the reduction or total elimination of physical models and the experimental methods associated with them. This is especially useful in task analysis under production conditions, where the utilization of purely experimental procedures is not always possible and often does not produce sufficiently accurate results. The major units of analysis in this study are cognitive and behavioral actions and operations and therefore in the same way as considered above laboratory studies examples this is morphological analysis (see page 66).

Two manufacturing operations performed at an assembly line were selected for analysis utilizing systemic-structural activity (Bedny et al., 2001). The first involves welding two brackets to the neck of a milk jug or flask (Figure 4.8). The second operation comprises the welding of two handles to the same neck (Figure 4.9). On the assembly line these production operations were performed sequentially (Figure 4.10).

Both operations were performed during a short time period and with high rates of recurrence. They appeared to be nearly identical in their physical demands and

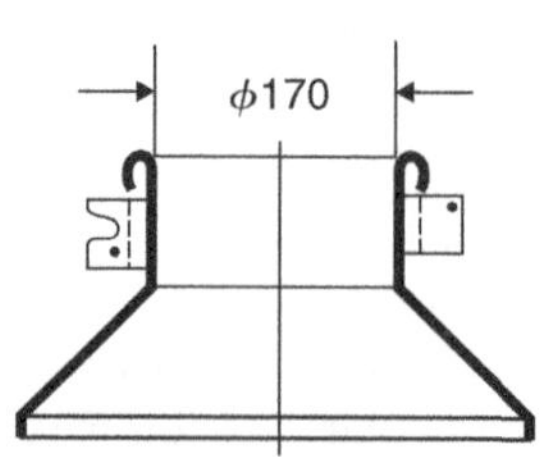

FIGURE 4.8 Neck with welding brackets.

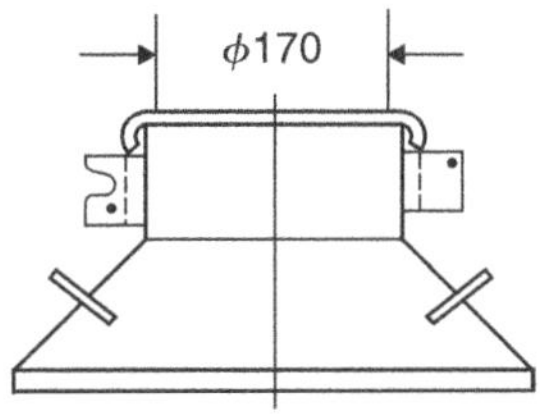

FIGURE 4.9 Neck with welding brackets and handles.

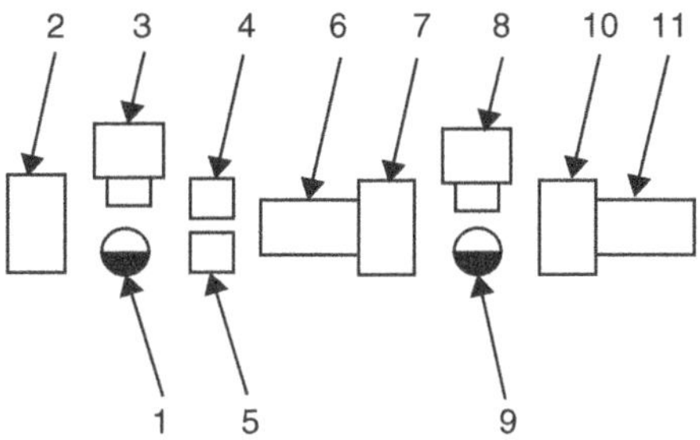

FIGURE 4.10 Planning of working places for operation of welding brackets and handles to the neck. 1 — Welder while welding brackets; 2 — bin with neck; 3 — welding apparatus; 4 — bin with rear brackets; 5 — bin with front brackets; 6 — conveyer's inclined plane; 7 — bin filled with necks having welder brackets; 8 — welding apparatus; 9 — welder while welding the handles; 10 — bin with handles; 11 — conveyer's inclined plane.

movements but quite different with regard to the cognitive regulation of motor activity. Thus, one of the more important theoretical and practical aspects of this study was to demonstrate how the quantitative measures of complexity described earlier can capture differences in the difficulty of cognitive regulation of human activity during the performance of two manufacturing operations with nearly identical behavioral characteristics.

The various stages and levels of analysis of these operations are described below.

4.3.1.1 Stage One: Qualitative Analysis

In general, this first stage of activity analysis includes many different methods. In the study discussed here techniques of objectively logical analysis were applied at the qualitative stage (Bedny and Karwowski, 2003).

(A) Welding brackets to the neck
The operation of welding brackets to the neck of the jug (see Figure 4.9) consists of the following steps.

The first operation involves welding two brackets to the neck of the milk jugs. The layout of the workplace for this production operation is presented in Figure 4.10. Storage bin 2 is filled to the neck. Bins 4 and 5 each contain different kinds of brackets. The brackets are placed to the right of the worker, while the bin filled to the neck

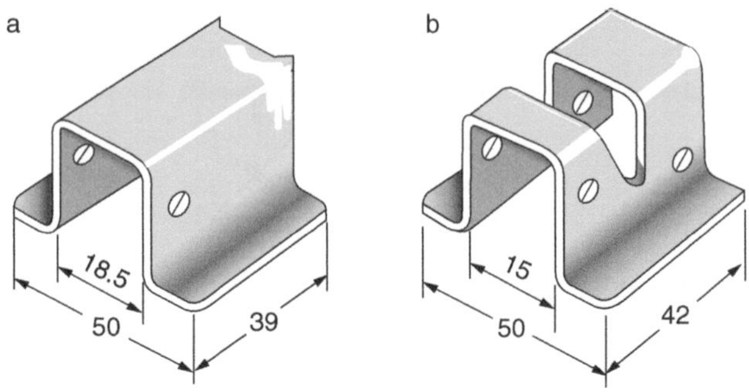

FIGURE 4.11 The rear (a) and the front (b) brackets.

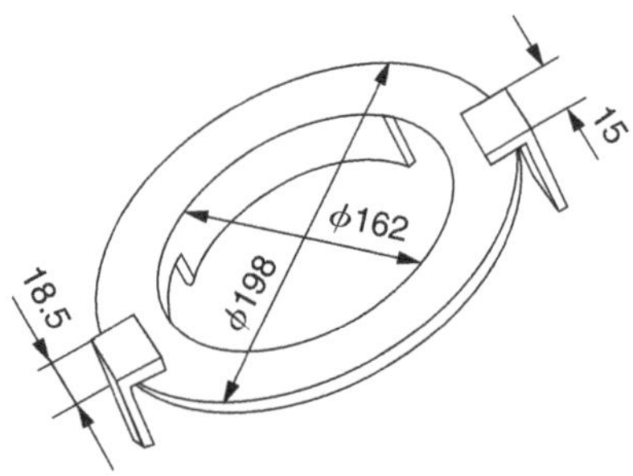

FIGURE 4.12 Jig for welding brackets.

is located to his left. Figure 4.11 shows two kinds of brackets, namely, the rear and front brackets, which are of different widths and shapes.

Moreover, the brackets are notched, which means that they can only be placed in a top-to-bottom orientation. The welding jig (shown in Figure 4.12) has two arms; one fits the front bracket, and the other is of an appropriate width to accommodate the rear bracket. The jig is used to align the brackets along an axis, guaranteeing that the brackets will be positioned precisely opposite one another. The brackets are installed into the required position over the arms of the jig, allowing them to be held steady while the worker welds their flanges to the neck of the flask. The arms of the jig only differ in their width, so that the size and shape of the arm correspond to the bracket to be held (see Figure 4.12).

The overall task requires the following steps.

1. Take a neck from bin 1 with the left hand.
2. Place a jig on the neck to align the brackets.
3. Put the neck (with jig) into the work position on a welding machine.
4. Hold the neck (with-jig) with the left hand; take one of the two brackets (either the front one or the rear) from bin 4 or 5 with the right hand.
5. Turn the bracket to the required position, weld the bracket to the neck with two sets of three spots on each flange of the bracket by operating a foot-switch.

6. Rotate the neck (with the jig still on it) 180° on the welding machine.
7. Take the remaining bracket, put it into the working position, and weld it.
8. Take the jig out and put it on the lap, and then take the neck with brackets and place it on the conveyer slide which drops it into a bin for the next worker (who undertakes the next operation "welding the handle to the neck").

Since the arms of the jig are of two different sizes, each calls for the corresponding bracket. Thus, in the passage from one complete welding cycle to another, the jig's arm (at the position to be worked on) alternates from one size to the other. Due to this alternation of the jig's arm, the operator must on one occasion begin the welding cycle process by taking a bracket from the front-bracket bin and on the next by taking a bracket from the bin containing the rear brackets. Thus, at the beginning of each welding cycle, the operator is required to decide whether the appropriate part to match the present position of the jig's arm is a front or a rear bracket. This decision requires attention to the width of the jig's arm and recollection of the last operation. In either case, an extraneous mental load is added to the task.

As the difference in widths between the arms of the jig is not highly salient, distractions or other sources of attention lapse typically result in a high frequency of mismatches between jig arm and bracket, leading to longer welding cycle times and greater error rates. In practice, the workers frequently "solved" this problem by simply cutting down the jig's wider arm to the same size as the narrower arm. This "solution" allowed them to begin each cycle by taking a bracket from the same bin. Although this clearly eases the mental load of the decision making, it also serves to introduce large variations in the post-weld position of the brackets, degrading the quality of the whole flask.

The six decision-making actions of the human operator are as follows.

1. If the narrow jig arm is in the work position, then reach for the front bracket.
2. If the wide jig arm is in the work position, then reach for the rear bracket.
3. Decide how to orient the rear bracket to place it on the jig arm after grasping it.
4. Examine and decide upon the acceptability of the rear bracket's quality while orienting it in relation to the jig arm.
5. Decide how to orient the front bracket in the appropriate position.
6. Examine and decide upon the acceptability of the front bracket's quality while orienting it in relation to the jig arm.

Analysis of the bracket-welding process indicated the following interrelated difficulties. Firstly, the production quota for each shift (up to 2000 flask necks to be completed) required that operators make a large number of decisions regarding the selection of the correct bracket, the acceptability of its quality, and its proper orientation with regard to placement of the notch. In addition to this quantitative requirement, the welders also needed to improve product quality by increasing the production of brackets and handles that were better aligned and therefore more durable. Thus, the selection, examination, and orientation of the brackets in the limited time available constituted the greatest difficulty for the welders.

The principal difficulty in the bracket-welding cycle resides in the number and nature of discriminating features involved. In selecting the fixing arm, the welder makes use of a single discriminating feature, the width of the fixing arms. In selecting the appropriate bracket, the welder must decide which bin it is located in. Neither of these features is salient, nor in any way makes self-evident which part should be selected, whether it is of the required quality, or how it should be orientated. Moreover, in those cases where the welders attempt to simplify the selection process by cutting the jig arms to the same width, another distinguishing feature related to the sequence of alternation between rear and front brackets is eliminated, requiring the welder to memorize which bracket was fixed in the previous operation. The problem is further complicated by the fact that if the front bracket was welded first on the previous neck, then the first rear bracket must be welded first on the next neck, otherwise, the welder must manually turn the jig through 180°. Typically, the welder controls the sequence in which the separate elements of the operation are performed, using a mental scheme of action that primarily depends on memorized information.

A fundamental aspect of this problem is that the workers are subject to severe time constraints. They must decide what type of bracket is appropriate for the current position of the jig arm very quickly. One factor confounding the workers' judgment is that the differences between the two arms of the jig are not at all salient, as they differ in their width by about 3 mm only. Thus, direct perception of the arms provides little inherent indication as to whether the welder should reach for a rear or a front bracket. Furthermore, the mapping of the width of the jig arm to the bin position contains no common perceptual elements. Thus, the welder must depend entirely upon nonconspicuous features of the width of the arm and the position of the bin.

As noted above, the problem is further complicated by the fact that if in the previous neck the front bracket was welded on first, then the rear bracket must be welded onto the next neck first; or, if the welder chooses to follow the same sequence as for the rear bracket, he must turn the jig — which is fairly heavy — through 180° with his hands. In practice, the welder frequently fails to select the correct brackets, and the mental strain generated by these subtle judgments causes an inconsistent product flow on the assembly line. Furthermore, the variability and magnitude of time needed for this manufacturing operation is also increased by these factors.

These demands on the welders' attention lead to premature fatigue. As we have seen, in order to defend themselves against tiredness, the welders would file the arms of the jig to the same size. This often results in a serious loss of quality control with regard to the positioning of the brackets, frequently requiring that the assembler who installs the flask cover at the end of the assembly line force the covers on with a mallet. This in turn means that the flasks covers are difficult to open and close and are not tight enough to hold the liquids for which they were designed. Additionally, erroneous selection of the brackets can lead to two rear (or front) brackets being installed onto the same flask. This means that the welder must also check as to whether the brackets are set up properly; this checking is, of course, also inhibited by fatigue.

Although the occurrence of unacceptable brackets is rare, their identification is made difficult by the lack of salient features by which they can be discriminated. Data concerning the number and types of defects in the bracket-welding process are shown in Table 4.16.

TABLE 4.16
Number and Types of Defects during Bracket Welding

	Number of cases	
	Psychological features are not taken into account	**Psychological features are taken into account**
Misalignment	590	0
Installation of two rear brackets	3	1
Installation of two front brackets	2	0
Wrong installation of brackets	5	2
Installation of false brackets	7	4

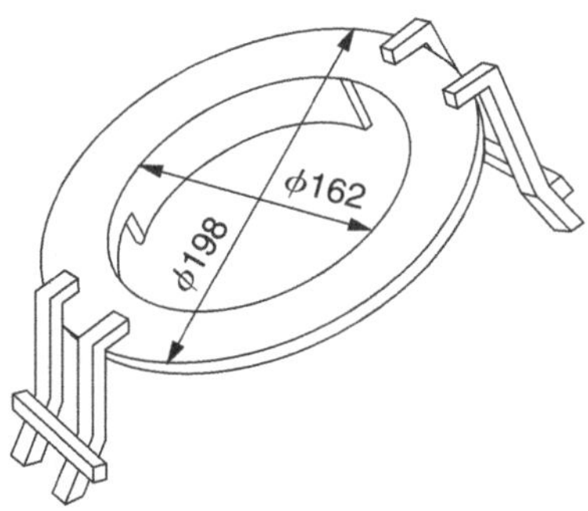

FIGURE 4.13 Jig for welding handles.

The defects shown in column two of Table 4.16 are relatively infrequent; their main drawback is that their detection at the completion stage of the operation in which the cover is installed on the flask provokes a contradiction between the requirements to maintain both the quality and quantity of task performance.

(B) Welding the handles to the neck
Immediately after the assembly operation discussed above, a second operation involving the welding of two handles to the same neck is carried out (see Figure 4.9). This operation consists of the following steps: bin 7 (as shown in Figure 4.10) is filled to the neck with the brackets already welded on. Bin 10 contains handles. The dimensions of this bin are similar to those of bins 3 and 5. In order to align the handles precisely, a jig similar to that involved in the previous operation is used. On this jig, the arms which align the handles opposite each other on the flask are the same size and shape (Figure 4.13). The handles are placed over the arms of the jig into the required position, allowing them to be held steady by the flanges.

The overall task involves the following steps.

1. Take a flask neck with welded brackets from bin 7, using the left hand.
2. With the right hand, place a jig on the neck to align the handle.

3. Put neck-with-jig into work position on the welding machine with both hands.
4. Hold the neck-with-jig with left hand, and take a handle from bin 10 using the right hand.
5. Weld the handle to the neck by depressing a switch with the right foot, using two sets of three spots on each flange.
6. Rotate the neck (with the jig still in place) 180° on the welding machine.
7. Take the next handle, put it into the work position, and weld it.
8. Remove the jig and put it on the lap, take the neck with brackets and handles and place it on the conveyer slide, which drops it into a bin for the next worker.

Two similar decision-making processes take place during this operation; when the operators put each of the handles into the work position they must examine the part in order to determine whether it is of acceptable quality.

An analysis of the workstation (shown in Figure 4.10) shows that the work space characteristics and organization are essentially identical for both operations. This implies that for the worker these two operations entail similar physical but different mental actions. In the second operation the work situation is symmetrical: equally sized and shaped handles from a single basket are welded onto the flask using a symmetrical jig, with the same size and shape of arm for either handle. This symmetry eliminates the mental actions involved in deciding which part to take in order to match it to a particular arm which happens to be in the work position. Further, unlike the brackets the handles are not notched. Their right-left, up-down symmetry eliminates any need to make decisions regarding bracket orientation.

Consequently, in the second assembly operation, the physical actions are unchanged from one cycle to another. The only decision the welder has to make is whether or not the grasped handle is of acceptable quality. The features of the handles that determine their acceptability (quality) are more salient than the features used to determine the quality of the brackets. Accordingly, interviews with the welders showed that the operation of welding the brackets is perceived by them as being more complicated and tiring.

4.3.1.2 Stage Two: The Algorithmic Analysis

The second stage of the study involves the delineation of an algorithmic description of the welding work process, where each entire operation is divided into elementary operators and logic conditions.

(A) The algorithmic description of operation "welding of the brackets to the neck

Table 4.17 shows the general descriptive form of the algorithm for the bracket-welding operation. As any operation may be analyzed at various levels of detail, algorithmic description is an iterative process. It begins with the development of an approximate description which can be amended and expanded as subsequent stages of activity analysis provide further information.

TABLE 4.17
Algorithmic Description of the Bracket-Welding Operation

Members of algorithm	Description of algorithm members
O_1^{ε}	To take a neck from the bin, to put a jig into the neck, and to install it on the welding machine
O_2^{α}	To discriminate the type of fixing arm
$l_1 \overset{1}{\uparrow}$	If the fixing arm is broad ($l_1 = 0$), perform O_3^{ε}, if the fixing arm is narrow ($l_1 = 1$), perform O_8^{ε}
$\overset{7}{\downarrow}\overset{4}{\downarrow}\overset{2}{\downarrow} O_3^{\varepsilon}$	While holding the neck and the jig with the left hand, take the bracket from the front bin
O_4^{α}	Determine whether the bracket is suitable or not
O_5^{α}	While taking the bracket from the bin to place on the neck, he determines whether the position of the notch on the bracket's end is facing toward or away from the worker and decides l_3
$l_2 \overset{2}{\uparrow}$	If the bracket is rejected ($l_2 = 1$), execute once more O_3^{ε}; if the bracket is suitable ($l_2 = 0$) then take into account l_3
$l_3^{\mu} \overset{3}{\uparrow}$	If the notch on the bracket's end is facing toward the worker ($l_3 = 0$), then perform O_6^{ε}; if it faces away from the worker (l_1), then leave it in this position and perform O_7^{ε}
O_6^{ε}	Turn the bracket 180°
$\overset{3}{\downarrow} O_7^{\varepsilon}$	Set up the bracket from the front bin and weld it on
$'O_7^{\varepsilon}$	Turn the neck with the jig 180°
O_0^{μ}	Recall what kind of bracket is welded
$l_4^{\mu} \overset{4}{\uparrow}$	If the bracket from the front bin is welded (the fixating arm is narrow $l_4^{\mu} = 0$ in the working position), then perform O_8^{ε}. If the bracket from the rear bin is welded (the fixating arm is broader ($l_4^{\mu} = 1$), then perform O_3^{ε}
$\overset{1}{\downarrow}\overset{5}{\downarrow} O_8^{\varepsilon}$	Take the bracket from the rear bin while holding the neck and the jig with the left hand
O_9^{α}	Discriminate whether the bracket is suitable
O_{10}^{α}	Determine simultaneously the position of the straps
$l_5 \overset{5}{\uparrow}$	If the bracket is rejected ($l_3 = 1$), repeat O_8^{ε} — 'if the bracket is suitable ($l_5 = 1$), then take into account l_6^{μ}
$l_6^{\mu} \overset{6}{\uparrow}$	If the strip connecting the right and the left side of the bracket that is facing the worker is wide ($l_6^{\mu} = 0$), then perform O_{11}^{ε}. If the strip connecting the right and the left side of the bracket that is facing the worker is narrow ($l_6^{\mu} = 1$), then leave it in this position and perform O_{12}^{ε}
O_{11}^{ε}	Turn the bracket 180°

(Continued)

TABLE 4.17
Continued

Members of algorithm	Description of algorithm members
$\overset{6}{\downarrow} O^{\varepsilon}_{12}$	Set up the bracket from the rear bin and weld it on
$l_7 \overset{7}{\uparrow}$	Condition of process termination or continuation up to operator O^{ε}_{11}; (not to turn the neck with the jig 180°) including O^{ε}_{13} (if $l_1 = 0$, then $l_7 = 1$)
O^{ε}_{13}	Pass on the neck for the next operation

Note: $l_7 \uparrow$ If operation starts from welding rear bin first ($l_1 = 1$) then go to O^{ε}_3 and perform up to O^{ε}_7 (not to turn the neck with the jig 180°), then go to O^{ε}_{13}; otherwise perform O^{ε}_{13}.

TABLE 4.18
Algorithmic Description of the Handle-Welding Operation

Members of algorithm	Description of algorithm members
O^{ε}_1	Take a neck from the bin, put a jig into the neck, and install it in the welding machine
$\overset{1}{\downarrow} O^{\varepsilon}_2$	While holding the neck and the jig with the left hand, take the handle from the bin
O^{α}_3	Determine whether the handle is acceptable
$l_1 \overset{1}{\uparrow}$	If the handle is rejected ($l_1 = 1$), repeat O^{ε}_2; if the handle is acceptable ($l_1 = 0$), perform O^{ε}_4
O^{ε}_4	Set up the handle and weld it on, turn the neck with the jig
$\overset{2}{\downarrow} O^{\varepsilon}_5$	While holding the neck and the jig with the left hand, take out the second handle
O^{α}_6	Determine whether the handle is acceptable
$l_2 \overset{2}{\uparrow}$	If the handle is rejected ($l_2 = 1$), repeat O^{ε}_5; If the handle is acceptable ($l_2 = 0$), perform O^{ε}_7
O^{ε}_7	Set up the handle and weld it on
O^{ε}_8	Pass the neck on to the next position

(B) The algorithmic description of operation "welding of the handles"

Table 4.18 and Table 4.19 present tabular algorithmic descriptions of the handle-welding operation. Table 4.18 shows a general description and 4.19 a revised, final, and more detailed version.

TABLE 4.19
Algorithmic Description of the Handle-Welding Operation (Detailed Version)

Members of algorithm	Description of algorithm members
O_1^{ε}	Take a neck from the bin and put the jig into the neck
O_2^{ε}	Install it in the welding machine
$\overset{1}{\downarrow} O_3^{\varepsilon}$	While holding the neck and the jig with the left hand, take the handle from the bin and bring the handle to the major working area
O_4^{α}	Decide if the handle is acceptable
$l_1 \overset{1}{\uparrow}$	If the handle is rejected ($l_1 = 1$), repeat O_3^{ε}; if the handle is acceptable ($l_1 = 1$) perform O_5^{ε}
O_5^{ε}	While holding the neck and the jig with the left hand, take out the second handle
O_6^{ε}	Set up the handle and weld it on
O_7^{ε}	Turn the needle with the jig
$\overset{2}{\downarrow} O_8^{\varepsilon}$	While holding the neck and the jig with the left hand, take second handle from the bin and bring the handle to the major working area
O_9^{α}	Decide if the handle is acceptable
$l_2 \overset{2}{\uparrow}$	If the handle is rejected ($l_2 = 1$), repeat O_8^{ε}; if the handle is acceptable ($l_2 = 0$), perform O_{10}^{ε}
O_{10}^{ε}	Set up the handle and weld it on
O_{11}^{ε}	Pass the needle on to the next position

This example demonstrates that, as was noted earlier, algorithmic description is an iterative process where subsequent, improved versions increasingly approach the optimal description possible. Specialists can begin with a global or enlarged description, subsequently decomposing it into more detailed analyses, or vice versa. For example, some members of the initial algorithm can be subsequently decomposed into several members of the algorithm. In cases where the sequence of actions during task performance must be kept in the operator's working memory, the members of algorithm will usually contain no more than three actions, and they will only rarely contain four. The number of included actions can be more in those cases where a member of the algorithm represents a sequence of simple actions that do not involve working memory. Usually, each member of an algorithm integrates tightly connected actions; it has a logical completeness, and is directed toward the achievement of the subgoal of the task.

From this discussion it can be seen that the notion of "level" of analysis should be distinguished from the notion of "stage" of analysis. The notion of "levels" refers to degree of detail of the description, while the notion of "stages" refers to the use of differing methods of description. These notions of levels and stages of analysis are important concepts of any systemic analysis.

With respect to the perceptual discrimination required, qualitative analysis of the handle-welding operation revealed an important difference from the bracket-welding operation. Unlike the brackets, the handles are symmetrical and are placed laterally on the flask; their welding does not require the operator to ascertain the part's left/right or top/bottom orientation. By the same token, the worker is not required to attend to the size or shape of the fixing arms on the jig which holds the handles in place. Furthermore, the discrimination of acceptable from unacceptable handles is easier, and the requirements for precision are less. Finally, stereotypical, repetitive activity dominates the whole process.

4.3.1.3 Stage Three: Temporal Analysis

Work activity is a process that unfolds over time. Therefore, temporal analysis provides the next stage in analyzing the production operations. The reader will recall that this stage of analysis uses typical elements of activity or psychological units of analysis. The notion of typical elements of a task refers to the elements of work; for example, the first member of the algorithm for the bracket-welding operation O_1^{ε} (Table 4.17) "To take a neck from the bin, to put a jig into the neck, and to install it on the welding machine" is an example of a typical element of the task or a technological unit of analysis.

This member of algorithm can be understood as including the following actions: left hand action — "move hand to the neck and grasp it" (the first action), "move neck to the welding machine" (the second action); right hand action — "move jig to the neck and install it into neck."

Describing these actions using technological units of analysis is fairly imprecise, and those who read such descriptions may encounter some difficulty in understanding exactly what actions are being performed by the worker. Therefore, it is preferable to translate this description into one using typical elements of activity or psychological units. We can achieve this by describing the performed actions as a system of standardized motions, for example, by utilizing the MTM-1. For example, the first left hand action includes the following standardized motions: "reach for new neck," $R50A_{BA}$, which signifies moving the hand to a particular object across a distance of 30 cm, using the simplest method (that which requires the lowest level of concentration of attention). This movement is accompanied by a body movement (AS30). The remaining 10 cm of movement is transferred into movement R10B, which requires an average level of concentration of attention. In the final stage of the action the worker performs the movement "Grasp" (G1A). We can describe the other two actions included in the first member of the algorithm in the same way.

Typical elements of a task are customarily used in the initial stages of analysis, particularly in the absence of good information regarding work activity. In the later stages of analysis, typical elements of a task are redefined in terms of typical elements of an activity. This translation allows the development of time structure descriptions of activity and also supports analyses of the task description in terms of understanding the worker's activity structure and evaluation of task complexity. "Taking the neck from the bin with the left hand" is a description in terms of typical elements of a task (technological units), whereas the same action described as "Reach" and "Grasp" utilizes typical elements of activity. It can be seen that the algorithmic descriptions outlined above utilized typical elements of task more often.

Therefore when one describes time structure of activity at the next stage of analysis he should transfer typical elements of task into typical elements of activity or psychological units. In the same way we can describe the other actions.

In general, in the absence of good information regarding work activity, initially typical elements of a task are customarily used. Later, typical elements of a task are redefined in terms of typical elements of an activity.

It is essential during time structure development to render the typical elements of a task in terms of the typical elements of an activity. This translation from typical elements of a task to typical elements of an activity enables one to develop time structure of activity. Further, it allows analyzing precise descriptions of a task and understanding what the worker does. Therefore during algorithmic description of a task we utilized first of all the typical elements of the task. When we describe the time structure of a task typical elements of activity (psychological units of analysis) are used. Time structure models use a hierarchically organized system of units of analysis (members of the algorithm, actions, and motor or cognitive operations) to describe activity as a coherent, structured process unfolding over time.

It should be noted that when we undertake a time structure analysis it is both a different stage of analysis, as it calls for a difference in the language or taxonomy of description, and also a different level of analysis, as it entails a more detailed specification of the process (that is, it is microlevel analysis). The notion of "level" is distinguished from the notion of "stage," which refers to different methods of description.

A tabular form of time structure description of the bracket-welding operation which makes use of the MTM-1 system is shown in Table 4.20. This description uses hierarchically organized units of analysis: firstly operators, then the sequence of motions that correspond to motor actions, and then microelements. The cognitive components of activity are presented in a similar fashion. This becomes obvious when we utilize the graphical model of time structure.

It should be noted that operators O_7^{ε} and 'O_7^{ε} contain very short, simple, and repetitive motions/actions. As discussed earlier, a greater number of such actions can be clustered into a single operator than is usually the norm when developing members of the algorithm.

Table 4.20 shows the time structure of the bracket-welding operation expressed in tabular form. It can be observed that any member of the algorithm can be translated into a temporal substructure of the assembly operation. For any member of the algorithm, comparing data from the algorithmic description with the time structure description can help to determine the logical sequence of shifting from one temporal substructure to another.

In addition to the tabular form discussed above, a graphical method of representing the time structure of work activity has also been developed, where the duration of any element of activity is illustrated by using an analogue system made up of horizontal lines. Activity elements are specified above the lines, either in terms of the MTM-1

TABLE 4.20
Time Structure Description of Bracket-Welding Operation (Units of Time — 0.01 min)

	Left arm			Right arm		
Operators and logical conditions 1	Description of movement 2	Type of microelement 3	Time 4	Description of movement 5	Type of microelement 6	Time 7
O_1^{ε}	Reach for new neck	$R50A_{BA}$ (AS30)	0.78	Release old neck (Hold on to Jig)	RL1 M44A	
	Hand makes contact with neck	mR10B	0.4	Move jig close to neck		
	Grasp neck	G1A	0.12			
	Move neck closely to the welding machine	$M40A_{BA}$ (AS30) T45M	0.95	Move jig to neck	M6C	0.35
	Move neck with jig to welding machine	M20B	0.63	Install jig into neck	P1SE	
	Sum of time units = 3.07; (1.84 sec)					

			Visual work load and decision making		
O_2^α	Identify type of jig arm	EF (dashed box)		Single glance required	
l_1			Decision making during the performance of M20B: O_2^α and RL1		
O_3^ε			Release neck with jig	RL1	0.12
			Reach to the front bin	R65B	1.36
			Take bracket	G1A	0.12
			Move bracket toward jig arm	M60B / ~~T75S~~ (bracketed)	1.22
				Sum of time	2.82 units (1.7 sec)
			Visual work load		
O_4^α	Evaluate the quality of the bracket	EF / ~~E~~F (bracketed, dashed box)	Relying on a single glance evaluate bracket's defects and jig arm for shape		# #

(Continued)

TABLE 4.20
Continued

Operators and logical conditions	Left arm: Description of movement	Left arm: Type of microelement	Left arm: Time	Right arm: Description of movement	Right arm: Type of microelement	Right arm: Time
1	2	3	4	5	6	7
O_5^{α}	Identify the position of the bracket			One glance		##
L_2	While executing P1NSE and glancing EF referenced logical conditions can be determined					##
L_3						
				Install the brackets on the arm of the jigs	P1SE	0.33
				Move neck with jigs and brackets to start position for welding ($S_l = 0$; $D_l = 1.0$)	M5C2/2	0.27
				Press pedal with right foot	LM	0.43
O_6^{ε}	Move into next position for welding ($S_l = 0$; $D_l = 1.0$)	M1C 2/2		Release pedal	LM	0.43

			Press pedal with right foot	LM	0.43
	Move into next position for welding	M1C 2/2	Release pedal	LM	0.43
			Press pedal with right foot	LM	0.43
			Release neck with jig and brackets (after welding on the flange at three points)	RL1	0.12
O_7^{ε}	Move into the next position for welding	M6C2	Move arm (distance about 8 cm)	R8A	0.32
			Take neck with jig	G1A	0.2
‘O_7^{ε}			Simultaneously with *M6C2*, R8A and G1A lift foot	LM	
			Press pedal with right foot	LM	0.43
	Move to next position for welding ($S_l = 0$, $D = 1$)	M1C 2/2	Lift foot	LM	0.43
	Move to next position for welding	M1C 2/2	Press pedal with right foot	LM	0.43

(Continued)

TABLE 4.20
Continued

	Left arm			Right arm		
Operators and logical conditions	Description of movement	Type of microelement	Time	Description of movement	Type of microelement	Time
1	2	3	4	5	6	7
				Lift up foot	LM	0.43
	Release neck with jig	RL1		Press pedal with right foot	LM	0.43
				Lift foot	LM	0.43
				Move arm tangentially along circle defining the neck, 10 cm and	M10A	0.37
				release neck	RL1	0.12
	Take neck with jig	G1A	0.12			
	Move neck with jig down	M8A	0.32			
	Release neck with jig	RL1	0.12			
	Move arm	M8A	0.32			

	Grasp neck	G1A	0.12			
	Move neck with jig down	M8A	0.31			
	Release neck with jig	G1A	0.12			
	Move arm	R8A	0.32			
	Take neck	G1A	0.12			
Sum = 8.22 units (4.93 sec)						
L_4	Decision making during period of turning the neck with the jig			EF		
O_8^{ε}				Move arm toward rear bin	R65B	1.36
				Take bracket	G1A	0.12
				Move bracket to arm of jig	M60B / T75S (T crossed out)	1.22
Sum = 2.7 units (1.62 sec)						
O_9^{α}	Determine quality of bracket	EF / EF (E crossed out)	Glancing in the same way as O_4^{α} and O_5^{α}			

(Continued)

TABLE 4.20
Continued

	Left arm			Right arm		
Operators and logical conditions 1	**Description of movement** 2	**Type of microelement** 3	**Time** 4	**Description of movement** 5	**Type of microelement** 6	**Time** 7
O^{α}_{10}	Determine position of brackets					
l_5	Making decision in the same way as l_2 and l_3					
l_6						
O^{ε}_{11}	Execute in the same way as O^{ε}_{6} and O^{ε}_{7}					
O^{ε}_{12}	(This time was not included in computing total)					
Total of time= 5.9 units (3.54 sec)						
O^{ε}_{13}	Release neck with jig	RL1	0.12	Move neck to next operation ($S_l = 0.11$; $D_l = 1.04$)		M35A
Totaltime = 0.12 units (0.07 sec)						
Grand total of operation = 22.85 units (13.7 sec)						

rules or using other predetermined time systems for cognitive and physical activity. Figure 4.14 uses this method to show the time structure of the bracket-welding activity. In this example, the length of the horizontal lines denotes the duration of the element of activity in terms of microelements as defined within the MTM-1 system. The table also uses the following symbols: Right Hand (RH), Left Hand (LH), Mental Processes (MP), Body (B), Right Leg (RL), and Left Leg (LL). Any elements in a section that refer to another unit of the algorithm are indicated by a dashed line.

In some cases, the graphical description also presents the probability of occurrence of a particular element of activity in the time structure. This provides a tangible means for representing the logical structure and temporal sequence of an activity, facilitating both qualitative, and quantitative evaluations of work activity. This method is especially valuable in cases where a large number of activity elements are executed simultaneously.

The time structure of the second production operation "welding of the handles" will not be described here, as the time structure of its motor activity is the same as in the first production operation. Differences between the first and second operations can be related to the absence of those cognitive actions associated with receiving of information about width of the arms, and with decision making about what kind of handle should be selected and the orientation of the handle before its installation in the required position. Due to the very short duration of these cognitive processes and the fact that they are performed simultaneously with the motor actions, the performance time of these production operations is insignificantly different.

It should also be noted that in the example algorithmic description operators O_7^{ε} and 'O_7^{ε} are so closely interconnected that their division into two discrete members of the algorithm is somewhat arbitrary.

4.3.1.4 Stage Four: Quantitative Complexity Evaluation of the Production Operations

The final stage of analysis involves the quantitative evaluation of the complexity of the production operations. Task complexity is one of the more important integrative characteristics of the task. This characteristic is critical for evaluation of mental effort during task performance. The complexity of a task is the major cause of mental workload during task performance (Hancock and Caird, 1993). The quantitative analysis of the complexity of production operations can be performed only by using psychological units of analysis and only after having developed a time structure of the activity under study. The quantitative evaluation of our sample production operations is performed using the methods described in the previous sections.

(A) Evaluation of task complexity of the bracket-welding operation prior to intervention

All the major elements of activity during the performance of this production operation have a probability of 1. Only those elements of activity associated with detection of defective brackets or handles have a lower probability of occurring, this being approximately 0.01. The detection of one of these less probable events takes 0.26 sec, and thus the time taken up by such an event is $0.01 \times 0.26 = 0.0026$ sec. As these

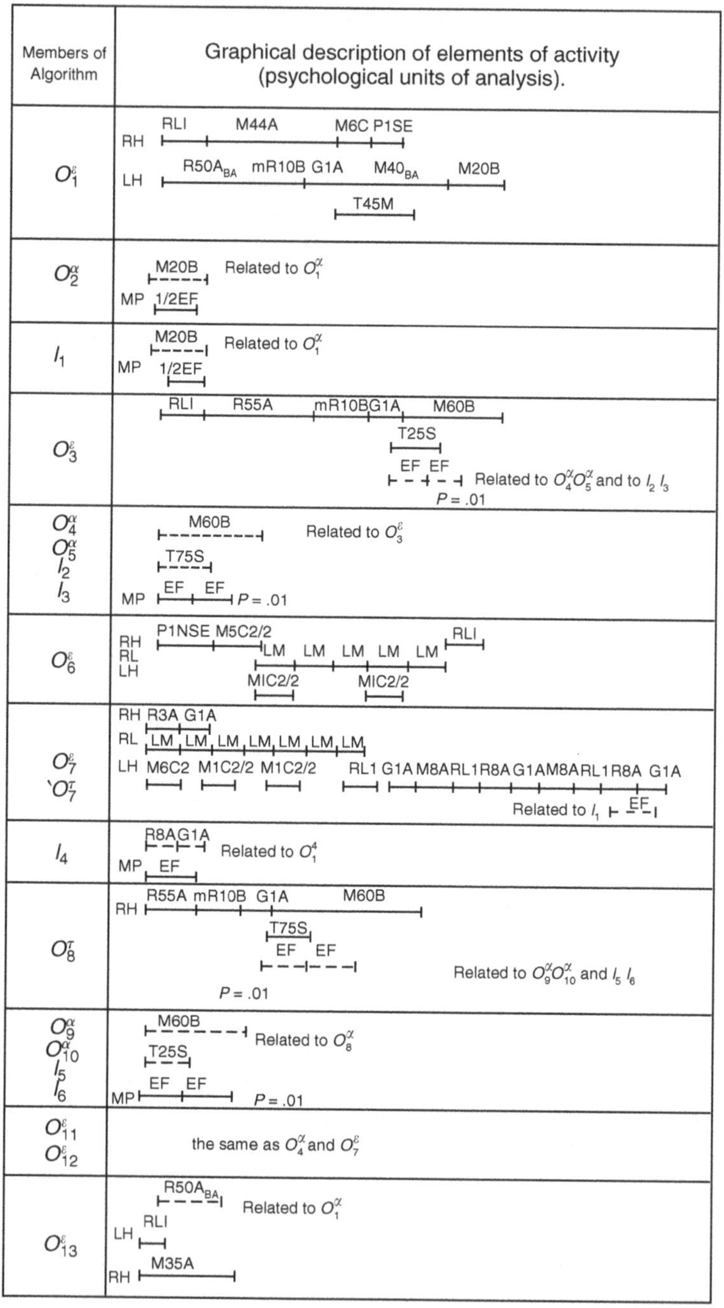

FIGURE 4.14 Time structure of the production operation "Brackets welding on."

events can happen twice during the operation, their total contribution to the overall task time is $0.0026 \times 2 = 0.0052$ sec. This very short duration can safely be disregarded when calculating the complexity of task performance, and this facilitates calculating the mean time of performance, which is more straightforward when the probability of all elements of the activity occurring is one. In this operation all cognitive elements of activity are performed simultaneously with the motor components. Therefore, according to formula (4.1) the total time for performing the production operation is:

$$T = O_1^{\varepsilon} + O_3^{\varepsilon} + (O_7^{\varepsilon} + {}^{\prime}O_7^{\varepsilon}) + O_8^{\varepsilon} + O_{11}^{\varepsilon} + O_{13}^{\varepsilon} = 1.84 + 1.7 + 4.93 + 1.62 + 3.54 + 0.07 = 13.7 \text{ sec.}$$

In our calculation we can also ignore the additional time required to perform those operators (O_3^{ε} and O_3^{ε}) which describe the use of defective brackets, as these events have a probability of 0.01 and therefore the time period involved (0.03 sec) is insignificant.

The procedure for calculating the complexity of the cognitive components of the task can be explained as follows. All the cognitive elements of activity are regarded as simple decision-making actions executed at the sensory–perceptual level (Bedny et al., 2000). Each of these actions consists of two mental operations: receiving information and making a decision. We use only microelement EF to determine the duration of these cognitive elements. According to the rules developed for complexity evaluations, half of this duration is related to afferent operators and half to logical conditions (see page 203).

Consider logical condition l_4^{μ}. This logical condition describes a decision-making process based not on external information but rather on information drawn from long-term memory, as there is no perceptual information (O^{α}) immediately present that can guide decision making. This is why there is no afferent operator preceding l_4^{μ} in the algorithm. Instead of externally presented information a worker utilizes information from memory. Therefore instead of O^{α} we introduce operator O_0^{μ} that stands for "Recall what kind of bracket is welded". In order to simplify calculation of complexity measures, we do not utilize conventional rules according to which half of EF is related to O^{α} and half - to logical condition l_4^{μ}. All this time was assigned to l_4^{μ}. Hence, duration of l_4^{μ} is 0.26. The duration of l_4^{μ} is 0.26 sec (the complete time for one EF element), whereas the time required to perform each of the other logical conditions is 1/2 EF. The overall time spent on logical conditions (L_g) is calculated according to Formula 4.2:

$$L_g = l_1 + l_3^{\mu} + l_4^{\mu} + l_6^{\mu} = 0.13 + 0.13 + 0.13 + 0.26 = 0.65 \text{ sec.}$$

The total time taken by afferent operators is

$$T_{\alpha} = O_2^{\alpha} + O_5^{\alpha} + O_{10}^{\alpha} = 0.13 + 0.13 + 0.13 = 0.39 \text{ sec.}$$

As noted above, in our example we use only microelement EF to calculate the time needed for afferent operators and logical conditions. Thus, in this situation L_g should be equal to T_{α}. However, L_g is actually found to be greater than T_{α}, leading to the conclusion that at least a portion of some of the decision-making operations is performed on the basis of information extracted from memory. This conclusion is also confirmed by the calculation of L_{ltm}, those logical conditions based on the retrieval of information from long-term memory.

Let us consider in more detail how the various logical conditions are performed. Logical conditions l_1 and l_4^{μ} are associated with making the decision as to which bin the hand should be directed. In the situation where the workers have cut down the jig's wider arm, they must make this decision solely on the basis of the bin's position.

It has been experimentally established that discriminating the spatial position of different signals is complicated and thus makes the decision-making process more complicated. Therefore, in this case we can conclude that the position of the bin is not sufficient to provide a sound basis for consistent decision making. As a result, the workers must mainly perform their decision-making actions on the basis of rules drawn from memory, with little external support. This means that not only l_4^{μ} but l_1 as well very often are performed using information extracted from memory.

The next problem is connected with logical condition l_3^{μ}. The worker's decision on how to position the bracket depends on the position of the notch at its end. This decision is also made on the basis of a memorized rule, namely, that the notch should face away from the worker's body. Logical condition l_6^{μ} is even more complicated. As we have seen, when the worker violates the technological process by reducing the dimension of the jig's wider arm, he must thenceforth always remember which kind of bracket was fixed previously when beginning a new welding cycle.

Thus, all the logical conditions (decisions) in the task were preformed primarily on the basis of information derived from memory. Thus, according to Formula 4.11 we can conclude that $L_{ltm} = 1$. The other measure associated with memory is N_{wm}, the duration of time for which information must be sustained in working memory. The worker must remember which kind of bracket was just welded until he decides to which bin the hand should next be directed. During this period the jig-with-neck must be rotated through 180°. Once the decision has been made the worker can forget this information. Thus, the period during which the worker keeps the information in his working memory starts when the rotation of the jig is begun (see operator O_7^{ε} when the left-hand microelement "take neck with jig" begins). The time period ends when O_7^{ε} is finished. The duration is 1.1 sec. Therefore, according to Formula 4.12 N_{wm} will be 0.08. When attempting to interpret these memory-oriented measures (L_{ltm} and N_{wm}) we must take into consideration the fact that the production operation is repeated up to 2000 times during each shift. So for each shift the total time of working memory load can be as much as 36.6 min, and the total time taken up by decision-making processes based on information extracted from memory up to 21.4 min per shift.

All the logical conditions we have been considering are performed in a habitual sequence and are thus considered as stereotyped. Only the logical conditions l_2 and l_5 which are associated with the decision as to whether or not the brackets are of suitable quality can be considered as nonstereotyped (i.e., changeable) elements of activity. However, as we have already discussed we can safely discount l_2 and l_5 when calculating the complexity of the task. Thus, according to 4.7, $L_{st} = 1$ and $L_{ch} = 0$.

In this production operation only O_1^{ε} and O_{13}^{ε} can be considered to be stereotyped efferent (executive) components of activity. All the other efferent operators represent essentially changeable executive components. From this we can calculate that the time for $t_{st} = 2$ sec, and for $t_{ch} = 11.7$ sec. Therefore, according to 10 (9) and 10(10), $T_{st} = 0.15$ and $T_{ch} = 0.85$.

There are both repetitive logical and efferent components of work. For example, those efferent operators such as O_3^{ε}, O_7^{ε}, O_8^{ε}, O_{11}^{ε}, O_{12}^{ε} include some motor components which are repeated more than once during the production process. Values for those measures such as Z^l, Z^{ε}, Z^{ex} that describe the repetitive components of the activity under analysis are presented in Table 4.21 (here we refer to the column showing values for the bracket-welding operation before intervention).

According to the five-point ordered scale, the cognitive components of activity in this task are related to the third category of complexity (the minimal level of complexity for cognitive components). The efferent (behavioral) members of the algorithm are related to the first (the simplest) and second (average) levels of motor activity complexity. The overall complexity category for this algorithm is 2, which is considered as being an average or moderate level of task complexity.

If logical condition $l_1 = 1$ additional turning the neck with the jig at 180° is required.

During performance of production operation once l_4 or l_7 is used because they are the same logical conditions (if $l_4 = 1$ then $l_7 = 0$; or $l_4 = 0$ then $l_7 = 1$).

(B) Evaluation of task complexity for welding the handles

Although this production operation — which is performed immediately after the one described above — has a nearly identical organization in terms of the workplace and the workers' movements, its cognitive components are significantly different. Therefore, for the purposes of this study it is important we demonstrate that not only qualitative but also formalized and quantitative methods can capture these differences, helping us to obtain the kind of information required to improve the design of the work process. The quantitative evaluation of the complexity of this operation is shown in Table 4.21 (here we refer to column four, showing values for the handle-welding operation). The time allowed for logical conditions (L_g) is 0.004 sec. The same duration is allotted to the afferent components of activity, T_{α}. As can be seen, both $L_{ltm} = 0$ and $N_{wm} = 0$, as the memory load during task performance is negligible. As there is neither a significant perceptual nor logical workload, it is the executive components of activity that are important in the handle-welding operation.

As can also be noted from the relative values of the afferent operators $T_{\alpha} = 0.004$ sec, and the stereotyped efferent components of activity $N_{st} = 0.999$, this operation is characterized by a high level of stereotyped executive activity. In cases like this, such measures as complexity of algorithm and its separate members as well as measure Z^{ex} are critical.

4.3.1.5 Comparative Analysis and Interventions

(A) Pre/post intervention qualitative analysis for welding brackets

The models we have developed of the flask-neck assembly process, and their subsequent quantitative complexity evaluation, make it possible to conduct further, more detailed qualitative analyses of these two production operations without resorting to experimental methods. This section presents this stage of analysis.

Consider the microelements "reach to the front (or back) bin" and "install the brackets on the arms of the jig." The motion "reach to the front (or back) bin" if carried out when the parts are heaped in the bin, can be considered as commensurate to MTM-1 microelement RB, which denotes a situation where the worker attempts to reach a

TABLE 4.21
Quantitative Evaluation of Complexity of Manufacturing Operation

Measures (all time measures in sec)	Operation of welding brackets: Before intervention	Operation of welding brackets: After intervention	Operation of welding handles
Time for algorithm execution (T)	13.70	13.70	12
Time for performance of logical conditions (L_g)	0.65	0.42	0.004
Time for performance afferent operators (T_α)	0.39	0.52	0.004
Time for performance of efferent operators (T_{ex})	13.40	13.40	12.0
Time for discrimination and recognition of distinctive features of task approaching threshold characteristics of sense receptors (T'_α)	0	0	0
Proportion of time for logical conditions to time for executive activity (N_l)	0.05	0.03	0.003
Measure of stereotyped logical components of activity (L_{st})	1.00	1.00	0
Measure of variable logical components of activity (L_{sh})	0	0	1
Measure of stereotyped executive component of activity (N_{st})	0.15	0.18	0.999
Measure of variable components of executive activity (N^{ε}_{sh})	0.85	0.82	0.003
Category of complexity:			
1. algorithm	2	2	2
2. members of algorithm	2;3;3;2; 3;3;3;3; 3;1;3;2; 3;3;3;3 3;2;1	2;3;3;2; 3;3;3;3; 3;1;3;2 3;3;3;3; 3;2;1	2;2;3;3; 1;2;3;3; 2;1
Proportion of time for repetitive logical components of work activity (Z^l)	0.41	0.25	0.50
Proportion of time for repetitive afferent component of work activity (Z^α)	0.003	0.25	0.50
Proportion of time for repetitive efferent component of work activity (Z^{ex})	0.38	0.38	0.20
Proportion of time for logical components of work activity depending largely on information selected from long term memory rather than exteroceptive information (L_{ltm})	1.00	0	0
Proportion of time for retaining current information in working memory (N_{vm})	0.08	0	0
Proportion of time for discrimination and recognition of distinctive features of work process approaching threshold characteristics of sense receptors (Q)	0	0	0

single object at a location which may vary from cycle to cycle. According to the rules of the MTM-1 system, it is the fact that the brackets are already sorted into distinct bins that limits the action to RB; another relevant MTM-1 microelement, RC, can only apply when the worker attempts to reach an object that is jumbled with other objects in a group so that search and select occurs (Maynard, Stegemerten, 1948). Thus, if conventional MTM-1 rules are used, the decision making involved in this process is simply ignored. Our argument is that not only should we take the decision-making action involved into consideration, we should facilitate it by choosing an appropriate container for the brackets, simplifying the decision-making process. Only after such an intervention could one with some approximation classify the microelement "reach to the bin" as RB. Some innovations at this step of performance will be outlined below.

Next consider the microelement "install the brackets." The MTM-1 system provides three categories of installation (position). Which category is applied depends upon the degree of symmetry of the objects to be installed. The first category, PS applies to those situations where the object to be installed is symmetrical, and thus no consideration of the orientation of the work object is required, and no rotations need be carried out. The second class, PSS, is called "semi-symmetric;" this refers to situations where two or more alternative orientations are acceptable (PSS). Finally, the third class, PNS, applies to nonsymmetrical installations, where the object can be installed in only one position. A graphical representation of the concept of nonsymmetrical installation according to the MTM-1 system is presented in Figure 4.15.

According to MTM-1, a nonsymmetrical installation operation consists of two microelements; the microelement "turn," designated as T75S and the microelement "symmetrical position" (PS).

In the bracket-welding production operation under consideration, the brackets may only be installed in one position; this is clearly a nonsymmetrical installation situation. In this operation time for the microelement "orient the brackets" overlaps with the element "move brackets toward jig arm"; only for microelement PS (symmetrical position) is time nonoverlapping. It can be seen that the MTM-1 system presents a nonsymmetrical installation in terms of the relationship between the shape of the part to be installed and the shape of the installation location; matching

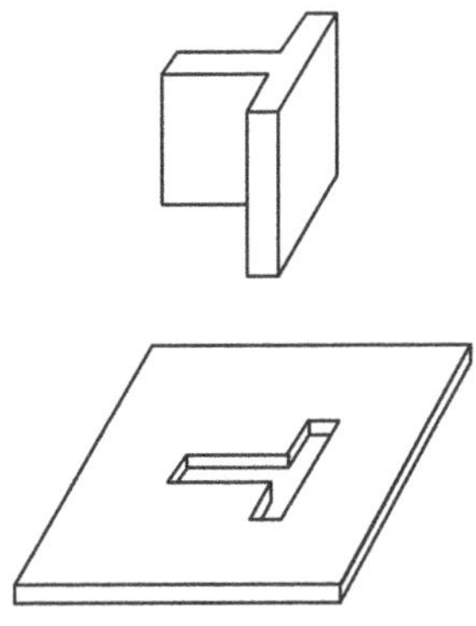

FIGURE 4.15 Nonsymmetrical installation of objects according to the MTM-1 system.

these shapes determines what category of installation or positioning can be applied (Figure 4.15). This implies that according to MTM-1, the worker relies upon exteroceptive information (external information) to guide his rotation of the part prior to installation. However, in our example the worker must recall instructions residing in long-term memory in order to orient the brackets. We can see that although the microelements "installation or position" in our example can be referred to the same category of nonsymmetric positioning, this microelement is actually executed by different psychological mechanisms than those postulated by the MTM-1 system. According to MTM-1, the operator utilizes exteroceptive information when performing installation; in contrast to our example he utilizes information from memory. This analysis demonstrates one of the most important principles of the systemic–structural approach: the combination of established microanalytical methods of work study with the study of strategies of performance and the integrated structure of activity. This approach allows the analyst to identify the character of the interrelationships between different elements of work activity, as well as the mutual influences operating within its structure. We can make use of existing formalized models of activity to facilitate our psychological analysis of the structure of activity, organize our thinking, and eliminate needless experimentation.

Analyses of work operations such as that carried out above provide a rigorous basis for intervention aimed at improving the design of the work process. In this case, such an intervention should aim to facilitate the decision-making processes involved with determining in which direction the workers should move their arms and how they should install the brackets on the jig. Thus, this redesign focuses on the logical conditions l_1, l_3^{μ}, l_4^{μ}, and l_6^{μ}. Conditions l_1 and l_4^{μ} are connected with the decision-making process regarding the direction of the arm movement, where the worker makes use of situational features such as the width of the jig arms and the position of the bins.

It is well-established that discriminative and decision-making processes which use position as a cue are more complicated than those which use color as a basis for the same kinds of discrimination or decision-making. This means that when executing operators l_1 and l_4^{μ} during the decision-making process, the positional features of the work situation cannot adequately relieve the worker from the necessity of having to continually recall in which direction he should move his arm. Based on this and the other considerations outlined above, redesigning the work procedure should include the following steps.

1. Redesign the body of the jig in such a way that it consists of two halves of unequal width. That is, by increasing the width of one half by 3 mm and decreasing the other half by 3 mm, produce a 6 mm difference between the halves that still preserves the overall weight of the jig. The wider section should be associated with the wider arm, and the narrower section with the narrower arm.
2. Cover the wider section of the jig ring with yellow plastic, and the narrower section with a dark green plastic. These color cues enhance the discriminative properties of the two jig ring sections: the yellow color tends to enhance the impression of largeness, while the darker color tends to create an impression of diminished size. Thus, these two features intensify the

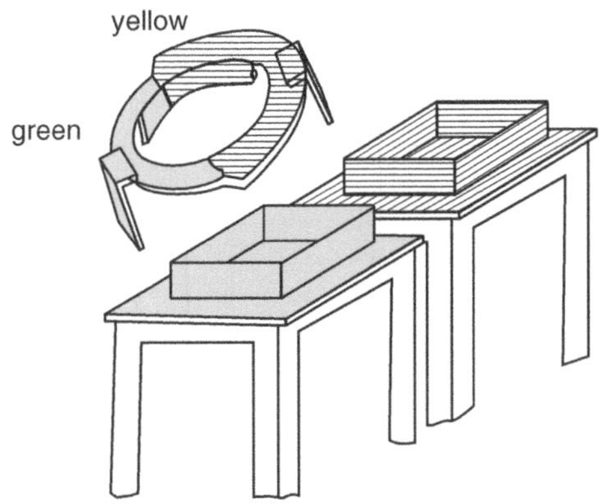

FIGURE 4.16 Demonstration of external association between jig position and position of the bins with brackets.

intrinsic relationship between the appropriate jig arm and its associated bracket.

3. Cover the bins which contain the rear, wider brackets with yellow plastic, and the bin with the front, narrower brackets with dark green plastic. Correspondingly colored coverings should also be applied to the table surfaces on which the bins rest (see Figure 4.16).
4. Reconfigure the shape of the jig's arms so as to produce a notch on one and an aperture on the other, this providing an additional cue as to the relationship between each of the arms and the orientation of the corresponding brackets.
5. Reduce the overall weight of the jig by perforating the metal part of the jig's ring.

When the changes set out above were actually implemented in the workplace, they were found to substantially improve the salience of the discriminative features of the task, simplifying the workers' decision-making processes. Not only were the visual cues enhanced by the color, but haptic cues were introduced through the alteration of the jig's ring sections, reducing the visual workload.

With respect to orienting the brackets through operations l_3 and l_6 reconfiguring the shape of the jig's arms to produce a notch on one, and an aperture on the other, provided intrinsic cues as to the relationship between the jig's arm and the orientation of the corresponding brackets (see Figure 4.17). This also eliminated any necessity for the workers violating the technological requirements of the work process by cutting down the wider jig arm.

In essence, the redesign outlined above produces a closer convergence between those microelements of the work process which involve recall and decision making and the descriptions offered the MTM-1 system, which only consider exteroceptive information. Therefore logical conditions l_3^{μ}, l_4^{μ}, and l_6^{μ} can be considered as l_3, l_4, and l_6 because they are performed based on exteroceptive information. Logical condition i_1 that in most cases has also been performed based on

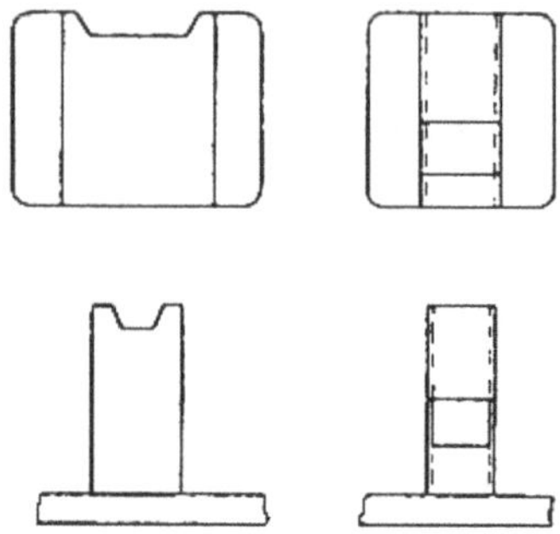

FIGURE 4.17 Association between jig's arms and position of the brackets.

information from memory, has been performed based on external information after the innovation.

(B) Revised task complexity for welding brackets

In closing, the complexity of the redesigned operation can also be analyzed as follows (see Table 4.21 after intervention). Postintervention, the time taken up by logical conditions (i.e., decision-making elements) is reduced to 0.52 sec. Correspondingly, the time for afferent operations is increased. If one compares the values of T_{α} and L_{g} before and after the intervention, it can be seen that the proportion of time for logical conditions is reduced from 0.05 to 0.03, showing that the decision-making process is simplified. At the same time the perceptual workload is increased, but only negligibly. It should also be noted that perceptual processes are less difficult for the operator than decision-making processes. Post-intervention, the value of L_{ltm} is reduced from 1 to 0.0, entirely eliminating the memory workload during the decision-making process. In other words, before the intervention, the workers largely relied on information from long-term memory when making decisions. After the intervention, choices can be made wholly on the basis of exteroceptive information, thus substantially reducing the complexity of the decision-making process.

The extent of the load on working memory is indexed by N_{wm} (the proportion of time the worker maintains information in working memory), which preintervention had a value of 0.08 sec. Following the intervention, N_{wm} equals zero. The intervention also reduces L_{g}.

In this task executive components (N_{ex}) predominate. Following the redesign the proportion of time dedicated to logical components of activity, N_{l} is further reduced, from 0.05 to 0.03 sec. Analyzing measures such as Z^{α} and Z^{l} shows that post-intervention the proportion of time for repetitive logical conditions is decreased, while the proportion of time for repetitive afferent operators is correspondingly increased. Thus, although overall the afferent components of activity are increased slightly, the more complicated logical components of the activity are simplified and the load on working memory is markedly reduced. The other complexity measures do not show any significant changes. These results demonstrate the precision with which the quantitative measures of complexity can be applied to common manufacturing operations.

The important practical results of this research were a reduction in the overall defect rate, as well as a marked reduction of the cognitive workload and thus the mental fatigue of the workers. However, this case study of manufacturing operations has important theoretical as well as practical implications. Many of the methods currently in use fail to capture the psychological dimensions of work design and improvement especially their quantitative and analytical aspect. In this case, such a failure would result not only in the setting of inappropriate time parameters for completing the assembly operation, but also in inadequate training being provided for the welders undertaking this psychologically complex task. From the theoretical perspective, a special significance attaches to these operations which on the surface appear to be nearly identical in terms of their physical demands and movements but exhibit marked differences in terms of the degree of difficulty involved in the cognitive regulation of that motor activity. Thus, one of the more important theoretical aspects of this study was to explore how quantitative complexity measures derived from systemic–structural activity theory can identify these important differences.

4.3.2 Application of Systemic-Structural Activity Theory to the Design of New Equipment

In this section we briefly examine the application of systemic–structural activity theory to an example of new equipment design. This study concerns the design of an Unmanned Underwater Vehicle (UUV), a semiautomated robotic system used for the exploration of the sea bed or to perform underwater construction work. Three design alternatives were considered for the UUV; the first was suggested by engineers, the second by specialists in the field of human factors (Yur'nev, 1975; Yastrebov, 1977), and the third is our own design solution, based on SSAT. In what follows we will consider, compare, and contrast all three versions.

4.3.2.1 Stage One: Qualitative Analysis

(A) General psychological description

There are various methods which may be adopted when seeking to optimize the efficiency and safety performance of an underwater exploration system. Here, the main idea is to replace a manned underwater vehicle with an unmanned, remotely controlled one — a UUV. One approach is to place the machine's control equipment on the shore, on a pontoon, or, when the vehicle must operate at a considerable depth, on a surface vessel stabilized in relation to underwater beacons. However, operating robots in this manner is highly complicated, particularly when we consider the use of UUVs for construction purposes. Our exploration of the continental shelves and ocean depths have involved the development of various underwater machines for laying deep-water pipelines and cables on the ocean bed. In this work the movement of the UUV from one position to another is especially important, but the control process is made more difficult by the bad visibility which results from the agitated and murky mixture of water and sand particles associated with the construction activities.

Using remote controls substantially complicates the control of underwater vehicles as — when compared to using an onboard pilot — it distorts the correlation

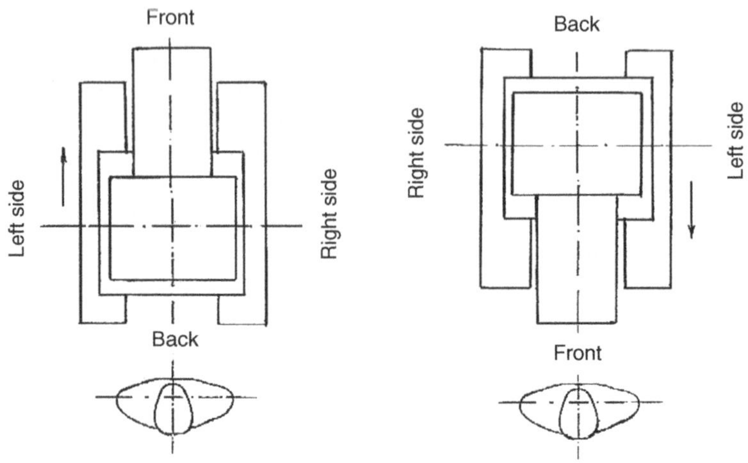

FIGURE 4.18 Example of relationship between axis of operator body and axis of UUV.

between the axis of the operator's body and that of the vehicle. This distortion means that the vehicle operator must make constant mental adjustments between these two axes, producing considerable mental fatigue. As an example of this problem, Figure 4.18 illustrates two possible relationships between the body position of the UUV operator and that of the vehicle with regard to four positional beacons on the seabed.

As can be seen in this figure, any 180° rotation of the UUV violates the initially established relationship between the orientation of the operator's body and that of the vehicle. In this situation, the visual information on the screen comes into contradiction with the operators' motor manipulation of the controls. Since the operator handles the UUV by remote control using data from a display, his or her coordinate system is almost entirely independent on that of the UUV. Thus, the operator may encounter problems in selecting a right- or left-turn, or backward or forward movements. Yur'nev (1975) and Yastrebov (1977), based on the analysis of human factors in the literature, suggested that such remote control processes could be made easier by using a rotating control panel that changed its position concurrently with changes in the UUV's position. Rotation, inclination, and declination of the control panel would be automated, based on positional feedback from the UUV.

Our approach to this design suggests that the 360° rotation of the control panel should be determined not by feedback from the UUV but rather by the active control actions of an operator who, simultaneously with the movement of the UUV on the seabed, also rotates his display panel in the same direction and to the same degree. With manual control the operator has greater knowledge about what influences she or he exerts on the UUV. Because of this, operator can more easily predict changes in the vehicle's movement pattern. A rotating display mask was also proposed. This display screen has a scale mask and a marker icon showing the current position of the UUV (see Figure 4.19).

When the operator's panel rotates as the UUV turns, the display mask should rotate at the same angle but in the opposite direction, making a direct association between the operator's coordinate system and that of the UUV. In this design the operator's vertical

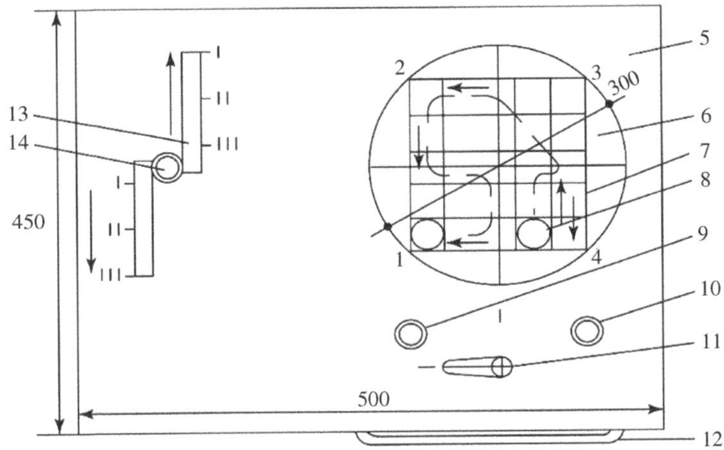

FIGURE 4.19 Plan view of control board. 1–4 Points indicating position of underwater electronic beacons; 5 — control board's horizontal panel; 6 — control board's display; 7 — coordinate mask; 8 — code image of UUV; 9 — left turn button; 10 — right turn button; 11 — UUV's engine switch; 12 — operator's hand rest; 13 — stick slot; 14 — stick.

and horizontal axes will always align with their counterparts on the control panel. This way of representing the spatial information is much closer to a real situation and thus does not require any recoding of information on the part of the operator.

Using this approach (and also drawing on an analysis of the design of bulldozer control panels carried out by the Komatsu Company, Japan) we developed a new design for the UUV's control and display panels (shown in Figure 4.19). In our version, when a button is pressed the control panel rotates to the same degree, and in the same direction as the UUV; the mask rotates by the same degree as the UUV, but in the opposite direction.

Thus, a qualitative analysis of the UUV's control principles allows the designer to consider three possible alternatives: a stable panel with a moving mask coordinated with the UUV's movements; a stable panel with a stable display mask; a rotating panel and stable display mask; rotating panel and rotating in opposite direction display mask. Practical constraints meant that choosing between these three possibilities required that the complexity of each alternative be evaluated using only analytical procedures. First of all, let us discuss the initial conditions. The UUV is moving along a given trajectory (Figure 4.19) that includes all typical cases of its movements between four beacons in a given square. This selection of the more representative movements of the UUV reflects the principles of constraint-based design. By predefining the possible area of the UUV's movements and selecting the more representative and critical of them we can predict the advantages and drawbacks of any version of the control panel design for any possible trajectory of movement. For the purposes of our study, the speed of the vehicle is taken as being constant and equal to 2.5 km/h. The seabed beacons are located at points 1–4, bounding a square which measures 50 m on each side. Each quarter of a square is designated by the number given to the seabed beacon in that quarter.

The UUV operator's task is described below. According to concept of design described in this chapter, the first stage requires that we develop verbal descriptions of the task at various levels of detail. In this first stage we use task elements or technological units of analysis, subsequently translating them into activity elements or psychological units of analysis.

(B) Broader verbal description of task
According to the trajectory of movement (the dashed line Figure 4.19) the task consists of the following steps:

1. Move the UUV to the turning point from its initial position in the fourth square
2. Stop the UUV and return it to its initial position using reverse gear
3. Move the UUV to the turning point again, and then stop it
4. Turn right 90° and move the UUV to the second turning point
5. Turn the UUV left by 135°
6. Move the UUV along the given path to the third square
7. Turn the UUV left 45° and move along the given trajectory to the second quarter
8. Turn the UUV 90° so that its front side faces the operator
9. Move the UUV to the next turning point
10. Stop the UUV and place it in reverse gear
11. Return the UUV to the previous turning place and stop it
12. Move forward again and then stop the UUV at the position of the first turn
13. Turn the UUV 90° and move it to the end point
14. Stop the UUV and turn off the engine

Based on the verbal description, a symbolic model was developed (see Figure 4.20).

In the presented figure only a part of the described above task is presented. In the symbolic model the task ends by the first right turn at 90°, that is step 4 of the description given above.

(C) Detailed verbal description of the task
A more detailed verbal description of this stage of task performance was developed on the basis of the verbal and symbolic descriptions presented above.

1. Turn on the engine with the right hand and move the shift stick to a moderate speed position using the left hand
2. Return hands to the initial position
3. Wait while the UUV moves to the first turning point
4. Move the shift stick to a moderate reverse speed position using the left hand
5. Wait until the UUV gets to the turning point
6. Turn the UUV through 90° by pressing the right button
7. Stop the UUV by placing the shift stick in the neutral position using the left hand

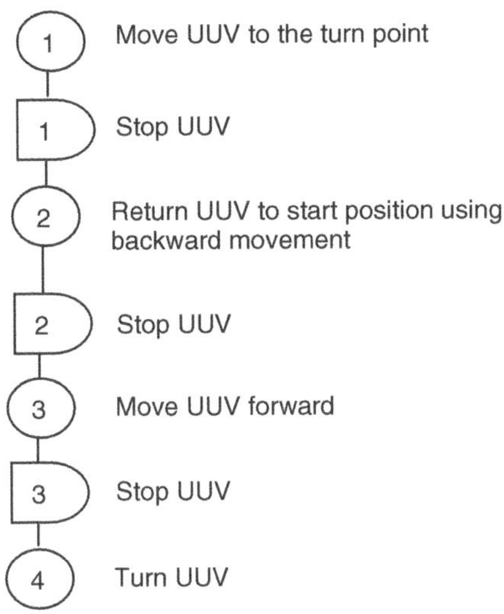

FIGURE 4.20 Broader symbolic model of the task (fragment).

This example demonstrates that task description can be performed at several different levels of detail. This means that the various symbolic models produced can be understood as forming a hierarchically organized system in which we use the same symbols to describe the same activity at different levels of detail. Both symbolic models and verbal descriptions are related to the qualitative stage of analysis; these methods supplement each other. The qualitative stage of analysis includes various levels of description, each of which has its corresponding units of analysis.

4.3.2.2 Stage Two: Algorithmic Analysis

An algorithmic description of the task is presented in Table 4.22. This algorithm requires some explanation. As the sequence of its elements is unambiguously determined by the trajectory of the UUV, no logical conditions (or their associated arrows) are shown in the left column of the table. While afferent operators are also not included, an active waiting period is designated here as $O^{\alpha w}$. This is because the visual afferent activity involved in the task almost entirely overlaps the active waiting period. Those components of activity associated with memorizing the required trajectory of the UUV movements are also ignored, as the trajectory of the vehicle remains the same in all three versions of the design and thus does not materially impact on those aspects of the design under discussion. Algorithmic description of the task is presented in Table 4.22.

In the example of algorithmic description presented above, each of those members of the algorithm associated with motor activity usually comprise 2–3 motor actions integrated by a higher order goal. For example, O_1^{ε} includes the following motor actions: turn on the engine with the right hand; move the stick control to the moderate speed position with the left hand; return the hands to the resting position. In a similar

TABLE 4.22
Algorithmic Description of Task Performance "Movement Control of UUV"

Members of algorithm	Description of members of algorithm
O_1^{ε}	Turn on the engine with the right hand, move the stick control to the moderate speed position with the left hand and return it to the hand rest position
$O_2^{\alpha w}$	Wait until the UUV moves to the first turning position
O_3^{ε}	Reverse the stick control to the moderate backward speed position with the left hand and return it to the hand rest position
$O_4^{\alpha w}$	Wait until the UUV returns to the first starting position
O_5^{ε}	Move the stick control to the moderate forward speed position with the left hand and return it to the hand rest position
$O_6^{\alpha w}$	Wait until the UUV moves to the first turning position
O_7^{ε}	Move the right hand to the right turn button, press and hold until the UUV turns right at 90°
$O_8^{\alpha w}$	Wait until the UUV turns right at 90°
O_9^{ε}	Release the right button and return hand to the rest position. Stop the UUV movement by moving the stick control to the neutral position
O_{10}^{ε}	Turn on the UUV movement ahead by moving the stick control to the moderate speed position
$O_{11}^{\alpha w}$	Wait until the UUV gets to the turning point
O_{12}^{ε}	Move the left hand to the left turn button, press and hold until the UUV turns at 135°
$O_{13}^{\alpha w}$	Wait until the UUV turns left at 135°
O_{14}^{ε}	Release the left button and return hand to the rest position
$O_{15}^{\alpha w}$	Wait until the UUV gets to the turning point
O_{16}^{ε}	Move the left hand to the left turn button, press and hold until the UUV turns at 45°
$O_{17}^{\alpha w}$	Wait until the UUV turns left at 45°
O_{18}^{ε}	Release the left button and return hand to the rest position
$O_{19}^{\alpha w}$	Wait until the UUV gets to the turning point
O_{20}^{ε}	Move the left hand to the left turn button, press and hold until the UUV turns at 90°
$O_{21}^{\alpha w}$	Wait until the UUV turns left at 90°
O_{22}^{ε}	Release the left button, press pedal "stop." Move the stick control to the neutral position
O_{23}^{ε}	Move the stick control to the moderate forward speed position with the left hand and return it to the hand rest position
$O_{24}^{\alpha w}$	Wait until the UUV gets to the turning point
O_{25}^{ε}	Press pedal "stop," moves the stick control to the moderate backward speed position with the left hand, and release pedal
$O_{26}^{\alpha w}$	Wait until the UUV gets to the turning point
O_{27}^{ε}	Press pedal "stop," moves the stick control to the moderate forward speed position with the left hand, return it to the rest hand position and release pedal

(Continued)

TABLE 4.22
Continued

Members of algorithm	Description of members of algorithm
$O_{28}^{\alpha w}$	Wait until the UUV gets to the turning point
O_{29}^{ε}	Move the left hand to the left turn button, press and hold until the UUV turns at 90°
$O_{30}^{\alpha w}$	Wait until the UUV turns left at 90°
O_{31}^{ε}	Release the left button and return hand to the rest position
$O_{32}^{\alpha w}$	Wait until the UUV gets to the turning point
O_{33}^{ε}	Move the right hand to the right turn button, press and hold until the UUV turns at 90°
$O_{34}^{\alpha w}$	Wait until the UUV turns right at 90°
O_{35}^{ε}	Release the right button and return hand to the rest position
$O_{36}^{\alpha w}$	Wait until the UUV gets to the turning point
O_{37}^{ε}	Move the right hand to the right turn button, press and hold until the UUV turns at 90°
$O_{38}^{\alpha w}$	Wait until the UUV turns right at 90°
O_{39}^{ε}	Release the right button and return hand to the rest position
$O_{40}^{\alpha w}$	Wait until the UUV gets to the finishing position
O_{41}^{ε}	Press stop pedal with right leg, move the stick control to the neutral position with the left hand and turn off engine with the right hand

fashion the active waiting period ($O^{\alpha w}$) also integrates several different cognitive actions, although we have not considered these separately during our analysis.

Symbolic description of the members of the algorithm are example of standardized psychological units of analysis. Verbal description of members of algorithm should be considered as technological units of analysis (typical task elements).

In this study the same design configuration was selected for both the controls and the display, as in each version of the task the UUV was required to move along the same trajectory. The only significant differences between the versions were whether a stable panel and mask or a rotating panel and mask were used. As a result, all three versions of the design produced the same algorithmic description. This suggests that in order to produce useful results the study needs to examine the designs in finer detail, indicating the need to move on to the next stage of analysis, time structure design.

4.3.2.3 Stage Three: Time Structure Analysis

In the previously described stages of analysis, task elements or technological units were used as units of analysis. From that point of view, only typical elements of activity were used as units of analysis. This means, for example, that instead of using task element "move a stick" one uses the element of activity "move-M," or instead of task element "to take a meter reading," one uses the typical elements of activity "eye focus-ET," "simultaneous perceptual action," etc. (the same task elements can be performed by different elements of activity).

The waiting period of time is an important component of task that is considered in this study. The waiting period of time, according to developed rules, is classified only by the level of concentration of attention. Therefore, all measures concerned with afferent operators have a value of 0. Hence, all measures of logical conditions also have estimated numerical values equaling 0. At the same time some quantitative measures concerned with estimates of active waiting time are also included. They will be discussed shortly.

As can be seen during algorithmic description of a task, typical elements of the task (technological units) were used as units of analysis. At this stage of analysis, they will be transformed into typical elements of activity (psychological units). Each member of the algorithm will be considered a subsystem of activity, comprised, in turn, of smaller units. Hence, this stage of analysis involves decomposition of activity into more detailed levels.

Time activity structure can be done in the form of either a table or a graph. If the elements of activity are implemented consecutively, we can limit descriptions to a table. For our purposes, accordingly, a tabular representation is sufficient, since, in most cases, all elements of activity are performed in an ordered sequence.

In Table 4.23 we present time structure of activity for the third version of the task (rotating panel and rotating in opposite direction coordinate mask).

Complexity of task performance when the UUV is moving in the first square on the ocean bed (before turning of the UUV) is the same for different versions of the UUV. This part of the activity was eliminated from further consideration (this fragment of task is described by operators $O_1^{\varepsilon}-O_8^{\alpha w}$). Time performance of efferent operators from $O_9^{\varepsilon}-O_{41}^{\varepsilon}$ (without waiting period) is 14.2 sec; time for waiting period from operators $O_8^{\alpha w}$ to $O_{40}^{\alpha w}$ is 295 sec; and total time of task performance is 309.2 sec. It was discovered that the time structures of task performance for the first and second versions of the task were identical; because of this we present only one table for both versions of the UUV in Table 4.24.

For the reasons described above, we do not consider in following discussions, that part of the activity that is described by members of algorithm $O_1^{\varepsilon}-O_8^{\alpha w}$ (before turning of UUV).

Analysis of time structure of activity did not indicate a preference for either the first or second versions of the UUV as the time structure for both versions is the same. However, evaluation of the time structure of task for the third version of the UUV compared with the time structure of the task for the first and second versions of the UUV demonstrated the advantage of the third version. Let us consider a few examples. All turnarounds of the UUV, except for the first turn (O_7^{ε}), are complicated by interfering information from an display, as there is no support for a relationship existing between the operator's body coordinate axes and that of the UUV. This leads to more complicated cognitive regulation of movements associated with hand displacement toward either the left or the right button. For example, when an operator performs O_{12}^{ε} he presses the left button and the UUV starts to turn at 135°. However, the control board does not exactly indicate a left turn as the operator usually expects to see it. This becomes particularly obvious when an operator attempts to perform either

TABLE 4.23
Time Structure of Activity during Task Performance for the Third Version of the UUV

Algorithm members	Motor activity: Right hand, right leg	Time (sec)	Left hand, left leg	Time (sec)	Mental activity: Description	Time (sec)
O_1^{ε}	R13A + G1A+ M2.5A + RL1+ R13E + G5	0.7	R32A + G1A+ M2A2.5+ M10B2.5 + RL1+ R43E + G5	1.65	—	—
$O_2^{\alpha w}$	—	—	—	—	Active waiting	21.4
O_3^{ε}	—	—	R43A + G1A+ M10A2.5+ M4A2.5+ M10B2.5 + RL1+ R24E + G5	2.0	—	—
$O_4^{\alpha w}$	—	—	—	—	Active waiting	21.4
O_5^{ε}	—	—	R24A + G1A+ M10A2.5+ M4A2.5+ M10B2.5 + RL1+ R43E + G5	1.86	—	—
$O_6^{\alpha w}$	—	—	—	—	Active waiting	21.4
O_7^{ε}	R13A + G5 + AP2	0.72	—	—	—	—
$O_8^{\alpha w}$	—	—	—	—	Active waiting	10.0
O_9^{ε}	RL2 + R13E + G5	—	R43A + G1A+ M10A2.5+ M2B2.5	1.24	—	—
O_{10}^{ε}	—	—	M2A2.5+ M10B2.5 + RL1+ R43E + G5	1.0	—	—
$O_{11}^{\alpha w}$	—	—	—	—	Active waiting	20.0
O_{12}^{ε}	—	—	R13A + G5 + AP2	0.6	—	—
$O_{13}^{\alpha w}$	—	—	—	—	Active waiting	15.0
O_{14}^{ε}	—	—	RL2 + R13E + G5	0.24	—	—
$O_{15}^{\alpha w}$	—	—	—	—	Active waiting	39.0
O_{16}^{ε}	—	—	R13B + G5 + AP2	0.6	—	—
$O_{17}^{\alpha w}$					Active waiting	5.0
O_{18}^{ε}	—	—	RL2 + R13E + G5	0.26	—	—
$O_{19}^{\alpha w}$	—	—	—	—	Active waiting	25
O_{20}^{ε}	—	—	R13A + G5 + AP2	0.6	—	—
$O_{21}^{\alpha w}$	—	—	—	—	Active waiting	10

(Continued)

TABLE 4.23
Continued

	Motor activity				Mental activity	
Algorithm members	Right hand, right leg	Time (sec)	Left hand, left leg	Time (sec)	Description	Time (sec)
O^{ε}_{22}	LM (dotted box)	—	RL2 + R13A+ G1A + M10A2.5 M2B2.5	1.0	—	—
O^{ε}_{23}	—	—	M2A2.5+ M10B2.5 + R43E +G5	1.15	—	—
$O^{\alpha w}_{24}$	—	—	—	—	Active waiting	27.0
O^{ε}_{25}	LM LM	0.5	R43A + G1A+ M10A2.5+ M4A2.5+ M10B2.5	1.55	—	—
$O^{\alpha w}_{26}$	—	—	—	—	Active waiting	27.0
O^{ε}_{27}	LM LM (dotted box)	0.25	M10A2.5+ M4A2.5 + M10B2.5 +RL1 + R43E+ G5	1.49	—	—
$O^{\alpha w}_{28}$	—	—	—	—	Active waiting	27.0
O^{ε}_{29}	—	—	R13A + G5 + AP2	0.6	—	—
$O^{\alpha w}_{30}$	—	—	—	—	Active waiting	10.0
O^{ε}_{31}	—	—	RL2 + R13E + G5	0.26	—	—
$O^{\alpha w}_{32}$	—	—	—	—	Active waiting	20.0
O^{ε}_{33}	R13A + G5 + AP2	0.62	—	—	—	—
$O^{\alpha w}_{34}$	—	—	—	—	Active waiting	10.0
O^{ε}_{35}	RL2 + R13E + G5	0.26	—	—	—	—
$O^{\alpha w}_{36}$	—	—	—	—	Active waiting	20.0
O^{ε}_{37}	R13A + G5 + AP2	0.62	—	—	—	—
$O^{\alpha w}_{38}$	—	—	—	—	Active waiting	10.0
O^{ε}_{39}	RL2 + R13E + G5	0.26	—	—	—	—
$O^{\alpha w}_{40}$	—	—	—	—	Active waiting	20.0
O^{ε}_{41}	LM LM (dotted box) R13A + G1A+ M2.5A + RL1+ R13E + G5	0.7	M10A2.5+ M2B2.5 + RL1+ R32E + G5	1.1	—	—

TABLE 4.24
Time Structure of Activity during Task Performance for the First and the Second Versions of the UUV

	Motor activity				Mental activity	
Algorithm members	**Right hand, right leg**	**Time (sec)**	**Left hand, left leg**	**Time (sec)**	**Description**	**Time (sec)**
O_1^{ε}	R13A + G1A+ M2.5A + RL1+ R13E + G5	0.7	R32A + G1A+ M2A2.5+ M10B2.5 + RL1+ R43E + G5	1.65	—	—
$O_2^{\alpha w}$	—	—	—	—	Active waiting	21.4
O_3^{ε}	—	—	R43A + G1A+ M10A2.5+ M4A2.5+ M10B2.5 + RL1+ R24E + G5	2.0	—	—
$O_4^{\alpha w}$	—	—	—	—	Active waiting	21.4
O_5^{ε}	—	—	R24A + G1A+ M10A2.5+ M4A2.5+ M10B2.5 + RL1+ R43E + G5	1.86	—	—
$O_6^{\alpha w}$	—	—	—	—	Active waiting	21.4
O_7^{ε}	R13A + G5 + AP2	0.72	—	—	—	—
$O_8^{\alpha w}$	—	—	—	—	Active waiting	10.0
O_9^{ε}	RL2 + R13E + G5	—	R43A + G1A+ M10B2.5+ M2B2.5	1.30	—	—
O_{10}^{ε}	—	—	M2B2.5+ M10C2.5 + RL1+ R43E + G5	1.24	—	—
$O_{11}^{\alpha w}$	—	—	—	—	Active waiting	20.0
O_{12}^{ε}	—	—	R13B + G5 + AP2	0.66	—	—
$O_{13}^{\alpha w}$	—	—	—	—	Active waiting	15.0
O_{14}^{ε}	—	—	RL2 + R13E + G5	0.24	—	—
$O_{15}^{\alpha w}$	—	—	—	—	Active waiting	39.0
O_{16}^{ε}	—	—	R13B + G5 + AP2	0.66	—	—
$O_{17}^{\alpha w}$					Active waiting	5.0
O_{18}^{ε}	—	—	RL2 + R13E + G5	0.26	—	—
$O_{19}^{\alpha w}$	—	—	—	—	Active waiting	25

(Continued)

TABLE 4.24
Continued

	Motor activity				Mental activity	
Algorithm members	Right hand, right leg	Time (sec)	Left hand, left leg	Time (sec)	Description	Time (sec)
O_{20}^{ε}	—	—	R13B + G5 + AP2	0.66	—	—
$O_{21}^{\alpha w}$	—	—	—	—	Active waiting	10
O_{22}^{ε}	LM (dotted box)	—	RL2 + R13A+ G1A + M10B2.5 M2C2.5	1.01	—	—
O_{23}^{ε}	—	—	M2B2.5+ M10C2.5 + R43E +G5	1.17	—	—
$O_{24}^{\alpha w}$	—	—	—	—	Active waiting	27.0
O_{25}^{ε}	LM LM	0.5	R43A + G1A+ M10B2.5+ M4B2.5+ M10C2.5	1.77	—	—
$O_{26}^{\alpha w}$	—	—	—	—	Active waiting	27.0
O_{27}^{ε}	LM LM (dotted box)	0.25	M10B2.5+ M4B2.5 + M10C2.5 +RL1 + R43E+ G5	1.63	—	—
$O_{28}^{\alpha w}$	—	—	—	—	Active waiting	27.0
O_{29}^{ε}	—	—	R13C + G5 + AP2	0.72	—	—
$O_{30}^{\alpha w}$	—	—	—	—	Active waiting	10.0
O_{31}^{ε}	—	—	RL2 + R13E + G5	0.26	—	—
$O_{32}^{\alpha w}$	—	—	—	—	Active waiting	20.0
O_{33}^{ε}	R13C + G5 + AP2	0.72	—	—	—	—
$O_{34}^{\alpha w}$	—	—	—	—	Active waiting	10.0
O_{35}^{ε}	RL2 + R13E + G5	0.26	—	—	—	—
$O_{36}^{\alpha w}$	—	—	—	—	Active waiting	20.0
O_{37}^{ε}	R13A + G5 + AP2	0.72	—	—	—	—
$O_{38}^{\alpha w}$	—	—	—	—	Active waiting	10.0
O_{39}^{ε}	RL2 + R13E + G5	0.26	—	—	—	—
$O_{40}^{\alpha w}$	—	—	—	—	Active waiting	20.0
O_{41}^{ε}	LM LM (dotted box) R13A + G1A+ M2.5A + RL1+ R13E + G5	0.7	M10A2.5+ M2B2.5 + RL1+ R32E + G5	1.1	—	—

a left or right turn when the UUV is moving in the opposite direction. For example, consider the area of the UUV movements associated with beacon 2. When an operator presses the left button (#9), the UUV turns to the right and when he presses the right button (#10), it turns to the left (Figure 4.19). This complicates the displacement of right- or left-hand movement, respectively, toward right or left buttons. Consequently, according to previously described rules (see page 202, rule 6), the movement of the hands to the required buttons, instead of more simple version of elements RA, should be classified as elements RB or even as RC. The last two require more concentration and performance time. In the third version, where stereotypical right and left turns are performed, the movements of the hand toward the right or left buttons always requires less concentration and also can be related to the element RA, according to the developed rules and classification suggested by system MTM-1. Pressing of the left or right buttons is always associated with the same turn of the UUV on the screen. The same applies when an operator shifts the control stick for moving the UUV ahead or back or deciding to stop the UUV. When an operator moves the control stick into the neutral position, it is a situation equivalent to moving an object "against stop" (Barnes, 1980). However, this movement is associated with information contrary to the actual movement of the UUV on the screen (the first and the second version of UUV) and therefore increases the concentration of attention. This is why this movement should be transferred from the first to the second category of complexity. In the third version of the UUV, when this contradiction is eliminated, this movement should always be considered as RA (low-level concentration of attention). When an operator shifts the control stick into a position that corresponds to the particular speed of UUV movements, it is RB (average level of concentration of attention is required in this case). However, in the first and second versions of the UUV, these control movements are associated with contradictory information of UUV movements on the screen. For example, when an operator shifts the control stick forward, the UUV is moving backward on the screen. Hence, a higher level of concentration is required and RB is transferred into RC. In the third version, the control movement coincides with the movements of the image of the object on the screen. In general, it is clear that the time structure of activity in the first and second versions of the UUV is more complicated than in the third version. The existence of interfering information regarding the UUV's position in the first and second versions means that the time to perform motor responses equals (time performance of efferent operators from $O_9^{\varepsilon} - O_{41}^{\varepsilon}$ without waiting period) 16.8 sec, as compared with 14.2 sec in the third version (see Table 4.23 and Table 4.24). The time for the waiting period from operators $O_8^{\alpha w}$ to $O_{40}^{\alpha w}$ is 295 sec for all three versions. The total time of task performance for the first and the second versions of the UUV is 311.8 sec. and for the third version it is 309.2 sec.

4.3.2.4 Stage Four: Quantitative Analysis of UUV Task Complexity

As the elements of activity do not overlap each other in time, we can estimate the complexity of all three versions of the UUV without developing graphical models of time structure. At this point, all qualitative and temporal data are transformed into quantitative measures, which describe the complexity of the UUV's task performance for all three versions of design (4.25). Let us estimate the complexity of the waiting period. During movements of the UUV along a straight line, an operator should remember the position of objects in relation toward his own coordinate system (in the first and second cases). This causes an additional workload for the working memory. Due to this, we assign all waiting periods, when a screen's marker is moving along a straight line on the screen, to the second category of complexity. In a preliminary study, it was discovered that a bulldozer operator's levels of attention in real conditions begins to rise at 2–3 m before the turn. When the UUV is moving, an operator should predict the moment of a turn on a coordinate scale, taking into account the position of the UUV toward the coordinate axes of his own body. Because of this the last 2–3 m along a straight line, before a turning point, demands an even higher level of concentration. It means that the last 3 sec of the movements before turning, corresponds to the third category of complexity. The most complex are turns with waiting periods associated with $O_{28}^{\alpha w}$, $O_{32}^{\alpha w}$, and $O_{36}^{\alpha w}$. In these cases, an operator starts to predict his action earlier. Because of this, we should claim that the last 5 sec, instead of the last 3 sec, could be related to the third category.

After an operator presses the "turn" button, he should hold his hand stationary until the turn has been completed. When the first type of panel is used, an operator needs to be vigilant and stop the turn by interrupting it to press a button according to the position of the screen marker. Due to this, turn processing has been assigned to the second category of complexity, except for the last 2 sec that are assigned to the third category (high level concentration of attention).

In the second type of panel, when an operator can navigate the rotating object not only by a moving screen marker, but also with the help of visual and vestibular senses, less control functions that allow one to transfer the second category interval to the first category are needed. In the third type of a panel, when both panel and scale rotate, object orientation corresponds to the operator's body coordinate system. This decreases the workload of working memory during waiting intervals when the UUV moves along a straight line. It also simplifies navigation during turnarounds. Due to this, the period of waiting time associated with straight line movements and turning around, as previously assigned to the second category, may be transferred to the first one. We also do not need to allocate the special 5-sec intervals of the third category, when an operator is involved in waiting periods described by $O_{28}^{\alpha w}$, $O_{32}^{\alpha w}$, and $O_{36}^{\alpha w}$. In these three cases, waiting periods have been reduced to 3 sec.

As the majority of waiting period time in the first and second versions is related to the second category (more than 70% of waiting period), the whole waiting period is assigned to the second category, done in accordance to a scale ordered on five levels. When the third version of the UUV is used, the first category takes the most time, so the waiting period is assigned according to the five-level scale. The total time for the waiting period is equal to 295 sec. Analysis of the percentage of waiting period as related to different levels of complexity shows that in the first version of the UUV 84% of waiting period is in the second category of complexity and 16% is in the first category. In the second version, 19% of time is in the first category, 65% is in the second, and 16% is in the third. In the third version of the UUV 86% is related to the first and 14% to the third category. It means that that second version of the UUV has a simpler waiting period than in the first version, but the difference is insignificant. Accordingly, the third version of the UUV has the obvious preference. It is also possible to evaluate the complexity of waiting periods related to separate members of the algorithm (from $O_8^{\alpha w}$, and up to $O_{40}^{\alpha w}$). However, we did not calculate these measures. It should be noted that the waiting period associated with stress can be assigned to the fourth or even fifth category of complexity. As logical conditions are absent and waiting periods and performance intervals (related to efferent operators) follow each other in a well-predicted, regular manner, the measure of stereotyped components of waiting period is 1. Accordingly, the measure of variability of this component of activity equals 0. The other interesting measure in this example is the measure which describes the proportion of time for repetitive waiting periods Z^w. If an operator is processing the same information during different waiting periods, then these periods of time are considered repetitive. The proportion of time for repetitive waiting in the first version of the UUV is zero, because different mental actions are behind apparently similar waiting periods. This can be explained by the different orientations of the UUV in relation to the operator's body during different parts of the UUV trajectory. As a result, different mental actions are required in order to maintain awareness of the correct orientation of the UUV.

Consider this measure for the second version of the UUV. The operator selects appropriate action to push either the left or the right button before moving the hand to the button. In the same way as in the first version of the UUV, the waiting period that precedes pushing the button and turning the UUV to the required position is variable (it differs in mental content). However, an active waiting period of time associated with the turning of the UUV to the required angle includes repetitive cognitive actions. In the second version when the control board can turn together with the UUV, the operator simply ignores information from the display. After pushing the button, the operator waits until the control board turns to the required position and then releases the button. The operator performs turning of the UUV at 90° angles five times. In all cases, the operator uses the same external information and the position of the control board. Therefore, time for the first turning is not repetitive, but all others periods of time for turning are repetitive. The time for turning of the UUV at 90° is 10 sec. Hence, the total time for four similar turns is 40 sec. This is the overall time for the repetitive waiting period. Hence proportion of time for repetitive waiting period Z^w for the second version of the UUV the time will be 0.14.

In the third version of the UUV, all straightforward movements connected with repetitive active waiting periods, coincide with each other. The first straightforward movement was already performed before the first right turn of the UUV. In other straightforward movements, the operator used exactly the same mental operations as he did during the waiting period. Only the first turning of the UUV was at 90° and turnings at 135° and 45° can be related to the variable waiting period of time. All these three turns require 30 sec. Hence, the proportion of time for the repetitive waiting period Z^w in the third version of the UUV is 0.90. This measure demonstrates that cognitive activity becomes simpler the during active waiting period in the third version of the UUV, in comparison to the first and second versions because the variability of the cognitive actions in this period is reduced. Part of the active waiting period in the whole works process ΔT_w is significant for all three versions and changes occur in a range of 0.94–0.96. This means that the basic characteristics of an active waiting period as described above are important during performance of a considered task.

Let us consider quantitative measures related to the executive components of activity. The time when an operator is actively involved in task performance for the first and second versions of the UUV are the same and equal to 16.8 sec (see Table 4.24). For the third version of the task, the time is 14.2 (see Table 4.23). As explained before, the operator performs practically the same actions. However, they are simpler and this causes a reduction in time performance. According to the developed rules of algorithmic description, logical conditions can be introduced only when the operator after receiving information or extracting it from memory, decides what kind of action should be chosen. In considered task trajectory the UUV movement is determined in advance. Receiving information and decisions about the UUV's straightforward movements were performed during the active waiting periods. Therefore, time performance of afferent operators and logical conditions are zero. Active waiting periods and their executive components (efferent operators) follow in the same order because there are no logical contradictions. Hence, all efferent operators, in the same way as an active waiting period, are stereotypical and the variability of executive activity is zero.

The category of complexity of efferent activity in general, according to the five-level scale as a whole, is equal to 2 for the first and second versions of the UUV (see Table 4.25). For the third version of the UUV, this measure is equal to 1. The complexity of different efferent operators, according to the five-order scale is presented in the same table. For example, moving the left or right hand to button 9 and 10 in the first and second versions corresponds to the second or third categories of complexity. This can be explained by the fact that there is contradictory information coming from an indicator and the real position of the operator's body. This contradiction is eliminated in the third version. As a result, the same motor actions in the third version always correspond to the first category of complexity. Let us consider the complexity evaluation of the efferent operator O^{ε}_{12} (move hand to the left button and press it) in the first and second versions of the UUV. In this example, the member of the algorithm consists of one motor action that includes two motions — move hand and press button. According to the MTM-1 system, these motions can be described as R13B + G5 + AP2 (Reach, distance 13, type B; Grasp, type "contact" symbol G5; Apply Pressure, type AP2). The time for R13B is 0.276 sec and for AP2 is 0.38 sec. For G5, according to MTM-1, time is equal to zero. Microelement R13B requires an average level of concentration and is relegated to the second level of complexity. Microelement AP2, according to the same criterion, can be relegated to the first level of complexity. The time for R13B requires 41% of the total time performance of O^{ε}_{12}. Therefore, time for the first category of complexity element AP2 is less than 70% and for this member of the algorithm, according to above described rules, we assign the second category of complexity. In the third version of the UUV, the same member of algorithm O^{ε}_{12} includes microelements R13A + G5 + AP2. Microelement R13A requires less concentration of attention (related to the first category of complexity). Therefore member of algorithm O^{ε}_{12} is assigned to the first category of complexity. The complexity of the other efferent members of the algorithm for different versions of the UUV is presented in table. From this table one can see that twelve members

TABLE 4.25
Quantitative Measures of Evaluating Task Complexity for Manipulation of UUV on Underwater Surface

Name of measure	First version	Second version	Third version
Time for algorithm execution (seconds) (T)	311.8	311.8	309.2
Time for performance of logical conditions (L_g)	0	0	0
Time for performance of afferent operators (T_α)	0	0	0
Time for performance of efferent operators (T_{ex})	16.8	16.8	14.2
Time for discrimination and recognition of distinctive features of task at threshold level ($'T_\alpha$)	0	0	0
Proportion of time for logical conditions to time for executing activity (N_l)	0	0	0
Measure of stereotyped logical components of activity (L_{st})	0	0	0
Measure of variable logical components of activity (L_{ch})	0	0	0
Measure of stereotyped executive components of activity (N_{st})	1	1	1
Measure of variable executive components of activity (N_{ch})	0	0	0
Scale of complexity X_r (except active waiting period)			
(a) algorithm	2	2	1
(b) member of algorithm			
O^ε_9; O^ε_{10}; O^ε_{12}; O^ε_{14}; O^ε_{16}; O^ε_{18};	2; 2;2;1;2;1	2;2;2;1;2;1	1;1;1;1;1;1
O^ε_{20};	2;	2;2;2;2;2;2	1;1;1;1;1;1
O^ε_{22}; O^ε_{23}; O^ε_{25}; O^ε_{27}; O^ε_{29};	2;2;2;2;2	1;2;1;2;1;1	1;1;1;1;1;1
O^ε_{31}; O^ε_{33}; O^ε_{35}; O^ε_{37}; O^ε_{39};	1;2;1;2;1;		
O^ε_{41}	1		
Proportion of time for repetitive afferent components of work activity (Z^α)	0	0	0
Proportion of time for repetitive logical components of work activity (Z^l)	0	0	0
Proportion of time for repetitive efferent components of work activity (Z^{ef})	0.13	0.13	0.85
Proportion of time for logical components of work activity depending on information selected from long-term memory rather than exteroceptive information (L_{ltm})	0	0	0
Proportion of time for retaining current information in working memory (N_{wm})	0.83	0.83	0
Proportion of time for discrimination and recognition of distinct features of task approaching threshold characteristics of sense receptors (Q)	0	0	0
Proportion of active waiting periods in total work process (ΔT_w)	0.94	0.94	0.95
Scale of complexity of active waiting periods (W_r)	2	2	1
Proportion of time for repetitive waiting periods of work activity (Z^w)	0	0.14	0.90
Measure of variable of waiting periods in work process (W^{ch})	0	0	0
Measure of stereotyped of waiting periods in work process (W^{st})	1	1	1

of the algorithm had a second category of complexity in the first two versions of the task. In the third version, all members of the algorithm have the first category of complexity.

Another measure to be considered is the proportion of time for repetitive efferent components of work activity (Z^{ex}). In all three versions, the operator performed externally identical actions, but in the first two versions, the same actions required the operator to use totally different information. This means that cognitive regulation of the same motor actions was changed, since the UUV's axes continuously changed their position in relation to the operator's own axis. Because of this, Z^{ex} equaled 0.13 for the first two versions. In the third version, because the operator's axis and that of the UUV coincided, the same motor actions are regulated by the same cognitive operations and the proportion of time for repetitive efferent components of work activity increased to 0.85. This significantly simplifies the control of the UUV (see Table 4.25). In the first and the second versions of the UUV, we can relate O^{ε}_{14}, O^{ε}_{18}, O^{ε}_{31} (left hand movements), and O^{ε}_{35} and O^{ε}_{39} (right hand movements) to identical or repetitive members of the algorithm. This means that the left hand performs RL2 + R13E + G5 twice. Another action performed twice is "press pedal," which also can be described as repetitive. In sections of the ocean bottom restricted by the second beacon, the UUV performs forward and backward movements and then moves forward again. As a result, O^{ε}_{27} repeats O^{ε}_{23}. If we know the total time of performance of repetitive efferent members of the algorithm and the total time for executive components of activity, we can easily determine the proportion of time for repetitive efferent components of activity. In the third version, the axis of the UUV coincides with the axis of the operator's body; thus most motor actions become identical not only in their external or observable manifestation, but also in their internal cognitive regulation. The only nonrepetitive actions are those that require performance of O^{ε}_{9} and O^{ε}_{41}.

The next important measure of complexity is N_{wm} (proportion of time for retaining current information in working memory). In the first and second design versions, an operator must remember the position of the UUV's axis toward his own. This leads to a significant load on the working memory. An operator does not need this data except when he performs O^{ε}_{14}, O^{ε}_{18}, O^{ε}_{31}, O^{ε}_{35}, and O^{ε}_{41}, returning his hands to the initial position. The performance of O^{ε}_{41} does not demand memory when the left hand returns. This time equals (RL + R32E + G5) plus the right hand's time (turn off the engine). These elements of activity take 2.91 sec. Therefore, the time when the operator retains information about the position of the UUV is 16.8 − 2.91 = 13.89 sec. From this, the proportion of time for retaining current information in the working memory for the first and second versions of the UUV, is shown by $N_{wm} = 13.89/16.8 = 0.83$. In the third design version, the operator does not need to keep the UUV's relative position in mind. Because of this, N_{wm} is zero. This is evidence of the sharply decreasing workload of the working memory for the last version. Similar calculations were made for other components of work activity (see Table 4.25). This demonstrates that the third design alternative was superior to the other two versions.

Proposed quantitative measures of complexity allow us to figure out even insignificant changes in design decisions. Figure 4.21a shows the transmission slots in a "Kamatsu" bulldozer. Figure 4.21b gives our proposal of the slot suggested for the UUV control board.

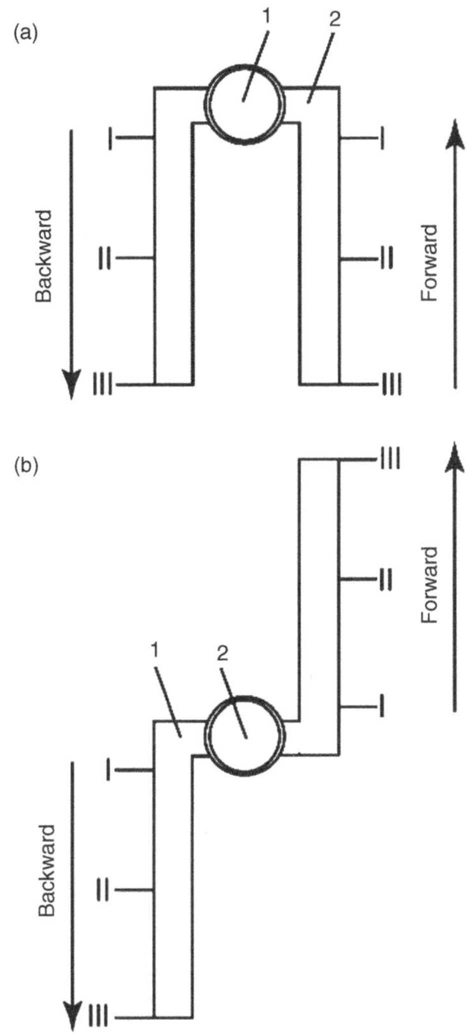

FIGURE 4.21 (a) Transmission slots in a "Kamatsu" bulldozer. (b) Proposed transmission slots in UUV. 1 — Stick; 2 — transmission slot; I–III speed of movements.

To do this we need to assume some ways of using this stick. Assume that we need to make nonconnected movements of the stick from the neutral position; all three speeds of moving forward and backward have an equal probability of being used. The difference in design solutions can be found by using the following measures: "category of complexity cognitive or executive members of the algorithm according to the five-level ordered scale" and "proportion of time for logical components of work activity depending on information selected from long-term memory rather than exteroceptive information" (L_{ltm}). In conditions when the operator has to decide, based on interpretation of seabed conditions, that he should move the UUV forward or backward (configuration A), we can describe two different situations. In one case, the UUV is moved backward and the operator shifts the control stick inside of the stick slot back (slot located farther and left from the operator's body). In the second situation, it was decided to move the UUV forward and this also required movement of the control stick back inside the other, more closely located slot. In the second situation, the decision is performed based on contradictory information (the UUV movement forward is associated with movement stick backward). In this latter case, according to the five-level ordered scale, the decision-making process should be transferred from the third category of complexity to the fourth category. Moreover, there are two slots and deciding which one should be selected is based on data in memory, rather than on externally presented information. If interpretation of information and associated decision making is not required, another measure of complexity can be used. This measure represents the level of complexity of executive components of activity according to the five-level order scale. When the operator moves the control stick backward and when the UUV moves forward, then this movement should be transferred to a higher level of complexity. Hence, depending on the specificity of using the control stick, different measures can discover deficiencies of the design solution according to version A. As an example, we can consider the method of calculation of complexity of stick manipulation activity according to the five-order scale. Concerning the size of the slot, the length of horizontal movement from neutral point left or right is 40 mm, the vertical movement to the first speed position is 50 mm, to the second 100 mm, and to the third position 150 mm. An existing variant (version A) can be described in the following way. Movements of stick from a neutral position for different UUV speed movements backward can be described by microelements M4A, M5C, M10C, and M15A (where numbers give the distance in centimeters and MA means — move object against stop, MC — move object to exact location). The category of complexity for them is correspondingly 1, 3, 3, 1.

When a UUV moves forward, an "interfering factor" appears. This factor is linked with displacement of the shift stick backward to force the UUV to move forward. According to developed rules this leads to an increase in the level of complexity of movements. The microelement M4A (horizontal movement) does not change. The movements M5C and M10C remain the same (in MTM-1 system does not exist more complicate category of complexity). The movement M15A, according to developed rules, is transformed into M15B (from a simple version of movement to more a complex one). At the same time, when we assign the category of complexity for M5C and M10C, instead of selecting the third category, we give them the fourth category. Therefore, category of complexity for all movements is 1, 4, 4, 2 (when the UUV moves forward).

The next step involves calculation of time performance of different movements. M4A requires 0.19 units, M5C −0.31, M10C −0.48 and M15A −0.48 units, respectively. In the forward movement of the UUV, instead of M15A we have M15B. The last requires 0.53 units. Therefore the total time for moving the UUV backward is 1.46 units and the total time for moving it forward is 1.51 units. At the next step, we calculate the time for performance of M4A and M15A (the first category of complexity). This time is $0.19 + 0.48 = 0.67$ units. The time for performance of the third category of complexity is $0.31 + 0.0.48 = 0.79$ units (M5C + M10C). Next, we define portions of the first and the third categories of the total time of movements.

The first category — 0.67/1.46 = 0.46; the third category – 0.79/0.54 = 0.54. Hence, no one category reaches level 0.7. Therefore, when we move the UUV backward, the category of complexity of this executive activity has level 2. When the UUV moves forward, because of the existing "interfering factor," we can obtain the following results. The time for the first category is 0.19 (M4A), the time for the second category is 0.53 (M15B) and the time for the fourth category is 0.31 + 0.48 = 079 ($M5C + M10C$). Therefore, the proportion of time for different categories when the UUV moves forward is:

The first category—0.19/1.51 = 0.13; the second category—0.53/1.51 = 0.35; the fourth category —0.79/1.51 = 0.52 (M5C and M10C were transferred into the fourth category). According to developed rules, the fourth category is applied to coefficient 2. Hence, the relationship of the fourth category to the total time manipulation is more than one. Therefore, when we move the UUV forward, the category of complexity of this executive activity has level 4.

If we consider another design solution (Figure 4.21 version b), these deficiencies can be eliminated and the complexity of moving the UUV forward and backward has category of complexity 2. Finally, we need to consider whether or not the design solution which recommends use of the stick moved inside a slot is based on the desire to preserve the natural relationship between the movement of the hand and the UUV.

Utilization of the third version of the UUV significantly reduces the complexity of the task and simplifies not only the executive stage of the task's performance, but also the active waiting period when apparently not being involved in any visible activity. The third version of the UUV reduces the diversity of performed cognitive and motor actions; workload of memory and attention and makes actions more atomized. This gives an opportunity, when necessary, for the operator to switch to other easier ongoing tasks.

The systemic–structural principle of design can be efficiently used in training processes. For example, it is very useful to use a special simulator for developing skills that require handling the UUV movements along the seabed (Figure 4.22).

This simulator helps to outline the task that should be performed and secure realization by students. In real situations it is dangerous to do this because control of

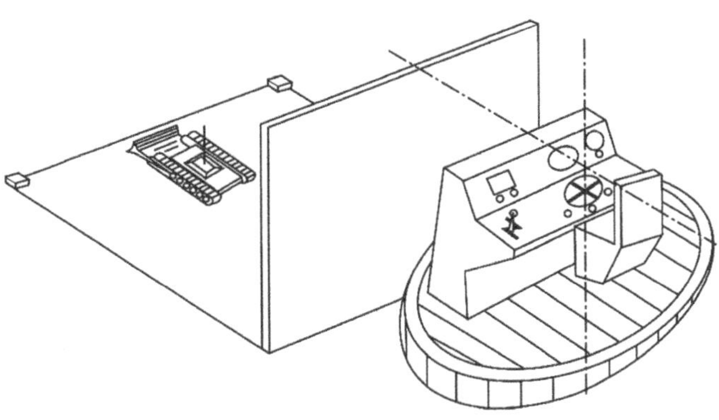

FIGURE 4.22 Simulator for underwater, unmanned vehicle.

the UUV movements along the seabed requires complicated skills and an inexperienced student may overturn the UUV. Utilization of a simulator helps overcome these obstacles.

The simulator that is presented in Figure 4.22 consists of rotating panels described earlier. Before the panel is a screen and behind it is an area that models various seabed profiles. A self propelled proper sized cart can move in this area. Data about the cart's position is transmitted to the rotating display and the student can change the position of the cart using corresponding controls. In this manner, a simulator helps to recreate the tasks involved in control of the UUV's movements along the seabed. The structure of this task is very close to a real one. The imitated shelf of the seabed can be changed during training processes from a simple one to a more complicated one.

Data, which is presented in this chapter, demonstrate that the systemic-structural analysis of work activity can successfully solve complicated design problems using only analytical procedures. In the earlier chapters, we did not discuss task complexity evaluation of computer based tasks. This problem will be considered later.

In Chapters 3 and 4 we widely utilized MTM-1 system. However we've used this system in a way that is totally different from traditional methods. Motions and cognitive elements in the MTM-1 system relate to the concept of operation in activity theory. This system offer a very precise language of activity description. In conjunction with other units of analysis and methods developed in SSAT MTM-1 system can be very useful. Time characteristics of activity can be corrected during experimental study if it is required. Activity is a process and without well defined units of analysis and their corresponding performance time it is very difficult to develop models of activity. At the same time very often we can restrict our analysis to qualitative and algorithmic analysis or even only to just qualitative analysis. In the following chapter we'll consider the functional analysis of activity which is a qualitative systemic analysis.

The analysis of the preceding material shows that SSAT is much more concerned with the temporal aspects of human performance than the cognitive approach is. Time is not just an important criterion for the evaluation of work productivity and efficiency. Failure to function within the time limit is considered a failure in operator's performance in man-machine systems. Finally, time in combination with the qualitative analysis is one of the most objective ways to measure not only behavioral but also cognitive aspects of human performance.

5 Foundation of Functional Analysis of Activity

5.1 INTRODUCTION TO FUNCTIONAL ANALYSIS OF ACTIVITY

5.1.1 DIFFERENCES BETWEEN THE CONCEPT OF SITUATED ACTION AND SELF-REGULATION OF ACTIVITY

Functional analysis of activity describes activity in relation to the context in which it occurs. It shows how activity is developed in a particular context based on the mechanisms of self-regulation. Situated aspects of human performance are also described in Suchman's (1987) "situated concept of action." According to this concept, purposeful actions are inevitably situated actions occurring in the context of particular circumstances. In contrast to traditional cognitive psychology, where purposeful actions are determined by plans, in the situated action concept, actions while systemic, are never planned (Suchman, 1987). She did not describe precisely what action is in her theoretical concept and how it differs from the term "action" in activity theory. Hence, we will not go into a deep discussion of this term in her work. However, we can only infer that action in Suchman's work can be understood as purposeful behavior involved in task performance and that it includes cognitive components. Action in Suchman's concept with some approximation can be considered to be similar to the term "activity" during task performance in activity theory. This approximation is required for comparison of the relationship between stable and situated components of human performance in situated action concept and in the systemic–structural theory of activity. According to Suchman, cognitive science overemphasizes the importance of the plan and considers it a major mechanism that guides actions. According to functional analysis of activity, a plan is only one possible mechanism of actions and activity regulation.

By plan or program of performance, we understand the content and sequence of the different components of an activity or separate actions (mental or behavior) by means of which an activity or separate actions should be performed. We use the term "plan" when the subject deliberately and consciously determines the sequence of the elements of an activity in a particular situation. The term "program" is used in situations when planning is unconscious and has a very short duration. Without a plan or program, goal-directed activity cannot be performed. At least some components of a plan or program should be developed before an activity or action starts. This outstripping can happen even when we talk about an unconscious program that is triggered automatically. In a more detailed way a plan or program is developed during an activity or action execution. A plan or program in some degree passes ahead of the activity. Bernshtein (1947, 1966) demonstrated that the motor action program consists of different subprograms that have a hierarchical organization, where high-level programs can be conscious and low-level programs unconscious. There are two kinds of behavior. The first is conscious goal-directed activity and the second reactive behavior. In the second case motor programs can be triggered automatically in the same way as in cognitive actions.

According to Bernshtein (1947) there are four main levels of self-regulation of motor actions. They are associated with different levels of the nervous system. Therefore self-regulation has a hierarchical organization. The concept of hierarchy is based on the idea that the nervous system has superordinate and subordinate components. Higher order subsystems provide guidance for the subsystems below them in the hierarchy. It is also involved in verbally logical thinking. This level of self-regulation is also important at the beginning of the development of motor skills. Bernshtein (1996) considered skill development a constructive process with different stages. Not only cognitive but also motor skill development can be considered a problem solving process. Bernshtein describes four main levels of motor action regulation: the first level is responsible for the meaningful symbolic aspects of movement regulation. (e.g., I need to describe some idea). It is an abstract level movement regulation. This level is particularly important at the first stage of skill acquisition and is also associated with the sense components of movements. For example, I need to write the following word that includes the following letters. Therefore the highest level of self-regulation is involved in the symbolical coordination during speech and writing. The second level of movement regulation is involved in special aspects of movement performance. This level provides the body movements in space. For example, this level provides movement of the hand in space during writing. The third level is associated with the kinesthetic sensitivity of movements. The kinesthetic level utilizes information about our body parts from receptors in joints and ligaments and in the muscle fibers. In our example, this level determines some handwriting features. The lowest level provides control of the muscle group tonus. This level works in cooperation with the equilibratory senses, which deal with the body position. The relationship between levels of regulation and their importance can be changed during the skill acquisition process. The first level described above is responsible for the conscious regulation of actions. The lower levels, which are developed during the training process, are responsible for the unconscious level of regulation. The conscious level of motor action regulation performs auxiliary functions as a kind of "scaffold" at the first stage of skill acquisition. At the farther stage of the motor skill acquisition this level of motor regulation is abbreviated or even becomes redundant.

Later this idea was assimilated by others (Broadbent, 1958). It was discovered that self-regulation includes the hierarchically organized mechanisms of planning. For example, a model of the movement regulation developed by Gordeeva and Zinchenko (1982), which we described in our work (Bedny and Meister, 1997), includes an integral program and a differential program. Zinchencko et al. (1978) also developed a model of a perceptual action which includes a microblock "program-formation mechanism." Konopkin (1980) developed a model of self-regulation of sensory–motor activity that included program-formation mechanisms. This is more apparent for complex holistic activity. A'model of self-regulation of activity developed by Bedny and Bedny, Karwowski also included the program-formation mechanisms (Bedny and Meister, 1997; Bedny and Karwowski, 2003,2004).

There are evaluative mechanisms (comparators) that check the executive stage of activity against the program. Corrective impulses can change the program of performance if the neural system detects discrepancies. This aspect of regulation can be unconscious. There are different possibilities of using feedback (Bedny and Karwowski, 2003), which we do not discuss here. We only want to pay attention to the plan and program that is sometimes a very complex, hierarchically organized system where unconscious components have a subordinate relationship to conscious planning of actions. It is well-known in gymnastics. For example, when the coach explains a sequence of movements to a gymnast, consciously controlled movement can automatically trigger an unconscious level of program regulation of separate muscles. Therefore the plan or program is an important mechanism of the self-regulation of motor actions or activity in general. During the skill acquisition process the importance of the high level of movement regulation decreased.

There are a number of other important mechanisms of regulation of activity. Activity becomes adapted to a situation based on the process of self-regulation (Anokhin, 1962; Bernshtein, 1966; Bedny and Meister, 1997; Bedny and Karwowski, 2004d; Konopkin, 1980). Self-regulation includes the mechanisms of planning and plan execution; however never reduced to these mechanisms. An activity is always considered a combination of prespecified and situated components.

Bernshtein (1947) demonstrates that when a subject attempts to repeat the same movement multiple times, careful registration of each movement reveals that each of them is unique to some degree. This is repetition without repetition. Each repetitive action, is not only performed based on data derived from past experience, but is also constructed and adapted for a particular situation. According to Bernshtein, movement of different segments of the body can be considered a result of the collective influence of both internal and external forces. The external force field acts outside the human body. The internal force field operates within the body and depends on the interaction among segments, muscles, and other internal organs.

Human movement is executed by various limbs, which can be thought of as multiple kinematical chains. These movements are executed within complicated force fields, which are never static. Therefore, it is very difficult to coordinate a specific movement within these dynamic force fields. No in advance developed program can supply precise directions to the body segments by themselves. Moreover, the same neural program can result in different effects because of the changeability of the external and internal force fields. Thus, a one-to-one interconnection between central neural impulses that are organized as programs and movements cannot exist. In this situation, coordination of movements is provided by the coordination of central neural impulses and dynamic phenomena with peripheral body segments. This occurs because of feedback; the ongoing performance of the action and the state of the muscle enable one to introduce specific corrective impulses during the execution of movement. A model of the motor movements regulation developed by Bernshtein (1966) contains six basic mechanisms:

1. Effectors (motor) Apparatus, the work of which is governed according to specific parameters
2. Order Apparatus, which is based on a specific program to introduce the required meaning of control parameters
3. Receptor Apparatus, which perceives the ongoing meaning of parameters and their stimuli
4. Comparison Apparatus, which evaluates any discrepancy between the desired and actual parameters
5. Translating Apparatus, which translates data from the comparison apparatus into corrective impulses used by the Regulator
6. Regulator Apparatus, which governs the effectors

According to this model, the essence of coordination is overcoming superfluous degrees of freedom in joint movement. The secret of coordination is not expending neural impulses organized as a program to depress reactive forces but using them to perform actions.

Object-oriented activity always includes the same planning in advance. For this reason, activity can be referred to as a self-regulation system. Systemic-structural theory of an activity states that plans and goals of an activity can be developed in advance. However, these plans and goals can be corrected or changed altogether, depending on the ongoing situation. The plan as the program of performance can be developed on a conscious or unconscious level of self-regulation of an activity. The plan cannot be regarded only as a retrospective reconstruction of actions. The plan or program-formation stage of actions can be formed on conscious and unconscious levels and can proceed so quickly that we are unaware of it. This plan can be modified during the execution of an activity. The plan or program of performance can be considered an important anticipatory mechanism of activity regulation. In emphasizing improvisation and response to a contingency, the situated action de-emphasizes the study of more durable, stable phenomena that persist across situations (Nardi, 1997). At the same time in activity theory one should distinguish goal-directed activity from reactive behavior. In the last case situational stimuli can trigger already existing programs of action almost automatically.

Another outstanding scientist Anokhin (1962) develops the theory of self-regulation of a conditioned reflex. According to his model of self-regulation, performance of actions always requires an evaluation of relevant and irrelevant information and comparison of these data with past experience. Based on these data an organism develops or sometimes simply selects an adequate program of performance for a particular situation and corrects this program based on feedback influences. Anokhin and Bernshtein regarded the organism as an active system that can not only be adapted to a situation but can also change the situation according to the purpose of the activity or behavior. A more elaborate model of movement regulation suggested by Gordeeva and Zinchenko (1982) also contains a different kind of program/plan mechanisms of motor action regulation. Suchman's "situation concept of action" ignores all data that are developed in the field of self-regulation of an activity.

Those concepts such as dynamic mental models, dynamic reflection, subjectively relevant task conditions (Bedny, Karwowski and Jeng, 2004), and situation awareness (Endsley, 1995, 2000) are an example of desiccation of the situated aspects of cognitive activity and behavior. The meaning of the situation has stable and dynamic aspects. Even in the same relatively stable, externally presented situation, the subject can extract totally different meanings. This becomes possible only when the subject can use a flexible system of mental actions or operations, which is adapted to a required situation and goal of activity. An interrelationship between the situated and stable components of an activity can be found in the desiccation of an object meaning and categorical meaning. Object meaning has a situated character (Zinchenko, 1995). For example, while playing, a child freely assigns meanings to different objects depending on the situation. Categorical meaning is a more stable phenomenon. It is a result of convention and sociohistorical developments. Object meaning can be transferred into categorical meaning. The meanings of different phenomena are the building blocks of mental models of situations. Mental models of situations, meanings, comprehension and forecasting are the result of flexible, adaptive mental actions. Actions and operations are integrated in more complicated situational mechanisms of self-regulation. They are always stable and adaptable to situation components. Planning as a mechanism of self-regulation can be rigid, flexible, and adaptive. In the last case we can use the term "strategy." In contrast, the concept of the plan or program as an important mechanism of self-regulation strategy implies flexibility and plasticity of the activity in general, based on the evaluation of the situation or activity outcome and the internal state of the human. The major purpose of self-regulation is the process of continuing reconsideration of cognitive and behavior strategies when internal and external conditions have changed. In a different kind of activity plan and strategy have different importance.

Not only cognitive but also motivational aspects of activity regulation are important in the situated activity. A person's motives can be divided into two groups: sense-formative and situational (Leont'ev, 1971). Sense-formative motives are relatively stable and determine a person's general motivational direction. Situational motives are connected with immediate ongoing activity and the solving of specific tasks. As a result, situational motives are more flexible. The sense-formative motive is connected with individual features of personality. The situational motive is more involved with task performance. The content of situational motives, their hierarchical organization, and their relative weight can be changed, depending on the character of the tasks to be solved, the temporal stages of task performance, and informational feedback about task solution. Motives can be needs, attractions, sets, desires, etc. Emotions may also accompany motivated behavior. According to Hilgard et al. (1979) there is no clear-cut distinction between these notions, and it is an unresolved issue in psychology. However, in activity theory this notion is clearly distinguished.

Reykovski (1979) wrote that motivation is emotions plus the directness of action to a specific goal. Emotional–motivational components of activity have inducing and regulatory functions (Zarakovsky and Pavlov, 1987). Inducing components have only one function: to direct the person to achieve a specific goal. The regulatory components have four functions: switching, reinforcing, compensation, and organization.

If an individual has several concurrent motives, the switching function enables him to concentrate on the behavior most closely related to the goal of the activity which has more subjective value for the person. The reinforcing function provides rewards and, thus, reinforces desired behavior. For example, reward increases response whereas punishment decreases response. The compensation function enables emotion to be transformed into increasing levels of motivation. For instance, time constraints increase the emotional intensity of the operator and motivate him to mobilize effort to achieve the desired goal. They increase the speed of task performance. The organizing function of emotion promotes recognition of any conscious discrepancy between existing and required methods of achieving a goal, and, thus, tends to a more correct organization of activity. This function is connected with the selection of correct strategies of activity depending on the emotional state of the person and the requirements of the task.

The emotional–motivational components of activity are associated with the concept of will. The will contains mechanisms that sustain activity under conditions in which obstacles to goal achievement appear. The willing process can appear when a conflict exists between motives in the operator. In such cases any change in the activity can be mediated by a conscious act of will.

5.1.2 Basic Characteristics of Functional Analysis

Functional analysis regards activity as a goal-directed self-regulated system. Therefore self-regulation is a major concept of functional analysis. The major purpose of functional analysis is the description of activity strategies that are considered a result of the self-regulation process. At the general stage of functional analysis one can describe different strategies in general terms. At a more detailed stage of functional analysis, specialists can describe strategies of performance in relation to different functional mechanisms or functional blocks. The more functional mechanisms involved in the description and analysis of activity strategies, the more detailed is the functional analysis. In the previous section we did not consider general principles of self-regulation of activity. In this section we describe psychological aspects of self-regulation as a basis of functional analysis. The concept of self-regulation in psychology is very often used incorrectly. "Self-regulation" as currently used in psychology has been made synonymous with such notions as "will-power," ego-strength and volition (Kuhl, 1992). Heckhausen (1991) and Kanfer (1996) regarded self-regulation as a motivational processes. Any self-regulative system is defined in the first instance by its goal orientation (Ackoff, 1980). According to Wiener and Rosenbluth (1950), a system is goal-oriented if it continues to pursue the same goals through variable behavior, as environmental conditions change. This system can also change its goals and methods of their accomplishment as a result of processing other system functions.

This permits a more precise definition of self-regulation. Self-regulation is an influence on a system that derives from the system in order to correct its behavior or activity. It is necessary to differentiate between "self-regulation" and "regulation." Regulation is an external influence on the system whereas self-regulation is internal. Self-regulation is an intrinsic self-organizing tendency of the system itself.

Self-regulation is not a psychological notion; rather it is a cybernetic concept. One fails to see what is gained by trying to translate these traditional notions into modern cybernetic idioms under the rubric of self-regulation. All psychological functioning and behavior can be construed as a self-regulation process. The concept of self-regulation becomes meaningful only when the self-regulation model is developed. One defines it in terms of a functional block model. In the absence of the specification of functional blocks and feedback mechanisms, the use of the concept of self-regulation is little more than a metaphor.

Traditionally the human information-processing approach describes mental processes as a sequence of elemental stages. Typically these stages are organized in linear order. Depending on the purpose of study, decomposition can be performed with different levels of detail. An example of a more detailed analysis is decomposition of reaction time latency into different stages. Sternberg (1969, 1975) was one of the first who discovered these stages by utilizing the additive factor method. This is an important achievement in cognitive psychology because it demonstrates that even the more simple cognitive process that is performed in a very short period of time is not a homogeneous phenomenon. In activity theory these stages are considered a components of cognitive actions that sometimes have a very short duration (Gordeeva and Zinchenko, 1982). However some of these stages are not stable and can be changed or totally be eliminated depending on the specificity of cognitive actions. Further, these stages are not only organized in a linear manner, but also have a loop-structure organization. Therefore, feedback influences are important during the organization of these stages (Bedny and Meister, 1997).

The stages of information processing are also described in analyses at the macro-level. Usually these stages are not precisely defined. An example of a broader or general level of description of these stages in cognitive psychology is the work of Norman (1986). He described seven stages of human activities, in which some can appear out of order, some skipped and some repeated. These stages are organized in a single loop and also performed in sequence. These stages include perception, interpretation, evaluation, intention, action specification, and execution. Norman designates this description of activities as approximate action theory. In this theory, the term "action" is more closely related to the concept of purposeful behavior and should not be mixed up with the concept of action in activity theory. The described stages are not precisely defined. For example, the concept of a goal in this scheme is not explained as done in activity theory. It is difficult to understand how the goal is related to intention and evaluation stages. What is the relationship between intention, goal, and concept of motivation? How is the executive stage related to physical activity, etc? All steps in the processing of information and stages of performance of activities are organized in a linear sequence. In general, cognitive approaches have limitations because the described stages are not precisely defined; they are not organized into a sequence and there is no precise understanding of how these data can be applied in practice. In systemic-structural activity theory all stages or mechanisms of information processing during the performance of cognitive or behavioral components of activity are described very precisely. They are organized into a self-regulated system with feed-forward and feedback interconnections. In particular, all important situation mechanisms and the relationship between them can be considered during task analysis. Hence systemic-structural activity theory eliminates the gap between theory and application.

There are two types of self-regulation processes in human beings: physiological and psychological. They are interconnected, but each of them has its own dynamics. The physiological self-regulation model is based on homeostasis. There is a tendency for physiological disturbances to be reduced and for departures from an optimal state to be corrected in order to restore balance. Many physiological imbalances are corrected automatically. The structure of physiological self-regulation processes is wholly predetermined.

The psychological self-regulation system can change its own structure based on its experience. Such a system can form its own goals and subgoals and its own criteria for activity evaluation. This is a goal-directed self-regulative process. Self-regulation provides the integration of cognitive, executive, evaluative and emotional aspects of activity. Given the complex set of variables involved, people exhibit an infinite diversity of activity. Internal changes in the psychological aspects of self-regulation emerge not only from external influences, but also from those changes that are prefigured in previous experiences. Thus, in systems of self-regulation prespecified experience from memory that can be adapted to a situation assumes fundamental importance.

It is important to take into consideration that during any individual self-regulation process, the program of self-regulation, the criteria of evaluation, and even the goal of self-regulation may be changed. Self-regulation functions in time; it is based on past experience and on evaluation of the present situation and anticipated future. From a psychological point of view, self-regulation may be considered a process that supplies coordination among the various psychological functions in accordance with a specified goal.

Strategy is fundamental to self-regulation. Broadly speaking, strategy is a method of taking into consideration information during planning as well as throughout the processes of achieving the goal. However, between plan and strategy, there are significant differences. A plan is something that is stable, even rigid, because it describes a more specific situation than does strategy, which has a more dynamic character. "Strategy" implies flexibility, plasticity, and variability of means, the capacity to change a program based on task outcomes, changing conditions, and the internal state of the human.

The major problem facing the self-regulation system is the process of continuing reconsideration of activity strategies when internal and external conditions or situations have changed. Sometimes this results not only in changes in the methods of achieving the goal, but a change in the goal itself.

When we talk about action, operations, motions, etc. a major consideration was temporal–spatial localization of the different elements of activity. All this refers to the morphological description of activity; in other words, its construction. When we talk about self-regulation, we describe activity functionally. In this case, during the analysis, functional mechanisms or functional blocks, through which one can achieve a specific goal, become basic units of analysis. This means that we attempt to describe activity not so much in terms of cognitive processes or cognitive and behavioral actions as in terms of more complicated integrative functional mechanisms.

The purpose of the deployment of these units of analysis is connected with obtaining and enhancing data regarding the crystallization of cognitive processes into more complicated mechanisms and reviling their role in the regulation of activity as a holistic system. The same psychological processes are included in a particular activity with a specific organization and carry out a specific function. Moreover, the same function may be realized by the different content of cognitive processes. The content of the functional mechanism can also be described in terms of cognitive and behavioral actions. When the functional mechanisms are presented as components of the model of activity self-regulations with their feed-forward and feedback interconnections, they are defined as function blocks. Different psychological phenomena included in activity may be considered from the aspect of their functional role in the regulation of activity. Under these circumstances the function block can be defined as a coordinated system of subfunctions, which has a specific purpose in the regulation of activity. For example, some function blocks can be responsible for creating an image of a situation or a program of execution and complex functions of control and corrections. The function block at a later stage of analysis may be decomposed into more detailed subfunctions.

From this it followed that functional analysis was developed in several steps. The first step entailed the identification of cognitive processes (cognitive analysis) involved in performance, determination of cognitive actions, and operations. The next step involved integration of these processes or actions into particular subsystems with specific regulatory functions within a structure of activity. They are called functional mechanisms. However this is not a strictly sequential step. Functional mechanisms do not exist in isolation. They are integrated into a holistic self-regulated system or a model of self-regulation. At this stage of analysis functional mechanisms can also be called function blocks. Each function block mediates a particular function in the regulation of activity. The interrelationship among the functional blocks is critical to the understanding of activity regulation.

Every function block can include the same cognitive processes. For example, they can include perception, memory, imagination, thinking, etc. However their integration can be performed in different ways depending on the specificity of the task in hand. Function blocks can be defined in an unvarying manner, but their content may, and indeed will, vary. Task context is also extremely important in determining how function blocks are used, their degree of development, and how they perform. The content of the function block can change but the purpose of each function block in a self-regulation model will remain the same. The meaning of the function blocks in any specific activity can be understood only in relation to other function blocks.

The concept of function block was used in the first section of this chapter when the self-regulation of motor actions was described. In cognitive psychology when one tries to describe psychological microprocesses, the notion of the function block is also used. In this case, it describes psychological microprocesses such as iconic memory and mechanism of scanning information.

More precisely, these blocks can be called functional microblocks. However, when we apply the concept of the function block in the model of self-regulation of activity as a whole, the function block has a much more complex architecture and describes activity at the macrolevel. At this stage of analysis we do not apply chronometrical studies of very short duration cognitive processes. Each function block requires much more time for its realization.

The functional analysis of activity at a particular stage enables the distillation of explanation in terms of functional mechanisms and blocks from the study of cognitive processes. However, if required, researchers can deconstruct the function blocks into lower- level cognitive processes embedded in these function blocks. Any cognitive process included in activity may be approached in terms of its role and position within a particular function block or mechanism. From this it follows that cognition may be studied from the perspective of its functional role in the regulation of activity, which in turn implies the unity of cognitive and functional analysis.

A fully elaborated model of self-regulation has the following characters:

1. All the function blocks are interrelated.
2. Each block is a functional subsystem directed to achieve specific subgoals of the activity.
3. Each function block is part of a flow of activity with multiple entry points and exits.

Functional models composed of such blocks should be distinguished from multiple figures and schemes that also consist of boxes that are informally introduced by researchers to present their ideas. To introduce a new block into a functional model, one has to prove experimentally and theoretically that a certain functional mechanism does exist and then determine its interconnections with other functional mechanisms of the developed model.

5.1.3 Meaning as Functional Mechanism of Activity

With increasing reliance on automation to support human performance scientists attempt to find more efficient methods to study human work in complex socio-technical systems. These systems are characterized by cooperation of people with technical components and tend to be composed of many different elements and diverse interaction between them. One important trend in these systems is computerization when computer-based technology is tailored with traditional technology and creates computer-based information systems. Computerization of man–machine systems increases the role of intellectual components of work. As a result, human activity becomes more complex and variability of work strategies increases.

The problem with computer-based information systems is not lack of information but finding what is needed when it is needed (Endsley, 2000). Comprehension and interpretation of data in a dynamic world is a major problem of these systems. The concepts of a stable and dynamic model, a conceptual model, subjectively relevant task conditions, situational awareness, etc. become critical in the study of these systems. Therefore great emphasis should be placed on analysis of the semantics of the work domain (Rasmussen and Goodstein, 1988). The theoretical foundation of these aspects of analysis is the study of the concepts of meaning and sense and their role in work activity.

The evolution and development of human culture has depended on the human ability to use sign systems. These sign systems have continuously changed, evolved and become more and more complex. Currently, labor is increasingly dependent upon sign systems, with the information presented to the operator usually encoded in these systems. In light of this the idea of sign and its meaningful interpretation, while having a longstanding history, has gained renewed importance. With the rapid dissemination of computer systems the role of sign systems has and will continue to increase. In activity theory meaning is studied in connection with the concept of sense. Meaning is an objective phenomenon, which can be transformed into a personal, subjective sense. Anything that is significant for the subject has personal sense. The study of meaning and sense within activity theory are grounded in the work of Vygotsky (1962), Leont'ev (1978) and others. According to Vygotsky, man acts on nature indirectly through the use of special tools which serve a mediating function. When an individual interacts with objects he uses external tools. During the process of mental development the individual internalizes various sign systems and uses them as internal tools for thought. From this it follows that there exist two kinds of signs, one of them in the external world and the other in the mind of the subject. The signs in the mind of the subject fulfill the role of an internal psychological tool. An understanding of how these two types of sign systems interact makes it possible to understand how the meaningful interpretation of presented information by the subject takes place.

Commonly three semiotic aspects of signs are distinguished: syntactic aspects, semantic aspects, and pragmatic aspects. Syntax considers the nature of the relationships between signs, semantics considers the relationships between signs and referent objects, and pragmatics describes the relationship between signs and the individual interpreting them (Morris, 1946). The semantic aspect of signs is related to the concept of meaning. While the relationship of the sign and its referent is of extreme importance, it is also important to consider that this relationship is the product of human activity. In order to obtain knowledge about the object one has to perform certain actions and operations with it. They can include discovering the specificity of that object's interaction with other objects, transformation of that object from one state to another, etc. Signs can be manipulated in the same way that other objects are manipulated. However, in order to obtain the required knowledge of a given sign system it is not sufficient to manipulate the material form of its sign.

The meaning of a sign is the most important aspect in the subject's interaction with that sign. A sign is not a regular object and cannot be manipulated as such. A sign does not function according to regularities of real object but rather according to laws applied to the sign's meanings. In order to understand the sign it is important to consider the sign in relation not only to its referent, but also to the activity of which it is a part and which grants it meaning and sense (Shchedrovitsky, 1995). A symbol is a sign only because people relate to it as such. However this does not mean that the interpretation of a sign is a purely subjective process. The meaning of a sign has an objective character in that it is the result of sociocultural development. This sociocultural development is what gives a sign a standardized method of its interpretation. The sign exists because there exist individuals who are a part of a culture, which uses this sign and is able to assign a certain meaning to this sign.

The study of sign systems is impossible in isolation from the study of sociocultural principles of the formation of individual consciousness and the ways in which these principles influence the means and norms of the culture. Activity theory in its study of the relationship between the sign and meaning considers not only the individual but also the world of culture created by human activity.

The fact that people can interpret signs in the same way is proof of the objective existence of meaning, which is independent of the subject interpreting the sign. However, this objective meaning forms within the activity that has not only an individual–psychological but also a cultural–historical nature. From this, it becomes apparent that an attempt to determine the meaning of a sign exclusively on the analysis of the interrelationship between the object and the sign is impossible. It is also important to consider that not all signs are related to objects; for example, some signs refer to abstract concepts such as increases in speed, or the concept of energy; others may refer to the interrelationship of other signs such as mathematical symbols and verbal syntactic markers. The meaning of some signs is determined by the functional relationship between signs. The meaning of others is uncovered through referents that are real objects.

The objective meaning of a sign is discovered by the human activity directed toward interpretation of the sign. Currently, the study of sign meaning and interpretation focuses on the sign's relation to the object or denotation. This approach derives from the work of Frege (1948,1892). However, from the perspective of activity theory, sign meaning should be studied not only in relation to the object or other signs but also in relation to human activity. The relationship between an object and a sign and between different signs exists only in the context of human activity. In order to understand a sign system it is important to consider it in relation to activity in which this sign system exists and from which it acquires its meaning and sense.

The process of understanding involves discovering the relationship between objects and signs and internalizing the meaning of a sign. When through this process a symbol comes to carry meaning it becomes a sign. This sign in turn, according to Vygotsky (1978), is a tool used in the thinking process and in the interpretation of a situation. Furthermore an individual's information-processing capabilities are associated with his/her creation and manipulation of signs. The sign functions as a tool or instrument of cognition for human cognition, which is not only a process but also a system of mental actions. One of the most important systems of signs employed in mental actions is natural human language, which is used as a tool for the development and carrying out of higher order cognitive functions.

This natural role of the sign in human mental activity leads to its importance in the work of the operator. In the operator's activity information is presented directly as a real object and indirectly in an encoded form, through various sign systems. In Section 5.1.4, we see how the individual can interpret these signs and extract meaning from the situation through different systems of mental actions.

The meaning of an object is determined by the relation of the action with that object to the situation in which the action is performed. Here we are referring to object-related meaning, one of whose characteristics is its dependence on the situation.

Object meaning is derived from an individual's practical experience. Object meaning has situational character. Object meaning or situational meaning of objects is determined through the relations of an action to a situation and exist from this point of view, only during the performance of a particular action. In this case the meaning of an object depends less on the object's properties and more on the subject's task and goal (Genisaretsky, 1975).

This was demonstrated early on by Vygotsky (1978) in his study of preschool-aged children. Vygotsky pointed out that while playing with objects, children ascribed functions to them that were far from their true functions in the adult world. For example, during play children ascribe meanings to objects a in very flexible manner. They can apply the word "car" to the chair because they mentally manipulate the chair as a car.

A different kind of meaning is the categorical or idealized meaning, which is part of the verbal categories that one masters (Gordeeva and Zinchenko, 1982). Categorical meaning has a stable character and does not depend on the situation. In the process of an activity with others, object meaning can be transformed into categorical meaning. The situation-dependent nature of an object meaning is one of the features that distinguish it from categorical (verbal) meaning, which is relatively stable and fixed in a word (Zinchenko, 1978). However, though distinct, categorical and object practical meanings are interdependent. One of the most important aspects of human activity development is the integration of symbolic and object practical components of activity.

Categorical meaning has an objective sociohistorical character. This objective property of signs, for example, allows a subject interacting with a computer to know the meaning of the various symbols on the screen and also permits him to know the meaning of the various indicators before him. This necessary constancy in sign meaning is the result of the connection between the sign, thinking, and culture. The meaning of a sign concentrates the experience of many generations and determines the specific nature of thinking characteristics of a given culture. The constancy of meaning and its relationship to culture allows us to view culture as a semiotic system, or net of meanings, which is superimposed by the individual on the surrounding natural environment and artifacts (Sokolov, 1974). Usually meaning is considered in relation to the conscious aspects of activity, and in most cases can be verbalized. However in activity theory meaning can also be associated with nonverbalized unconscious activity. This aspect of meaning will be considered later. Different kinds of meaning constitute the building blocks of an operator's mental model of reality.

Meaning is the result of a particular kind of activity, the goal of which is not the transformation of a situation but rather its understanding and interpretation. Based on the analysis of this situation the subject interprets its meaning in terms of the goals and actions which he can perform. In this activity of comprehension symbolic systems have a special role. Human activity in general and comprehension in particular involves the substitution of real objects by sign systems, each of which requires a specific system of actions and operations. Here the sign begins to serve the function of the real object of activity. The subject performs different actions not only with the material form of the sign but also with the meaning of the sign. During an activity the sign systems allow the subject to extract and fixate relevant aspects of objects and phenomena. Having expressed certain aspects of the objects in sign form we simultaneously determine a system of actions and operations that correspond to that particular sign form.

For example, the number as a sign makes quantity an entity independent of the objects to which it refers. The number sign system determines the system of operations or actions, which allows us to reflect the quantitative aspects of the objects at hand. In a task with numbers the subject fixates a particular content in sign form and then performs actions with these signs, actions that are in turn determined by the sign system being used.

Furthermore, the meaning and function of the objects under consideration are closely interconnected. For example, the shape of a chair determines how we can use it or what kind of actions we can perform with this object. The relationship between shape and function has logical and associative interconnections. Changing shape, therefore, can result in a change of function. However, very often minute changes in an object's image do not influence our understanding of that object's functions. On the other hand, when an object's shape is significantly changed, a reconsideration of the object's function must often follow. These data are important for the separation of perceptual and thinking actions during task analysis.

While the interpretation of the object's meaning by the subject is often influenced by the object's shape very often in operator activity, the image and function of objects do not match each other. Not only does this make the interpretation of information difficult, but it also hinders the regulation of executive actions performed by the operator. The operator's interpretation of the object's meaning in turn partially determines the activity of the operator. In conclusion, the meaning provides not only an orientation in a situation, but also regulates the executive actions of an operator. The importance of the notion of an action to the meaning becomes obvious in noting the relationship between the meaning and the function of an object which are a determining factor in operator activity.

To further elucidate the relationship of meaning and action let us consider the meaningful interpretation of a situation in comparison to Gibson's (1979) concept of direct perception. In the same way that perceptual features can lead to direct perception, semantic features of the situation can lead to direct interpretation. However, in cases when the semantic identifying features of situations are hidden, deliberate thinking operations and actions are required for interpretation of the situation. The more hidden these essential indicative features are from the perspective of the goal activity, the more complicated the gnostic actions and operations involved in the process of interpretation and comprehensions.

5.1.4 Sense and Its Interaction with Meaning

Along with the objective meaning of an object, sign, or word (a verbal sign) there is also the subjective interpretation of that meaning (Leont'ev, 1978). This interpretation of the objective meaning depends on past experience, current goals, and the motives of activity. The interpretation of the objective meaning is its transformation into the subjective sense.

The acquisition of meaning by the subject allows for the adequate comprehension of the objective characteristics of the external world. Sense, on the other hand, has a more personal character and allows for the adjustment of the individual to more specific situations and problems. These problems are in turn defined by the goals and motives of the subject, which are influenced by the current situation.

Meaning and sense often overlap; they diverge in cases where there exists the possibility of variable interpretation of the same facts and data. When one considers the notion of meaning as it exists in the consciousness of the individual, its relationship to the external world becomes central. On the other hand, when considering the notion of the sense, we focus on those aspects of meaning specific to a given subject. Meaning determines the position or role of an object among other objects. Sense, on the other hand, determines the relationship between objects and the needs of the individual (Gal'perin, 1966). Notably a connection exists between the psychological concept of meaning and sense within activity theory and the earlier philosophical concept of meaning and sense developed by Frege (1948,1892). He considered denotative meaning as an objective characteristic of the object, which should be distinguished from its idiosyncratic interpretation.

In cognitive psychology there are logical or objective meaning and psychological or subjective meaning. According to Ausubel (1968) logical meaning "... is inherent in certain kinds of symbolic material by virtue of its very nature ... psychological meaning, on the other hand, is a wholly idiosyncratic cognitive experience." In contrast, in activity theory, objective or logical meaning is a result of sociocultural development of human activity. A symbol becomes a sign because people assign meaning to it during the course of their activity and social interaction. Therefore, logical meaning cannot be considered a certain kind of symbolic material that posseses inherent meaning. Meaning does not exist without human activity, culture, and historical development. Similarly, psychological or subjective meaning cannot be reduced to purely cognitive experience. Meaning and sense include not only cognitive but also emotionally evaluative, motivational, and goal-related components of activity. Therefore, meaning involved in the context of different situations goes through certain modifications.

The notions of meaning and sense in activity theory allow us to analyze why the same meaning, in the context of a different situation, acquires a different sense for the subject/s. This difference between objective meaning and subjective sense is most apparent in social interaction. An example comes from a book by Kuzemchenko (1982) "*Funny stories*." A teacher asks a student "Tell me how long do mice live?" The student responds, "That depends on the cat!" In theory, the student gave the correct answer because often mice fall prey to cat's, as a result, the length of the mouse's life is determined by when and if it meets a cat. However as Tikhomirov (1984) says, one has the feeling of a paradox. The teacher was referring to the biologically determined duration of a mouse's life, while the student was referring to the living conditions imposed by its environment, specifically its encounter with a predator. Consequently the same phrase (in this case, question) has several correct interpretations or meanings. That which the subject views as the correct interpretation is the sense of the phrase.

Such paradoxes result when a given phrase has a different meaning for different subjects. In reality the sense is made concrete in the context of the activity and is often referred to as the contextual sense. From this example we can conclude that meaning and sense are connected by the activity of comprehension and interpretation. Further, meaning and sense emerge as different components of this activity. These notions are distinct components of a sign, which develop through the comprehension process. In this aspect meaning and sense are elements of activity and the social interactions that involve comprehension and interpretation.

Sense can be viewed as a structural organization with different images and representations of the same object. Each of these reflections represents separate functional characteristics of situational elements which in turn relate to other elements of the situation. For example, the same verbal expression has several interpretations and meanings. However, in the process of communication the speaker is able to convey a given sense to the listener, who on his part comprehends not any number of meanings but that sense which is being communicated. The listener often extracts the appropriate sense from any number of possible meanings associated with a given expression and ascribes the same sense to externally different verbal expressions when appropriate.

Sense is a dynamic psychological entity, which is developed by involving the same sign or object in different systems of functional interactions. While there is a natural commonality between concept formation and sense formation, the two are distinct. A concept determines stable nonsituated features and characteristics, objects, or situations. Sense, on the other hand, is a dynamic entity which includes the extracted features and attributes of a sign and situation that are critically important for a particular time or stage of interpretation.

Another important property of sense is that it combines within itself affective and intellectual components (Vygotsky, 1956). As a result, in the analysis of activity it is critical to consider not only the cognitive aspects of sense but also the emotions it engenders. Regarding this Bassin (1973) asserted "sense disconnected from emotions — is a logical construct, the emotions disconnected from sense — are physiological reactions." Hence, the sense of interpreted events is defined by their relationship to the goal of activity and its motive/s. The notions of sense and motivational processes are intimately related but distinct. Sense refers to the cognitive–emotional components of activity. Motive, on the other hand, is related to its activity-inducing components; it lends activity directness and is the energetic component, which drives activity towards the achievement of a specific goal.

Words, images and nonverbal symbols can be organized as categorical semantic systems that provide objective interpretations of external phenomena and reality as a whole. Objective information that is presented to a subject through sign systems is understandable to the subject because of the previous acquisition of meanings and related senses. The increased ability of a subject to interpret properly a given sign system hinges on the increasing diversity of meanings and related senses that the subject acquires.

Comprehension and understanding of a sign system can be viewed as a flow of information in the form of a system of meanings and senses, which correspond to the given information. This flow of information further results in the development of new concepts and meaningful connections between them. Rubinshtein (1958) considered comprehension to be the process of analyzing objects from different perspectives as well as connecting and relating them to other objects. Through this process of analysis through synthesis, the object under consideration allows the subject to discover new features and relations with new objects. This process is facilitated by extracting an

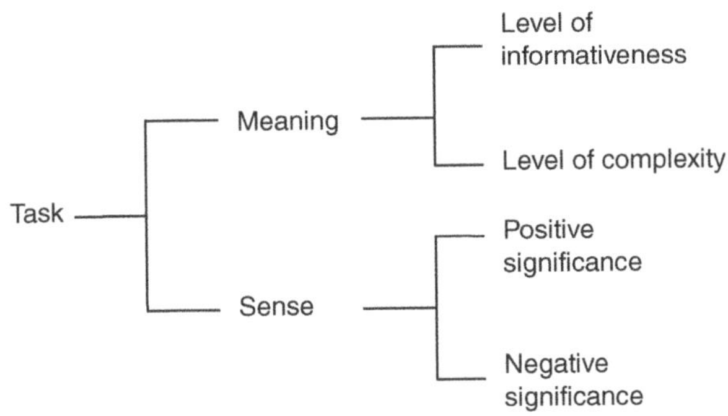

FIGURE 5.1 Relationship between meaning and sense.

unknown and connecting it to something already known, one of the major strategies of the thinking process.

Meaning has two important characteristics: level of complexity and informativeness. To what extent is the meaning of the task or situation informative to a person? The level of complexity reflects the other aspects of meaning connected with a situation; how comprehensible is the meaning of the task to a person?

Sense also has two characteristics: the values solving a task has for a person, and the positive emotions accompanying the achievement of a goal. This type of situation has positive significance.

Goal achievement and task solution threatened by obstacles, such as danger, are accompanied by a negative emotional state. This situation has negative significance.

The foregoing can be described graphically as in Figure 5.1.

5.1.5 Functional Model of Goal Formation Process

Theoretical data presented in the preliminary sections demonstrate that the psychological mechanisms described above can be presented as function blocks of the process of self-regulation. However in addition to describing above the major characteristics of goal, meaning, and sense, here we focus on the self-regulative aspects of this problem. It highlights the regularities of the formation and modification of the goal and meaning in accordance with the context of the particular situation and transformation of meaning into subjective sense. Self-regulation emphasizes the interrelationship between these mechanisms and their mutual influences (Bedny and Karwowski, 2003). As a first step we consider meaning and sense from the functional point of view.

One of the important distinguishing characteristics of considering meaning from the functional perspective is the connection of meaning with motivation. We can elicit not only inducing but evaluative factors of motivation. Actions are not only produced and events are not only perceived but are also evaluated by subjects in relation to their motives in the form of their personal sense. Therefore, a motivational process has two components: those related to "sense" and those related to "motives." These two components of motivation are intimately interconnected within activity but are nevertheless distinct. Sense refers to cognitive- emotionally evaluative components of activity, while motives determine directness and are energetic components for achieving a specific goal.

Emotionally evaluative aspects of the senses are connected with the significance of events and actions or situations in general for the subject. Cognitive aspects of the senses are associated with idiosyncratic interpretation of objective meaning. Objective, commonly accepted, meaning is transformed into the idiosyncratic sense for each individual. The personal significance within a goal-directed activity leads a person to interpret the meaning of the presented information and transform it into the subjective sense. This individual sense creates a predilection of human consciousness by predisposing the individual to make certain judgments and decisions, and take certain actions. This notion of sense is of critical importance in the cognitive evaluation of the situation, and the evaluation of the significance of the situation.

Meaning emerges as a more objective phenomenon. Through meaning, which is an essential component of human consciousness, we apprehend phenomena in the environment. Meaning is reflected in consciousness as an image or concept and the relationship between them. Meaning provides categorization of a situation from a finite number of potential alternatives that are relevant to the goal of activity. Sense provides the transfer of commonly accepted meaning into idiosyncratic sense for each individual. Individual sense creates a predilection of human consciousness (Leont'ev, 1978).

As the motivational state changes within the same situation it alters the meaning of that situation. Consequently, meaning and sense are not rigid, fixed structures, but flexible, self-adjusted systems. For example, a "knife" in one situation might be perceived as an instrument for cutting bread; however, in a different situation the same knife would be looked upon as an instrument for killing. Interpretation of activity always has a goal-directed character. In the context of the subject's activity, meaning cannot be independent of motivation because subjects never accept or formulate unmotivated goals. (Shmelev, 1983). Furthermore, it is personal significance that transfers objective meaning to subjective sense.

Personal significance impacts the comprehension and interpretation of a situation. From the multitude of potential semantic features that can be included in meaning, real categories include only those features which are relevant to the goal of activity, the significance of the situation, and motivational state of the subject. The significance of the categorical features of an object and sign can change through the self-regulation of activity. This, in turn, can result in a change in the algorithms of categorization or interpretation and strategies of activity in general. As a result, the same information can be interpreted in a different way. A person's inherently stable motives determine the preferable strategies for selecting relevant semantic features in the evaluation of a situation. This results in a predisposition towards interpreting the meaning of the situation in a particular way.

Depending on the preexisting emotionally motivational state and the specifics of the situation, a particular kind of mindset is created which manifests itself as a tendency to trigger particular strategies for gathering and interpreting information. These strategies can be either conscious or unconscious and depend on the self-regulation mechanisms of activity (Bedny and Karwowski, 2004). There exist both unconscious and conscious strategies. These strategies consist of flexible mental systems of conscious actions and unconscious operations. However, the flexibility of unconscious strategies is restricted. Conscious strategies, which proceed according to the goal of activity and consist of a system of gnostic actions, are a more flexible method of selecting information and interpreting it meaningfully. Both these kinds of strategies can be considered dynamic algorithms, which are stochastic in nature and based on which subjects develop sense.

These above described strategies of gathering and interpreting information have both emotionally evaluative and motivational features as well as cognitive features, which are an integral part of the thinking processes. These dynamic strategies of activity, which are controlled by the mechanisms of self-regulation, lend a dynamic nature to meaning and sense. Furthermore, as the strategies to which they are related, meaning and sense have two aspects: semantic and emotionally motivational. As a result, a person in the same situation can extract a totally different sense.

Functional analysis distinguishes between two kinds of meaning: representative and functional. Representative meanings perform only cognizable functions and have no practical value. A subject does not use such meanings in practice. Objects, which possess this type of meaning, are only potentiality significant to the subject and impact activity only indirectly. Functional meaning is directly involved in practical performance and regulates human activity. Representative meaning is static in nature while functional meaning is dynamic and adaptive. In light of this distinction all functional meanings are representative; however, not all representative meanings are functional (Shmelev, 1983). For example, for many people, data concerning the makeup of atoms possess only representative meaning. However, for physicists this information is both representative and functional.

When the distinction between functional and representative meaning is applied to operator activity one can see that as the amount of information that needs to be processed by an operator grows, there is an increased probability that meaning will become only representative or declarative and not functional. An operator begins to develop a system of representative meanings, which cannot be used in practice. A discontinuity between representative and functional meaning in operator activity can lead to the failure of an operator to correctly interpret information relating to a problem or task.

Finally, we need to briefly describe the important components of activity from past experience. The general background of the subject also influences the strategies of performance and therefore can be regarded as a functional mechanism. It includes general and professional knowledge of the subject, knowledge of culturally accepted norms of behavior, customs that describe how the community functions, and past experience acquired through activity that evolves over time within a culture. The interaction of past experience and new input information results in the assessment of the meaning of the immediate input information. When the input information corresponds to the background of the subject objective input information is transferred into similar subjective meanings or personal sense. When past experience significantly differs, then substantial variations in interpretation of the same information between subjects can be observed. Specificity of goal formulation, personal sense, and interpretation of meaning depend in significant degree upon past experience available to the performer at the present time.

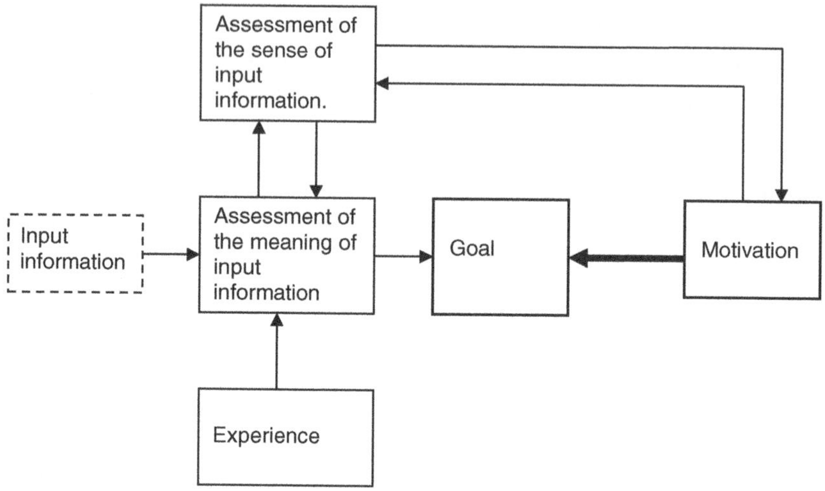

FIGURE 5.2 Functional model of goal formation process.

In activity theory goal and motive create a vector "motive → goal" that lends activity goal-directed character. The more intensive the motive, the greater the effort expended to reach the goal. The systemic-structural theory of activity suggests a model of formation and goal acceptance that describes this problem from the standpoint of self-regulation, which is one of the major principles of activity functioning (Figure 5.2). Goal formation is a dynamic subsystem consisting of the following function blocks: (1) assessment of the meaning of input information; (2) assessment of the sense of input information; (3) experience; (4) motive; (5) goal.

In a developed model those aspects of sense that are involved in emotionally evaluative aspects of activity have a personal significance associated with function block "sense." For example information that is perceived subjectively as more important for achieving the required goal becomes more significant. Actions associated with overcoming obstacles become more significant. Actions that permeate to avoid failure in risk-taking behavior also become more significant. As can be seen in the presented model, function block sense is not regarded as idiosyncratic meaning but as emotionally evaluative mechanism for cognitive data. Sense is the tension and emotional coloring that the activity in a particular situation has for the subject. Sense reflects the relationship of the subject to an ongoing situation. From this model one can see that interaction of past experience with new input information results in the formation of the meaning of the information. Simultaneously, assessment of the sense of information and possible actions is being formed.

Function block "sense" is connected with the evaluation of the significance of input information. Feedback from function block "sense" from which the personal significance derives influences the interpretative process and formation of subjective meaning. Therefore the interaction of function blocks past experience, meaning and sense is important in the formation of subjective meaning. Meaningful interpretation of input information (can be extracted not only from external data but from memory as well) is an important source in the goal formation process. The function block "sense" also interacts with the motivational (inducing) block. The more significant the information or situation in general the more intensive the motive, the more effort expended by a subject to reach the goal. If the motivational process is negative the subject will avoid imposition of the goal. Motivation can also influence sense and therefore meaning. Hence individual sense creates a predilection of human consciousness. The sense aspect of motivation and its inducing aspects may be in conflict Personal significance, value, valence, or utility of the goal bears on the affective (emotional) evaluation of the task or goal. One may consciously be aware of the importance of a task for another, can understand that the task is complex and solving it is prestigious for others and for oneself, and that one has enough ability to solve this task, but for the individual (I personally) this task may not be significant or valuable. In this situation one will not be motivated to solve the task.

From the self-regulative point of view, these two aspects of motivation are treated as distinct functions: "sense" and "motivation" (see Figure 5.2).

Finally in our model functional block, meaning will be considered first of all as a cognitive mechanism, which provides transformation of objectively presented meanings of different elements of situation into subjective situational meanings. In contrast, functional block "sense" will be considered first of all from an emotionally evaluative perspective. Therefore the concept of "significance" becomes the major focus of analysis in this case. How significant the different elements of a situation are for the subject are the major aspects of analysis associated with this function block. Of course cognitive and emotionally evaluative aspects of activity cannot be totally separated. In the same way we will consider these function blocks in other models of self-regulation.

In the presented model vector motive $\rightarrow$ goal is a result of the complex interaction of different blocks. In this model the vector is depicted by a bold line. The model describes not only cognitive aspects of the goal formation process but also evaluative and inducing factors of motivation.

The "motive-goal" relationship provides direction to the self-regulative process. In trying to understand the self-regulative process, we need to determine precisely the goal of the activity. Studies show that different individuals may have an entirely different understanding of a goal, even if objectively identical situations or instructions are given (Konopkin, 1980; Bedny, 1987). As a result we distinguish between "subjective" and "objective" understanding of a goal. Let us consider a practical example, which demonstrates the importance of a goal in human performance. Among pilots controlling a flight, cases of goal deviation were discovered. Sometimes pilots would change the goal of an activity involuntarily. The pilot would be unaware of it. Instead of controlling the aircraft, they would start to control pointers on the different instruments. In other words, instead of using the display patterns to control the aircraft, they focused their attention on controlling instrument pointers. This resulted in the pilot losing orientation of aircraft position during flight (Beregovoy et al., 1978). Instances of goal switching frequently occur during automatic flights when the pilot, functions as a passive observer performing a backup function. In a stressful situation when the pilot urgently must mobilize his resources for performance, he loses sight of the goal of the activity. In such cases, his behavior exhibits chaotic explorative features. Pilots do not simply receive information but actively extract and evaluate it according to the goal of activity. It was discovered that functions of the display apparatus can change as a function of the goal of the activity. The goal of a task is not only interpreted in different ways but also formulated by pilots and this influences the specificity of strategies of performance. A goal accepted by a subject emerges as the leading mechanism in the regulation of activity. Its regulative function may be defined as systemic integrative. Owing to this mechanism the entire process is integrative as a directive process mobilized to achieve a conscious goal. A goal emerges as a conscious determinant that specifies the selection of information relevant to goal realization. Thus, a goal directs the functioning of other functional mechanisms. Goals are also an important anticipatory mechanism reflecting what must be achieved during the performance of diverse actions.

A goal has both objective and subjective aspects, which interact to influence the manner in which the activity is performed. Goals can be self-induced or imposed in instruction. In a task when a goal is formulated in advance based on instructions the acceptance of the goal becomes important. In a self- initiated task the goal is formulated by the subject independently. Even when goals are presented to the subject by means of instructions or external requirement, interpretation of the meaning of the goal can vary depending on the context of the idiosyncratic and situational factors. From the self-regulation point of view, we can distinguish the following stages of acceptance or formulation of a goal. In self-initiated activity there are the following steps:

Goal formation → Goal selection → Goal acceptance.

In "instructed" activity (the task is prescribed) the sequence of stages is

Goal recognition → Goal interpretation → Goal acceptance.

All these stages are considered a self-regulative process. They include conscious and unconscious components.

The presented model brings a different understanding of goal in comparison with the prevailing western theory. For example, goals in activity theory do not have such attributes as intensity (Lee et al., 1989; Kleinback and Schmidt, 1990). A goal may be precise, clear, and conscious, totally or partially. Motives, on the other hand, as an energetic component, may be more or less intensive. The more intensive the motive the more desirable the goal and the more effort expended by a subject. In the work of the above listed scientists, the goal "pulls" activity. In activity theory, the goal "pushes" activity.

The study of goals as a functional mechanism or blocks of self-regulation embodies a number of specific aspects of activity analysis. These begin with goal specification in terms of a task or discrete action and include the mechanisms representing the goal, the position it occupies in a hierarchy of goals, coordination with other goals, and the specific integrative role of the goal with respect to other function blocks within a holistic system. Major influences of the goal on other function blocks in varying contexts, and the precise formulation of subgoals and final goals are studied. Further, specific relationships among verbally logical and imaginative components of the goal, a subject's conscious awareness of the goal and the extent to which he can verbalize the components of the goal or freely select a goal has also been addressed. On the other hand, for imposed goals, as opposed to autonomously selected ones, researchers have examined the extent to which such imposed goals are accepted and the process of their acceptance. Other questions arise for imposed goals including the variances and failures of subjective correspondence to objective requirements for imposed goals, possible versions of the subjective interpretations of the goal, effects of particular interpretations of the goal on selecting and interpreting information, deployment strategies and task performance. To what extent the subjectively accepted or developed goal corresponds to the objectively required goal should be studied. Similarly, the effects of individual differences of personality and past experience on the formation, interpretation and acceptance of a goal, as well as the effect of the form of presentation of external instructions and presented information have been studied. Finally, research has attended to the subjective significance of the goal to the subject, correspondence between the subjective significance of a goal and its objective value, tolerances for deviation from the formally presented the goal etc.

5.2 GENERAL MODEL OF SELF-REGULATION OF ACTIVITY

5.2.1 Unconscious Level of Self-Regulation

Data presented in previous sections and those given below are the foundation for the general model of self-regulation of activity. The self-regulation model comprises closely interconnected but distinct functional mechanisms or function blocks (see Figure 5.3).

The number of function blocks in this model is constant, but their content varies and the degree of their involvement in a particular activity may be varied. Not all function blocks should be considered in any particular task analysis. Depending on the features of the task that the operator performs, an ergonomist may prefer certain function blocks and their relationships and ignore or treat as background other function blocks and their relationships.

The same parameters of activity at different stages of self-regulation can be used in different ways. For example, separate parameters of the goal can be used as criteria of success at the final stage of self-regulation. A goal can change based on the feedback influences during self-regulation of activity which provide the dynamics of the goal-formation process. Transformation of goals during the self-regulative process is an example of important strategy for solving a complicated problem. This means that the same components of activity depending on their role in the process of self-regulation can be included in different function blocks. In contrast to the previously discussed models of self-regulation (Bedny and Meister, 1997), in the present model we describe two channels of information-processing. Channel 1 involves mostly conscious, voluntary processing of information. Channel 2 involves mostly unconscious or automated processing of information.

Their interaction and relative importance change based on the specifics of the situation and the past experience of the subject. Any practical situation presented to

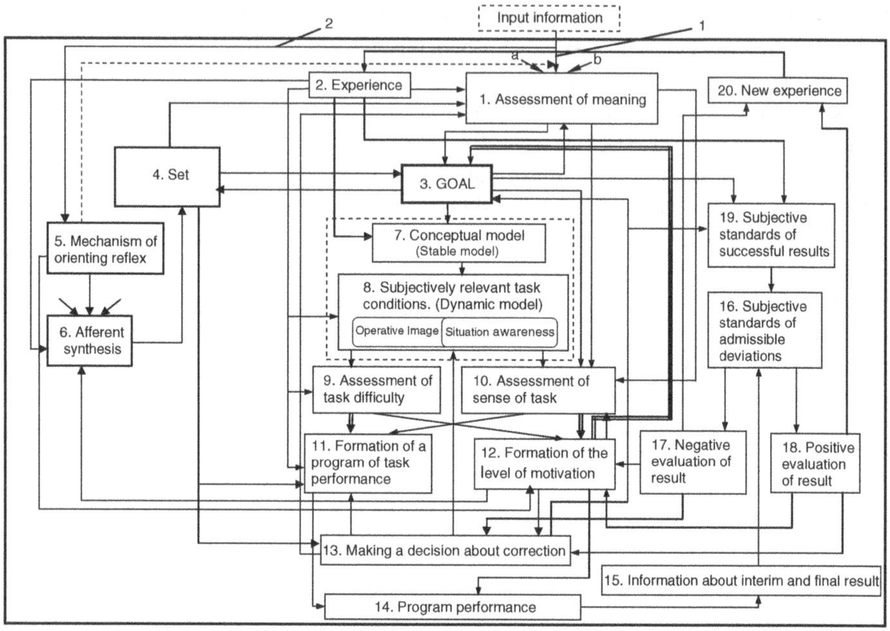

FIGURE 5.3 The general model of self-regulation of activity.

the subject has not only relevant but also irrelevant information. The subject must actively extract the required information. Relevant information is presented by vertical arrows and irrelevant by inclining arrows a and b. Let us briefly consider automated channel 2. Incoming information activates block 5 (orienting reflex). Orienting reflex, introduced by Pavlov (1927), creates conditions for a heightened receptivity of the organism to sudden changes in the situation. This is accomplished through the development of a complex, shortlived and transitory physiological processes, the change in the activation level in the neural system with a general inhibition of conquering ongoing activity. This reflex is conveyed by such components as eye movement reaction, head movement, blood vessel reaction, heart rate, concentration of attention, etc. In general, this mechanism provides automated tuning to external influences and is involved in involuntary processing of information at the first stage of interaction with the situation. If the stimulus is repeated, the orienting reflex weakens.

Sokolov (1960) proved that the orienting reflex is not a result of direct stimulation but is produced by signals of discrepancy. This discrepancy is developed when neural patterns from incoming signals or information is compared with the trace formed in the brain by preliminary stimulation. The greater the differences between these neural patterns, the stronger the orienting reflex. Orienting reflex can activate both conscious and unconscious channels of information processing. In both cases, this activation suggests involuntary activation of these channels. However, when the conscious channel is activated the involuntary stage of activation is transformed into the conscious voluntary processing of information. In those cases when only the unconscious channel is activated, information processing is performed at an unconscious level. In the first case, function block 5 (orienting reflex) activates block 1 (assessment of meaning). This situation is represented in Figure 5.3 by a dashed line that connects block 5 with block 1. In the second case, function block 5 influences block 6 directly and activates unconscious channels of information processing. Orienting reflex also directly influences motivational block 12 and activates the motivational processes.

Motivational block 12 interacts with block 6 (afferent synthesis). This mechanism provides analysis, comparison, and synthesis of all data that the organism needs in order to perform an adaptive response in given circumstances (Anokhin, 1962). A major stimulus that causes a reaction never exists in isolation. Anokhin differentiated between a major stimulus that causes the response, supplementary environmental stimuli that influence the response, information extracted from memory that is relevant to this response, and the current motivational state. The organism must be tuned to all of these influences including major and environmental stimuli. The current motivational state is also very important. This motivational state depends on the current needs of the organism. It selectively activates a relevant neural structure that becomes sensitive to specific stimulation. Selection of an adequate major stimulus is achieved by comparison of different stimuli with dominant motivation and any specific responses also associated with the past experience. Therefore relevant information that is extracted from memory also influences the formation of response. Afferent synthesis integrates information from the above listed sources. It is able to select the appropriate stimuli that are important to the temporal needs of the organism from any external and internal influences and provides a holistic integrative evaluation of the situation. This mechanism functions automatically.

A goal-directed set is formed (block 4) based on the evaluation of the situation. The set is characterized by the role it plays in the formation of purposeful behavior. The set is responsible for the creation of an internal state of the organism that determines the purposefulness of human behavior but this state is not conscious. The set creates a predisposition to processing incoming information in a particular way or a predisposition to performing particular actions (Uznadze, 1961). There are different kinds of sets. Some are relatively stable and depend on individual features of personality. On the other hand a goal-directed set depends on the situation. In this work we mostly consider the latter.

A goal-directed set manifests itself as a dynamic tendency to the completion of interrupted goal-directed activity (Zeigarnik, 1927). The set receives emotionally motivational coloring through afferent synthesis. It determines the stable and sequential course of activity and allows the activity to retain its goal-directed character in constantly changing situations without the subject's awareness of the conscious goal. In general, the set to a large extent is an unconscious regulator of activity and performs the same functions as a conscious goal at the conscious level of self-regulation. Hence motivational processes are critically important to the three function blocks considered above.

A set interacts with function block 1 (assessment of meaning) as well. This means that a set influences the way a subject interprets different components of a situation. One of the major aspects of meaning is the relationship between an object and the sign that presents the object. A subject internalizes different sign systems and uses them as internal mental tools for the thought (Vygotsky, 1978). Therefore, function block 1 (assessment of meaning) is responsible for the interpretation of the meanings of the different sign systems.

A long line of research has studied the verbal and conscious aspects of meaning. When one considers these aspects of self-regulation, function block 1 (assessment of meaning) is associated with conscious goal (function block 3). At this stage of analysis we consider how block1 (meaning) works together with function block 4 (set). The aspects of self-regulation that are related to the extraction of "nonverbalized situational meaning" are associated with the process of interaction of meaning and set. Different forms of nonverbalized meaning were described by Puskin (1978) and Bedny and Meister (1997). With the help of nonverbalized meaning, the subject can extract distinct and essential characteristics of the situation that are germane to the solution of the particular problem. This process is carried out using nonverbalized unconscious thinking operations. In cases when function block 1 is associated with a conscious goal (block 3) conscious thinking actions and verbalized level of interpretation of information become more important. Therefore function block 1 (assessment of meaning) can be involved in conscious and unconscious levels of information processing. When block 1 is involved in unconscious processing of information it is directly associated with function block 10 (assessment of sense of task). The material presented above demonstrates that the subject is able to interpret separate elements of a situation on the unconscious level.

A set can directly influence executive function block 11 (formation of a program of task performance) and function block 13 (making a decision about correction). Through these function blocks the set can associate with other function blocks. Presented material demonstrates that Leont'ev's (1978) statement that activity only consists of consciously regulated actions that are in turn comprised of unconscious operations is not adequate. Activity also includes unconscious reflection of reality. Hence psychic reflection consists of not only conscious actions and their operations but also unconsciously performed operations that are not a part of consciously performed actions. A non-verbalized meaning or situational concept of thinking are also important components of unconscious reflection of reality (Bedny and Meister, 1997; Pushkin, 1978; Tikhomirov, 1984). Activity can not be reduced to the consciously performed actions. Psychic reflection is not only a system of cognitive actions but also a process of unconscious and conscious reflection. Reflection process and cognition are organized as a self-regulative system. In this section we considered the unconscious level of self-regulation in an abbreviated manner. Later we consider this problem in more detail.

5.2.2 Interaction of Conscious and Unconscious Levels of Self-Regulation

Channel 2, considered above, is connected with the unconscious level of self-regulation (5.2.3). When the conscious level of self-regulation is used, the information is processed along channel 1. Usually both channels will coordinate with each other. The same information can be transformed from one channel of information processing to another. This is provided through interactions of block 4 (set) with block 3 (goal). If instead of block 4 (set) the leading role in activity regulation is performed by block 3 (goal) it will provide transformation from the unconscious level of self-regulation to the conscious level of self-regulation. This demonstrates that the unconscious set can be transformed into the conscious goal and vice versa. For example, after a person finishes a workday and gets into his car to drive home, he has to formulate a goal to go home before he is able to do this. However, after the formulation of this conscious goal a driver can shift his attention to other tasks. As a result the conscious goal to drive home is transformed into an unconscious set. It permits the driver to center his attention on the on-going tasks that are correlated with driving a car and at the same time "keep driving home." The driver also has the possibility of talking with a passenger. This also requires a formulation of conscious goals that are associated with conversation and connected with the function of block 3. At the same time the goal to drive home does not disappear. It is simply transformed into an unconscious set and continues to influence the driver's performance through the unconscious channel of self-regulation. At a certain time the driver is aware that he needs to take a certain exit to reach home; the unconscious set is then transferred back into a conscious goal. The content of this goal is the awareness that the driver is going home and not to any other destination. This example demonstrates that the unconscious set regulates the driver's behavior in a way similar to a conscious goal. The transformation of the conscious goal into an unconscious set when attention shifts to new tasks allows the subject to return quickly to the formerly interrupted task. Goals are the only components of self-regulation that have to be conscious during self-regulation. Other stages of self-regulation may be only partly conscious or unconscious.

We have described above how two levels of self-regulations interact. Let us now consider briefly some function blocks that are involved at the conscious level of self-regulation. More detailed information can be obtained from other sources (Bedny and Meister, 1997; Bedny et al., 2000). At the conscious level of activity regulation information is entered along channel 1 to blocks 1 and 3. Based on the interaction of these blocks, the primary interpretation of separate components of the situation and developing of conscious goal of activity is provided. At the next stage the subject formulates a stable or dynamic model of the situation with the help of interdependent function block 7 (formation of task) and block 8 (subjectively relative task conditions). If the situation changes slowly and does not require its continual mental transformation only block 7 is activated. If the situation is dynamic and the operator mentally transforms the situation then function block 8 is involved. Both blocks contain imaginative, verbal, conscious, and unconscious components. For simplicity, in the model described in Figure 5.3 these components are only presented in block 8. Thus, blocks 7 and 8 are responsible for the creation of the holistic mental model of reality. However, block 7 is associated with the development of stable models and block 8 with the development of dynamic models. Block 8 provides mental manipulation of the inner images and symbols in order to create an internal dynamic model of events that are progressive in time. The subblock "operative image" to a large extent provides unconscious dynamic reflection of the situation. The subblock "situation awareness" includes a logical and conceptual subsystem of a dynamic reflection of the situation. These two subsystems of dynamical reflections overlap. Conscious and unconscious components of dynamic reflection can to some degree transform into each other (Bedny and Meister, 1999). Hence, a person can mentally manipulate inner images and symbols to create an internal model of events progressive in time. This dynamic reflection can be enriched with additional data from internal and external sources that are necessary for each particular period of time. The imaginative manipulation of the situation can be, to a large extent, unconscious and easily forgotten owing to the difficulty of verbalization. SA as a component of function block 8 includes a logical and conceptual subsystem of dynamic reflection in which the operator is very conscious of information processing. Imaginative and conceptual subsystems of dynamic reflection partly overlap. The operator is also very conscious of the information processing in the overlapping part of the imaginative subsystem. The nonoverlapping part of imaginative reflection can be considered also as containing a preconscious reflection. With the shifting of attention, an increase of will, and a change in the situation, a preconscious reflection can become conscious or, alternatively, what was conscious earlier can become unconscious. All this may be reflected in the individual by "vague feelings," which can also affect conscious components. Therefore, function block 8 (subjectively relevant task conditions) is involved in a dynamic reflection of the situation and creation of a dynamic model of a situation. It also provides a constant transformation of information on conscious and unconscious levels according to the goals that arise before an operator. Usually what is subjectively significant to the operator is presented in the dynamic reflection of the situation. However, these elements of the situation are not always objectively important. This can lead to faulty orientation in the situation and distortion of the internal model of reality. All data that are contained in SA can be verbalized. Data which are in the subblock operative image can be verbalized partly or cannot be verbalized totally because some aspects of these data associated with unconscious processing of information (there are no verbal equivalents) and others can be very quickly forgotten. Therefore a dynamic reflection of the situation cannot be reduced to verbal protocol analysis. The conceptual model, image-goal, and subjectively relevant task conditions together form the "mental model of reality." This is a "mechanism whereby humans generate descriptions of system purpose and form, explanation of system functioning, observed system state, and prediction of the future system state" (Rouse and Morris, 1986). An individual may create a mental model of reality through the sequence of performance of mental actions or operations. Mental operations are associated with unconscious aspects of the creation of this model. In those situations when only unconscious mental operations are involved in the creation of this model this process is very often perceived is as simultaneous. In more complicated situations direct recognition may be impossible, and gnostic activity may involve a system of explorative conscious actions.

Function blocks assessment of task difficulty (9), assessment of sense of task (10) and formation of the level of motivation (12) are involved in the activity regulation later on. These two function blocks together with function block set (4) and goal (3) are basic mechanisms involved in the motivational processes. We consider them in the following section.

The function block "experience" reflects the past experience of subjects. The interaction between past experience and new input information is an important stage in the assessment of the meaning of immediate input information. When the input information has optimal complexity and matches past experience, the subject may then infer the same meaning from identical input information. Here objective input information is transformed into similar subjective meaning for the subjects who have had a similar past experience. When input information is very complex and does not match past experience, then substantial variation in subjective meanings will be observed. Therefore interaction of past experience and input information is an important aspect of meaningful interpretation of a situation. Past experience includes not only cognitive but also emotionally motivational components and evaluation of task difficulty. What we have described is the first subsystem of self-regulation. This subsystem may be called "goal formative and orientating components of self-regulation." At this stage the subject orients to the situation, develops goals and a subjective representation of task, and develops a dynamic model of the situation.

5.2.3 Motivational Aspects of Self-Regulation

The study of motivation helps to understand how a subject accepts or formulates goals and acts in pursuit of them. Here we attempt to demonstrate how self-regulation help us to understand the motivational process. Self-regulation, which is the foundation of functional analysis, links motivation, cognition, and behavior in a unitary system. Traditionally, motivation is considered an important concept in the personnel/organizational area of industrial/organizational psychology. At the same time motivational factors are practically ignored in ergonomics study. However, from a self-regulation point of view, motivation influences the interpretation of the meaning of an event. This factor demonstrates that motivation is critically important for a correct understanding of a human information processing system. Cognitive processes should be considered in unity with motivation. Motives and goals create, in a self-regulation process, goal-directed tendencies until the self-regulative cycle is completed.

Functional analysis, which considered activity as a self-regulative and goal-directed system, pays attention to situational specific aspects of motivation where conscious and unconscious motivational components interact with each other. General needs and conscious and unconscious motives influence human activity through their integration with situational specific conscious goals or an unconscious set and reflection of situational requirements.

Motivational processes in our model are considered in close connection with cognitive processes and goal formation (Figure 5.3). The goal carries out integrative functions. Thanks to this mechanism all function blocks are combined into a holistic system. Motive as an energetic component and goal as a cognitive component create a vector that makes self-regulation a goal-directed process (see feedback from block 12 to block 3, Figure 5.3). Motives can also be associated with set through afferent synthesis. Without a goal or set motives are transformed into reactive emotional impulses. The other mechanism that is important in the motivation process is function block "assessment of task difficulty" (9). One can distinguish between objective complexity of a task and subjective evaluation of task difficulty. We hypothesize that the more complex the task the greater the probability of the task being difficult for a subject. The subject can evaluate the same task as more or less difficult depending on his past experience or the individual features of the task. Therefore cognitive effort depends on task difficulty. The essential purpose of block 9 is to evaluate how difficult the task will be for the person who performs it. The individual may under- or overestimate the objective complexity of the task. An individual can, for example, overestimate the difficulty of the task and as a result the task can be rejected in spite of the fact that objectively the subject is able to perform it. On the other hand, the individual can underestimate the difficulty of the task and, as a result, select inadequate or inappropriate strategies for its performance, thereby failing to solve the problem.

The concept of difficulty can be approached from two different perspectives. It can be studied as characteristic of the task or as a functional mechanism of self-regulation. In the first case the concept of difficulty becomes important for task complexity evaluation. It is the evaluation of the objective characteristics of the tasks. In the second situation the concept of difficulty is considered a functional mechanism of self-regulation. In this situation it is important to evaluate when a person believes that he possesses the necessary abilities and experience to accomplish the goal of the task. Therefore concepts of one's abilities, self-efficacy etc. become important during functional analysis. For example (Bandura, 1982) described self-efficacy as an important mechanism of motivation. This means that the concepts of one's abilities, self-efficacy, etc. in their relation to task become important for function block "assessment of task difficulty." At the same time evaluation of task difficulty or difficulty of goal attainment does not predetermine the motivational processes. Because the function block "assessment of task difficulty" has complex relations with other function blocks, no simple derivation about motivation levels may be made. Incorrect assessment of difficulty can result in an inadequate personal sense or motivation to sustain the effort towards completing the task. This causes us to consider other function blocks.

Function blocks 10 (assessment of sense of task) and block 12 (formation of the level of motivation) play a leading role in the motivation of activity. Evaluation of the sense of the task or goal is the cognitive–emotional component of evaluation. In contrast, the motivation block determines the inducing component of motivation. These two components are intimately interconnected. Sense is linked with the subjective significance of goal attainment. There is a transfer of commonly understood meaning into an idiosyncratic sense for each individual. The individual's sense creates a predilection of human consciousness (Leont'ev, 1978).

In our model, sense is connected with the subject's evaluation of the significance of situation or task. Thus, sense of activity is an evaluative mechanism. In contrast, the motivation that derives from sense is more action-oriented. Motivation determines the intensity of our induction to the goal. The interconnection of factors of significance and motivation was studied not only on psychological but also on physiological levels. Sense interacts with objective meaning and influences the interpretive process.

Block (12) is also connected with the block "making decision about correction"; therefore, the goal can be corrected or replaced. The presented model shows that the motivation block also influences the program of performance. The last is related to the executive aspects of motivation. Motivational block 12 is connected with block 19 (subjective standard of successful result). Under motivational influences the subjective standard of success can significantly deviate from the objective requirements. This fact explains why the quality of performance very and often depends not on cognitive abilities but on the motivational state of a person. Finally, it may be noted that positive and negative evaluation of results (blocks 17 and 18) directly influence the motivational process.

5.2.4 Motivational Stages of Self-Regulation

The model of self-regulation (Figure 5.3) gives us the opportunity to extract different stages of motivational processes. Each stage has its own specific features. Our model will outline five stages of motivational processes. Let us briefly consider these stages.

Motivation is important not only at the conscious but also at the unconscious levels of self-regulation. At the unconscious level information about external situations can interact with the present need that governs motivation as well as with set, particularly with those aspects of set that depend on instructions (goal-directed set). This emotional–motivational state precedes meaningful interpretation of different aspects of a situation and formation of a conscious goal. This motivational stage is triggered by external stimuli that are not sufficiently conscious. Information received through the orienting reflex mechanism activates dominant motivation that is adequate to a particular situation. Afferent synthesis integrates this motivational state with information from major stimuli, information from environmental stimuli relevant to the situation and information from memory. Based on the analysis and integration of this information a mental set can be formed. An important aspect of this mechanism is the formation of a motivational tendency that makes activity a goal-directed process. This motivational tendency, which is also called a preconscious motivational stage, can trigger executive aspects of self-regulation or initiate the formation of a conscious goal.

The second stage of motivation is involved in the formation or acceptance of a conscious goal. This stage can be developed after the preconscious stage of motivation is formed and the set is transformed into a conscious goal (see relationship between blocks 4 and 3). This motivational stage can also be developed through channel 1, bypassing the preconscious stage of motivation. Channel 1 is associated with conscious processing and can directly be connected with blocks 1 and 3. Function blocks 3 (goal) can be connected with block 10 (sense). Based on this connection the goal initiates motivational block 12. Block 12 influences goal formation or he goal acceptance process through feedback. This stage of motivation is also called the "goal related stage of motivation." The preconscious stage of motivation cannot only precede other motivational stages but can also function in parallel. Further the goal-related stage of motivation could be transformed into the preconscious motivational stage. This stage involves transformation of the conscious goal into an unconscious set (see connection of block 3 with block 4).

The third motivational stage is related to the evaluation of difficulty and significance of task. Through feed-forward and feedback influences between blocks 9 (difficulty), 10 (sense) and 12 (motivation) and their interaction with blocks 7 (conceptual model) or 8 (subjectively relevant task conditions) the motivational stage related to task formation, its evaluation and acceptance of the task is activated. This motivational stage is called the "task- evaluative aspects of motivation." The fourth stage is related to the executive aspects of motivation and is associated with the goal attainment process. We call this stage "the executive or process-related stage of motivation." At this stage of motivation interaction among functional blocks 9 (difficulty), 10 (sense), 12 (motivation) and blocks 11 (formation of a program of performance), 13 (making a decision about correction) and 14 (program performance) is of the utmost importance. The fifth stage of motivation is related to the evaluation of the activity result (result-evaluative stage of motivation). At this motivational stage, analysis of the interrelationship between function blocks 17 and 18 (negative and positive evaluation of result) and their connection with motivational block 12 plays a leading role. Analysis of this stage of motivation also involves the study of the interconnection between blocks 12 (motivation) and 19 (subjective standard of a successful result).

All stages of motivation are intimately connected and can be in agreement or in conflict. For example, the positive or negative evaluation of an activity result has meaning for a person only in a situation where the person is motivated to achieve the goal. Similarly, motivation to achieve a goal may be unrelated to the executive or process related stage of motivation. In this case positive motivation to achieve the goal may be combined with a negative motivation for task performance. This may be considered a conflict or contradiction between different motivational stages. One often encounters this in both work performance and learning activity. Positive motivation for goal attainment may be combined with negative motivation attached to the assessment of the task difficulty. This can lead to rejection of a desired goal. Overestimation of one's own result can lead one to ignore external evaluation, thereby losing valuable information.

According to the presented model of self-regulation, a complex relationship exists between difficulty (block 9) and motivation (block 12). An increase in the difficulty of a goal or task does not always increase the level of motivation as stated in goal-setting theory (Lee et al., 1989) . The level of motivation depends on a complex relationship between the function blocks of "assessment of difficulty" (block 9) and "assessment of the sense of task" (block 10). The relationship between difficulty and significance is presented in Figure 5.4.

Task difficulty and significance can be changed from a very low level to a very high level, hence the relationship between function blocks 9 (difficulty) and 10 (sense) can vary. For example, task difficulty can be evaluated as very high and significance as very low, as designated in Figure 5.4a.

In this situation a person will not be motivated to perform a task because he or she does not have any reason to waste a great deal of effort on a task that has no significance to him or her. If the difficulty is very low and significance is very high this can produce an optimal level of motivation (see Figure 5.4b).

If "difficulty" and "significance" are low but not extremely low, the person is still motivated to perform the task but with very low motivation (see Figure 5.4c). This kind of work is usually perceived as monotonous and boring. We can see that the different correlation between function blocks difficulty (9) and sense (10) can produce different motivational states.

We can also consider the construct of self-efficacy developed by Bandura (1982) according to which all motivational manipulations are effected through self-efficacy.

From the self-regulation point of view, if a person evaluates a goal as not being significant and very difficult due to his low self-efficacy, the resulting negative influence on motivation increases the probability that the goal will be avoided. On the other

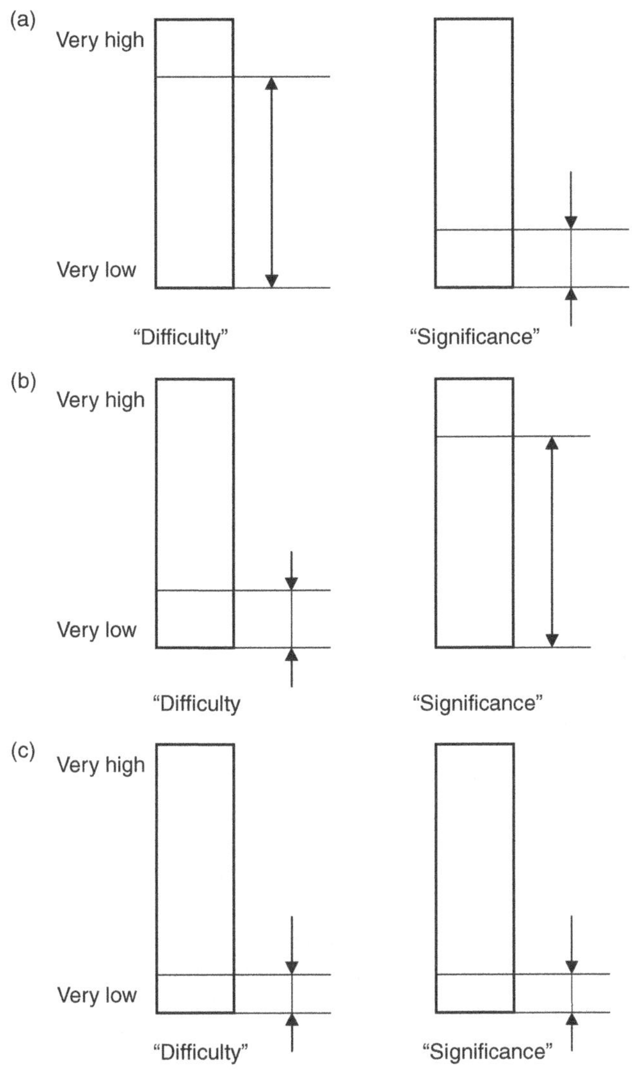

FIGURE 5.4 Relationship between difficulty and significance.

hand, if the goal is significant or highly desirable, those with low self-efficacy can nevertheless be motivated to strive for the goal. In certain situations, high self-efficacy can have negative effects. For example, athletes who regard their self-efficacy as very high may underestimate the strengths of their opponents, diminishing their motivation and consequently their performance. We can see that self-efficacy beliefs are not enough to explain motivational processes. Similarly it is not sufficient to explain motivation by specificity of the goal-formation process. Motivation can be explained as a dynamic process, which depends on the complex relationship between different mechanisms of self-regulation. Finally, a motivational process cannot be understood or analyzed outside of the social context in which it occurs. Once accepted by a subject, goals become more significant to him if they match the socially developed norms and standards of the community in which the activity is performed. Therefore, goals that match social norms and standards are strong motivational factors.

5.2.5 Executive and Evaluative Aspects of Self-Regulation

Function blocks 11, 13, and 14 are involved in the executive (implementation) stage of self-regulation (Figure 5.3). Block 11, "Formation of a program of task performance," involves the development of a program of execution of actions directed to achieve the accepted goal. This mechanism represents information regarding the method to be used in achieving the task goal and may or may not be conscious. This program is developed prior to the performance of the task or action performance and can be modified during that performance. A program of performance can comprise hierarchically organized subprograms. Some of them can be conscious, others unconscious. Some of them can be responsible for holistic activity, others for separate components of activity. The program formation stage can be complex and unfold in time or be simple and proceed subjectively almost simultaneously. The program depends on past experience (block 2), "Subjectively relevant task conditions" (block 8) and "Assessment of task difficulty" (block 9). At the same time function block 11 can very often be under the influence of function block 10 (Assessment of sense of task). This means that the emotionally evaluative component of activity can be an influence on the cognitive mechanism responsible for developing the program of task performance. The program-formation stage can be modified and in this process involves block 13. This block can correct the dynamical model of situation (block 8) and can even change the goal of the self-regulation process.

The next stage is associated with the realization of the program. Here we only stress that there are two types of executive programs: a rigid and a dynamic one. The first type is associated with stereotyped and automated kinds of activity. In such cases function block 14 is triggered almost automatically and blocks 11 and 13 are not involved. Dynamic programs are very plastic and changeable. They can be adapted through urgent reconstruction of the program in accordance with situational requirements through function blocks 13 and 11. The unconscious level of self-regulation is associated more with rigid programs. The conscious level of self-regulation is more involved when flexible and dynamic programs are in use.

The last stage of self-regulation is an evaluative stage that involves function blocks 13 through 19. This stage of self-regulation enables the individual to obtain new data regarding deviations that the individual can take into the next performance cycle. A subjective standard for successful results is critically important at this stage of self-regulation. This standard can significantly deviate from the objective standard presented through instructions. A subjective standard of success has a dynamic relationship with the goal and past experience. This standard can be modified during performance. Modification can be done through feedback from function block 13 (making decisions about corrections). In some situations, different components of a subjectively accepted goal may be utilized as a subjective standard of success, although by itself the goal does not contain sufficient information for the evaluation of the results of activity. Sometimes the formation of this function block occurs almost immediately; in other cases its formation emerges as a complicated process associated with the evaluation of interim and final results of the task performance.

In the literature, more attention is paid to proving that a subjective standard of success exists than to how this standard is formed. Konopkin et al. (1983) developed an experimental method that made it difficult for the subjects to develop such a standard. The subjects had to measure the length of an object using special instruments. Because the precision of the measuring instruments was not very great, the subjects could not conclude anything from one trial. Thus, multiple repetitions of the measurement was required. Based on observation and verbal protocol analysis, the experimenters could determine how the subjects changed their standard and what they considered a "good" standard. It appeared that they used various strategies to discover the best standard. A common strategy was to develop a preliminary standard based on the frequency of obtaining similar results in measurement. The subject progressively refined the value of the standard by discarding the results that sharply deviated from the standard because this result was considered erroneous.

In the study of positioning actions Bedny (1987) also used two conditions. In one condition when subjects miss targets, they only received information about erroneous actions. In other experiments, the subjects additionally received a shock on the left hand. It was discovered that in neutral conditions without the shock, the subjects preferred the use of most of the width of the target as a parameter of success (risk strategy). They also did not often correct their strategy after erroneous actions. The subjects in the experimental "shock" condition changed their strategy of activity, narrowing the acceptable width within the target (unrisky strategy). At the same time, they immediately corrected their own action when they approached the edge of the target. As a result, the speed of action decreased under the experimental "shock" conditions.

Social learning theory has emphasized the significance of an individual's setting of a subjective standard of success. This standard depends upon standards established by other people through a process of social comparison (Bandura, 1977,1982). Bandura showed that during the process of creating this standard, the individual's evaluation of the standard established by other people who had performed the same or similar tasks is important in self-regulation theory. The development of subjective standards of success has a dynamic relationship with the goal and past experience. Such standards can be modified during goal formation and the decision-making process.

A subjective standard of admissible deviation (block 16) is also an important evaluative mechanism. In Morosanova and Stepansky's (1982) study, subjects had to stop a pointer moving in a circle, at a specified objective stop position. Errors varied to the left and to the right of the stop position. Errors made in one area (left or right) were considered by the subject only as chance variations and the subjects did not correct their subsequent actions if the pointer stopped in this area. This area was regarded by subjects as admissible deviation. Errors in the other area (with more significant deviation) were regarded by the subjects as critical and they corrected their actions if the pointer went into this area.

Morosanova et al. (1980) studied sharp shooting and found much the same thing. It follows from these studies that the subjective standard of success, while necessary, is not sufficient for self-regulation. Personnel must define which errors are significant and which are not. This definition derives three function blocks in the model of self-regulation: "Subjective standards of successful results," "Subjective standards of admissible deviation," and "Information about Interim and Final Result." These criteria are, as was pointed out previously, not only influenced by objective criteria but also affected by past experience, individual differences and level of aspiration. This process may occur either consciously or unconsciously.

Any output that varies from the "Subjective parameters of a successful result" and exceeds the "Subjective standards of admissible deviation" will be evaluated negatively or positively (functional blocks 17 and 18). Of course, if the output is exactly what is required, it will be evaluated only positively. As part of this process, individuals must define which deviations are significant and which are not.

Most approaches to self-regulation establish criteria for the correction of influences that register only when the person obtains negative results. The model presented in this chapter, on the other hand, also enables modifications to be made in response to positive evaluations of interim or final results. For example, "Positive Evaluation of Result" may increase achievement motivation through a connection with the block "Formation of the Level of Motivation" and be reflected at a cognitive level in the functioning of "Making a Decision About Correction." This psychological process where success motivates further success marks the distinction between psychological and homeostatic self-regulation. For example, the subject's standard of successful results can deviate significantly from the objective requirements. This standard can be modified during the self-regulated process. The function block "new experience" (20) demonstrates that in a process of activity the subject gains new experience.

The mechanisms of self-regulation are not developed evenly. Multiple interactions engender various rates and synchronicities within the development of various function blocks. Some of these blocks may be developed first, others later, the rest simultaneously. Initially in the development of particular functional systems of self-regulation, the function blocks of "assessment of meaning of input information," "goal," "assessment of the sense of the task," and "formation of the level of motivation" assume great importance. Due to feedback and feed-forward, they become more precise and differentiated. Function block "Assessment of the Sense of the Task" evaluates the subjective significance not of the task (which does not yet have an objective existence), but of the significance of the goal and the meaning of input information. At later stages, other function blocks begin to develop.

In particular cases, some function blocks are not developed in precise form, nor are they all of equal weight. For example, dynamic model of the situation may be reduced in the self-regulation process. In this case we may only attend to the function block "Conceptual model." Thus, it may be seen that in practical applications professionals may emphasize certain function blocks and their relationship as well as neglect or treat as background other function blocks and their relationships.

According to Landa (1976), Platonov (1982) and others, we can extract a threefold level of regulation of activity:

1. Level of stereotype or automaticity of performance.
2. Level of consciousness of the regulation of activity in terms of acquired rules and familiar strategies.

3. Level of regulation of activity based on general knowledge, principles, and heuristic strategies.

All these levels have a hierarchical relationship. This classification can be useful, for example, when we conduct human-error analysis. Errors more frequently occur when the level of regulation of activity and conditions of performance are poorly matched. For example, in the face of unpredictable changes during performance, stereotyped methods of activity may result in errors. In conditions of low levels of predictability with complicated problems, the ability to use appropriate knowledge and skill as well as adaptive new strategies assumes greater significance. As may be seen, the ability to exploit different levels of activity and transfer from one level to another is germane to controlling the number and gravity of errors.

Here one can see that suggested levels of self-regulation have some similarity with skills, rules and knowledge taxonomy suggested by Rasmussen (1986).

The discussed above model of self-regulation of activity demonstrates that cognition is not linear sequence of the information processing steps but rather a self-regulative system. This system gives us the insights into the activity process and helps to conduct activity analyses. A model of self-regulation can be interpreted as an interdependent system of windows (function blocks) from which one can observe human performance. For example, a researcher can open a window called "Goal" and at this stage pay attention to such aspects of activity as goal interpretation, goal formation, goal acceptance, etc. At the next step s/he can open another box called "subjectively relevant task conditions". Here a researcher would study such aspects of activity as "an operative image", and "situation awareness" and their relationship. Similarly, other windows or function blocks can be opened selectively depending on the activity peculiarities. Some function blocks might be skipped all together. Interrelation of the data obtained using different function blocks, its contradictions and coordination should be also taken into account. Therefore, a self-regulative model is a very flexible tool of activity analysis. It should be applied to perform qualitative systemic analysis of activity.

5.3 APPLICATION EXAMPLES OF FUNCTIONAL ANALYSIS OF ACTIVITY

5.3.1 Selective Examples of Functional Analysis of Activity

Let us consider some practical examples. It is well-known that the apparatus most frequently used in the study of human visual perception in cognitive psychology is the tachistoscope. This is a device to present visual information for very brief periods of time. The subjects should react in different ways according to the presented information. The same way a scientist conducts experiments in engineering psychology while designing visual displays and instrumental panels, the engineering psychologist uses time reactions and errors as major criteria for evaluation of the pilot's instrument panel or different displays. However, this procedure is often not sufficient, because, it ignores the mechanisms of self-regulation.

In an experimental study conducted by Dobrolensky and colleagues (1975) scientists tried to determine how pilots interact with a display showing a failure signal. This display presented signals, by chance, between others instruments. It was discovered that there was a delay in the reaction to this failure signal. As a first step scientist concluded that this delay resulted because the existing failure signals instrument was insufficient. However, comparing different data about eye movement, response time, and the pilots subjective opinion proved that this delay had reason. The pilots voluntarily delayed their responses to these particular signals because these signals had higher subjective significance in comparison with other signals. Delayed responses to emergency signals increase the precision and reliability of the pilots' actions. This means that the delayed response cannot be regarded as evidence that the display is insufficient.

Another applied research study demonstrates the importance of the subjectively accepted goal of activity and task significance (Dobrolensky et al., 1975). The purpose of this study was to evaluate the adequacy of four instrument configurations (A, B, C, D) for landing an aircraft in zero visibility. The methodology of the study involved complex technical equipment for the experiment and sophisticated statistical methods for data analysis. The initial experimental trials suggested that A was a superior configuration and allowed pilots to land with high accuracy relative to the other configurations. However, subsequent trials indicated no significant improvement in performance with A. Landing performance with configurations B, C, and D improved significantly after repeated trials, while performance with configuration A did not improve after repetition. Specialists made the preliminary conclusion that configuration A was unacceptable. However, in subsequent debriefings pilots reported that A was much easier to use than other configurations. It was later found that during the experimental trials pilots concluded that the purpose of the study was to evaluate their ability to perform complex tasks. Pilots reformulated the goal of the experiment, concluded that task A was not complex, and therefore focused their attention on tasks B, C, and D. They downgraded configuration A because according to them only performing with configurations B, C, and D could demonstrate their ability as pilots. Here the factors of significance and goal changed over the course of the experiment. This study also demonstrates that in psychological research instructions given during an experiment can be subjectively interpreted by the participant to mean something different from what the researcher intended. These interpretations influence the strategy assumed by the participant in task performance and consequently must be taken into consideration during the design of any experiment.

Let us consider another brief example. For increasing the reliability of flight a correctly designed emergency instrument is very important. If the flight regime is disturbed the pilot must change the regular sequence of actions. He should distribute attention between the ongoing task and the emerging situation. The pilot should extract the more important components of emergency problems. However, in this situation very often correct extraction of required information, its interpretation, and a distribution of attention between ongoing and emergency tasks are violated. This can first of all be explained by insufficient functioning of the self-regulating mechanisms, "goal," "assessment of meaning of input information," and "subjectively relevant task conditions." Insufficient functioning of these mechanisms provokes inadequate response actions. This data was proved in special experiments (Beregovoy et al., 1978). Very often engineering psychologists pay attention to modality or the physical intensity of an emergency signal. However, this recommendation is not sufficient. In some cases, for example, the increasing intensity of signals can bring forth the opposite results, worsening the operator's performance.

Let us consider other examples. It is important that in different conditions flight speed should not go past particular limits. While approaching the critical parameters of speed it is important that this information is correctly reported to the pilot. Sometimes, for these purposes, additional acoustical or visual signals are recommended. But in special experiments it was proven that the introduction of these signals and regulating them according to their intensity is not correct for the considered emergency situation. So in these cases it was discovered that pilots shifted attention from the ongoing task to emergency signals. The time for interpretation and comprehension of emergency signals significantly increased. Transformation of the attention from one goal to another resulted in the pilot forgetting information about the ongoing task.

For the purpose of efficient functioning of mechanisms "assessment of meaning of input information" and "Subjectively relevant task conditions," the goal of task performance in an emergency should not be destroyed. Based on this data it was recommended that the designer place emergency signals directly into the speed indicator. When the pilot approaches the critical speed a particular area of the scales lights up in red. The intensity of light and the area of the light increase as speed increases. In such a situation the pilot is more likely to react quickly and precisely to the emergency. It was discovered that transformation from one goal to another has negative consequences particularly in those situations when the ongoing task is more difficult and dynamic.

Expectations are an important component of the dynamic model of a situation. In one experiment it was studied how expectations influence strategies of performance. The failure of the autopilot was studied under laboratory conditions and the time required for detecting the malfunctioning was measured. Later the laboratory study was compared with the same study in real flight. It was unexpectedly discovered that in laboratory conditions, when the pilot evaluates only one situation, he spends more time than in real flight. This can be explained by the fact that in real flight the failure emerges within the context of expectation. Preceding events help the pilot to develop a dynamic model of the situation in the context of an adequate conceptual model of the flight. However, in the same situations a laboratory study very often helps to develop an adequate system of expectation which cannot be observed in real flight. As a result this factor helps the pilot to detect malfunctioning much quicker than in real flight. Therefore we should always take into consideration the possible strategies of performance in real and laboratory studies.

From these experiments we can see that an understanding of the mechanisms of self-regulation permits scientists to more correctly interpret data obtained from experiments and make design decisions more efficiently. This also proves that we cannot regard an operator as a device for processing information. A person actively selects information and interprets it in accordance with the goals, significance of the activity and mechanisms of self-regulation. Even very rigorous instructions in experiments can be interpreted in different ways. This cannot be considered merely as an incorrect design of the experiment. This is explained by the fact that activity is organized in accordance with the mechanisms of self-regulation. An experimental psychologist must take into consideration possible strategies of performance during the design of the experiment. This fact cannot be passed over in silence, as is generally done.

The first example pertains to the study of goal and motivation as mechanisms of activity self-regulation. This study examined the effects of the introduction of a time standard during training on trainee performance (Bedny, 1981). The participants were blue-collar workers in a manufacturing industry in the former USSR. Vocational training methods in the former Soviet Union were centralized and standardized. According to existing requirements instructors could use time standards for performance of production tasks only during the final stages of training. These requirements were based on the assumption that introducing a time standard early in the training would reduce the quality of trainee work. However, the vocational school found that when trainees completed the program and moved on to work at a plant they were unable to perform the work at the required pace. The lower-than-expected productivity resulted in reduced salaries of the young workers, causing in turn, dissatisfaction on the part of the workers and increasing the turnover rate. After long training without time limits trainees were not able to increase the pace of task performance as required under production conditions. Introducing a time standard at the end of the training process did not produce a significant effect. The purpose of the study was to create a training method that would increase the productivity of young workers and increase their capability to perform the job at the required pace.

The hypothesis of the study was that functional blocks such as goal and the evaluative and inducing components of motivation were critically important in this problem. During several observations of the training process and chronometrical studies the researchers found that a goal without time requirements is ambiguous and imprecise. A time requirement as to when a job must be completed gives precision to the goal. However, if trainees work for long periods of time without a time standard, introducing a time standard at a later stage does not improve performance. During training without a time standard the group of trainees forms its own standards and social norms in relation to the task being performed. Time has no personal significance for the trainees and they are not motivated to turn their attention to temporal requirements at the end of the training. In other words the objectively given goal with time requirements, which is provided at the final stage of training, is not accepted by trainees. The trainees develop their own subjectively accepted goals, which do not include time parameters. Based on these findings, the researcher proposed that a precise goal of performing a task at a required quality standard and within a required time should be introduced starting with the first training session. This ensures a precise formulation of the goal (cognitive function) which will be significant for the trainee (motivational) function. Multiple experiments with different groups of students confirmed these predictions. The students changed their performance strategy and began to regulate their activity not only according to qualitative but also temporal parameters. Special training methods were developed, which facilitated among the trainees a feeling of time during work and the organization of work in time-restricted conditions. The researchers further developed methods for determining the appropriate time standard for various stages of the training process.

The dynamic model associated with functional block "subjectively relevant task conditions" is an important mechanism in the study of various other kinds of activity in addition to human work, such as sports activity, particularly, team sports. However, even in the individual kind of sport this model plays a significant role. For example, in target shooting the shooter can consistently hit the bull's eye even when he cannot see different areas of the target. Success comes from mental interpolation performed by the shooter. Interpolation in this case involves the shooter performing mental operations which find the outer points of the target (opposite points on the biggest circle of the target) and mentally computing the middle between them. Through this process the experienced shooter can hit the bull's eye even without seeing it. Thus, the shooter creates a dynamic model of the situation, which can be modified during the shooting.

Let us consider another example with gymnastics. Suppose the gymnast is performing a layout on the parallel bars. The gymnast does not focus on perceiving the exact locations on the bars where he must place his hands. The gymnast attempts to grasp a general picture of the bars position. He may not see the exact location on the bars where he must place his hands. He places his hands on the bars he does not see by extrapolation of the information from parts of the bars he does see. Through extrapolation the gymnast creates a mental image of the situation appropriate for the task. This is a dynamic model of the situation which is created quickly and just as quickly disappears. Memory mechanisms are important in the creation of a dynamic model of a situation. Templates of situations stored in memory are modified and adapted to the situation. The factor of significance is important in the formation of these dynamic models. This demonstrates that dynamic models are influenced by emotionally motivational factors.

5.3.2 Functional Analysis of Process-Related and Result-Related Aspects of Motivation

5.3.2.1 Laboratory Experiments

Studies of the motivational aspects of self-regulation permit the differentiation of process-related and goal-related aspects of motivation. In the first case, the process of work stipulates motivation. In the second case, it is connected with the achievement of the required goal. According to the model of self-regulation, the goal-related motivation could be associated with the second stage of motivational regulation. Process related motivation, on the other hand, is related to the fourth stage of motivational regulation.

Contradiction or conflict between these stages is often encountered in the study of motivation during the performance of a monotonous job. In some situations the work process itself does not produce positive emotion. Moreover, the process of work can be conveyed with a negative emotional state. In these cases in order to sustain positive motivation during the performance of a task commitment to the goal-related aspects of motivation should be activated. Another important aspect of sustained work motivation is reducing the negative motivational state connected with the process of performance.

As the object of the study we selected a monotonous task where process-related aspects of activity cause negative emotional and motivational states. During the performance of this type of task the pace of performance and productivity can be significantly decreased. Our prediction was that reducing the positive motivation and the appearance of the negative motivation is connected with violations of the self-regulation process.

The first series of experiments was conducted in the laboratory. We studied how ongoing information about performance influenced the motivation of activity and pace of performance. Subjects, who worked on special devices, were involved in the performance of the required task. They were required to perform the same task multiple times when ongoing information about the quantity of the performed operation was absent and then when this information was present.

5.3.2.2 Participants

Five subjects 25 to 32 years old participated in this experiment. All of them were male. Four of them were engineers and one was a musician. The participants were motivated to take part in the experiment because they were interested in the study of the efficiency of performance.

In order to conduct this experiment a special device was developed (see Figure 5.5).

This device was a physical model of the production operation that imitated a monotonous job. The model consisted of panel 1, which contained two positioning switches 2 and 3 and bulb 4 that lit up after manipulation of the switches in a particular order. A hand push button 5 was placed to the left. It could turn off bulb 4. A special counter 6 that registered the amount of performed operations was placed to the left of the hand push button. The count increased by one every time bulb 4 lit up. Switch 2 had number 1 above it and number 3 below it. Switch 3 had number 2 above it and number 4 below it.

5.3.2.3 Procedure

According to the instructions, the subject had to turn switch 2 forward (to number 1), then switch 3 forward (to number 2), then turn switch 2 backward (to number 3) and

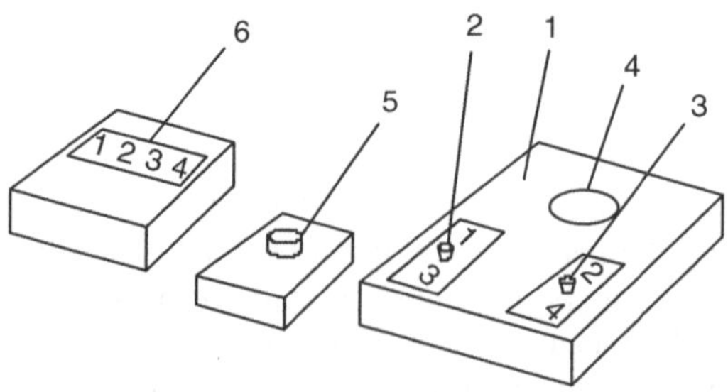

FIGURE 5.5 Physical model of production operation that imitates a monotonous job. 1 — panel with two positioning switches; 2 — left switch; 3 — right switch; 4 — bulb that informs subject about correct performance of actions; 5 — button which can turn off the bulb 4; 6 — counter that registers the number of performed operations.

switch 3 backward (to number 4) using his right hand. After that bulb 4 turned on. The subject turned off the bulb by pushing button 5 using his left hand. The operation was completed and then repeated. Therefore the goal of this task was to light the bulb and then turn it off. The operation comprised four motor actions with the right hand and one with the left hand. To perform the switching in a particular order, concentration on the sequence of the performed actions was required and did not give the subject the opportunity to count the number of the performed operations. On the first day of the performance of the task the counter was placed in a position not observable by the subjects. This information was available only to the researcher.

The lighting of the bulb could also be achieved by moving the switches in the opposite direction (from 4 to 3, then from 2 to 1). However this information had not been given to the subjects. Any other sequence of switching would not light the bulb and the counter would not count that operation. The subjects were to follow the required order of switching or the opposite order if they discovered it. They had to perform 1800 operations. For this job they had 2 h. After 2 h or after the performance of 1800 operations, the experiment was interrupted. The instruction included a demonstration of the required pace for the task's performance. The subjects were informed that if they performed at this pace they could complete 1800 operations in 2 h. They had an opportunity to take 2 or 3 min breaks several times. On the first day of the experiment all the subjects were asked to surrender their wristwatches.

5.3.2.4 Design Experiment

In the preliminary study it was discovered that on the second day productivity was reduced by about 15 to 25% if the information about the quantity of produced operations was not presented to the subjects. This was due to the increasing negative attitude toward performing the tasks.

The obtained data allowed us to design the following experiment, which was conducted over 2 days. The subjects performed actions in the required order. On the first day they did not receive ongoing information about their productivity (the counter was removed from an observable position). After 2 h the experiment was completed. On the second day the subjects performed the same task. This time the counter was placed in an observable position as demonstrated in Figure 5.5. The subjects could obtain ongoing information about their productivity any time. During both days the subjects were not allowed to wear their wristwatches. Therefore, they were not able to observe how long they had been working. The experiment was conducted with one subject at a time.

We registered the amount of tasks performed by the subject, not only during the two-hour periods but also over each 30-min period. We observed the subjects' behavior and their emotional state during the task's performance. The amount of breaks taken by the subjects as well as their verbal and emotional expressions were registered. After 1 h and 10 min of performance we asked the subjects how long they thought they had worked. When the experiment was completed we interviewed them.

The important issue under investigation concerns the extent to which the ongoing information about productivity influences different motivational stages. An analysis of the obtained results showed that on the first and second days we could observe a negative emotional state during the task's performance. However, this state was more apparent on the first day of the task's performance. As a result we observed an increasing negative motivation during the task's performance. Negative motivation produced "psychic saturation." Karsten (1928) first described psychic saturation as a psychological state connected to a strong desire to stop working. A person demonstrates a variation in the task's performance, the quality of work is reduced, the pace of performance slowed down, and finally there is a complete inability in performing a task. If the subject switched to a new task this emotional state disappeared. An analysis of the experiment allows for the hypothesis that this state is a result of the violation of the self-regulative processes. Let us analyze the results of the experiment in more detail.

On the first day, when the subjects did not receive ongoing information about their productivity, the state of psychic saturation appeared after 30 min of work. The subjects varied the pace of performance which became very quick and then very slow. They also changed the sequence of switching. During these changes the subjects discovered they could use switching in the reverse direction. After this finding the subjects started to change the sequence of switching. The amount of breaks during job performance increased (see Table 5.1, first day).

The subjects demonstrated a desire to stop working. They continued to perform tasks only after multiple persuasion and the insistence of the experimenter. Some subjects started to lose concentration and focus on other data nonrelated to task events. They attempted to examine devices, and tried to converse with the researcher. All subjects demonstrated curiosity to know how long they had been working and how many operations they produced, etc. The quantity of production operations during each 30 min constantly decreased (see Table 5.2).

A single factor analysis of the variance for each group taking 30 min works as the factors for four subjects was done. Statistical data demonstrate that changes in productivity were statistically significant (F-value $= 5.85 > F_{\text{crit}} = 3.9$ and $p < .05$). The fifth subject was excluded from statistical analysis in this case because he worked on the first day for only one hour thirty minutes and on the second day

TABLE 5.1
The Amount of Breaks during Job Performance

	Without information about ongoing productivity (first day)		With information about ongoing productivity (second day)	
Subject	Quantity of work breaks	Duration of work breaks	Quantity of work breaks	Duration of work breaks
1	13	30 min 15 sec	3	2 min 30 sec
2	18	38 min 40 sec	9	10 min 20 sec
3	15	34 min 26 sec	7	10 min
4	14	34 min	5	7 min 5 sec
5	5	7 min 30 sec	2	2 min
Average	65	144 min 60 sec	26	31 min 55 sec

TABLE 5.2
Quantity of Production Operations during Each 30 min

	Without information about ongoing productivity				With information about ongoing productivity			
Subject	**1**	**2**	**3**	**4**	**1**	**2**	**3**	**4**
1	380	340	282	248	492	628	680	—
2	456	402	382	301	586	564	650	—
3	350	365	293	203	501	612	687	—
4	400	417	370	311	486	614	709	—
5	720	820	560	—	1057	1043	—	—

TABLE 5.3
Quantity of Operations and Time of Performance

	Without information about ongoing productivity		With information about ongoing productivity	
Subject	**Quantity**	**Time**	**Quantity**	**Time**
1	1250	2 h	1800	1 h 35 min
2	1441	2 h	1800	1 h 45 min
3	1205	2 h	1800	1 h 30 min
4	1508	2 h	1800	1 h 28 min
5	2100	1 h 30 min	2100	1 h 5 min

for only one hour before he finished his work. During the experiment all subjects attempted to calculate the amount of tasks performed. However the sequence on switching did not allow them to achieve this. We could observe the deterioration in feeling, as time went by. Only one subject was able to relatively precisely determine how long he worked. Most subjects expressed their negative attitude to work. They also reported that during the task's performance they had the impression that time went by very slowly. One subject described the work in the following way. "The next operation is not different from the previous operation, it seems to me that I perform the same operation all the time." Similar explanations were given by other subjects. Because the level of motivation was low and because of psychic saturation, four subjects did not perform the required amount of production operations. Subject five was the exception; he worked almost without interruptions and after one hour and thirty minutes he had performed 2100 operations. After that he expressed the desire to stop working and the experiment was interrupted (see Table 5.3).

Analysis of the subjects' behavior showed that they tried to obtain any information about the quantity of the operations that they had performed. For example, they often asked how many operations they performed, how long they worked, how many operations they needed to perform, although the subjects knew that the experiment required the withholding of this information from the subjects.

On the second day, when the subjects received the information about ongoing productivity, (the counter was installed in an observable position) psychic saturation was expressed less frequently and increased only slowly. Psychic saturation became observable after approximately one hour of work. The general amount of breaks and their duration also were reduced (see Table 5.1, second day). When analyzing data in Table 5.1 we need to take into account that on the first day of the experiment subjects often started to work again after the breaks only after multiple persuasion and insistence of the experimenter. On the second day the subjects started to work after the breaks on their own. Productivity of work, in most cases, increased over each 30-min period. (see Table 5.2, second day). Statistical analysis demonstrates that an increase in productivity over each 30-min period was statistically significant (F-value $= 15.53 > F_{\text{crit}} = 5.1$ and $p < .05$). All the subjects noticed that the last operations were performed more easily than the previous operations. Hence, at the last stage of work the subjects were in a more positive emotional state, and the evaluation of the required time intervals became more precise. Most subjects mentioned that on the second day "time did not pass as slowly as on the first day." The task that was formulated on the second day, when ongoing information about productivity was presented, motivated the subjects much more to perform the job than on the first day. On the second day, all the subjects performed with the required standard production rate. It took them less then 2 h to perform 1800 operations.

Observation of the subjects and further discussions with them showed that the subjects checked the counter when they felt the need to sustain their motivation and regulate the pace of work. At such periods of time the executive stage of activity self-regulation was followed by the evaluative stage of self-regulation. The obtained information helped to eliminate uncertainty in the regulation of quantitative and temporal parameters of activity.

During this study, we recorded all events obtained during the observation (protocol analysis). Data obtained during protocol analysis are not presented.

A comparison of protocols obtained during our study demonstrated an advantage in work performance when a person obtained information about the dynamics of his productivity. On the second day improvement in performance was observed for all the other subjects. All subjects expressed their opinion that on the second day it was much easier and less monotonous to work. Even the fifth subject, who worked very quickly on the first day almost without psychic saturation, increased his productivity on the second day. He also stated that on the second day he felt less monotony.

5.3.2.5 Field Experiments

(A) Experiment in production environment

The experiment was conducted at a plant that produced milk churns. There was an auxiliary shop at this plant that used the scrap metal to manufacture lugs for the metal buckets. This work could not be automated because the workers used scrap metal. It was a highly repetitive task that was characterized by inherent job monotony and single production operation is very simple. The worker takes a sheet of iron scrap with her left hand, puts it in her right hand and then puts it into a blanking press tool. Then she presses the foot pedal with her left foot. A finished part is dropped into the bin. She turns the metal scrap in other required positions and presses the pedal again. One piece of scrap usually yields no more than two or three parts. Therefore, she repeats simple movements with her left and right hands and the pressing of the pedals during one cycle. By using her right hand, the worker threw the remaining piece of scrap into a special bin kept for collecting the remaining scrap. Producing one piece of a product required approximately 2–3 sec. Usually, the workers produced 18 thousand pieces of product per shift. The simplicity of the production operations and the high level of repetition caused a feeling of extreme monotony on the job. This resulted in a high level of turnover, in spite of the good salary.

During the production process workers counted the number of produced parts. After producing one hundred pieces, the workers put one part in a special collecting place. During and at the end of the shift, the workers counted the amount of pieces in the collecting place to determine their productivity. Counting 15–18 thousand pieces during each shift caused strong neural tension that resulted in mental fatigue. All of the workers felt tired at the end of the shift and complained that they were having headaches and counted when they slept during the night, causing them not to feel rested when they woke up. This was also proven by special studies of mental fatigue.

The administration at the plant attempted to evaluate productivity based on the weight of the products during each shift. However, all workers continued to count the amount of products. The workers preferred to count, in spite of fatigue, because this method of evaluation of productivity was accepted as standard. When the researcher suggested that workers perform their job without counting, most of them rejected this method of performance after two hours of work and began to count the produced pieces.

A preliminary laboratory experiment brought us to the conclusion that workers used counting because the information about ongoing productivity sharply decreased the feeling of monotony and psychic saturation. We concluded that it is impossible to perform a tremendous amount of repetitive operations without information regarding ongoing productivity. Based on this assumption it was suggested that a special counter be installed on the press which could inform workers about the number of produced pieces at any time. Workers would receive information about ongoing productivity and as a result they could stop counting the number of pieces produced. This innovation proved our assumption that for performing highly repetitive tasks information about ongoing productivity is important. For the evaluation of ongoing output workers used not only the counter but also the big wall clocks in the shop. They evaluated their productivity by comparing the quantity of product with time performance. Each worker made his assessment whenever he felt fit. Therefore, the specifics of the feedback and its timeliness were determined by the workers. After this innovation all the workers stopped counting, which sharply decreased complaints about being tired and they also stopped counting during their sleep. A special study showed that after the innovation worker's mental fatigue was drastically reduced.

(B) Time study and monotony during training process

Here we briefly describe another field experiment. This experiment studied the formation of professional skills required for performing work at a standard pace. As an example we chose the lathe process in which young workers were trained to perform turning operations on cylinder bushing. The purpose of this study was to find out how the dynamics of productivity changed during the training process when trainees were informed about a required standard production rate and when the standard production rate was not presented. We selected two groups of trainees who had similar educational backgrounds and success in performance. Each group contained six 16–18-year-old males and worked for 3 days. The experiment was conducted individually with each trainee. They worked in a real workshop and did not know they were being observed. The control group was not informed about the standard production rate. The trainees in the experimental group were informed that they needed to produce 60 bushes per day. This standard production rate was not difficult for the trainees. Nevertheless, because of the repetitiveness of the operations the trainees felt monotony. This kind of work is very often encountered in machining operations. Our suggestion was that a precisely formulated goal to perform the required number of pieces of product could have motivational influences on the trainee's performance and they could increase their productivity. The result of the trainees' performances is presented in Table 5.4. Analysis of the obtained result demonstrated that in the control group where the trainee worked without information about the standard production rate, productivity on average was 32.2 pieces per shift. In the experimental group their productivity increased to 82.3 pieces per shift. A difference in productivity of performance between the two groups during the three days according to the Student's t criterion was statistically significant ($P < .01$). It is also interesting to compare the dynamics of productivity during the 3-day period. In the control group that worked without the time requirements productivity was gradually reduced over the 3 days. In the experimental group where the trainees received time requirements productivity increased from day to day (see Table 5.4).

A single factor analysis of the variance for each group taking days as the factor was applied. The F value for the control group was not statistically significant

TABLE 5.4
Number of Turned Parts

	Control group			Experimental group		
Subject	First day	Second day	Third day	First day	Second day	Third day
1	21	20	36	80	85	85
2	40	35	35	75	80	90
3	33	30	29	85	92	90
4	48	46	33	74	72	83
5	40	42	11	70	90	85
6	29	30	17	75	80	90
Average	36.1	33.8	26.8	76.5	83.1	87.1

($F = 1.67 < F_{crit} = 3.68$ and $p < 0.05$). However, the experimental group significantly increased their performance each day, which is evident from the F-value $= 5.67 > F_{crit} = 3.68$ and $p < 0.01$. Therefore, statistical changes in the dynamics of productivity were discovered only in the experimental group.

We concluded that introducing the time standard requirements not only increased productivity, but the dynamics of productivity improved for the monotonous job performed.

(C) Discussion on obtained results and conclusion

In our experiments we selected the types of work where goal-related (1st stage) and process-related (3rd stage) motivational stages are in conflict. On the one hand, the subjects accepted the given goal (goal-related stage of motivation) that was a part of their duties and responsibilities related to their job requirements. On the other hand, the work performance itself produced a negative motivational state. If a negative motivational state during work performance outbalances a goal-related stage of motivation, the person could not only stop working but also quit the job altogether. Turnover, which was clearly observed in our field studies, has been induced by this situation.

Studies demonstrate that psychic saturation that emerges during the performance of monotonous work is caused in a significant degree by the violation of the mechanisms of self-regulation. The distinguishing features of repetitive tasks are their position in relation to the final goal. Information about ongoing productivity permits workers to distribute their energy evenly during the shift, plan the pace of performance, and evaluate their success in approaching the final goal. Therefore, receiving ongoing information about productivity plays an important role in positive motivation. The more intense the negative emotional state during performance, the more important the ongoing information about approaching the required goal. Each following result or group of interdependent results are evaluated in relation to the preliminary results obtained and in relation to the final goal of activity. This explains why intermittent goals have a positive emotional influence. Information about the dynamics of productivity improves the feeling of time during task performance and gives workers an opportunity to consciously and voluntarily regulate their activity over time. Information about the dynamics of productivity sharply decreases the negative aspects of the process-related motivational stage.

Experiments conducted with trainees also demonstrated the necessity of studing the relationship between goal-related and process-related stages of motivation. A precisely formulated goal (including its temporal and quantitative aspects) helps trainees to increase positive motivation of activity. It shows that positive goal-related motivation can predominate over negative process-related motivation. In contradiction with the opinion of Lee et al. (1989) we've concluded that at the first stage of work performance it is not very important whether the goal is challenging or not to the subjects. What is more important is how accurately this goal is presented to the subjects, in relation to the task in hand. This can be explained by the fact that subjects cannot predict whether this goal is difficult or easy for them to achieve. Existence of a precisely formulated goal (block 3) and understanding of its importance (block 10) are major factors that increase the significance of the goal and activate the subjects' goal-related motivational stage. A factor of significance (block 10 sense) emerges as an independent emotional-evaluated mechanism that is critically important in the motivational process. During the performance of monotonous work an externally precisely formulated goal and its significance become vital to the subject. Subjects can accept an externally set goal only if the goal is perceived as subjectively significant.

It is interesting to find out why productivity increases in time limited conditions and why it decreases in the absence of these conditions. The data obtained can be explained by the mechanisms of self-regulation. Trainees do not know in advance that they will attempt to increase their productivity. The formation of a more difficult goal, namely, "to produce more than is required on the first day and more than on the previous day" emerged during work performance. In order to sustain interest in the task's performance (significance of task) trainees attempt to increase the difficulty of work. Maintenance of motivation can be considered a cyclical process. This cycle can be explained in the following way. Even on the first day students achieve the required productivity fairly easily. To maintain the evaluative stage of motivation in these circumstances students increased their subjective standards of success (block 19). Through the associated mechanisms (function blocks 16 and 18) this leads to the formation of positive motivation. Self-increasing of productivity up to a particular quantitative level during all the three days emerged as a dynamic criterion of success. Students do not know its value ahead of time, and it is gained at the last stage of job performance. The level of aspiration, feeling of tiredness, etc. is very important in the formation of a subjective standard of success. This standard is subjectively associated with a self-command such as "It is enough for me," "I can stop now".

On the second day students realized that the job was performed not difficult for them. Moreover, students acquired the necessary skills during performance. This led to decreases in subjective evaluation of task difficulty on the second day. The decrease in perceived difficulty (block 9) resulted in a decreasing significance of the task (block 10 "sense"). This in turn was followed by the degradation of motivation (block 12). In order to maintain the required level of motivation, on the second day students increased the difficulty of the goal (task) again. However, even this higher result was achieved relatively easily on the second day. The subjective evaluation of the task's difficulty went down again due to further enhancement of skills, which in turn led to the reduction of significance and motivation. As a result students "raised the bar" (increased the difficulty of the task or goal) again and the cycle repeated itself. Due to this dynamic the subjects were aware that they were capable of better performance. This increased the self-esteem of the subjects because they proved that they could perform more difficult jobs. Therefore motivation (specifically goal formation, process-related and evaluative stages of motivation) is sustained and coordinated as a result of the dynamic relationship between blocks 3, 9, 10, 12, 18, and 19.

When a precisely formulated goal is absent, achievement of any acceptable result (without time standard requirements) during a simple task performance causes reduction of goal significance. This has a negative influence on the goal-related stage of motivation. Process-related aspects of motivation become particularly important here. The difficulty of the task gradually decreases due to its relative simplicity, high repetition, and automation of skills. This in turn reduces the significance of the task and then reduces the motivation (see relationship between function blocks "difficulty," "significance," and "motivation", Figure 5.3). In general, in our experiment when trainees receive a precise goal they form higher subjective criteria of success. When time-limited conditions were absent, the dynamic standard of success gradually deteriorated.

Experiments demonstrate that such function blocks of self-regulation as goal, difficulty, sense of task (significance), level of motivation, and subjective standard of success have a dynamic relationship. They are major mechanisms that determine the dynamics of productivity. The interaction of these mechanisms allows the required level of motivation to be sustained. Changes in productivity cannot be simply explained by the concept of feedback. Feedback alone cannot explain why in some cases productivity increased gradually from day to day and in other cases decreased. Self-regulative mechanisms and their relationship can explain this phenomenon. Feedback can in some cases may be inadequate to the goal of activity. This in general can lead to reduction of motivation and efficiency of performance. Furthermore, the concept of feedback is a complicated system of interdependent functional mechanisms of self-regulation. Our study demonstrates that work motivation should not be considered a separate mechanism, but a dynamical organization of different mechanisms. This study also demonstrated that cognitive and motivational mechanisms function together in the process of activity regulation.

In this section, motivation is construed as a dynamic system, which depends on the complex relationships between various mechanisms of self-regulation. We presented theoretical and empirical proof that motivation is not a homogeneous process but rather is composed of different stages. These stages play a specific role in the regulation of activity. The study of work activity has revealed five motivational stages: (1) a preconscious motivational stage; (2) a goal-related motivational stage; (3) a task evaluative motivational stage; (4) an executive or process-related motivational stage; (5) a result-related motivational stage. These stages are organized as a loop structure and in any practical situation some of these are more important than the others.

In this work we discuss the interrelationship between the set and goal and how it determines the transformation of the preconscious motivational stage into the conscious, goal-related stage and vice versa. The preconscious motivational stage can be activated involuntarily through the orienting reflex and can then be transformed into a conscious, goal-related motivational stage. This process leads to the transformation of the unconscious set into the conscious goal.

Conversely, when the current task is interrupted and attention shifts to a new goal, the former goal is transformed into an unconscious set. As a result, the goal-related motivational stage is transformed into the preconscious motivational stage. The availability of the former goal in the form of a set allows the individual, if necessary, to return to the formerly interrupted task.

The discussed above stages of motivation can be in agreement or in conflict with each other. An analysis of these stages and the illumination of their agreement and contradictions allow one to more precisely describe and analyze motivation in the context of a particular work activity. This makes the formulation of practical recommendations possible, facilitating positive motivation during human performance.

5.4 SELF-REGULATIVE CONCEPT OF LEARNING

5.4.1 Basic Characteristics of Self-Regulative Concept of Learning

At present, no single theory has been accepted as a paradigm of learning for psychology. Depending on the situation, psychologists may use one or another theory of learning. Here we present a self-regulative concept of learning derived from functional analysis of activity. The self-regulative concept of learning derives from the model of self-regulation of activity. This approach rejects the study of the stimulus–response paradigm as basic units of analysis during human learning. At the same time activity theory does not consider the learner to be an information-processing system, but as a subject who actively interacts with the situation according to the goal of the activity. The subject develops his knowledge and personality by acting or in other ways performing cognitive and motor actions. From this it follows that basic units of analysis during learning are cognitive and motor actions. Cognitive processes and actions are organized according to the principles of self-regulation. From this follow other units of analysis during learning. These units of analysis are function blocks. During the self-regulative process the subject not only comprehends or transforms the situation according to the required goal but also acquires knowledge and skills associated with the performed activity. Self-regulation is the real "machinery" of the learning process. In our model of self-regulation we have the function block "new experience," which is the result of the self-regulative process. It interacts with past experience (function block "experience"). Therefore a new experience reconstructs our past experience.

Thus, learning is treated as a recursive process which a goal-directed character governed largely by control, corrections, and regulations. In some situations learning is governed by the unconscious set. In this situation learning can, to a significant degree, be unconscious. Associative learning is an example of unconscious learning. Associations are the result of the self-regulative process. Learning derived from associations without conscious knowledge of the logical interrelationships between phenomena is based on reward and punishment, which perform informational and motivational functions. The informational function of reward and punishment is limited because there is no understanding of the stimulus–response relationship. On the one hand, reward and punishment are signals, and on the other, they play the role of motivation. When we talk about reward and punishment the motivational aspects are more important. Trials and errors observed during animals' learning is an example of simple explorative or gnostic behavior. In human learning a stimulus does not force us to react when it appears. Stimuli only promote the formulation of a goal. Later due to this, a task or problem is formulated. The activity during conscious learning unfolds in time as a sequence of distinct phases associated with formation or choosing of a goal, formation of a dynamic model of situation, formation of motivational stages, program of performance, developing subjective criteria of success, etc. From this perspective the concept of reinforcement is meaningless.

The model of self-regulation of activity is simultaneously a model of learning that emphasizes the study of the internal cognitive processes and their relationship with behavior and energetic (motivational) aspects of activity which are studied in unity.

In ambiguous situations when past experience does not match requirements, learning may be performed by way of trial and error when subjects can use conscious actions or unconscious operations and follow the evaluation of their consequences. In other situations, learning is a result of consciously regulated actions organized into strategies which gradually change during learning. Often, the purpose of instructional design is to provide a transition from learning by trial and error to goal-directed, planned learning. Through the working process and social interaction people changed the social environment and changed themselves. Here social interaction is considered a particular kind of activity, which is closely interconnected with object-oriented activity. As we demonstrated before, these two kinds of activity cannot exist independently and transform into each other. In this transitory process self-regulation is a basic process. The evaluative stage of self-regulation can explain learning much more precisely than the concept of feedback. It is the final stage of self-regulative sickle which includes different functional mechanisms. Interactions between these mechanisms are what we regard as feedback. As already stated before, self-regulation cannot be described as a linear sequence of stages as described by Bandura (1982) and presented in educational psychological texts (see Gibson and Chandler, 1988). This process should be described as a recursive loop structure with definable functional blocks that have both forward and backward interconnections.

One can view the self-regulation process as having several stages. The first is the stable stage of self-regulation, which is associated with well-developed skills. Next is the transition stage of self-regulation, which is of two types. One type is connected with switching from a well-known strategy to another well-established one. In another transition, the strategy must be transferred from the accustomed one to the new strategy with which the performer is not familiar. As one can see from the position of self-regulation, learning is a transition from one well-known strategy to a new strategy or adaptation of a well-known strategy to a new situation. The more difficult the task, the more intermediate strategies should be used. The number of trials and errors increases. This dynamic of strategies implies that learning can be described as a sequence of stages. Learning activity is constructed through a sequence of these stages.

The notion of strategy is fundamental to the self-regulation theory of learning. In this section, strategies are treated as plans for goal achievement that are responsive to external contingencies, as well as to the internal state of the subject. The notion of strategy is intimately associated with the notions of algorithmic and heuristic methods of performance. The concept of strategy is more comprehensive than that of either algorithm or heuristic. Strategies are typically a complex integration for deploying different algorithmic and heuristic methods of performance. Strategies have a dynamic and adaptive character, enabling changes in the approach to goal attainment as a function of external and internal conditions of the self-regulation system.

According to our model of self-regulation an unconscious level of self-regulation is associated with an unconscious set and unconscious operations. This learning is performed as blind trial and error. Explorative operations are triggered by the existing set. At a later stage the set can be transformed into a conscious goal and explorative operations can be transformed into conscious goal-directed actions. If a set transforms into a conscious goal self-regulation becomes conscious. It is possible opposite process during skill acquisition. Both levels can be involved in the same learning performance.

Analysis of the interrelationship between a goal and a result is also useful in understanding relationships between conscious and unconscious components of learning. Tikhomirov (1984) outlined three kinds of outcomes:

1. A conscious goal is achieved
2. During goal-directed activity persons achieve an inadvertent accessory outcome which is originally unconscious but can become conscious
3. An outcome is achieved based on involuntary behavior that is involved in the goal-directed activity; the accessory output always remains unconscious

Accessory outcomes that are not associated with the goal and cannot be expressed verbally are always unconscious to the individual. However, this result still influences the process of regulation of activity. Very often it is possible to provide transformation of an outcome into an outcome directly associated with a goal. In this case the outcome becomes conscious (Ponomarev, 1976). The attainment of a conscious outcome is considered a conscious level of self-regulation. When one achieves an accessory outcome, its results can become conscious by reorienting one's attention. In this case, the incidental learning is subsumed in the content of the goal-directed activity. This is transformed from an unconscious level of self-regulation into a conscious level. In the third case, outcomes and processes of self-regulation are always unconscious. In the first case, persons always make inferences about logical and functional interrelationships between functions and goals. In the second case, it becomes possible only after reorienting one's attention to these accessory results.

According to activity theory, one can talk about a cognitive level of learning only when the learning process is directed to achieving a conscious goal. From this it follows that learning can be voluntary and involuntary and conscious and unconscious. For example, a child can play with toys and acquire different skills and knowledge unconsciously without any "voluntary" efforts. However, major types of learning, by humans, are considered to be conscious. This type of learning suggests the existence of consciously established goals. It implies a conscious level of self-regulation that has complex interrelationships with an unconscious level of self-regulation. This last level of self-regulation can be considered an association's level of learning. As an example of a transition from an unconscious to a conscious level of learning, there is a transfer from unconscious explorative behavior (associative level of learning) into conscious explorative activity (cognitive level of learning). This activity is connected with one's conscious goal via the comprehension of a particular situation which emerges in explorative activity. At the conscious level of self-regulation exploration is performed with the help of conscious actions (cognitive or motor). At the unconscious level of self-regulation exploration is performed with unconscious operations, which also can be cognitive or motor. Classical conditioning with dogs and operant conditioning with rats are examples of the unconscious level of self-regulation. In human beings this level of self-regulation is more complicated and associated with the unconscious set.

Association can also appear as a result of a consciously developed action during its automatization. In this case, we are talking about the cognitive level of learning. From the self-regulation point of view, association is one of the results of the learning process. In order to understand why one or another association was formed during the learning process, it is necessary to study the self-regulation process that creates those associations.

One method that permits consciousness of the different components of a problem in different ways is known as reorientation. Introducing different instructions, prompts, reformulation of the goal, etc., results in the student becoming conscious of different components of the problem in different ways and developing different strategies for solving the problem. In one study Gal'perin and Sachko (1968) analyzed the student's skills in dealing with the use of a vice for fixing wooden pots. If the student secured the vice too tightly to the pot, the pot was damaged. If he secured the vice too loosely, the pot was unstable. The instructor used a verbal expression such as "Fasten the pot more tightly or loosen the grip on the pot, by turning the lever on the vice." These instructions were ambiguous. The student could not transfer the verbal instructions to muscular efforts and the instructor could not find the appropriate words to explain what the student should do. In such a case, the student acquired the correct action only after a long period of trial and error. The use of a special measuring device with a pointer eliminated the difficulty after a few trials. The researchers interpreted these results as the process of internalization but it was actually self-regulation. In this case, the student developed a subjective criterion of success based on the objective criterion presented by the measuring device. When conscious strategies are inefficient and one does not know how to change them, one starts to use unconscious strategies of self-regulation. The student without the special measuring device unconsciously clutched the pot either too tightly or too loosely and again unconsciously corrected his/her actions until the required muscle effort was attained to secure the pot appropriately. This required a long period of training. When the instructor used the measuring device with the pointer, unconscious motor processes were transferred to the conscious level.

There are other examples from gymnastics. The skills involved include many cognitive components, which may be conscious or unconscious. As in the last example, the trainer and the student had difficulty translating verbal descriptions into muscle coordination and required space orientation. The difficulty is exacerbated because of the holistic nature of the gymnastic act. Since the components of the movement are closely interconnected, it is difficult to deconstruct the act to enable the gymnast to practice an individual component.

When the trainer uses instructions such as "Straighten arm upward" or "Turn head left" it influences the gymnast's behavior on a conscious level of self-regulation (the gymnast consciously performs the action) while at the same time exerting an influence on an unconscious level. For example, when the gymnast turns his head, this immediately influences muscle tension, which by trial and error is corrected unconsciously.

Sometimes the learner acquires an incorrect technique which becomes automatized. Because of this, the orientating and the evaluative components of the self-regulative process are weakened. The learner cannot change the method of performance because the programmed components of self-regulation become dominant, which enable the learner to consciously control and regulate movements. One procedure to solve the problem is to "weaken" the automatization. The instructor starts to employ varying recommendations that may even be incorrect but that induce the learner to perform the particular action in such a way that it destroys the embedding. This is an example of influencing the unconscious level of self-regulation by means of conscious mechanisms.

According to the concept of self-regulation learning involves the acquisition of different strategies. These strategies can be described in algorithmic or quasi-algorithmic ways. Algorithms can serve as models of activity to be learned (Landa, 1976). They usually describe the general strategy of activity relevant to a particular class of tasks or problems. Algorithm-based learning pays attention to only behavioral but also mental actions. Therefore one of the major objectives of algorithmization in learning is the development of general strategies of thinking.

The more complex a task, the longer it takes the individual to find a truly effective strategy, which means that there may be a series of attempts to develop one, only to revise or discard it later on. These initial attempts are called intermediate strategies.

Venda and Ribal'chencko (1983) also conceptualized learning as a process of transition from one strategy to another. However, he did not describe psychological mechanisms that provide these transformational strategies. The presented data suggests that learning should be organized in accordance with the phases of skill acquisition, each of which in turn represents the development of preliminary strategies and the transition to more refined ones.

From the specificity of the learning process, it is very important to pay attention to orienting components of activity. The orienting component of self-regulation of activity includes function blocks such as "goal" and "conceptual model" 'subjectively relevant task conditions," "assessing the task's difficulty," and "assessment of sense of task." There is a complicated relationship among them. For example, function block "subjectively relevant task conditions" demonstrates that the system of information presented to the students should be changed depending on the stage of the learning process. This means that the orienting basis of an action continually changes during learning.

The student uses different temporal components of an activity which cannot be considered erroneous. These components have an intermittent character and are not included in the final stage of performance. All this is evidence that the system of instructions should be changed depending on the stage of the learning process. Based on this, Bedny (1981) formulated the principle of dynamic orientation at different stages of skill acquisition. Dynamic orientation includes changes in the method of presentation of indicative features or reference points in the task, and new material, changes in feedback, and conscious control of performance. This requires a dynamic system of instruction to be applied to the learning process. We consider this aspect of teaching and training in Section 5.4.2. Units of analysis of activity during learning can also change. This can be explained by the fact that the content of action performed by a student changes.

Motivational aspects of learning can be studied through analysis of the relationship between function blocks such as "goal," "assessment of task difficulty," "assessment of sense of task," and "formation of the level of motivation."

The learner is particularly sensitive to the influence of feedback in the initial phases of learning. However, in the last stages of learning, the role of feedback sharply decreases and the learner begins to work in accordance with the program of activity he has developed. To introduce conscious feedback into automated skills can weaken them. The evaluative stage of learning, designated simply as feedback according to the self-regulative model of activity, has a complicated structure. It includes different functional mechanisms such as "subjective standard of successful result," "subjective standard of admissible deviations" etc. Therefore, the concept of feedback is not sufficient in the study of learning. The role of different function blocks involved in the evaluation of performance during learning becomes important.

The theory of internalization plays an important part in the conceptualization of learning (Piaget, 1952; Gal'perin, 1966; Talizina, 1975) According to the self-regulation concept of learning, external practical and internal cognitive actions are interconnected. Because of the existence of feed-forward and feedback interconnections between these action types, the possibility of their comparison and evaluation enables active formation of both external and internal actions. External motor actions can be considered a support for the performance of internal mental actions. During the gradual acquisition of mental actions, external support is less necessary and, finally, they can be performed only on a mental plane. During the simultaneous performance of external and internal actions, they actively shape and correct each other but are not transferred from an external to an internal plane. This causes changes in the strategies of mental actions or operations. The character of these changes is, of course, influenced by external activity. The existing cycles of feedback influences regarding external motor activity influence the structure of mental operations, which, in turn, influences external motor activity. Because of this, we can use the term "comparison" of mental and motor actions during the learning process. This is one of the important self-regulative mechanisms of transforming the strategies of activity during learning. The transformation of strategies of activity during learning is not an internalization in traditional meaning, as stated by Gal'perin (1969), but rather a process of construction. From this it follows that learning is regarded not as an internalization of a readymade standard, but rather as a process of actively forming separate actions and strategies of performance. In this dynamic process, the relationship between the conscious and unconscious components of learning continually changes. In general, the learning process can be described as the individual stages through which a learner must pass. The more complex the acquired activity, the more the intermediate strategies are utilized by the student. Learning as a transition through stages and strategies requires the student's dynamic orientation. Only in simple situations, when the student has adequate past experience, can he acquire knowledge and skills, bypassing intermediate strategies.

In the foregoing, learning was described first of all as the result of the individual activity of the learner, organized according to the principle of self-regulation. However, the learning activity carries not only an individual–psychological character but also a sociopsychological character. The learner acts in situations that always include social relations. Rubinshtein (1959) noted that the psychology of the individual is always social. Through the organization of individual activities society shapes individual consciousness. Engaging in activity, the individual not only changes the situation but also forms himself. From this follows Rubinshtein's famous phrase "External causes act through internal conditions" (Rubinshtein, 1957). Even during isolated, individual learning the social aspects are present. But they always exert their influence on the learner through his activity. Individual-psychological and sociopsychological aspects of activity are always interrelated and are transformed into one another during the learning process.

At the same time, in social or group learning the strategies of learning activity are different from isolated activity. These strategies become more complex due to the greater significance of group norms, standards and goals. The self-regulation of the individual's activity is organized taking into consideration group norms and standards. Not only cognitive but also emotionally motivational aspects of activity are changed. The self-regulative processes of activity of each individual are coordinated with other individuals by a variety of methods which vary according to the specifics of the social interaction. In the course of learning groups form their own norms and goals of activity. Each individual within group learning analyzes his goals depending on the goals of the individuals and group goals. Norms and goals set by instructions in the process of learning can contradict socially accepted goals and norms of the group. As a result, in social learning the objective norms and goals given to the individual by instructions can be significantly modified under the influence of social relations. The evaluation of the achievements of one's own activity is done in the context of information about the achievements of other members of the group. Therefore, in group learning there are changes in the criteria for evaluating the success of activity, norms and standards of activity, and even the goals of individual activity. This leads to changes in the regulatory mechanisms of activity for each individual and the formation of new strategies of activity.

The process of the self-regulation of activity of each individual is modified depending on the sociocultural setting in which the individual's activity is embedded. Through the coordination of self-regulating processes of different individuals during learning, activity becomes socially and culturally situated. At the same time, activity is a result of individual construction. The social and individual aspects of learning activity are always present and interdependent.

5.4.2 Application of Self-Regulative Concept of Learning

Some specialists in learning point out that between the psychological theory of learning and practice, there exist essential differences. As Goldstein (1974) said, "There is a wide gulf separating learning theory and principles from what is actually needed to improve performance." According to the principles of learning developed in activity theory, application of theory is a major criterion for its validation. Below, we present two simple examples that demonstrate the inefficiency of the behaviorist approach.

One can evaluate the behaviorist paradigm against practical situations. Ormrod (1992), in her textbook, "*Human Learning*," provides the following example of the behaviorist approach, "Many newer cars sound a loud buzzer if the keys are still in the ignition when the driver's door is opened; removal of the keys from the ignition is negatively reinforced because the buzzer stops." What would happen if, instead of a buzzer, one used beautiful music? Maybe in this case the driver would close the door and "key leaving" behavior would be maintained by the positive reinforcement of the beautiful music. Taking the key would, under the behaviorist approach, entail self-punishment.

Whether one uses the buzzer or music, the issue is having the driver take the key. Here one can see that behaviorism fails to understand that the essential element in this concrete action is the goal of removing the key from the ignition, regardless of the means of doing so. The buzzer only recalls what the driver has to do. But in behavioristic learning theory conscious goals do not exist.

Another example is found in Wade and Tavris (1990) in their test bank from the second edition of their textbook, *Psychology*. On page 185 they provide the following question and answers.

A prisoner is released from jail early because of good behavior. This is an example of:

1. Positive reinforcement
2. Classical conditioning
3. Negative reinforcement
4. Punishment

For these authors the correct answer is negative reinforcement of proper behavior. But prisoners who are being released not only remove the negative stimuli of prison but at the same time have an opportunity to reenter society, all of which are very powerful positive incentives. In such cases behaviorist theories fail to capture the causal dynamics of the practical situation. Even in those cases when the behaviorist approach can explain with some approximation simple examples in practice, they can be more precisely described based on the concept of learning which is derived from activity theory. Any theoretical concept of learning, in order to prove its existence, should demonstrate its ability to be used for teaching and instruction.

According to functional analysis of activity the basis of even simple learned responses are self-regulative processes. For example, the salivary conditioned response of Pavlov's dog is a result of the self-regulation process. This simple conditioned response can be explained according to the model of self-regulation of conditioned reflex developed by Anokhin (1969). The model of self-regulation of conditioned reflex according to Anokhin (1962) was also described in Bedny and Meister (1997). In all these examples the self-regulation process provides formation of the desired responses unconsciously. Therefore, the basic assumption of our concept of learning is that any learning is based on principles of self-regulation. In human learning it is a much more complex process, which cannot be reduced to the concept of feedback or homeostatic self-regulative process. Our psychological model of self-regulation includes conscious and unconscious levels of self-regulation and eliminates the problem associated with ignorance of emotion and motivations. The evaluative stage of self-regulation demonstrates that those functional blocks such as "negative evaluation of result" and "positive evaluation of result" closely interconnect with "evaluative and inducing components of motivation." Therefore, reward and punishment perform not only motivational but also informational functions. The first who paid attention to the informational aspects of reward and punishment was Tolman (1932). The concepts of reward and punishment are not very productive in the study of human learning. Instead of these terms we use the concept of result. Informational aspects of evaluation of the result is always associated with its motivational evaluation. Skinner (1974) introduced the concept of reinforcement because he attempted to eliminate subjectivity in psychology. However, psychology as a science should always study the subjective world of a person in an objective way. Reinforcement can be regarded merely as a motivational factor which forces the subject to do something. In our view, it does not add anything new in comparison with the terms of reward and punishment. At the same time, the term reinforcement contradicts the principles of self-regulation where the evaluative stage of activity performance includes complicated mechanisms such as subjective standards of a successful result, subjective standards of admissible deviations, etc. All these function blocks are complex cognitive mechanisms that interact with goal, meaning, and other mechanisms of self-regulation. Two examples that we have already considered demonstrate the weakness of the behavioral approach to the study of human learning.

From a self-regulative point of view there are two levels of learning: the associationist level of learning and the cognitive. The associationist level of learning is connected with unconscious processes and set, cognitive with conscious, goal-related process. Both levels are interconnected and depend on self-regulation. The more difficult the task in the learning process, the more intermediate strategies are required and therefore the number of trials and errors are increased.

Let us consider the practical application of the self-regulative concept of learning. Vocational teaching (training of operators and blue collar workers) can be divided into theoretical and practical components. Vocational teaching, at the first step, is realized in the form of theoretical teaching and then in the form of training. Theoretical teaching targets the formation of general theoretical and special knowledge. Training is considered a process of developing new skills, knowledge, attitudes and abilities which can be shaped through practices. The relationship between these components gradually changes during vocational teaching.

In activity theory there are two basic systems of vocational teaching, the operationally complex system and the problem-analytical system (Batishev, 1977). In the first system major units of the training process are production operations, in the second system task-problems. In the operationally complex system the student as a first step acquires actions associated with separate components of tasks. At the final stage of training, the student performs a holistic production operation. He also studies how to perform different tasks under different conditions. In the problem-analytical system, which is adapted for semiautomatic and automatic systems, the most important task-problems are extracted. Each problem is divided into smaller units called "situations." The trainee studies this problem first theoretically and then during practice. Training and instruction are intimately connected with the theory of learning.

Let us consider some practical examples.

By operating different things, a person discovers essential, but hidden, connections and relationships between different phenomenan and their features. The learner performs actions not only on different things, but also on the content of his own psyche-sign, concepts and images. In learning by observation the student does not perform externally observed actions. However, he performs internal, mental actions. The student performs an observation in accordance with the goal of observation which can be regarded as a particular task-problem. The learner can actively interpret observed events and, in a similar situation, perform in a totally different way or use the same method in totally different situations. Observers develop interpretative strategies of performance. Learning by observation as an independent method is not effective in vocational training. In the study done by Bedny (1979), subjects performed tasks that involved inserting pins into the holes of a pin-board. The behavior in this task is overt. The actions and their sequence as performed by subjects may be precisely observed. Subjects were divided into two groups. One group used individual training by direct experience; the other used training through observation. Learning by direct experience was significantly superior to that of learning by observation. Two components of skill acquisition were discovered. One group of components was connected with a certain sequence of actions. The other was associated with strategies of performance of particular actions (strategies of attention, method of grasping of pin, sequence of the decision-making process, etc.). The first group of components may be learned by observation. The second group of components is cognitive in nature and cannot be learned by observation. Among the second group of components are those that are potentially subject to direct observation only through performance and direct supervision. Without supervision, the subjects during long periods of time "overlook" efficient strategies and "blind trials and errors" take place. Therefore, learning by observation should be combined with practical performance. During training and learning it is important to provide transfer from actions with real things to internal mental actions and vice versa. This methodic recommendation is derived from the principle of unity of cognition and behavior. The concepts of motive, goal, and method of performance are critical in learning. The technology of instruction in some instances may be similar. However, their theoretical basis and method of explanation may differ. The successful technology of instruction is often based on the intuitive decisions of teachers. Theories of learning in such instances enable us to explain these successes or failures and facilitate dissemination or corrections.

Sometimes orientating components of task performed by students can be very precise. However, this information can be redundant and contradict the natural mechanisms of regulation of activity. In one experiment (Novikov, 1986) a trainee learned how to apply the exact effort to a particular control when using additional orienting points in an oscilloscope. The trainee very quickly performed the required motor actions without errors. However, when the information from the oscilloscope was removed, the trainee could not perform the required motor actions correctly. When the instructor started to use discreet feedback, which included turning on a bulb only during the period when the trainee performed errors, it resulted in a significant improvement in real performance. From the position of self-regulation of the activity it can be explained in the following way: in the situation with oscilloscope performance the trainee uses only visual external feedback (external contour of self-regulation). The internal counter of self-regulation associated with muscles feedback is not activated. However, the internal contour of self-regulation in this task is critically important. In the second situation when discreet feedback with a bulb was used the internal contour of self-regulation is activated. This causes increasing efficiency of performance in a real situation.

In other situations a special simulator for acquiring filing skills was developed. This simulator can present information about the required trajectory of the file on the oscilloscope's screen. Two training methods were used. In one method the trainee immediately used the simulator. In the second method the trainee worked without the simulator and then worked with the simulator. It was discovered that the second method was more efficient. In the second method the trainee preliminarily performed incorrect actions. Later he could compare erroneous and correct actions. This helped the trainee to form a subjective standard of successful results and a subjective standard of admissible deviations. It was discovered that activation of explorative activity which was adequate to a performed task permitted more successfully developed required mechanisms of self-regulation. In general, the activation of explorative activity helped to evaluate adequate strategies of performance.

In order to increase the efficiency of learning, according to the process of self-regulation it is important to provide a comparison of different actions by the student (verbal, sensory, cognitive). For example, when a student acquires a sensory skill connected with discriminating between different colors and shades, he can use different color samples and compare them with their verbal designation. The more verbal designations that can be used by the student, the greater the discrimination between colors can be made. The principles of comparing different actions and verbalization in learning cognitive and motor skills permit the student to transfer unconscious components of self-regulation to the conscious level. The student is able to perform conscious self-control by introducing conscious components into the evaluative stage of self-regulation. This example demonstrates that the association between sensory–perceptual and verbalized components is the result of the self-regulative process. Verbalization during the training process provides a more efficient memorization of instruction and transformation of instructions into self-instruction and comparison of verbal actions with cognitive ones. Here, we can also mention the classical example in the developing of sonar-man skills. The training process involved discrimination of underwater sound, and comparison of this data with the sound reproduced on an oscilloscope. Acoustical stimuli varied and the student could compare them by using acoustical and visual information. He attempted to interpret different stimuli. During trials he received feedback and corrected his responses. Gradually the student formed an association between nonverbal sound and its meaning. Therefore an association is the result of a complex self-regulative process which includes cognitive (meaning), sensory–perceptual and motor actions, their comparison and corrections. In this example it is very important not only in the development of technical devices but also in the development of the correct exercise methods. Practitioners developed a complex system of exercises which included complex cognitive activity. This method of training cannot be reduced to a simple pairing of conditioned and unconditioned stimuli as interpreted from the behaviorists' point of view. Even Pavlov's classical experiment with the salivation response of dogs can be more precisely explained as an unconscious self-regulative process (Anokhin, 1962, 1969; Bedny and Karwowski, 2004).

Instructions, including explanation and demonstration, are not sufficient in the training process. Many components of a skill cannot be explained verbally, and aspects of acquisition of these skills are found on the unconscious level. Very often real equipment cannot be used at a particular stage of training because it is associated with dangerous practices in the production environment. In this situation training simulators are very useful. As an example we consider the simulator for the development of separate isolated skills of processing a curved surface by manipulating the longitudinal and cross feeds during training of the lathe operator (Bedny, 1981). This kind of work can arise in maintenance, repair, job-lot production and, serial production. In an apprenticeship, the novice student imitates the different trajectories of the cutting instrument (lathe tool) by turning two handles. The purpose of this exercise is to prepare the student to move a lathe tool along a certain trajectory correctly. However, in this situation, when a lathe is turned off, the exercise has little meaning for the student. The absence of orienting components of activity, or criteria for evaluation of the actions reduces the efficiency of training and the student's interest in this task. This is why a special simulator to develop this particular skill has been suggested. This simulator presents orienting components of activity, provides the required feedback, introduces motivational factors, etc. All these factors provide more effective strategies of self-regulation. A picture of it is shown in Figure 5.6. The simulator is installed on the lathe. It includes frame 9, which is fixed horizontally in a spindle and quill (11) A brass plate (10) is installed in this frame. It has a special configuration of milled flutes in it. A support device (8) was installed in the tool post. At the end of the support devices vertical followers (7) was installed. The free end of this follower was introduced into the brass plate flute. The task required the student to move the follower along the flute by turning the two handles without touching the sides of the flute. A special dashboard, with a red flashing bulb (1), was placed at the right bottom corner of the dashboard. The flashing red light of this bulb indicates an error in movement by the student (the student can see the signal about errors by using his peripheral vision). At this time a special noise was also emitted. The slower the student corrected his movements, the more errors were recorded on the meter (4). In order to count the number of passes, a second meter (5) was installed on the dashboard. When the student finished one pass, a green bulb (1) flashed. Performance time was recorded on a special clock. After a given time, the bell would ring, and the system would shut down. This meant that the performance time for the task had expired.

The experimental study and the practical application of this training simulator demonstrated that performing two or three tasks in a period of thirty minutes sharply increased the quality of processing. This simulator demonstrated that introducing the orienting and evaluative components of activity significantly increased the efficiency

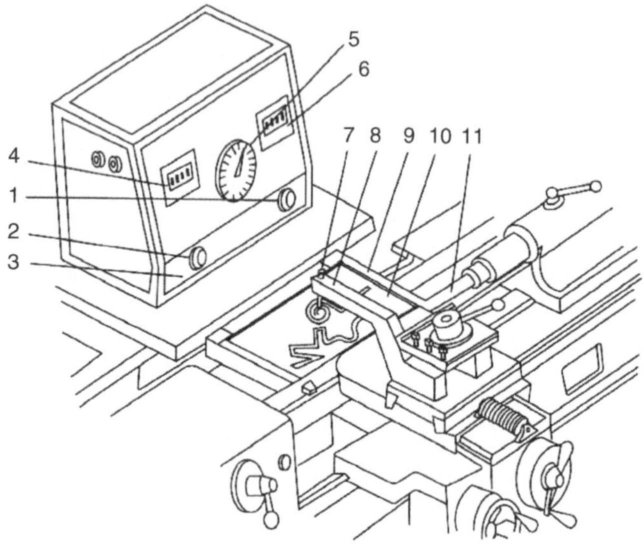

FIGURE 5.6 Simulator for development of separate isolated skills of processing a curve surface during training of a lathe operator. 1 — Red flashing bulb which indicates an error in student's movements; 2 — green flashing bulb which indicates that student finished one pass; 3 — dashboard; 4 — meter which counts number of passes; 5 — clock which measures time performance; 6 — meter which counts number of errors; 7 — vertical followers; 8 — support devices; 9 — frame; 10 — brass plate with flute; 11 — spindle.

of the training process. We need to pay particular attention to the fact that this simulator provided the precise goals of the task and increased the significance of the task. It, in turn, increased the motivation of students during training.

Krukov and Kremen (1983) have developed a method of training that illustrates the importance of the orienting components of activity in pilots' training. They developed a method called "Method of Critical Points." Critical points of a flight may be treated as a certain imaginative spot of the flight profile associated with the transition from one stage of flight pattern to another. Each critical point has its own logical structure of indicative features. The noninstrumental information such as vibrations, acceleration, etc. also play a important role. Depending upon the task performed by the pilot, the logical structure of indicative features in the same stage may be modified. Trainees study how to extract appropriately indicative features (relevant instrumental, as well as, noninstrumental information) and constitute a coherent structure. Essentially, this process is a diagnostic one. At the first stage, training is performed on ground conditions. For this purpose the instructor can use slides, movies, actual signals, vibration, noise, etc. The trainee should interpret the situation and describe the image of light verbally, and manipulate simulation materials (a combination of perceptual, imaginative and verbal actions). At the second stage the trainee studies how to perform similar tasks in real flight together with instructors. The strategy of the pilot's behavior is described as a logically organized system of perceptual, mental, and behavioral actions in which he is involved. Based on the evaluation of correct/incorrect actions and their organization, pilots acquire strategies for performing this diagnostic task efficiently and developing an adequate dynamic model of the flight.

5.4.3 Dynamic Orientation in Training

A fundamental role of training is to develop and maintain the students' adequate conceptual and dynamic models and formation of adequate strutegies of performance. In the self-regulative theory of learning the process of developing skills is represented as a series of stages of searching for and developing suitable strategies of activity. These strategies require the organization of individual activity components into multidimensional activity structures which may not all be acquired simultaneously. Sometimes the process of acquiring a skill must be divided into individual components and must be mastered separately. However, it is not always possible to divide a complicated skill into separate components. In this case the instructor should use the holistic method of training. Contradictions emerge between the necessity of using the holistic method of training and the inability of the student to comprehend all the components of skill mastery. Dynamical orientation is also important in this case. The instructor should emphasize student attention on those components of skill which are more important in a particular phase of training. In this case the holistic method of training is combined with instruction that emphasizes student attention on those components of skills which are more important in a particular phase of training. (Bedny, 1981; Bedny and Zelenin, 1988). From the self-regulation point of view it is associated with an adequate goal formation process and associated with it is the process of extraction of "subjectively relevant task condition." These functional mechanisms influence significantly the strategies of dynamic orientation. Based on specially developed instructions, the teacher can change the focus of attention to different components of the task while neglecting in some degree the other components.

Depending upon the stage of skill acquisition, different reference points in new actions are required. The reference point of activity or actions together presents an orienting basis of activity. During the transformation from one stage to another, the student changes attention to different parts of the same element of activity by extracting new reference points of actions, which correspond to particular stages of skill acquisition. Therefore orienting components of the actions have dynamic features.

An orientating basis of action can be changed depending on the stage of learning. This also causes the executive components of action to change. Based on the evaluation of the results of the actions, the learner performs the subsequent corrections, as well as necessarily adjusted orienting and executive components of actions. As a result of this loop-structure process, the structure of the acquired actions is gradually modified in the appropriate direction. All this permits the conclusion that the basis of the learning process is self-regulation. Therefore, the orienting basis of actions or activity is not abbreviated and internalized as stated in the concept of learning suggested by Gal'perin (1969) but is reconstructed based on the mechanisms of self-regulation.

Dynamic orientation is closely interconnected with the dynamic evaluative stage of self-regulation. For example, feedback about errors can be introduced based on dynamic criteria of success. These criteria can be changed depending on the stages of the training process. At this time, the student also does not have any feedback regarding errors being made in those parts of the task that should not be controlled at that stage. Dynamic orientation provides a combination of partial and holistic methods of learning together. This method is holistic because the student performs a holistic task. At the same time, it is part learning because the student concentrates on particular aspects of the task.

Novikov (1986) used this principle during the training of blue collar workers in a vocational school. He studied the hand chipping of metal. The parameters involved in this skill included the precision and force of a hammer used on a flat chisel, the method of grasping the chisel, the pressure exerted, the duration of work without pauses, muscular bioelectricity, and flexibility of the elbow. He discovered that all individual skill parameters could not be mastered simultaneously and that the interrelationships of those parameters to each other sometimes worked against each other. Thus, mastering one skill parameter sometimes resulted in the weakening of other parameters. In spite of the same instructions given to all students, they actively searched for more personally suitable methods of task performance.

Changes in the performance of skills are explained as resulting from the reconstruction of activity strategies. One recommendation from this study was to use dynamic instructions, which change according to the stage of skill acquisition: the students' attention can be reoriented and recommendations about self-control can be introduced. For example, in the beginning the student was afraid that he might hit the left arm with which the flat chisel is held. Because of that the recommendation was made to install a rod in the vice at the same angle as the flat chisel. This device prevent potential injury of the left hand. The student could regulate hammering by the use of visual and kinesthetic information. In the second stage, a firing-pin with a special surface is used. This results in a sharper deviation of the hammer after incorrect hitting. This information is more helpful in directing the students to evaluate hammering more kinesthetically. In the third stage, students use a protective device which protects the left hand when holding the rod with the regular firing-pin. This enables the students to hammer the rod with the same force and transfer their attention to the point of the rod fixed by the vice. (expert workers should look at the object fixed by the vice, but not by the firing-pin). Similarly, other stages of the training process were developed. Applying a dynamic method of orientation to the students with a holistic method of training sharply increases the efficiency of the training process.

In another experiment Khodikina and Portnoy (1985) taught students how to use a keyboard. They used different methods; the student alternated between typing separate letters and their combinations; they were required to transfer attention from precision to speed and vice versa, and from free pace to a strictly required pace. Because of these variations the speed of mastering skills increased to more than twice that of the traditional method.

The study illustrates the importance of a dynamic orientating basis of student performance. The activity goal may contain requirements that cannot be simultaneously taken into account by the students. In these cases, instructions should be changed at different stages of knowledge acquisition. These examples underline the importance of development knowledge and skills by utilizing dynamic orientation in learning situation relevant to the corresponding stage of training process. Therefore, the function block 8 on Figure 5.3 that represents mechanism of self-regulation responsible for creation dynamic model of situation is vital in learning and training.

Let us consider training in gymnastics. The whole method of training is ideally used with beginners in gymnastics, when learners perform uncomplicated skills; however this method is more and more widely used, even during the teaching of complex tasks. The advantage of this method is that the gymnast and the trainer do not have to overcome the problem of integrating separate parts into a holistic system. A disadvantage of this method is that a gymnast is often not prepared to perform the whole task especially at the beginning of its acquisition. This is why the instructors also use the part method of teaching. A disadvantage of this method is that the gymnast and instructor may not be able in the future to integrate these separate parts into an integrated system.

It is very difficult for gymnasts to pay attention to the whole learning element, especially at the beginning of the training process. A major concern of the learner at the first stage of learning of such a task is the fear of injury. In this situation the instructor orientates the gymnasts on the main parts of the elements at that particular stage of learning (different reference point). The focus of attention is thus directed to the different aspects of the movements at different stages of learning. During the explanation of the techniques of performance, the specificity of the performance is not emphasized for all the components of motor actions. Depending upon the stage of skill acquisitions, different reference points in new actions are required. During the transformation from one stage to another the gymnast changes attention to different parts of the same element by extracting new reference points which correspond to particular stages of skill acquisition. At this time, gymnasts do not have any feedback regarding errors being made in those parts of the movements that could not be controlled at that stage. Therefore, orientating basis of action can be changed depending on the stage of learning when the student performs a holistic task.

Because the holistic task performance at the first step of training is difficult for the gymnast the method "through guided assistance" is used. In this case the coach physically helps the gymnast correctly move different body parts through specific movements and, at the same time, prevents incorrect aspects of performance. While acquiring these elements, the assistance is gradually reduced and changed until the gymnast approaches independent performance. The acquisition of the proper technique "through guided assistance" should be distinguished from spotting, the major purpose of which is safety. The training through guided assistance changes according to stage of the training and the specifics of dynamic orientation of the gymnast in the holistic task. Therefore, dynamic orientation in combination with the "through guided assistance" method helps the trainer combine the whole and part method of training in sufficient complexity for the gymnastic tasks.

Parts can be extracted objectively as a separate element which becomes the task for the gymnast. At the same time, by using the notion "dynamical orientation," parts can be extracted subjectively by formulating particular goals for the gymnast and focusing his attention on specific subparts of the task. The gymnast continues to perform the holistic element, yet alters his attention to separate components. As a result of this method the instructor eliminates any difficulty associated with the further integration of separate parts of the complicated movements into a whole.

As a last example, we consider the principle of dynamic orientation of pilots during the training process. This example demonstrates the importance of the orientating components of activity. One of the important components of orientation for a pilot is his image of flight (dynamic model of situation). The pilot can of course fly, based on displayed (instrument) information, without having a sufficiently developed dynamic model of flight, but under these circumstances, flight reliability is reduced. The flight dynamic model includes not only verbally logical components but also image ones. Moreover, the imaginative components are more important. Imaginative components become particularly important when the pilot acquires flight skills largely in a ground simulator. These components of a dynamic model are not sufficiently conscious. Although the pilot has a simulated visual display (external view of the environment) and certain physical stimuli, such as engine noise, the pilot in the simulator learns how to orient himself primarily by instrument information. Learning in ground simulators requires a method of training that produces a more valid image of flight and "a sense of feeling" about the aircraft. Formation of a flight image depends upon a particular system of actions as well as upon a system of indicative features that engender the image. At the first stage of learning it is very important to demonstrate to trainees how to select indicative features relevant to a particular situation. From the perspective of self-regulation the more important function blocks providing the image of flight include "Image-Goal," "Conceptual model," and "Subjectively Relevant Task Conditions" (Bedny and Karwowski, 2004). The specific relationship among these function blocks determines the relationship between the relatively stable components and the dynamical components of trainees orientation during the learning process (dynamic and stable model of flight). The relatively stable components are "goal" and "conceptual model." The dynamical components are "Subjectively Relevant Task Conditions."

By changing the system of verbal and written instructions and by bringing the student's attention to diverse aspects of activity depending on the different stages of the training process the instructor can help develop dynamic models of the situation adequate to stages of training and to transfer unconscious elements of activity to conscious levels. Comparing different actions and particularly comparing verbal and nonverbal actions are the most important principles in this process. The described methods of training lead to a change in the relationship between such subblocks of self-regulation of activity as "operative image" and "situation awareness," which are submechanisms of the function block "subjectively relevant task conditions."

One important aspect of the part and whole teaching is the correct separation of the whole task into its constituent parts. Separation of the whole into its elements can destroy the structure of performing the holistic task. One way is that of discovering the degree of interconnectedness and interdependence between different parts. Those parts of the elements which are closely interconnected and that very strongly influence each other should not be separated during training. The student should acquire these parts together. The degree of interconnections can be determined either by expert evaluation or by statistical methods. For example, based on the correlation method, the instructor can determine the degree of interrelations between different parts of the activity during task performance. However, in practice the instructor does not use statistical methods. Usually the selection of parts is based on expert analysis. Those parts of elements of activity which are closely interconnected and acquired together by the students as unitary segments, we call "didactical unit of teaching." Any didactical unit of teaching consists of one or several closely interconnected actions. Usually they are presented as members of the learning algorithms of task. One important criterion of the separation of these didactical units is that any change to one component strongly influences changes to other components of this unit. The interrelationship between these components makes it very difficult for the student to voluntarily correct separate components of didactical units. At the same time, large influences of different elements of activity can sometimes be observed between elements that do not follow directly after each other. Any didactical units have their own reference points to which a student must pay particular attention depending on the stage of skill acquisition.

We want to stress that reference points of activity or indicative features of situation are critically important not only for meaningful interpretation of separate element of situation, but also for creation of a holistic dynamic model of situation.

All material presented above demonstrates that activity is constructed during the learning process and the major machinery for this construction comprises functional mechanisms of self-regulation. The process of acquiring knowledge and skills should be considered a process of active formation of cognitive and motor actions based on the mechanisms of self-regulation, which can be performed at the conscious and unconscious levels. As a result of self-regulation, the strategies of activity performed by a student change during the learning process. The more complex the acquired activity, the more intermediate strategies will be utilized by students.

6 Functional Analysis of Orienting Activity

6.1 BASIC CHARACTERISTICS AND MECHANISMS OF ORIENTING ACTIVITY

6.1.1 CONCEPT OF ORIENTING ACTIVITY

As tasks have become more complicated, the flexibility of activity during task performance has significantly increased. This is particularly the case for computer-based tasks. In modeling and design of flexible activity, the concept of orienting activity is particularly useful. In the following discusion, we concentrate on orienting activity. This activity is explorative, or gnostic, in nature and, therefore, very flexible. The main characteristic of orienting activity is its dynamic reflection of the situation. The cognitive processes performed serve reflective, regulative, and evaluative functions. In orienting activity, cognition initially focuses on reflective functions. Reflection may be depicted by the following scheme (Platonov, 1982).

Reflected object → Reflected system → Reflected representation.

From this scheme it follows that the result of reflection depends on both the reflected object and the features of the reflected system. Reflection can be considered the mental representation of reality. Dynamic reflection of the situation is the major purpose of orienting activity. Activity, in general, has four stages: goal formation, orientation, execution and evaluation. In orienting activity executive components are significantly reduced. Therefore, one should distinguish between orientation as a stage of activity and orienting activity when reflection of reality is the major purpose of the activity. The major purpose of orienting activity is interpretation of the situation. The major purpose of activity, in general, is transformation of an object of activity, according to the goal of the activity. Of course, this transformation requires interpretation of the situation. Orienting activity is to some degree similar to the concept of situation awareness in cognitive psychology (Endsley, 1995). However, orienting activity is a much broader concept, which includes both conscious and unconscious components, as well as motivation and other elements.

Orientating activity is performed through explorative actions and operations. The goal of such actions and operations is not transformation, but to explore the situation. More often than not, orientating activity is performed via internal mental actions and operations. However, in some cases such explorative actions can be external-motor; the main purpose of such actions is generally the manipulation of the object for the purpose of cognition. Sometimes activity, which involves the comprehension of a situation, is called gnostic activity. In dangerous and uncertain situations, such orientating activity can be stimulated almost automatically. The greater the uncertainty and stress, the more marked such explorative action tends to be. In light of the fact that such activity is often not goal directed, this could lead to a considerable loss of time and the accumulation of mistakes in the work of the operator.

Orientating activity includes the processes of decision making. However, the specificity of decision making in orientating activity is directed to features such as the detection of a signal, categorization of phenomena, decision making about explorative actions, and so forth. Consequently, one should distinguish between decision making that precedes executive activity, and decision making that is an important mechanism of orientation activity.

Sokolov (1963) examined reflection from the neuropsychological point of view. In his research, Sokolov demonstrated that neural mechanisms modeled the external world by specific changes in the internal structure. The set of changes produced in the nervous system by external objects is isometric with the changes to the external objects. These changes comprise an internal model or image of external events. This model performs a vital function in shaping and modifying human activity, predicting dynamic changes in the environment, and adjusting activity to this environment. Reflection is a constructive activity process, in which a neural model of the environment is created by the nervous system. Sokolov demonstrated that an orienting reflex is produced as a result of discrepancies found in the nervous system when information about the incoming signal is compared with the trace of an earlier stimulus. Sokolov's research gave a neuropsychological basis to the concepts of reflection and psychological model, which previously were studied only from the philosophical and psychological points of view.

In activity theory, psychological reflection is a basic concept for describing various cognitive processes. One important function of psychological reflection is to provide a dynamic reflection for a situation. Reflection of the dynamics of a situation refers both to capturing changes in the situation as well as the capacity of a subject to perceive and interpret the situation from different perspectives. Through the dynamic reflection of objects and situations, representations in the human mind encompass multifaceted sets of features and relations. Subjects thereby forecast the future and revisit the past. Dynamic reflection entails an active interrelationship among the subject, the object and the context through the deployment of various mental actions and operations. Dynamic reflection of situations enables the subject to discover different aspects of the situation, their changes over time, as well as changes in the relationships between the conscious and the unconscious in human activity. The purpose of orienting activity is consequently not any transformation of an object according to the goal of activity. Rather, the purpose of orienting activity is interpretation of a situation, comprehension of its possible meaning, and forecasting the possible changes in the situation, according to the goal of the activity. Based on this reflection, a subject can develop a dynamic model of the situation.

There are two distinct aspects to the dynamic reflection of a situation. In one aspect, subjects reflect the dynamics of a situation without any active involvement in its transformation. In the other, the subjects are actively involved in either the transformation of objects, whether material or ideal, or in the transformation of the situation in general. In the transformation of the situation, subjects will illustrate interactions of those two dynamics. The dynamics, which are not a direct function of the subject's own activity, must be included with those he or she affects in an interactive and complex fashion to the situation. It, therefore, suggests that the dynamic reflection of a situation, according to the activity theory, emerges as a complex, multifaceted phenomenon.

Orientating activity proceeds next to the performance of executive actions. It can be either an independent activity or a particular stage of activity. In the previous section, we considered orienting components of activity. In this section, we consider orientation in a situation as an independent activity. The more complex tasks are, the more often a subject is involved in the performance of independent orienting activity.

From the perspective of activity theory (AT), Endsley's concept of situation awareness (SA) coincides with some aspects of orientating activity. The mental and situational models in the study of SA are similar to the conceptual and dynamic models introduced in activity theory in the mid 1980s (e.g., Zarakovsky et al., 1974). However, the notion of SA ignores the relationship between conscious and unconscious components of dynamical orientation and the relationship between verbalized and nonverbalized aspects of activity. Conscious dynamical orientation of the situation when examined only from the point of view of cognitive psychology resulted in neglecting such aspects of orientation as goals, motives, cognitive actions, and operations. The concept of goal in AT is different from that described in SA (Endsley, 2000). In AT, the conceptual model (stable model), and the dynamic model (situation model) are results of the functioning of particular mechanisms in a self-regulative system. Besides, SA overlooks the fact that comprehension of a situation frequently requires explorative actions. The decision-making mechanism is excluded from SA. However, decision making mechanisms are always involved in the final evaluative stage of SA. One should distinguish decision-making involved in orienting activity from that involved in executive activity. The model proposed by Endsley is not logically consistent. In her model, SA is treated simply as another box in a flow diagram of the human information processing system. Boxes such as SA, decision making, and performance of actions as stages of information-processing in such models suggest the involvement of various psychic processes involved in the functioning of each box. Therefore, the box labeled information processing mechanism in Endsley's model cannot be described as an independent mechanism (box) of information-processing. The Endsley model pays more attention to the functioning of memory and attention but does not concentrate on the functioning of thinking mechanisms, which are major components in the creation of a dynamic model of the situation. We consider these mechanisms in more detail in the following section. We do not intend to deny the interesting findings obtained in SA studies. We hold, however, that AT can make a significant contribution to the study of those phenomena that are currently addressed by SA. In general, the orienting stage of activity and orienting activity provide operative reflection of reality. It provides dynamic orientation in a situation and the opportunity to reflect not only the present but also the past and future as well as on both actual and potential features of a situation. This dynamic reflection contains logical–conceptual, imaginative, conscious and unconscious components. Based on these, an individual can develop mental models of external events. Mental models have operative features and include processing of meaningful interpretations of a situation. They enable operators to understand a system and predict future states through mental manipulation of model parameters.

6.1.2 Operative Thinking and Dynamic Reflection

Thinking has a leading role in the building of dynamic models of a situation. The most important type of thinking in the study of complex man–machine systems and human–computer interaction is operative thinking. The key features of operative thinking are (Pushkin, 1965):

1. Operative thinking is directed towards obtaining solutions to practical problems.
2. Practical actions are formed and immediately analyzed through operative thinking. The performance of these practical actions, in turn, allows the subject to immediately correct and change the thinking process.
3. Operative thinking is often performed under time constraints and can be accompanied by stress. Consequently, operative thinking is studied in connection with the emotional–motivational aspects of activity.
4. Operative thinking performs diagnostic, planning, control, and regulative functions.

The following are the most important components of operative thinking: creation of a meaningful situation, structure or dynamic model of a situation (the development of meaningful units of thinking, and their structuring based on the connection between different elements of the situation); dynamic recognition; anticipation of the final situation's effect based on an already developed mental structure; formation of an algorithm of situation transformation (developing principles and rules of task solution); and development of a plan for the performance of actions (developing and determining the sequence of actions in a particular situation).

The content of verbalized or nonverbalized thinking actions and operations is important in the study of operative thinking. Actions are the conscious elements of the thinking process, while operations are the unconscious elements. During operative thinking, problem solving and task performance very often are based on visual information, and thus eye movements become an important component of the thinking process. Consequently, use of eye movement registration is widely spreading in the study of operative thinking. AT distinguishes between perceptual and thinking eye movements. Eye movements involved in the thinking process are explorative in nature and are directed at extracting the meaning of the situation, as in mentally reconstructing the situation, and so forth. The contrast between eye movements involved in perception and thinking is discussed later in this work.

Thinking is an uninterrupted process of moving from one mental action or operation to the next (Rubinshtein, 1973; Brushlinsky, 1979). In creative processes, the sequence of such actions and operations cannot be predefined. As a result of the thinking process, there is a transition from one meaning of a situation to another, which leads to a deeper comprehension of a situation. Thinking actions perform the functions of analysis, synthesis, comparison, and generalization. Analysis is the extraction of the various facets, elements, properties, and relationships of the object or situation. Analysis breaks up the object or situation into various components and extracts those aspects of the object which are subjectively most significant. This subjective processing suggests thinking as the emotional–motivational aspect of activity. Analysis extracts specific units of the objective content of reality. As the action and operation content of analysis changes, the type of content extracted from reality changes as well.

The operations and actions of analysis are always related to the operations and actions of synthesis. Synthesis connects the extracted content of reality with previous knowledge stored in memory. Rubenshtein (1973) termed the interrelationship of analysis and synthesis as "analysis through synthesis." The operations of comparison and generalization are dependent on analysis through synthesis. Analysis through synthesis is a necessary condition for relating the objective content to a symbolic form, or relating two symbolic forms to one another. For example, to extract a specific property from an object and transform it into symbolic form that property is analyzed and then connected to another known property, which serves as the standard. Both real objects and the corresponding symbols can be manipulated in this way. The symbolic form itself becomes the object of examination, and its elements can later be analyzed and integrated as real objects. For a sign/symbol, however, the most significant aspect is not its physical characteristics but rather its meaning. The meaning of a sign is its objective characteristics, which are connected to the particular sociohistorical development of a specific society. The meaning of a sign is disclosed during comprehension activity, where thinking processes are of leading importance. Data obtained through symbol manipulation can then be fixed in new sign forms. Sign forms are then separated from the objective world and integrated into formal knowledge. As a result, along with the actions and operations with real objects, the subject is able to perform actions and operations with a sign system. The close interconnection between practical actions with real objects and symbolic actions with signs are one of the distinguishing characteristics of human activity.

Synthesis makes possible the mental operation of comparison, which allows for generalization. Generalization occurs based on two broad groups of features, those based on similar (nonessential) features, and those based on essential features. Generalization based on nonessential features can lead to incorrect solutions to a problem. In some cases, the notion of "indicative features of a situation" is used instead of the notion of essential/nonessential features. Essential indicative features of a situation or object are critically important in solving ill-defined diagnostic problems. Such tasks can be accomplished by using diagnostic algorithms or heuristics that are sometimes referred to as human identification algorithms or heuristics (Landa, 1976). The identification process can be performed at the perceptual level and at the conceptual level. In the study of operative thinking, the latter becomes critical. The essence of the identification problem involves a search for indicative features essential to particular problems and developing general rules for checking their presence or absence. Based on these procedures, the subject can make a conclusion about the connection of an object or phenomena to a specific category. Thinking actions and operations are always organized in accordance with the goal of the thinking process. Through these actions and operations, the subject forms a system of hypotheses about the situation. Each hypothesis is compared with the present situation and with past experience. Based on this comparison, the hypotheses are either rejected or accepted. If the subject does not reach the goal of activity during the problem solving process, the problem is subdivided into separate subproblems, each of which has its own intermediate goal. Attainment of success or failure is controlled by the influence of feedback about the course of the thinking process. This presupposes having criteria for the evaluation of the thinking process and its results that are often unavailable prior to actually solving the problem. These criteria must thus have a hypothetical character and need to be able to change throughout the course of the thinking process. As a result, feedback connections are dynamic and changes during the thinking process, and thinking as a whole, emerge as a dynamic and self-regulating process.

Research on operator activity demonstrated that a situation can be presented to a subject in the form of a group of discrete data. In such cases, the subject extracts a series of identification features from the situation. The subject determines the meaning of the situation and formulates the problem based on the interrelationship of the features discovered. The situation, which is objectively presented as a discrete group of data, is mentally reconstructed into a whole. The subject formulates a solution plan based on the received data. The thinking process reconstructs reality into a dynamic model of the situation. The function block "subjectively relevant task conditions" is responsible for the construction of such models. The functioning of this block involves not only thinking mechanisms but also perceptual processes and mechanisms of memory. Particularly important in this situation can be schemas and scripts which represent the cognitive structure of human memory. Schemas and scripts are developed through experience and exert specific effects on the way information is encoded, stored, and interpreted (Rumelhart, 1975; Norman, 1976). They are involved in the creation of mental models of a situation. With the help of mechanisms of operative thinking, schemas and scripts are combined with each other and with new information. These modified schemas and scripts, together with imaginative components, provide basic frameworks for the creation of a dynamic mental model, which is only developed for temporary use. Situational models can then disappear from memory after achieving the required goal. Gradually, during repetition of similar situations, schemas and scripts existing in memory are modified, or new ones can be developed from them. Schemas and scripts and mechanisms of operative thinking are important components in the creation of mental dynamic models.

The function block subjectively relevant task conditions allows the subject to form several representations of the situation that differ in their subjective significance based on the same objectively presented situation. As a result, the mental reflection of a situation is not merely the mirror reflection of reality. Rather, the mental reflection depends on the dynamic activity of the subject, which is organized according to the principles of self-regulation.

6.1.3 The Gnostic Dynamic

The study of dynamic reflection of reality has a long tradition in activity theory. Pushkin (1978) introduced the important notion of dynamic reflection, which is

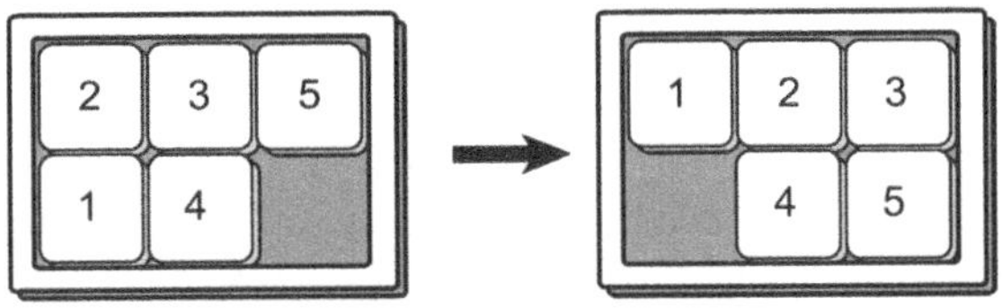

FIGURE 6.1 Combinatorial party game, "5."

called the situational concept of thinking. He concluded from his studies of visual operation tasks that such tasks embody problem solving components as their major content, the texture of which calls for more visual than verbal activity. The task conditions represent a stable system of elements, with some dynamic elements embedded. The goal of the task for the subject is to transfer initial placement of dynamical elements into the required pattern. As a laboratory model of this task, Pushkin (1978) selected a popular combinatorial party game, "5." The game consists of five plastic counters, fitted into a grooved plastic frame, to fill five of the six possible positions within that frame. The objective of the game is to rearrange the initial rectangular, random array of 5 elements into an array of elements in a natural numerical sequence. Subjects perform the rearrangement through the continual movement of the elements into the "open" position within the frame. For example, the initial position on the left should be transferred into the required position, as shown on the right (Figure 6.1).

The experimenter carefully observed the sequence of transformation from the initial to the final position. Pushkin videotaped the subjects' eye movements and supplemented the videotaping with observations of the subject's strategies of performance and interviews. As a result of this study, he was able to isolate eye movements that are components of perceptual actions from eye movements that are components of thinking actions. "Thinking" eye movements accompany the subjects' efforts to extract from a system of dynamic features those that reflect the relationship among diverse elements relevant to the required transformation of the initial situation, into the final situation. It corresponds to a required goal of solving the problem. At the stage of thinking movement, perceived elements are transformed into objects of thinking activity, which are termed situational concept. Qualitatively different "perceptual" eye movements accompany identification of colors, shapes, positions, etc., which are important for the perceptual process. In contrast to perceptual objects, situational concepts cannot exist in isolation. Each feature that is included in the structure of the situational concept may be discovered only in relationship to other components of the situation. Situational concepts do not have stable features but rather are adequate for a particular situation at a particular time. The situational concept has dynamic properties. At the first stage of the thinking process, a situational concept can be inadequate for the goal of problem solving and will affect the subject's structure during the solution process. The strategies for extracting different situational concepts, their reconstruction and transformation into other situational concepts, through which subjects develop a conceptual–dynamical model of the situation, is called a gnostic dynamic. It is important to note that the gnostic dynamic can be accomplished in the context of an external situation, which does not change. Features of the same elements may be actualized in various ways as a function of the particular situation, which enables operational specificity in representing a situation. Pushkin (1978) introduced a notion of gnostic self-regulation, which he discovered during his study of chess players. A situation is reconstructed on the basis of the results of a preliminary analysis. In a subsequent study, Pushkin (1978) used a Vietnamese chess game, consisting of counters with special hieroglyphs on them. The distinctive feature of the task in this game is the complexity of the required perceptual and intellective activity. Changing the instructions from asking the subjects to recall the position of the counters requires them to solve a more complicated, logical chess problem. It offers the opportunity to isolate perceptual actions from thinking actions. The researchers found that systems of situational concepts reflect dynamic relationships between the features of the situation elements. During the gnostic dynamic process, there is a constant extraction of different features of situation elements which is followed by determining the relationship between features of these elements and features of other situation elements. This constitutes the process of analysis through synthesis, described previously by Rubinshtein (1973). Thus, in intellectual, thinking activity, gnostic dynamic is a major aspect of activity with an external perceptual activity to accelerate the thinking process. The gnostic dynamic differs at the sensory–perceptual reflection of a situation, from conceptual logical thinking. When information about the situation is presented adequately to the operator, the gnostic dynamic is more efficient.

The results obtained in these laboratory studies were later adapted to explore the thinking processes of the operators in a practical situation. One example of such a practical task is that of a railroad dispatcher. The tasks that must be solved by an operator, similarly, have a problem-oriented character, and they are performed primarily on the basis of visually presented information, supplemented by verbal information. A visually observed situation, implemented on an instrument panel, is presented over time and in space, and may be represented as movement of dynamical units in a fixed structure. In a natural situation, opportunities to observe the field directly are strictly limited. Based on the dynamic information received visually on the panel, the dispatcher performs mental actions and operations that enable him/her to coordinate railroad movement. As may be seen, in this task one can extract movement of material objects (train) on a fixed structure (railroad), an internal psychological plan of the movement developed by the operator. It was discovered that successful performance of tasks depends on the operator's specifics of the gnostic dynamic. This task is distinctive, insofar as it entails remote control of a moving object, based on an informational model presented to an operator on a panel. The data presented on the panel is referred to as the informational model of the situation. Based on this model, operators develop their conceptual and dynamic models of the situation. The thinking process is fundamental in developing these models. Another distinctive feature of this task is that a train moves almost continuously, except for occasional, required stops. However, information about the continual movement is discrete in nature. Here, we observe a contradiction between continual movement of a controlled object and discrete information about this movement. In this situation, one important aspect is the protection against overload from excessive information, or, on the other hand, confusion because of insufficient information. The control panel is considered a specific external tool for operative thinking that interacts with the internal mental tools of an operator's thinking process. Strategies of the thinking process of different dispatchers were compared, from novice to expert, and from deficient to expert problems solvers, using verbal protocols, eye movement registration, error analysis, time performance, etc. Based on this data, different criteria and recommendations for control panel design and methods of operator's training were derived.

6.1.4 Conscious and Unconscious Components of Dynamic Reflection

Following Pushkin's death, Tikhomirov (1984) elaborated Pushkin's studies. He underscored that in gnostic dynamics, unconscious, nonverbalized aspects of activity are crucial. These involve diverse operations, which are outside of the subject's awareness. Based on these operations, subjects in the same situation unconsciously extract different situational meanings. Tikhomirov referred to such unconscious features of situation elements, and the situation as a whole as it is extracted by the subject, as the nonverbalized, situational meaning. This notion is similar to that of situational concept. He also discovered the existence of nonverbalized operational meanings of separate elements in a situation as well as the nonverbalized operational meaning of a situation in general. The specificity of extracted meaning affects subsequent dynamics of mental actions. The meaning of the situation, and its elements, develop as a result of a sequence of steps that configure the same elements in different systems of relations. The meaning of a situation may be comprehended through either conscious mental actions or through unconscious mental operations. Extraction of nonverbalized operational meaning, with the help of thinking operations, becomes propadeutic to problem solving. Such unconscious, propaedutic feature assumes relevance to subsequent steps that entail conscious thinking actions and conscious (logical) solutions to a problem. Pushkin (1978) discovered two stages of the complex problem solving process. The first is an unconscious stage, which involves the nonverbalized operational meaning; the second is a conscious stage, which involves promotion of a conscious hypothesis and problem solution.

There is also a stage directed to the exploration of the problem situation, which refers to understanding a situation as given, without changing the situation or solving it. Such explorative activity may also take place at a nonverbalized level outside of awareness. Following an unsuccessful attempt to solve a problem at the second executive stage, the subject typically returns through a feedback process to an exploratory phase. On the subsequent stages of the task solution, part of the nonverbalized meanings are transformed into verbalized and conscious meanings. Unconscious, nonverbalized meaning is always broader than its verbalized conscious equivalent. At the same time, the nonverbalized meanings of different situation elements, or the situation as a whole, do not capture the entirety of the objective meaning of the situation.

An analysis of the work by Brushlinsky (1979), Pushkin and Nersesyan (1972), Pushkin (1978), Tikhomirov (1984), Bedny and Meister (1997,1999), and others, leads to the conclusion that activity is generally regulated immediately, on both the conscious and the unconscious levels. These levels are interdependent and transformed into one another. The conscious levels of activity are mobilized around goal-directed actions, rendering goals of actions always immanent in consciousness, even if only for a very transitory duration. When actions are performed, they are often rapidly lost to consciousness. At the nonverbalized level, activity is mediated by unconscious operations that are not organized into independent actions. The particularity of these operations is not dependent on identifiable, discrete actions and their goals. While operations function automatically, their mobilization is affected only by the ultimate goals of a task. One way of transforming from the unconscious to the conscious levels of activity is the organization of automatic operations into holistic actions through the defining goals of actions. Conscious levels of performance are thereby mediated by goal-directed conscious actions, and unconscious levels are mediated by automated operations. In the latter case, iconic and symbolic transformation is performed without a verbal equivalent. Unconscious activities serve two general functions. On the one hand, the unconscious level of activities is immediately implicated in the conscious levels of actions, and these two levels are organized into a holistic activity structure. On the other hand, unconscious levels of activity are not implicated in basic conscious activity but are parallel processes providing alternate directions in the face of obstacles or distractions.

The study of an operator's visual thinking and unconscious components of activity affects both theoretical and applied studies of the imaginative processes and also development of the psychosemiotic approach in the study of nonverbal language of a visual nature. In this vein, Brodesky (1998) distinguished real space from psychological space. He cited a number of examples of verbal expression of psychological space such as "line of behavior" similar to the commonsense notions of "straight shooter" as opposed to a "crooked" actor. He also introduced the notion of a "circle of interest" or "area of activity," and several others. These are all verbal signs constructed as geometrical metaphors that seem to express the experience of reality. However, there is also a system of nonverbal visual signs that similarly express the experience of reality. Brodesky labeled this as "system toponem" after the concept of topologies of a set. Basic orientation in space entails a feeling of a position in space that is transformed into a sign system (toponem), which consists of nonverbalized meanings. Toponem possesses all the features of sign systems: (1) syntaxes — reflecting relationships among toponem in psychological space; (2) semantics — reflecting a particular complex of stereotypical associations connected with each particular toponem; and (3) pragmatics — connected with emotional influences of an individual. The distinctive feature of the toponem sign is that it does not possess a specific visual, material form but is rather a position within a psychological space. This position, or toponem in psychological space, determines the semantic meaning of such unverbalized visual signs. Orientation in space, formation of imaginative forms of thought, includes within itself a system of toponem as nonverbalized visual signs pertaining to psychological space. For example, one of the basic toponems in pilots flight is vertical position. Around this toponem, spatial images as a whole are developed as a complex of feelings defined by the relationship to the force of gravity. During take-off or acceleration pilots correct for the felt "toponem" induced by "g" forces in order to guide the aircraft properly in relation to the earth's gravity. This results in the changing of the coordinate system of psychological space, which accounts for many visual illusions. In such cases the interpretation of spatial positions is performed at an nonverbalized visual level. Pilots can overcome such illusions with voluntary effort and intellectual evaluation even in connection with verbalization. We conclude that information about space, where imaginative thinking plays a prominent role, calls for a different system of signs. Speech reflects one sign system, which has verbal meaning. Other systems of signs, such as toponem, have visual natures and are connected with nonverbalized meanings. Thus, while in his work Vygotsky placed emphasis on the importance of the verbal sign system and its connection with consciousness, in activity theory nonverbalized sign systems and their connection with intuition are, at present, also important. The connections between these sign systems needs to be considered in the discussion of situation awareness.

The above presented material demonstrates that a subject has the ability to develop a mental dynamic model of a situation. This model has conscious and unconscious as well as verbalized and nonverbalized, or imaginative, components. In orienting activity, the function block "Subjectively relevant task conditions" is the mechanism responsible for the final result of orienting activity and the creation of a mental model. Based on interrelationships with other blocks, this mechanism produces a dynamic model of the situation. The study of the function block "Subjectively relevant task conditions," addresses a number of specific aspects of the process of subjective representation. These include the role of dynamic models of the situation in an operator's activity, the relationship between verbally logical and imaginative components as well as between conscious and unconscious components in this situation. Further, this mechanism describes the basis for transforming the verbal and conscious to imaginative and unconscious processes as well as methods for representing information to an operator, methods of training to requisite transformations of imaginative into verbal ones, unconscious components into conscious ones, and vice versa. Further, it describes the relationship between extrareceptive and interreceptive source in the creation of the dynamic representation of the situation. The logic of transformation from one representation to another, the relevant inventory of dynamic images for a particular situation, is also described. The logical organization and structuring of the external features of the objective situation and its adequacy for an operator are also subsumed under this function. The role played by signals from instruments and noninstrumental signals in developing dynamical models as well as the role played by operative units of thinking and memory in the creation of models should also be studied. Is it possible to transform unconscious components of a dynamical model into a conscious verbal level? Does such a transformation resulting in deautomization and overloading working memory degrade performance? How can different stages of transformation from conscious levels to unconscious ones and vice versa be accomplished? Transformation into an unconscious level is connected with the development of skill and automatization. Can automatization catalyze undesirable consequences, because of rigidity and stereotyped recognition and interpretation, to novel and variable situations? What kind of explorative actions and operations are implicated in the development of a subjectively relevant task condition? Does interpretation of the situation call for executive actions, with risk of untoward outcomes in a trial and error process, before the situation can be clarified? How does corrective feedback correct this dynamic model, and how does the restructuring of the dynamic model increase the information about the situation? The described above material covers the most important points of the study of functional block "Subjectively relevant task conditions."

6.1.5 Emotional Motivational Components of Situational Reflection

Many studies in psychology, across numerous subfields of psychology, demonstrate that emotions affect cognitive processing. It is crucial to understand how emotions interact with higher mental functions, particularly thinking. Vasil'ev et al. (1980) experimentally studied the ongoing interaction between problem solving and emotions during chess play. They adapted a method for registering emotions, utilizing the galvanic skin response (GSR), to the index of emotional reactions to problem solving. The experiment used audio tapes, which recorded discussion of the strategies required to solve complicated chess playing tasks and, ultimately, achieve checkmate. The subjects were restricted from moving certain pieces, and their GSR was registered during problem solving activity. Verbal protocol data was compared with the GSR. Following the experiments, the experimenters discussed with the subjects their psychological states during the problem solving sessions, thereby comparing the problem solving task with the GSR. Emotional activation accompanied critical moments in task solution, and, moreover, emotional activation became more salient during the stage just preceding the performance of important verbal actions. In other words, emotional activation, as indexed by the GSR, precedes the verbal formulation of solutions. Actions and elements of situations that are crucial to the solution are seen as subjectively significant. Subjective significance, or subjective value, emerges as a reflection of the objective value of an element or a situation (Bedny and Karwowski, 2004). Thus, the factor of significance assumes profound importance in solving cognitive problems, and is accompanied by changes in the GSR. Emotional activation correlates closely with the nonverbalized level of situation reflection, preceding and preparing emotional activation to the nonverbalized solution. Only later can the nonverbalized solution be raised to a conscious, verbalized level. In one set of experiments, Vasil'ev and his colleagues (1980) introduced an instruction to remain calm during the problem solving session, which thereby reduced the efficiency of problem solving.

In one experiment, these psychologists measured the touch sensitivity of blind chess players using motion detecting devices. The subjects touched figures using only the index finger and thumb, while wearing a small ring on their index finger. The trajectory of their movement was registered on videotape. At any particular moment, a subject may touch only one piece, thereby precluding scanning of the chessboard. The data obtained was compared with the GSR. Figure 6.2 is presented as an index of these subjects' activity.

At the top of the figure are depicted strategies of sense of touch by fingers, using the symbolic code of chess pieces and positions. The middle of the figure indexes the GSR, and at the bottom, the player's verbal expressions are indexed. Emotions mediated the solution of the problem onto a conscious level. At the same time, this experiment revealed that the transition of nonverbal reflection of the situation into verbal expression occasionally takes place in the absence of emotional mediation. Whether or not emotional processes are involved depends on the complexity and significance of the task. In general, the study demonstrates the close interconnections between the conscious and unconscious aspects of cognitive activity as well as the close interconnection between the intellectual and emotional process and motivations

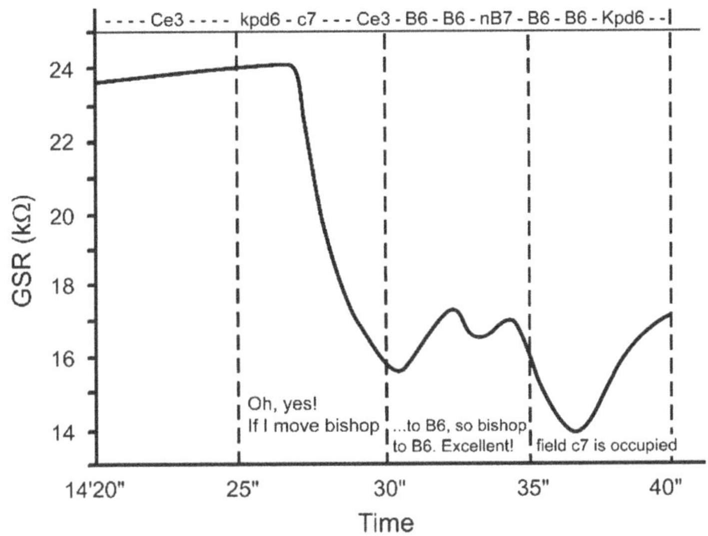

FIGURE 6.2 Galvanic skin response (GSR) during task performance.

in general. Under intellective emotions we understand how emotions contribute to intellectual processes.

Analyses of material presented in the previous work demonstrate the influence of motivational mechanisms for self-regulation of activity on orientation in surrounding reality. This is manifested in changes of afferent syntheses, goal formation process, dynamic orientation, etc. For example, Kotik (1978) demonstrated that the accuracy with which a pilot reads an aviation instrument often depends more on the significance of the instrument than on the visual features of the instrument. For example, the aircraft's altitude indicator has a rough scale; the distance between scale elements is about 5 mm and the distance between the numerals is about 15 mm. This instrument is very significant for pilots since they learn to read the horizon of the aircraft with an accuracy of about $\pm 1.3°$. The precision is even higher than that of their readings for other displays, which have a more detailed scale (Kotik, 1978).

Quite often, the criteria of activity evaluation are not given in advance in a ready unchangeable form. They can be developed and changed during performance based on a self-regulation process. This, in turn, leads to a change in orientation strategies. Not only cognitive, but also emotional evaluative factors (significance) and motivation influence the criteria of orienting activity. For example, the significance of a goal for an operator can be very low. Increasing task precision criteria in such situations may be ignored by the operator, who may develop subjective criteria of precision. These criteria may be lower than the objective requirements for precision, resulting in lowering the quality of task performance. One can influence the acceptance of precision requirements by increasing the significance of the goal. From this it follows that cognitive and emotional motivational processes are interdependent and this interdependence is critical in situational reflection.

6.2 FUNCTIONAL MODEL OF ORIENTING ACTIVITY AND REFLECTION OF REALITY

6.2.1 Automated Level of Self-Regulation of Oriented Activity

In this and the following section we consider some function blocks of self-regulation which we discussed before. However, they are now discussed in a more detailed manner. Moreover, we consider them in the structure of orienting activity, where they can function in a more specific way. In orienting activity there are also two levels of self-regulation.

The Automated Level of Self-regulation (ALSR) is that level in which the leading methods of self-regulation are nonverbalized and unconscious. At this level, the conscious and verbalized aspects of self-regulation play a subordinate role but are not excluded. This level of self-regulation is important in the analysis of the activity of the human operator, particularly where imaginative and nonverbalized strategies of activity are important.

In the same mode as proposed by Endsley (1995), we consider only those activity components that precede the performance of executive actions. In this section, we describe a functional model of self-regulation, where the unconscious level is dominant. The functional analysis allows formulation of models of self-regulations. These models are dynamically organized and directed towards a particular purpose (Anokhin, 1962). We avoid using the term goal in this situation because the final end state to which behavior is directed can be completely unconscious, or at least not precisely realized by the subject. The major unit of analysis is the functional block, which represents a coordinated system of subfunctions, with a specific purpose within the structure of activity. Arrows indicate the most significant connections between functional blocks. The functional model of orienting activity with a dominant automated level of regulation is presented below (Figure 6.3).

The dashed box "informational model" is not a function block. This is simply information presented to an operator. The informational model includes information organized in a particular fashion and transferred to the operator via specialized devices. The model presents the human operator with both information relevant to the particular situation and irrelevant information. In Figure 6.3, the relevant (incoming) information is designated by arrow a, while the irrelevant information is denoted by arrows b and c. The human operator must actively extract only that information which is relevant to the particular task.

We will discuss the concept "informational model" in more detail later. The situation is complicated by the presence of contradictory information. Slanted arrows 1 and 2 indicate the presence of noninstrumental information, which the operator must compare and connect with important information. Even the first stage of information reception is a complicated system of explorative actions and operations.

At this stage of information reception, the functional block "Reception mechanisms" (block 1) is of fundamental importance. This block is composed of two interdependent subblocks, the "sensory perceptual mechanism" (1a), and the "mechanism of orienting reflex" (1b). Block 1a is responsible for the formation of sensory

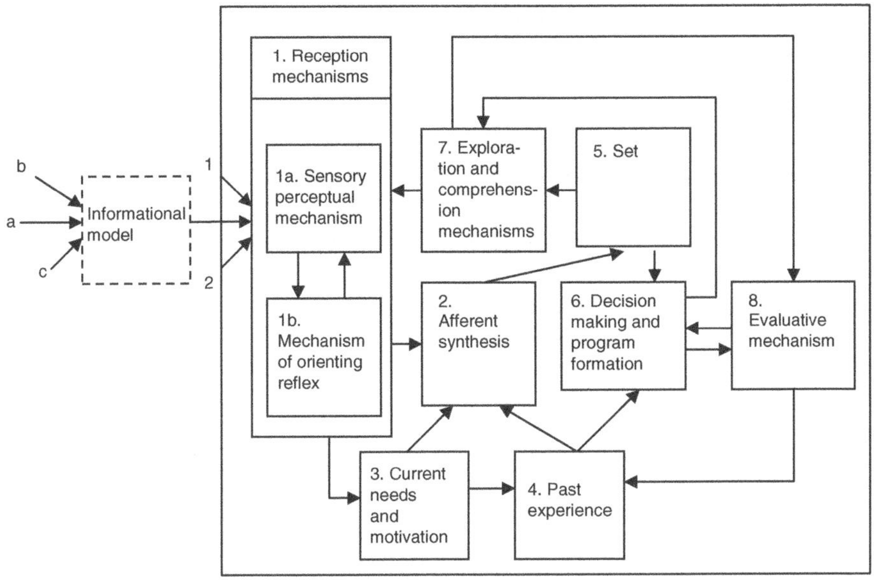

FIGURE 6.3 The model of automated level of self-regulation of orienting activity.

perceptual strategies for information reception. The processing of sensory–perceptual data can be accomplished through conscious sensory and perceptual actions or unconscious automated operations that do not have verbal equivalents. Through these actions and operations, the operator builds sensory and perceptual images of a situation. In this case, sub-block 1a is being analyzed as a system of sensory perceptual actions and operations. For example, signal detection includes decision-making actions at the sensory–perceptual level.

The human operator can use various strategies depending on the developed "set," the latter being related to unconscious tendencies in information reception, or purpose of activity. Evidence for the existence of complicated sensory–perceptual strategies is provided in research conducted by Swets (1964). This research showed that strategies chosen affect the ratio between hits and false alarms during a detection task. Bardin (1982) demonstrated that in a situation where two signals are very difficult to differentiate according to one dimension, subjects begin to use other dimensions to differentiate between the same signals. In each dimension, human operators can use different criteria. From these examples, it follows that even a simple task, such as signal detection, can be regarded as a complex problem-solving task, which includes decision-making actions, operations and strategies.

Functional block 1a is closely interrelated with functional block 1b, called the "mechanism of orienting reflex." Pavlov (1927) first described this reflex. We have already described this function block. Here we present some additional information.

There are three basic groups of the orienting reflex components: (1) motor, (2) vegetative, and (3) change in activation level of the central nervous system. The motor components include the eye movement reaction, head movement to the direction of the stimulus, auditory tuning, change of posture, etc. Vegetative components include blood vessel reaction, change in respiration or heart beat rate, and galvanic skin response. General activation of the neural system involves heightened attention and other factors. The source of the orienting reaction is the novelty of stimulus, change in stimulus and its characteristics, high stimulus complexity, uncertainty of stimulus, and so on. Sokolov (1963) discovered specific kinds of neuron detectors for novelty in the neural cortex.

The regularity of the orienting reflex should be taken into consideration during the design of computer displays that present information to the human operator. Sometimes such displays must attract our attention involuntarily. In other cases, involuntary attention can distract an operator from the ongoing work process through the orienting reflex, causing errors in performance. After attracting attention through the orienting reflex, information can be meaningfully evaluated with respect to its importance. The involuntary attention can be transformed to voluntary attention at the point where verbal–logical processes and consciousness begin to play an important role. The meaning of information input can be elicited directly when we refer to voluntary attention or indirectly when the orienting reflex mechanism is involved. Input information can come from some external sources and internally from memory. In general, the orienting reflex is an important mechanism in the involuntary automatic processing of information.

One of the critical functional mechanisms involved in automated information processing is called "afferent synthesis" (block 2) and was first described by Anokhin (1969). It was demonstrated that some neurons in the cerebral cortex could react to several stimuli from different modalities at the same time, processing light, sound, touch, etc. These neurons not only react to stimuli from different modalities but are also able to select, from many external sources, stimuli that are important to the needs of the organism. This can lead to the discovery of a complex functional mechanism of activity self-regulation called *afferent synthesis*. This functional mechanism involves various structures of the brain. In our model, this mechanism is represented by functional block 2 that integrates the following interactions (1) temporal needs and motivation; (2) mechanisms of memory that represent past experiences of reactions to particular situations; (3) effects of irrelevant stimuli; (4) effects of the most significant stimulus; and (5) effects of the orienting reflex.

As supported by neurological studies (Anokhin, 1969), the relevant signals should be analyzed and interpreted not only as a function of the external relevant stimuli but also in terms of the internal temporal state of the organism, its motivation (block 3), and past experiences (block 4). Based on the (internal) motivational state, past experience and comparison of the relevant and irrelevant information, an organism can react differently to the same external influences. Therefore, the specifics of explorative human activity are to a great extent dependent on functional mechanisms, "current needs and motivation" (block 3), and "past experience" (block 4).

Motivation has a particularly important effect on the processing of internal input from block 1. The change in motivational state can lead to change in the manifestation of integrative functions of afferent synthesis. Functional block 3 changes its motivational stage under the effects of temporal needs and external influences, which come from functional block 1. In this way, the external information not only informs the subject but also activates it. This interaction is also manifested when external information activates the orienting reflex mechanism, thus activating motivational processes. This was confirmed by psychophysiological experiments that showed that an orienting reflex is usually accompanied by a galvanic skin response and changes in the vegetative functions of the organism. In other words, incoming information not only affects cognitive processes but also activates motivational processes. All this indicates that when discussing automated responses to different signals, we have to consider the motivational state of the human operator. A person cannot be understood only as a logical mechanism of information processing, but he also has a bearing on or "attitude" toward the problem. Therefore in human information processing, the motivational and emotional states are significant. It should also be noted that both temporal states and permanent personality features affect afferent synthesis. These affect the afferent synthesis via the motivational block.

The next important functional mechanism (or block of activity) of self-regulation is a set (5). A set is formed through afferent synthesis and is defined as an internal state of the organism that is close to the concept of a goal but is not sufficiently conscious or completely unconscious and thus cannot be considered as such. A *set* can be characterized as the readiness or the predisposition of the subject, which arises with the expectation of a particular situation. A *set* gives a constant and goal-directed character to the course of unconscious components of activity and often leads to the formation of a conscious goal. A *set* creates a predisposition to processing incoming information in a particular way.

In activity theory, there are different kinds of sets. Personality features and personal sense determine the most stable kinds of sets (stable personal set). These shape the subject's view of the world, likes, dislikes, and such. A general, stable system of sets, which a person forms as a result of learning and experience, can become important characteristics of personality. There are also goal-directed sets, which form during the course of learning under the influence of instruction and specific situations. These sets are characterized by their role in forming goals of activity but are not sufficiently conscious to be the goal of an activity. Goal-directed sets manifest dynamic tendencies to the completion of interrupted goal-directed actions (effect) (Zeigarnik, 1981). In mental activity, a set determines the range of a hypothesis, which can arise for the subject. Various types of sets affect each other. The main functions of a *set* are the following:

1. A set determines the stable and sequential course of activity and allows activity to retain its goal-directed character in constantly changing situations, without consciousness of this goal-directedness by the subject.
2. A set provides for the adaptation of the subject to anticipated factors. It ensures the selectiveness and directness of our attention, and the corresponding organization of cognitive processes.
3. A set is to a large extent an unconscious regulator of activity and the process of set acquisition promotes the formation of personality features and character.

A set has a direct effect on functional blocks 6 and 7. Functional block 6 includes decision making and program formation submechanisms. We united these mechanisms for the simplification of the model, but if needed they can be viewed as separate functional mechanisms (Bedny and Meister, 1997; Bedny et al., 2000).

Functional block 6 is activated not only by a set but also by the mechanisms of memory connected to previous experience (functional block 4) and by evaluative mechanisms (block 8). One of the functions of block 4 is to activate and extract from memory information relevant to the specific situation. Block 6 is under the direct influence of blocks 4 and 8. Also, block 6 (decision making and program formation) affects a row of other blocks. This block significantly affects the change of activity evaluation criteria. The functional block "evaluative mechanism" (8) can also be subdivided into a series of functional subblocks (Bedny and Meister, 1997). However, this is avoided in this model for the sake of simplicity. This block becomes more complicated when the unconscious and conscious levels of self-regulation interact, and with the involvement of executive activity. In case of complication, functional block 8 can subdivide into a number of other functional blocks.

The functional block set (5) can affect the functional block "exploration and comprehension mechanisms" (7). A set gives goal-directness to the functioning of block 7, and maintains the stability of orienting activity already in progress that is carried out by block 7. Block 7 is the block that realizes the program that was activated by a set or was formed, or corrected, in block 6. The necessity for differentiating between the block of program formation and the block of program realization was analyzed by Konopkin (1980). The program (6) reflects information about the logical organization of actions and operations. In block 7, the given program is transformed into performance of explorative and orienting operations. Later, these operations can be transformed into goal-directed, conscious actions. As a result of current conditions of task performance, the execution of the program can vary within certain bounds. The execution is then evaluated by the evaluative mechanism (8), which then sends information to block 6, where the program is corrected.

The unconscious self-regulation of activity is considered here as a dynamical system, with numerous connections between the various mechanisms of activity. Despite the absence of a conscious goal, this system has a goal-directed tendency. Using the terminology of Bernshtein (1966), such a system aims to reach a "desired future." In such a system explorative operations, not fully conscious for the person, are dominant. Conscious, goal-directed actions are subordinate in this model of self-regulation.

In general, the system functions as described below. Instrumental and noninstrumental information as well as relevant and irrelevant information reaches block 1. This activates sensory–perceptual mechanism (1a) and mechanisms of the orienting reflex (1b). Reception mechanism (1) and, in particular, the mechanisms of the orienting reflex interact with current needs in block 3 and generate new motivational tendencies. This effects the activation of memory mechanisms relevant to similar situations, thus activating block 4. Information from block 1 (external information), block 4 (past experience), and block 3 (motivational influences), reach block 2 (afferent synthesis). In block 2 (afferent synthesis) these data are integrated and compared, and on the basis of this data, the *set* is formed (5). This *set* then serves as the basis for goal-directed explorative activity, which does not have a sufficiently conscious character.

The program of activity and decision making are formed in block 6, under the influence of a set. Simultaneously, the already formed program of block 7 is activated, and the results from block 7 are evaluated in block 8 (evaluative mechanisms). The evaluative mechanisms, in turn, influence block 6 (decision-making program formation) where programs can be corrected or changed. As a result, new influences arise on block 7, that is, the exploration and comprehension mechanisms. Simultaneously, the information goes to block 4 (past experience) from block 8 (evaluative mechanisms). From block 4, the information once again goes to block 2, and the entire program can be rebuilt.

6.2.2 The General Model of Self-Regulation of Orienting Activity

In the previous section, we discussed the model of self-regulation of orienting activity, which has a predominantly unconscious character. In this next section, we describe a functional model in which the conscious components of self-regulation dominate. This model also illustrates how the conscious and unconscious levels of self-regulation interact. The general model of the self-regulation of orienting activity includes a whole series of functional blocks that we analyzed in the previous section. Such a model is presented in Figure 6.4. The suggested model is adopted to the analysis of orienting activity and offers additional information about self-regulation.

From the previous model (Figure 6.3) we excluded the subblock 1a, as this subblock always closely interacts with the orienting mechanisms (1b) and, therefore,

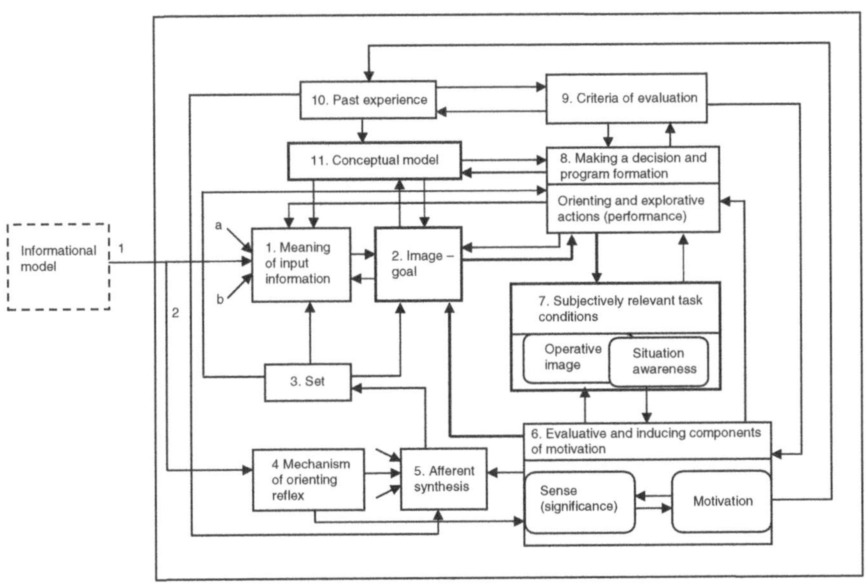

FIGURE 6.4 The general model of the self-regulation of orienting activity.

its inclusion would be redundant. Because we are describing an orienting activity, and executive actions are absent, the performance block carries out only explorative functions. This block, called an "exploration and comprehension" (performance block), is integrated with decision making into a single block 8. Exploration and comprehension are carried out through the orienting and explorative actions or operations (performance). The purpose of such actions and operations is exploration, and the interpretation of information is received by the operator.

In the proposed model, there are two channels of information processing. Channel 1 involves mostly conscious processing of information. Channel 2 involves mostly unconscious processing of information. Their interaction and relative importance change based on the specifics of the situation, past experience of the operator, and their personal characteristics. A brief discussion of the automated channel 2 is shown below.

The incoming information activates block 4 (orienting reflex) the function of which was described in the previous model. This mechanism provides automated tuning to external influences and effects the general activation and motivation of a subject through block 6. A series of additional mechanisms during the activation are necessary for transformation to the conscious level of self-regulation. Instrumental information is indicated by a horizontal line (line 1), noninstrumental information by diagonal arrows "a" and "b" (see functional block 1). Similarly, the influences from functional block 4 (orienting reflex) can be relevant (horizontal line) or irrelevant (diagonal arrows), these influences affect block 5 (afferent synthesis). This block is also affected by influences from motivational block 6 and block 10 (past experience).

Based on complex integrative activity, block 5 (afferent synthesis) develops influences that promotes the formation of block 3 (Set). Set (3) influences "meaning of input information" (block 1) and "image- goal" (block 2). In this way, the *set* interacts with the conscious channel of information processing. If activity regulation is predominantly unconscious, then Set (3) has a direct effect on "making decision and program formation" (8). In this case, all of the resulting information-processing has a predominantly automatic and unconscious character. The functional block "image-goal" (2) begins to form and only functions to a full extent during the latter stages of activity. If the developed set (block 3) does not reflect the reality, the interpretation of the situation can also become inappropriate. As a result, the explorative activity of block 8 can have a chaotic character. Adequate meanings of the input information (1) and goal (2) begin to form later. Such a situation is undesirable, especially under hazardous/risky situations.

It should be noted that when block 3 affects block 1 (meaning), with the lack of sufficient activation of block 2 (goal), the meaning in block 1 is primarily nonverbalized. This was studied by many authors, including Pushkin (1978), Tikhomirov (1984), and Brodesky (1998). In this case, meaning forms under the influence of the *set* and external information, not through the conscious explorative actions, but through unconscious explorative operations from block 8. Nonverbalized meaning is not sufficiently conscious but allows the subject to self-orient in situations requiring complex problem solving.

In general, during unconscious information processing (channel 2), goal (2) is not yet activated. Block 1 (meaning) is formed primarily under the influence of the *set* (3), conceptual model (11), and making a decision and program formation (8).

As a result, in block 1 the nonverbalized meaning becomes highly significant. Further, block 1 becomes responsible for the preliminary interpretation of input information from the signals of various devices that are not yet integrated into a holistic system. Their integration and interpretation as a holistic task or problem becomes possible only after the formation of the goal of activity (block 2).

A more detailed interpretation of the task, its reformulation, the extraction of subjectively relevant task conditions, and projection of the near future become possible through the interaction of blocks 2, 8 and 7. The interaction of these blocks is related to the transformation to the conscious level of self-regulation. At this stage, there is an intensification of the interaction between block 2 and blocks 1 (meaning of input information) and 11 (conceptual model). The interaction of the motivational components of block 6 with block 2 (goal) becomes especially important. This interaction which gives a goal-directed character to activity has been termed vector "motive-goal" and is indicated with a bold line. The formed goal gives regulatory feedback to block 8, and through it, to block 7 (subjectively relevant task conditions).

A bold line in Figure 6.4 also indicates this regulatory feedback. At this stage of information processing (channel 1) block 1 includes verbalized components and as a result there is formation of verbalized meaning of input. The meaning inherent in the objective information presented to the operator must be distinguished from its subjective interpretation. The first is logical meaning and the second is the psychological meaning (Ausubel, 1968). Information presented to the operator in a logical and meaningful way becomes potentially meaningful for the subject. To ensure the proper interpretation of the information, the operator must have professional experience and knowledge. These requirements are reflected in the model through functional block 10, "past experience," which represents the general background of the subject, including general and professional knowledge. The past experience forms a more specialized block 11, termed "conceptual model." Most significant in this block is the professional knowledge about possible situations and imaginative components of work activity related to a specific professional framework.

The conceptual model (11) has a direct influence on blocks 1 and 2. The goal of activity is formed as a result of complex interactions between blocks 3, 1, 8, and 11 (Figure 6.4). A possible situation may involve the automatic goal formation. This is observed when *set* (3) is adequate to the situation. In more complex situations, especially in those situations when the *set* contradicts the situation, a goal is formed gradually. Because the operator often develops the required future result of activity (goal), an image from functional block 2 on the model is presented as the "image-goal." The goal is not a component of the situation awareness in block 7, but rather, it has an independent and leading role in operator information processing. The goal accepted or formulated by the operator can correct and change the method of information interpretation and the development of situation awareness.

The reformulation of a goal can lead to significant changes in blocks 7 and 8 (Figure 6.4). This means that situation awareness can be significantly changed depending on a given goal. This fact has been confirmed by many experiments within activity theory, including study of a activity of the pilot. For example, Beregovoy et al. (1978) and Zavalova et al. (1986) demonstrated that reformulation of the goal leads to a completely different interpretation of the situation by the pilots and formation of totally different dynamic models. Analogous results were demonstrated by Konopkin (1980), and by Bedny (1981). This understanding of the goal in a structure of activity is different from the idea of goal in SA.

Block 8 is shown in the form of two integrated subblocks. These include subblock "making a decision and program formation" and subblock "orienting and explorative actions (performance)." This block also includes automatic, unconscious operations, and demonstrates that explorative activity is not simply information processing but also a system of explorative actions and operations. Explorative activity has a manipulative character, which includes both mental and external practical actions. This is because both the mental and motor actions carry information about the situation. The major purpose of explorative actions is not the transformation of the situation but rather an extraction of information about the situation.

Block 7 (subjectively relevant task conditions) becomes especially important in holistic situation interpretation and is dynamic in nature. This block includes two subblocks, namely, "operative image" and "situation awareness." These are both conceptual and imaginative components of activity, which provide a dynamic reflection of reality. Block 7 reflects the concept of situation awareness widely discussed in the literature of contemporary psychology published in English. By making a decision and program formation block (8), and the motivational block 6, block 7 has an impact on the goal and, if required, can change the strategy of performance. Block 7 not only orients the subject at any moment during the current situation, but also anticipates near future situations and, based on the current situation, infers what happened in the past. The human operator can mentally manipulate inner images and symbols to create an internal, dynamic model, which is progressive in time and reflects the current situation. From the cognitive point of view, operative thinking is particularly important at this stage of processing.

The mental, imaginative manipulation of the situation (subblock operative image) can be largely unconscious and is easily forgotten due to difficulty of verbalization. The subblock situation awareness is a component of functional block 7 and includes a logical, conceptual subsystem that dynamically reflects the situation in a verbally logical form. The imaginative and conceptual subblocks, which are included in the dynamic reflection situation, partially overlap as was described above. When unconscious processing dominates (channel 2), the subblock operative image becomes more important than the subblock situation awareness. In such a situation, the operator regulates activity based on "vague" feelings, or an intuition, but can often do this very successfully. Mechanisms such as "nonverbalized meaning," a "set," and "afferent synthesis" also become critical in this situation. This is why, for example, the term "feeling of aircraft" is important in the study of pilot activity (Ponomarenko and Zavalova, 1981).

The functional block subjectively relevant task conditions (7), in general, is responsible for developing a dynamic model of the current situation. In contrast, functional block 11 conceptual model is responsible for developing a relatively stable model, which includes different scenarios of possible work situations. Welford (1960) was the first to introduce the concept of conceptual model to psychology. Oshanin (1977), Zarakovsky et al. (1974), and Gordeeva, et al. (1975) infused such terms as conceptual models, stable model, operative image, and dynamic model into activity theory. The "image-goal" (2) is a model of the future result of the operator's own actions.

Functional blocks 2, 11, and 7 are outlined in bold (Figure 6.4), to indicate that they include image components. The modeling function of our brain is particularly obvious in the imaginative reflection of the situation. Usually, a model is treated as equivalant to some object or phenomenon. This model reflects the more essential features of objects of phenomena, from the perpective of the current task. Subjects may perform preliminary actions with these models, prior to interacting with real objects. Functional blocks contain imaginative components of real objects, or situations that stand for the real objects, with which a subject operates. Accordingly, these functional blocks are referred to as models. Thus, notions such as an image-goal are construed as a model of the future results of an action. Conceptual models are by the same token "constant models" reflecting diverse scenarios or situations germane to specific undertakings. Finally, we can understand subjectively relevant task conditions as a dynamic model, relevant to a particular situation, and to actions performed in this situation.

Functional block 9, criteria of evaluation, is important in the evaluation of the results of explorative activity. Through functional block 8, block 9 corrects the explorative activity. The results of this correction are then transformed into past experience. At the same time, the "criteria of evaluation" can be gradually developed and modified through the functional block 8. In more simple situations, the criteria of evaluation can be extracted from past experience.

Orienting activity contains both the informational and the energetic components. From the psychological point of view, the energetic components that are most important include motivational and related emotional components of activity. Goals cannot exist apart from motives, because there is no such thing as nonmotivated goals. Needs, wishes, desires, etc. become motives when they are connected with a goal, giving the activity its goal-directed character. Motivation affects both the conscious and the unconscious aspects of human activity at work. Without motivation there can be no "afferent synthesis" (5) or set (3). Thus, motivation is important in both conscious orienting activity regulations, where it creates the vector "motive-goal," and in the unconscious orienting activity regulation, where it is related to blocks 5 and 3.

Let us consider block 6 in greater detail. This block includes two subblocks, which mutually influence each other. One subblock is called the "sense" (significance), and the other "motivation." The subblock motivation of block 6 refers to the inducing components of motivation. In contrast, subblock sense includes the evaluative (cognitive-emotional) components of activity. The higher the level of motivation, the more mental and physical effort the person expands towards achievement of the goal.

The subblock sense (significance) and block motive in this model are equivalent to the blocks "assessment of the sense of task" and "formation of level of motivation" in the general model of self-regulation. Here, we are only concerned with the fact that sense influences the interpretation of objective (logical) meaning and that as a result of these influences subjects develop their psychological meaning.

The more significant the situation is for a subject, the higher the level of motivation. This increases the goal-directed tendency of activity during task performance.

The other functional block, which directly influences motivational block 6, is functional block 7. Functional block 7 is responsible for the formation of a dynamical model of a situation. This block switches attention from one feature of an object or situation to another, one that is more relevant for the actions performed at a particular time. In other words, the factor of significance is involved in extracting adequate "subjectively relevant task conditions," and can correctly (structurally) organize this information in a dynamic model. All this demonstrates that motivation is critically important in receiving and interpreting information.

The system of expectations also plays an important role in human information processing. The system of expectations activates different motivational tendencies and possible strategies of information processing. The orienting activity becomes even more difficult when incoming information is unpredictable and uncertain and contradicts existing expectations. This kind of information processing is briefly described below.

At the first stage of the orienting activity under uncertainty of information, one can distinguish a complicated interaction between the conscious and unconscious levels of information processing. Suppose that a human operator has already developed a system of expectations and then is suddenly presented with information that is unpredictable and hard to interpret. As a result of such a situation, the orienting reflex is immediately formed. The orienting reflex activates the central neural system; this, in turn, leads to the distortions of the already existing system of expectations and developed activity. The orienting reflex also influences the motivational state of the subject (block 6). This functional block activates "past experience" (block 10). The retrieved past experience does not always match the situation at hand. Conscious awareness of this process can be very low. The adequacy of past experience for a situation is then checked at the conscious level of self-regulation.

The influence of orienting reflex (block 4), motivation (block 6), and past experience (block 10) is integrated in afferent synthesis (block 5). The motivation dominant at a given moment plays an essential role in this integrative process. The afferent synthesis stage is completed by formation of the output, which influences the process activating functional block "set" (3). Based on the developed set, the adequate system of anticipations, expectations and goal-directed tendencies are shaped. The subject is not quite aware of this psychological state. The functional block *set* influences the blocks "meaning of input information" (block 1) and "image-goal" (block 2). Motivation (block 6) activates "making a decision and program formation" (block 8). Atomized operations are activated at the beginning stages of block 8 functioning. As a result of the following formation of blocks 1 and 2, unconscious operations in block 8 are transformed into conscious goal-directed actions. Thus, due to the influence of blocks 11, 6, 8 and external information coming out of channel 1, changes in blocks 1 and 2 occur.

Nonverbalized meaning is transformed to a verbalized meaning in block 1, and, at the same time, a conscious goal is formed in block 2. This, in turn, leads to block 7 being responsible for the creation of a dynamic model of the situation. Quite often at the first stage of receiving and interpretation of unexpected and contradictory information, the unconscious processing through channel 2 can dominate. Only at the following stage does the conscious information processing begin dominating through channel 2. Evaluative mechanisms and feedback influence self-regulation to facilitate subsequent corrections of the activity goal as well as more accurate interpretation of information, and so forth.

6.2.3 Informational, Conceptual, and Dynamic Models in Situational Reflection of Reality

In operator activity the main source of information about the situation and the object being controlled is a series of displays or instruments on a dashboard. The activity of the operator involves not real objects but their substitutes: signals, or external images, that imitate real objects. Information about the situation, instruments and tools that index the state of the object, and its external environment to the operator, mediate the control of the object. The assembly of information presented to the operator regarding the object of control and the external environment is referred to as the informational model. The informational model presents to the operator the most essential data about the control object and external environment. The better this model presents the essential information to the operator, the more effective the operator becomes in receiving and interpreting information. On the basis of the informational model, the operator develops an image of the real situation, performs an analysis and evaluation of the situation, plans activity, and makes decisions. The amount of information and its organization must be in accordance with the problem being solved by the operator. In each situation, the operator must extract the relevant information and distinguish it from irrelevant information. The operator must further integrate information from different instruments and analyze the possible dynamics of the situation on the basis of the indication of different displays and their relationships. Based on this analysis, the operator forms a problem-solving task.

Signals about the arising of a given situation can be instrumental and noninstrumental. When the operator has the opportunity to directly observe the object of control, he can receive not only instrumental but also noninstrumental signals. For example, during a flight the pilot receives information about the aircraft's position based on information he observes outside the cabin. In other cases, noninstrumental information is less definitive; for example, vibrations of the aircraft, resistance of controls, and changes in the background noise. Noninstrumental signals are particularly important in emergencies (Beregovoy et al., 1978). The informational model and the external situation includes both instrumental and noninstrumental information and relevant and irrelevant information. Irrelevant information often has an unpredictable, irregular character.

The operator must know how to select relevant information and distinguish it from irrelevant information. He must also be able to correlate the instrumental and noninstrumental information. An important aspect of operator activity is the ability to synthesize information to create a holistic, mental picture of the situation. Finally, the pilot must be able to predict the course of developing events and determine their causes. All of this demonstrates that the operator is often confronted with ill-defined tasks and problems, and such concepts as automated mental operations, situational concept, nonverbalized meaning, and toponem become important in the dynamic reflection of reality. From this it follows that not only verbally logical and conscious processes but also intuitive, nonverbalized processes are important in the activity of the operator. In dynamic stress situations, with time limits, the intuitive and imaginative components of activity are particularly important.

The information received by the operator can be classified as certain, uncertain, and contradictory (Beregovoy et al., 1978). The operator receives certain information from the output of various displays or verbal information. This is the information predetermined to be given to the operator in a specific situation. It is easily interpreted because the operator is trained to receive and comprehend this information. Contradictory information arises when the information of one set of displays contradicts that of others. Such a contradiction can lead to incorrect interpretation of the information. Uncertain information cannot be unequivocally interpreted and is characteristic of noninstrumental information. All of this makes it apparent that, in the operator's interpretation of information, not only conscious, verbally logical but also unconscious, intuitive methods of situation reflection are significant. This explains the importance of imaginative components in an operator's activity. From the perspective of AT, one can extract an unconscious and conscious level of activity self-regulation.

One of the major determinants of psychological development is human labor, characterized by the use of tools. Tools emerge as a product of sociocultural phenomena that prescribe practical actions and mental operations (Vygotsky, 1978). Based on external tools that mediate practical activity, humans developed internal tools to mediate mental activity. According to Vygotsky, these psychological tools are signs, which have their respective meanings. From this it follows that the informational model can be regarded as a particular kind of external tool for operator activity. During training and practical activity, based on the informational model, an operator develops his own psychological tools. These tools, which include images, verbal representations, goals, knowledge, motives, programs of action, non-verbalized sign, etc., are organized into a structure. This structure is the internalized, idiosyncratic world of the operator. The tools that make up this structure are past experience adapted for particular professional duties. The inner, idiosyncratic world of the operator is called the conceptual model. The conceptual model is distinct from past experience, insofar as it is more specific to particular duties and its imaginative components are critically important. The conceptual model includes an organized system of pictures of possible practical situations stored in long-term memory. The British psychologist Welford (1960) first introduced the notion of a conceptual model. However, it was later elaborated and refined by Zinchenko et al. (1974). The notion of conceptual model received rapid and widespread recognition in AT. Proper design of the information model as an external tool is a prerequisite for adequate development of the conceptual model as an internal tool. The content of a conceptual model is relatively independent of particular tasks and hence contains substantial redundancy in relation to the particular task performed by the operator.

The conceptual model is held in long-term memory and is generally resistant to change. This model is based on past experience as well as verbal or written instructions and is available to the operator prior to initiation of an action. The process of decoding the representation of the informational model involves the comparison of informational model elements with the elements of the conceptual model. At any particular time only those conceptual model components that are tied to the operator's specific actions are extracted and elaborated. These components which are adapted to specific actions make up the dynamic model. At different points during the history of activity theory, different terms were used to refer to this dynamic model. These differences in terminology reflected distinct aspects of this model and diverse approaches to its study. For example, Oshanin (1977), who emphasized the imaginative aspects of the dynamic model, introduced the concept of operative image. Konopkin (1980) introduced the concept of subjective model of significant conditions. Bedny (1981), in his work in vocational training, introduced the concept of dynamic orientation in the training process. Recently, Bedny and Karwowski (2004) elaborated the idea of the dynamic model in his study of the dynamic reflection of the situation. He introduced the concept of subjectively relevant task conditions as a functional mechanism, which is responsible for creating a dynamic model of the situation (Bedny and Meister, 1997). In contrast to the conceptual model, which is stored in long-term memory, information about subjectively relevant task conditions is held in short-term or working memory. The content of the dynamic model is determined both by information stored in short-term memory and by the specific character of operative thinking during the solution of a particular problem.

The functional mechanism of subjectively relevant task conditions is a function block in the model of activity self-regulation. This function block is responsible for building a dynamic model, which is adequate for the goal of the action and task. It includes both conceptual and imaginative components of activity. The block, subjectively relevant task conditions, consists of two subblocks (Figure 6.4) One subblock, called "situation awareness," includes a logical and conceptual subsystem of dynamic reflection that is connected with consciousness. Another subblock, in turn, is responsible for imaginative and unconscious reflection of information. The logical-conceptual subblock of dynamic reflection provides SA overlap, partly with subblock operative image. Therefore, subblock operative image includes conscious and unconscious components. Only those parts of the operative image that overlap with SA are conscious. The nonoverlapping part of imaginative reflection can be considered a unconscious reflection. With the shifting of attention, an increase of will and a change in the situation, a unconscious reflection can become conscious or, alternatively, what was conscious earlier can become unconscious. All this may be reflected in the individual by "vague feelings," which can also affect conscious components. Continual transformation of information, from one functional subblock to another, provides the transformation of conscious components to unconscious, and vice versa. Hence, this function block provides a more dynamic reflection or representation of reality, by switching from one feature of the object or situation to other features. An individual can manipulate inner images and symbols to create an internal model of events progressive in time. This dynamic reflection can be enriched with additional data from internal and external sources that are necessary for each particular moment. The imaginative manipulation of the situation can be, to a large extent, unconscious and easily forgotten, due to the difficulty of verbalization. Depending on the goal of activity, dynamic reflection can be changed. This, in turn, has an influence on executive strategies. Hence, a dynamic model of situation includes conscious as well as unconscious components that can be transformed, to some degree, into each other. Changes in the goal of the action, or activity, necessitate changes in the dynamic model. Such changes assure that the situation is adequately reflected in the actions performed. If the goal of action or activity changes, but the dynamic model does not, then this model becomes inadequate for the new goal and loses its dynamic nature. This disturbs not only the adequacy of the situation's reflection but also the adequacy of action regulation. The function block, subjectively relevant task conditions, interacts with the block for evaluative and inducing components of motivation. Subblock sense, or significance, becomes particularly influential. Often, what is subjectively significant to the operator, is presented in a dynamic model of the situation. However, these elements of the situation are not always objectively important. This can lead to faulty orientation in the situation and the distortion of the internal model of reality. The opposite can happen as well. The dynamic reflection of the task can also affect the subjective evaluation of the significance of the situation.

During the first stages of activity theory, development researchers outlined such basic elements of activity as the goal, motive, conceptual model, and operative image. These elements were not yet integrated into a holistic system. Through the works of Anokhin (1962) and Bernshtein (1966), it became clear that activity is organized as a functional system. According to Anokhin (1962), a functional system selects and integrates structures and processes for the performance of a precisely outlined act, behavior, or function of the organism. As described earlier (Bedny and Meister, 1997), this functional system was self-regulating and had a more physiological than psychological character. The need arose for a purely psychological concept of activity self-regulation. From the psychological point of view, a self-regulating system is a functional system which integrates various functional mechanisms of activity. These mechanisms are responsible for developing dynamic models of the situation and achieving the desired goal of activity. Researchers, like Konopkin (1980) and his colleagues, performed studies to develop a purely psychological theory of self-regulation. Konopkin proposed a model for self-regulation of sensory–motor activity. This model contains important elements of self-regulation, some of which are described in this volume. Bedny proposed a more general model of activity self-regulation (Bedny and Zelenin, 1988). More detailed models were published in English (Bedny and Meister, 1997; Bedny and Meister, 1999; Bedny et al., 2000). Recently, new elaborative models of self-regulation have been suggested by Bedny and Karwowski (2004). These models include not only conscious but also unconscious levels of self-regulation. Contemporary psychology accumulates tremendous amounts of data. This data is very often unrelated and sometimes contradicts itself. All of this makes it difficult to comprehend and interpret the data as well as apply it in practice. Different psychological phenomena do not exist in isolation. They influence each other and are organized as a system. Psychological mechanisms have systemic organization. Functional analysis of activity is a systemic method applied to study of human performance.

Human activity is a multidimensional system, and therefore, even with the most developed method of study, may not be sufficient for a comprehensive study of work. Activity as a complex system comprises different units and requires a systemic-structural method of analysis. Accordingly, the same activity during a task performance must be studied from different perspectives. This analysis presupposes the description of activity as a functional self-regulating system. Such a self-regulation system is not homeostatic, but rather goal-directed.

In the description of different components of the self-regulation system, we do not use the localization principles directed at the study of physiological mechanisms, but the functional principles that describe various functional blocks/mechanisms. On some occasions, the same elements of activity can serve different functions at various stages of activity regulation. Depending on these stages, the same element of activity can be analyzed from the point of view of different functional blocks. The systemic nature of the functional description is determined by the interrelationships of different functional blocks. This aspect of human *work activity* description allows it to be called functional in structure.

6.2.4 Application of Concept of Orienting Activity (Functional Analysis)

Here we will discuss a few applied examples. The speed with which pilot reads displays varies with the selected strategy. Kotik (1978) found that pilots in flight spend very little time (0.3–0.5 sec) reading quantitative aviation displays. To understand how pilots can achieve such speed he conducted an experiment with quantitative displays in laboratory conditions. In this experiment he utilized tachistoscope to present pilots with the pictures of quantitative displays. The pilots had to read the instruments as quickly and precisely as possible. Before the study began, subjects were told the kind of instruments to expect. Readings were considered erroneous if the error was more than one interval on the scale. Results indicated that, for the required accuracy, a pilot should spend 1.2 sec reading each instrument. The pilot's experience and skill level did not increase the speed of the readings.

In the real flight pilots develop strategy of reading such displays which are totally different from laboratory conditions. In laboratory study when pilots received information from an independent display they utilized quantitative readings. In real flight when pilots perceived non-isolated interdependent and logically organized data they utilize qualitative reading from quantitative displays. According to this strategy it is not important to read an exact position of individual scale pointer. It becomes important for pilots to determine how far the pointer position deviates from particular area on the display which is considered by pilot as a correct position of a pointer. Hence pilots developed the image of correct positions of pointers for a particular regiment of flight. They organize the information from the individual instruments into a holistic dynamic model and utilize a qualitative reading strategy. In experimental conditions and in real flight pilots develop different strategies without the full awareness of it.

Studies have indicated that the pilot organizes the information from his individual instruments into an interrelated, dynamic whole, or mental model. The pilot's dynamic model of flight (subjectively relevant task conditions) changes as the situation changes and as evaluation of the consequences of actions changes. In general, the orientating component of self-regulation during flight depends upon the functions blocks Goal, formation of task, subjectively relevant task conditions, assessment of sense of task and "motivation." The results of the pilot's actions through feedback can affect these function blocks. As a result, the dynamic model of flight can be changed.

Goal and motive are often leading factors in determining the dynamic orientation of a pilot during a given situation. For example, a goal and a persevering motive can completely change situation awareness. Let us review an example of a real flight during critical conditions (Ponomarenko, 1998). A pilot had little fuel remaining and he had no chance of making a second attempt at landing. Owing to bad weather and cloudiness the pilot lost sight of the airport. The pilot was in constant communication with the airport. According to the critical situation, and existing norms, the pilot was to abandon the aircraft. Despite this the pilot, who had a high degree of experience and fully understood that he had almost no chance of landing the plane, decided not to catapult but rather to find the airport and land the aircraft. Thanks to his skillfulness, and favorable circumstances, he achieved the set goal with little damage to the aircraft. Immediately after the flight, the pilot was debriefed. An analysis of the pilot responses demonstrated that during the flight, logical thinking directed at answering the question "Where am I in relation to the airport?" assumed leading importance. Based on this the pilot formulated the goal of landing the plane. The desire to land the plane became a persevering motive for the pilot, and other possible methods for escaping were eliminated. All the instructions received during radio communication, the comprehension of the situation, and evaluation of action results were evaluated from the position of the persevering motive and goal of activity. Through effort of will the pilot aimed to suppress the negative emotional state. The dynamic model of the situation formed on the basis of the persevering motive and the connected goal of activity. The decision of abandoning the aircraft dropped out of the pilot's consciousness. The negative emotional state violated adequate orientation of the situation. In general, the catastrophic state ended positively, thanks to the pilot's skill and favorable circumstances. However, research demonstrated that because of the malfunctioning of such mechanisms as the evaluative and inducing components of motivation and image-goal, apprehension and interpretation of the situation were inadequate. In other words, the function block subjectively relevant task conditions formed an inadequate dynamic model of the situation. An analysis of the verbal protocol demonstrated that the persevering motive and goal of activity determined the entire strategy for the selection an inappropriate interpretation of the situation.

Let us consider, briefly, other examples. Consider a pilot's activity during engine failure (Ponomarenko, 1998). Elimination of engine breakdown is a very complicated task. Breakdown of an engine is a task that requires different strategies of performance, depending on the specificity of the malfunction. Hence, this task includes a diagnostic stage. It is particularly important to develop an adequate dynamic model of the situation. In this dynamic model, nonverbalized imaginative components play a critical role. Both working memory and operative thinking, which includes nonverbalized aspect of thinking and intuition, are especially important. Research was conducted in conditions of real flight. There was an experimental panel in an aircraft. The experimenter introduced imitation of one engine failure by using the experimental panel. Time of performance of different elements of the task was then measured. Speech reaction, reporting what happened through radio exchange with the experimenter, was also utilized. Some physiological data were used. The study demonstrates that pilots use three different strategies during diagnosis of failure. It was also discovered that pilots utilize noninstrumental type of information. Pilots obtain this kind of information in the form of subjective feelings, which do not have precise verbal equivalents. In general, it was discovered that subjectively developed dynamic models

TABLE 6.1
Prescribed and Real Strategies of Information Gathering during Engine Failure

Sequence of information gathering according to standard procedural manual	Real sequence of information gathering in case of engine breakdown
1. Progressive pitch Turning of aircraft	1. Feeling of acceleration, resistance on control column and pedals, changing position of body, sound changes
2. Light indicator "ENGINE BREAKDOWN" is turned on	2. Decreasing of RPM
3. Decreasing of turning moment	3. Decreasing of turning moment
4. Pressure drop in gas system	4. Pressure drop in gas system
5. Decreasing RPM	5. Decreasing of exhaust gases temperature
6. Decreasing of exhaust gases temperature	6. Lighting of signal bulb (alarm signal) about engine failure
7. Pressure drop in oil system	7. Pressure drop in oil system

of a situation include, both verbalized and nonverbalized components. Imaginative components were particularly important in this model. Instrumental and noninstrumental information play a role in the development of this subjectively developed dynamic model. More interesting, in this study, was the fact that the standard, prescribed, and real strategy of gathering of required information during failure was different (see Table 6.1). Noninstrumental information was the major factor that influenced the development of an informal strategy of gathering information for diagnosis of failure.

In order to change diagnostic strategy of the pilot, a new design of the panel was suggested. Alarm signal "danger" was located directly over the instruments that reflect the parameter of engines functioning. Noninstrumental information automatically attracts the pilot's attention in this area. Ambiguous feelings about engine failure associated with noninstrumental signals were automatically transferred into conscious level of processing of information through instrumental signals. Preliminary ambiguous feelings and guessing is confirmed by information from instruments. This example demonstrates that functional analysis of activity permits, with high levels of precision, analyses and description of experimental data, from which can be developed applied design recommendations. In this experiment, the importance of the functional mechanism of self-regulation "subjectively relevant task conditions" was demonstrated. This mechanism is responsible for developing a dynamic model of the situation. Not only verbalized, but also nonverbalized, and particularly imaginative elements of the dynamic model, should be taken into consideration.

Let us consider the other example. It is known that the subject often selects a subjective criterion of success, which may differ from objective criteria. For example, a study demonstrated that in time restricted conditions, where the time standards of performance were subjectively significant, participants chose temporal criteria of success that led them to perform the task in a shorter time than required (Bedny, 1985). Divergences from these subjectively set criteria, during task performance, were viewed by the participants as a negative result. Objectively set criteria of success were almost ignored. This strategy of activity increases the reliability of task performance according to temporal standards.

Bardin's studies (1982) discovered different strategies during the performance of these tasks. For example, in experiments Bardin demonstrated that when two acoustic signals, difficult to distinguish on the basis of loudness were presented, the subject used additional discriminating criteria not given in instructions. The subjects began to perceive previously unnoticed qualities in the acoustical stimulus, and used these as discriminative criteria. The participants reported that the sounds became dimmed, brilliant, resonant, dull, etc. These features are new criteria of performance developed by the subjects. Such experiments demonstrate the logic behind understanding the sensory space as a multidimensional system according to activity theory. In these signal detection tasks subjects developed their own criteria of stimulus evaluation, thereby changing their dimensions of sensory space and strategies of activity.

These examples from applied research demonstrate that functional analysis of activity provides a unique lens for interpretation of a variety of experimental data. Such concepts as goal, motive, subjectively relevant task conditions (dynamic model), conceptual model (stable model), criteria of evaluation, etc. are complex mechanisms of activity regulation that determine the strategies of human performance. Consequently, functional analysis of activity is one of the most important methods of studying human performance.

One primary purpose of orientation activity is to create a situational reflection of reality, which is the process that constructs the dynamic model of the situation. Such a model must be adequate for the goals of the actions performed at a particular point in time. The dynamic mental model performs orienting and regulative functions in the structure of activity. The functional mechanism (block) called subjectively relevant task conditions is particularly important in the construction of this model. This mechanism has a complex structure and contains imaginative and verbally logical subblocks. It includes verbalized and nonverbalized sign systems. One of these subblocks is called situation awareness, a term introduced by Endsley. We regarded it as a submechanism that is responsible for conscious, dynamic reflection of the situation. This reflection includes verbal and imaginative as well as conscious and unconscious components. Operative thinking and working memory are particularly important in the functioning of this mechanism. The contents of the dynamic model constantly change and adapt to the set action goal. The dynamic nature of a human mind's reflection of reality is a central feature of this reflection and is necessary for situation comprehension and activity regulation. The dynamic model ensures a flexible transition from reflecting one set of object/situation qualities to reflecting others, and this transforms the process of reflection, in accordance with the goal of the activity.

The functional mechanism that generates the dynamic model does not work in isolation. One of the mechanisms that influence the function block of subjectively relevant task conditions is the conceptual model. The conceptual model is responsible for developing a broad and relatively stable mental model, which serves as a general framework for understanding different situations relevant to particular professional duties. Although this model is general, and is stored in long-term memory, it is more specific than past experience. Imaginative components are one of the distinguishing characteristics of this model and play an important role in its functioning. The data that correspond to the goal of activity are extracted from this model, adapted and transformed into a dynamic model of the situation. These dynamic models are enriched by external information actively extracted from the presented situation.

Of particular importance is the adequacy of the dynamic model for the goals of actions being performed at the time. A divergence between the intermittent goals of actions and the dynamic model requires either the transformation of the dynamic model of activity or change of action goals. If this does not occur, the interpretation of the situation is slowed, or becomes inadequate, and can lead to mistaken actions. Another mechanism important for the dynamic reflection of the situation is that it is responsible for the motivational aspects of activity regulation. The influence of this mechanism is the selection and interpretation of information, according to the factor of subjective significance. The predilection of the subject affects interpretation and comprehension of the situation.

The study of activity mechanisms, which are responsible for the construction of relatively stable and dynamic situational models, has a long-standing tradition in activity theory. This section describes SA as an important mechanism of orienting activity. The integration of the data presented here, and the materials gathered under the SA approach, will lead to further development of the functional analysis of activity.

6.3 EXAMPLE OF APPLICATION OF FUNCTIONAL AND MORPHOLOGICAL ANALYSIS IN THE STUDY OF HCI TASKS

6.3.1 Eye Movement Analysis in Activity Theory

In this chapter is demonstrated the possibility of combining experimental and analytical procedures during systemic-structural analysis of HCI tasks. When experimental and analytical procedures are combined it often allows to carry out an experiment in abbreviated manner. In some cases an experiment is reduced to a collection of some initial data for farther development of theoretical models of activity. Analysis and comparison of different models of activity eliminates necessity to use comparative analysis between experimental and control groups with following testing of the statistical hypotheses. After receiving initial experimental data a practitioner has an opportunity to develop theoretical models of activity during task performance and then switch to theoretical analysis and development. These analytical procedures enable a practitioner to abandon or discard incorrect solutions and develop a more appropriate one based on comparison of different models of activity performance. Below we will describe this method. In this study the computer is considered as "means of work" that generates various sign systems which can be used as tools or objects (Bedny and Harris, 2005). Computer generated sign systems first of all engages the visual organs. Hence, analysis of task performance HCl can be conducted by utilizing eye movement data. Very short duration cognitive actions, which are performed based on visual information, are dominant in computer mediated activity. As a result, microstructural level of analysis and eye movement registration will be introduced in this study. It is assumed that since cognition is not simply a process but also a system of cognitive actions, it can be studied from a systemic perspective. Systemic-structural theory of activity includes qualitative, algorithmic analysis, time-structure analysis and quantitative stages of analysis. Each stage can be performed with a different level of decomposition.

During qualitative analysis, a number of interdependent methods are used, which permit us to describe an activity from different perspectives (substages) and levels of detail. Any qualitative method of analysis should be directed to describe and discover specific aspects of multidimensional and multilevel systems of activity. Qualitative methods can also include experimental methods of study. However, any experimental method should be combined with a theoretical description of the structure of the holistic activity. In this work, eye movement registration is used as method that based on systemic-structural analysis principles. Most studies using eye movements follow a parametric method by obtaining and qualifying eye movement data as dwell time, gaze durations, scan path lengths, etc. However, these data are not sufficient for description and analysis of multidimensional systems of activity. In spite of the tremendous amount of data on the eye movement registration there are still methodological problems in the interpretation of eye movements data. Some scientists suggest that usability researchers do not always have a strong theory to perform eye movement analysis (Goldberg and Kotval, 1999). At present there is discrepancy between eye movement registration and eye movement analysis (interpretation). The significant progress has been made in the area of eye movement registration. However eye movement analysis is almost the same for a long period of time now. Difference in the interpretation of the eye movement in systemic-structural activity theory and cognitive psychology is presented below.

Systemic-structural theory of activity employs a different method of eye movement registration. Parametric method of eye movement registration should be combined with systemic principles of analysis, in which eye movements are associated with mechanisms of self-regulation or with actions performed by subject. This gives a better opportunity to describe the structure of activity.

The study of eye movements in activity theory is associated with one fundamental principle of activity: the principle of unity of cognition and behavior. According to this principle, motor activity not only provides transformation of the object but is also important in cognitive activity of human performance (Rubinshtein, 1973; Bedny et al., 2001). Motions transforming an object and changing its position in relation to the others at the same time give information about the object and situation in general. In spite of the tremendous amount of data on the eye movement registration there are still methodological problems in the interpretation of eye movements data. Some scientists suggest that usability researchers do not always have a strong theory to perform eye movements analysis (Goldberg and Kotval, 1999). There is discrepancy between eye movement registration and eye movement analysis (interpretation). The significant progress has been made in the area of eye movement registration. However eye movement analysis is almost the same for a long period of time now. Differences in the interpretation of the eye movement in systemic-structural activity theory and cognitive psychology is presented below.

Yarbus (1965) was the first one who developed the direct eye movement registration procedures. Theoretical background for this work was the study of unity of cognition and behavior in activity theory that has been advanced by Rubinshtein, 1973, Leont'ev, 1978. Study of eye movement in cognitive psychology and activity theory does share some similar principles. However, there are some differences in interpretation and application of the eye movement data in activity theory and cognitive psychology (Bedny, Karwowski, and Bedny, 2001; Ponomarenko, 2004). Yarbus (1965) proved that stabilization of the image on the retina results in disappearance of the object in 1.3 sec. Therefore micromotions involved in perceptual process and aid in the formation of perceptual image. Motions of eyes do not simply provide the position of eyes on different objects, and subsequent foveating, but also are involved in the formation or development of a perceptual image. For example, Yarbus showed that eye fixations are far from random. They depend not only on perceptual features of the object but also on the goal of the activity. Inducing different goals in the observation process can change eye movement pattern. Eye movements are reliable indicators of intellectual operations. Attention shifts, in most cases, are associated with shifting of the eye in a corresponding area. At the same time, direction of the eye sometimes does not coincide with attention. However this happens very rarely (Mit'kin, 1976). Zinchenko and Vergiles (1969) studied the process of visual thinking and discovered complicated micro-motions of the eyes during visual problem solving. These micro-motions are considered as an externalization of the thinking process. When eyes are involved in such micro-motions the subject does not perceive external information but rather manipulates visual images and performs mental transformation of the situation. Komishov (1968) discovered similar data during his study of eye movement of a pilot in the real flight. Pushkin (1978) and Tikhomirov (1984) demonstrated in their studies differences between perceptual and thinking eye movements during visual problem solving. It has also been discovered that mental imagination of objects increases micro-motions of eyes (Zinchenko and Vergiles, 1969). Mit'kin (1976) showed that different geometric figures presented on the screen lead to different saccadic movements. The study of the visual perception process and the sense by touch are examples that demonstrate the relationship between external and internal activity. Touch perception is accompanied by complicated hand and finger macro and micro-movements (Zaporozhets, 1969; Turvey, 1996). Micro and macro eye movements also was first discovered in visual perception (Yarbus, 1967). Comparison of hand and finger movements with eye movements shows that both kinds of movements perform similar functions in this perceptual process (Gordeeva and Zinchenko, 1982). Kochurova et al. (1981) performed a micro-structural analysis of motor actions and motions. They found cognitive components in motor activity. Motor and cognitive activity involve similar cognitive mechanisms. These studies demonstrates that movements involve cognitive functions. Eye movements can be used as reliable indicators of the operator's intellectual activity.

In cognitive psychology eye movement is usually attributed to attention processes, or perception (Sodhi et al., 2002; Rayner, 1992). Blind fixations are considered as discrepancy between eye movements and attention. On the other hand, in activity theory more attention is paid to the relationship between eye movements and higher mental functions. Eye movements/hand movements relationship in activity theory has been studied from the perspective of unity of cognition and behavior. SSAT advocates the use of the principle of unity of cognition and behavior as a key in the study of HCI (Bedny et al., 2001), It is recommended to utilize both eye and mouse movements in applied research because it helps to interpret eye movement data. It should be mentioned that in cognitive psychology scientists sometimes also utilize the combination of eye movements and mouse movements (Smith, et al, 2000; Harnof, 2001). However, these studies are based on totally different theoretical data. There are some similar data obtained in activity theory and cognitive psychology.

The basic characteristics of eye movement analysis are saccades and fixation. They usually include scan paths, frequency fixation, fixation duration and transition between areas of interest. There is no standard methods for identifying the fixations and saccades. At least three processes are assumed to take place within typical fixation with the duration of the 250–300 msec (Viviani, 1990). These processes include analysis of the visual stimulus in the fovea field, the sampling of peripheral field, and the planning of the next saccade. Yarbus extracted two kinds of eye movement, macro motions, or saccades, and micromotions, or microsaccades. Macro motions provide changes in gaze position of eyes on a stationary scene. Macro motions can be partly regarded as voluntary. However, in most cases a subject cannot give verbal description, in which case, he gazes differently at objects. This is why eye movements help us, in an objective way, to understand cognitive strategy. It takes about 100–300 ms to initiate a saccade, that is from the time a stimulus is presented till the eye starts moving, and another 30–120 ms to complete the saccade, depending on, among other things, the visual angle traversed (Mit'kin, 1974). The basic information is received by an eye during fixation. During a saccade, the eye can only detect an object, but cannot recognize it. Saccades can be initiated voluntarily but are ballistic: that is, once they are initiated, their path of motion and destination cannot be changed — which must be taken as an indication that visual attention in the peripheral area selects the next location for the eyes to move to. Pursuit motion is a much smoother, slower movement than a saccade; it acts to keep a moving object foveated. Nystagmus and drift are another kind of eye movement described in activity theory. They are also well described in cognitive psychology. Drift and microsaccades occur during fixations and consist of slow drifts followed by very small saccades (microsaccades) that apparently have a drift-correcting function, and like physiological nystagmus, are physiologically determined. Micromotions are always involuntary and unconscious. They are conveyed by processes of fixations on stable objects.

In activity theory, eye movement data are not used alone as a predictor of cognitive activity. Eye movement analysis is combined with other psychological methods of study. In activity theory, eye movement registration is combined with diverse qualitative methods, which include observation, debriefing, error analysis and registration of external behavior. This is important for the classification of cognitive actions, and particularly for the separation of perceptual and thinking actions, during eye movement analysis. A combination of eye movement and motor movement of the mouse, as objective methods with subjective data, is important at this stage of analysis. This is particularly relevant for separation of perceptual activity from intellectual activity. In this study, this principle is used to study human computer interaction tasks.

At present, observation of the subjects' field behavior have shown highly task-specific eye fixation strategies, and considerable regularity in fixation patterns between observers. This data demonstrates dependence of eye movement strategies on the features of the interface. Observations of natural behavior have demonstrated the highly task-specific nature of eye fixation patterns (Hayhoe et al., 2002).

6.3.2 Functional Analysis of HCI Tasks

From a functional analysis perspective, all tasks include some requirements and conditions. When subjects accept task requirements, they become subject to personal goal. Transformation of task requirements into personal goal is an important step of task performance. This process is always associated with the motivational aspects of activity, because a nonmotivated goal does not exist, in practice, according to activity theory. Motive-goal creates a vector, which gives a goal-directed character to activity during task performance. Differences may sometimes exist between specified task requirements and the subjectively accepted goal. This stage of task performance is called goal acceptance.

From a functional analysis point of view, one should distinguish between two aspects of goal emergence: goal acceptance and goal formation. Goal formation is associated with self-initiated tasks. Goal acceptance is connected with prescribed tasks. As explained later, goal acceptance is considered in this experiment.

This can be encountered in some practical situations. Therefore, this example has some practical implications. At the same time, in HCI tasks, the goal very often is formulated by the user independently. In this study, the goal was presented on the screen, in ready form. The goal concept in activity theory has a totally different meaning in comparison with cognitive psychology. For example, Preece et al. (1998) discuss goal concept in a very different way. According to these authors, the goal may be defined as a state of the system that the human wishes to achieve. The goal is regarded as a final state of the system. From this example, it can be seen that the goal is something externally given, in ready form, to the subject. The goal is regarded as a standard to which the subject approaches. In activity theory, on the other hand, a goal is always associated with some stage of activity. It includes stages such as goal recognition, goal interpretation, goal reformulation, and goal formation. Therefore, even in a simple situation, a goal requires recognition and some aspects of interpretation and acceptance. Later, it will be shown, that the goal is associated with the required active processing of information, even when the goal is presented as a ready standard. The concept of action in cognitive psychology is also defined in a totally different way. For example, Preece et al. (1998) regard action as a task that involves no problem solving or problem structure components. In activity theory, action is also defined differently. It can be cognitive and motor and include some cognitive and problem-solving aspects. This is precisely demonstrated in these studies. Thus, in our study, the goal is presented on the screen in ready form. The goal acceptance process consists of recognition, interpretation and acceptance of the goal. This stage of activity, during functional analysis is associated with functional mechanism, or function block, which is called "goal" in our model of self-regulation (Bedny and Meister, 1997; Bedny and Karwowski, 2003, 2004). These aspects of the goal are totally ignored during task analysis in cognitive psychology.

Such aspects of the task that are known to the subject, or presented as elements of a situation that should be taken into consideration during task performance, are known as conditions of the task. The conditions of the task include elements of situation, rules of transformation of situation, possible alternatives of solution or transformation of situation. Task conditions gradually change during task performance. Therefore, the initial situation, the intermediate situation, and the final situation should be distinguished. Task conditions include two major components, object and tools. In computer-based tasks, a situation is usually presented visually as a number of discrete elements, on the screen. In this case, past experience is also very critical. During task performance, subjects can activate the mental structure, which should be adequate to the external structure of the task. Based on this interrelationship, the subject can create a mental model of the situation. This model can be transformed and changed during the task performance. Under the same task conditions, people can develop different mental models of task, based on subjective understanding and interpretation of the task. It is important to understand that the same external situation on the screen can be constantly changed in the mind of the subject.

Mental models can be stable and dynamic. The functioning of these two closely connected models is associated with those function mechanisms, or function blocks, such as formation of task and subjectively relevant task conditions, in the self-regulation model of activity. Logical organization of stable and dynamic models in space and time provides a dynamic reflection of the situation.

The possibility of performing the same task in multiple ways in the HCI system creates a tremendous diversity of tasks during human–computer interaction. As a result, task complexity causes the initiation of explorative activity during task performance. Exploration is an important component of computer based tasks.

Interaction of tool with elements of orienting activity and past experience are important in the formation of "program of performance." It is another functional mechanism of activity.

In the following stages, the subject is involved in the execution of task performance. At this stage of analysis, it is important to ascertain the possible number of alternative activity strategies of performance execution that are significantly dependent on the orientation components of activity. Therefore, at this step of analysis, the executive stage of self-regulation is discussed. Any initial situation can be changed and evaluated from the point of view of approaching a final situation. The final situation is a result of activity during task performance. Results should be evaluated from the point of view of the goal of activity. Based on this, a subject develops subjective criteria of evaluating the success of performance. Subjective criteria of success can sometimes be derived from the objective criteria or objective requirements. This step of analysis is associated with the evaluative stage of self-regulation. The final step of activity is associated with evaluative mechanisms of self-regulation. Performance of a computer-based task has a lot of similarity with the solution of a problem. Methods of performance can be divided into a number of steps. The subject reorganizes the situation on the screen and then transfers the situation into subtasks. Based on this, the subject develops a new mental model of the situation and, associated with it, a program of performance. After execution, the results are evaluated and then the required corrections are introduced. Therefore, computer-based tasks should always be regarded as problem-oriented tasks. In activity theory, the task is inherently a problem-solving endeavor.

In computer-based task conditions, sometimes task requirements can be presented on the screen. Therefore, three basic areas on the screen can be extracted: goal area (if task requirement is presented on the screen), object area, and tool area according to activity terminology (please note this is not the same as object-oriented terminology in the case of object-oriented programming). Each area, if required, can be divided into subareas of interest. During task performance, a subject interacts with these areas and attempts to find relationships between them. For this purpose, he uses different strategies. From a functional point of view activity is studied as different strategies, which are derived from the mechanism of self-regulation. A more precise description of these strategies are possible, when task is presented as goal, object, and tool areas. Discovering relationships between these areas during performance is important aspect of functional analysis of HCl task.

Depending on the specificity of the task, and the purpose of the study, each of the above listed areas can be divided into more specific subareas of interest. For example, in our task, if it is required, tool area can be subdivided into subareas indicating functional groups or individual icons, depending on the granularity of purpose. Dividing task into areas and subareas (areas of interest) permits a more detailed analysis of the structure of the task depending on the functional purpose of each area and subarea of interest. Therefore, in this study the task is regarded as a system whose components are qualitatively different but interrelated. As can be seen in this section, the major purpose of functional analysis of a computer-based task was described. This is a systemic qualitative method of study. This method should be distinguished from traditional parametric methods of study in cognitive psychology. Parametric and systemic methods (functional analysis) should be combined.

Classification of elements of activity as goal, object and tool areas can be performed based on different levels of decomposition of activity. According to this, there is a need to differentiate those concepts such as goal, object and tool, when researchers study holistic tasks, or when they study separate actions. Each action has its own goal. Therefore, the concepts of goal, object and tool in this case become very dynamic. In the presented discussion, goal, object, and tool concepts are discussed at the level of holistic task analysis. At that level, those elements such as goal, object, and tool have more stable functions. This can particularly be explained by the fact that the goal of a task does not change until the task is completed. This permits us to extract three stable qualitatively describable areas of task: the goal area, the object area, and the tool area of the task.

6.3.3 Description of Experimental HCI Task[1]

Based on the previous description of important aspects of systemic-structural analysis of activity, the experimental method was developed using a hypothetical task and an interface. The interface of task with areas of interest is presented in Figure 6.5. The

[1] This section has been prepared together with Tirthankar Sengupta at New Jersey Institute of Technology.

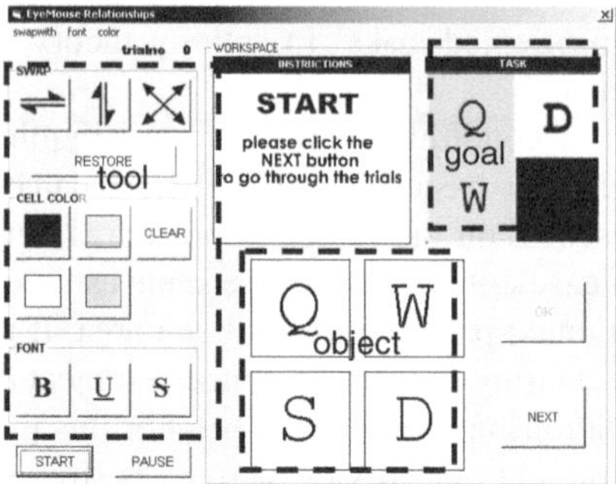

FIGURE 6.5 The interface with areas of interest.

purpose of this experiment was to develop methods and principles of creating activity models during performance of computer based tasks.

According to functional analysis of activity, the following areas of the screen were selected:

1. The tool area. This area of the interface display consisted only of elements of situation (icons), which could be clicked for imparting the feature desired and required by the arrangement.
2. The object area. This area consisted of the elements whose states were to be manipulated in order to achieve the final arrangement given in the goal area.
3. The goal area. This area is constant from the beginning of the trial to the end of the trial. This is the final desired arrangement the user must impart to the objects in order to successfully accomplish the task.

Tool was presented on the screen, designed as icons that perform specific functions. The functional purpose of the tool is presented in Table 6.2.

Subjects, based on a comparison of the goal and object areas, attempt to develop a mental picture or model of the initial situation of the task.

The features of the elements of situation that were manipulated in this experiment were:

1. Position: the location of the symbols with respect to each other
2. Color: the color of the cell containing the symbol
3. The format of the symbol

The task consisted of manipulating the objects for the features available through the tools. As a result, alphabets were altered in position, color and shape as the objects in the task. The main focus of the task is to alter the features of the objects, with available tools, according to the presented goal. The initial state of the objects was without any feature (position, color and format).

Task conditions imposed some constraints on task performance. Here, task constraints were introduced based on the sequence of the use of the tools on the interface.

TABLE 6.2
Tools for Task Designed as Icons on the Interface with Grouping

Position		Color		Format	
	Horizontal position switching		Red	**B**	Bold
			Green		
	Vertical position switching			U	Underline
			Yellow		
	Diagonal position switching			S	Striket hrough
			Blue		

The subjects initially had knowledge of the existence of a constraint but were not cognizant of the type of constraint in question. They were instructed to figure out the constraint while performing the task. Subjects can discover limitations or constraints in task performance only during independent exploration of possible strategies of task performance. In the case of computers and computer-based tasks, an initial phase of exploration is evident. The users used this phase in order to understand the task conditions and its relationship to task requirements. However, all the task conditions were not transparent in the initial exploration. In the course of regular use and different task structures, the user developed and improved the knowledge of the task conditions, and, finally, of the system as it interacts.

A total of 8 subjects completed 128 tasks (16 tasks each) on this interface.

The goal of the task was to impart the features to the different elements of the objects, so that the given arrangement was finally reached. The subject had to select an appropriate sequence of actions in order to impart the desired features to the objects.

The next aspect associated with the design of the interface is the representation of the system constraint. This was only imparted in the functioning of the tools or icons. Overall it is not visible to the user just by looking at the interface. The user only understands the constraint by interacting with the interface to perform the task.

The following methods of data collection were employed in the experiment at the qualitative stage.

1. Video recording of the complete session using interlaced video and eye movement analysis by using eye tracking system
2. Mouse event logging in terms of both the mouse movements and the actions carried out during the task
3. Debriefing and discussion with the subject and expert analysis

A combination of these methods and comparison of obtained data aid the process of description of the activity in a formalized manner and help us develop models of activity during the performance of HCI tasks. Studying eye movements, in combination with motor activity, in this study is derived from the principle of "unity of cognition and behavior" (Bedny et al., 2001). In this experiment, task researchers can shift to parametric, functional, or morphological analysis.

The eye movement data was collected using corneal reflection tracking system, and two video monitors. The first monitor was the eye monitor, for observing the pupil and the corneal reflections, and stayed within the field of vision (25° of visual angle for accuracy purposes of the equipment). The second monitor was the scene monitor, which recorded the complete task performance, along with an interlaced video of the point of regard. A schematic diagram of video data showing point of regards (POR) is given in Figure 6.6.

From functional analysis perspectives, it is important to find out the major possible strategies of task performance. Observation studies of the video of the task sessions suggested that the subjects used different strategies to obtain the required goal. Most of the subjects activated the positioning feature and then performed the color or font change. The subjects preferred to complete either feature by feature or using symbol by symbol. 55% of the subjects completed the task by first completing positioning, then the color and then the formatting features. This observed, and mostly preferred strategy, is analyzed here. Other strategies and comparison of the different strategies, is beyond the scope of this work. However, the final stage of algorithmic analysis requires analysis of different strategies and formulating the best strategy by combining the algorithms obtained. Based on the previous discussion, a complete task is

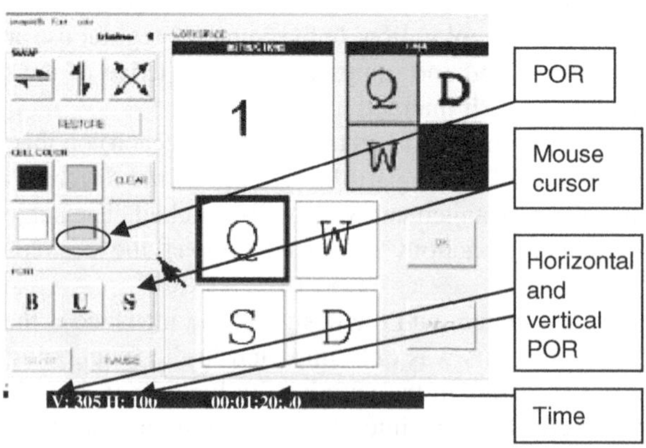

FIGURE 6.6 Schematic diagram of video data showing POR and mouse.

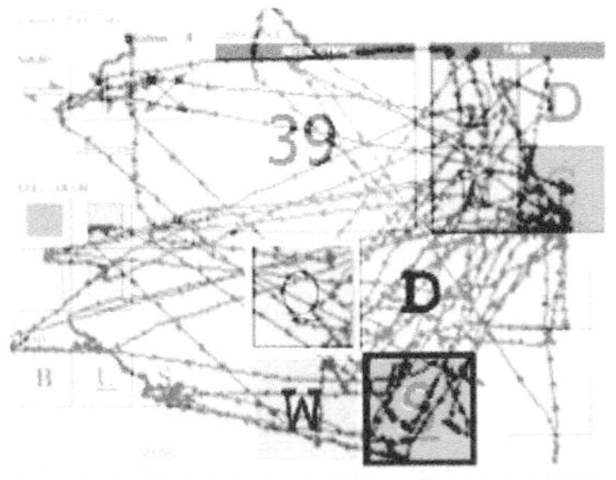

FIGURE 6.7 Cumulative scan path for the complete task.

chosen, and the cumulative scan path of one subject, who follows this strategy for the performance of that task, is demonstrated.

Finally, an example of the parametric method of analysis of eye movement registration data can be demonstrated. This is a traditional method of eye movement analysis, which is used as a preliminary step of data interpretation in the systemic-structural analysis of activity.

The cumulative scan path is obtained by using the point of regard data. The data are plotted on the interface, and the cumulative scan path is shown in Figure 6.7. Using the point of regard data, for this particular task, provides temporal and spatial measures; for example, scan path length (in pixels), as a measure of search behavior; cumulative dwell times or average fixation times, in respective areas of interest, as a measure of difficulty of information extraction; and other relevant measures of performance, such average movement times.

The total dwell time was 17.6 sec; the mean gaze duration for the task 367.45 msec. The total number of fixations was 58 and the total number of saccades was 54. The parameters, based on eye movement data, are useful for analysis of some aspects of design but are not sufficient for all stages of systemic-structural analysis of design. Analysis of a particular design solution supporting an activity, which is a multidimensional system, requires the systemic approach, and hence other methods as well as other associated data are required to explain this activity. In activity theory, it is important to associate all types of behavioral (motor) and cognitive actions in their logical and structural organization sequence. This will be discussed later.

6.3.4 Action Classification Analysis of HCI Tasks[2]

Before going to the next stage of analysis, there is a need to discuss the basis of extraction of unitary action from eye movement and mouse movement data as observed. During task performance, users execute a series of eye movements and gazes in the visual field.

Any saccade is contingent upon the preceding cognitive process, which is assumed to be a portion of the preceding gaze. This gaze includes the program of performance for the following saccade. This assumption is based on the work of different scientists,

[2] This section has been prepared together with Tirthankar Sengupta at New Jersey Institute of Technology.

including Just and Carpenter (1976). In their work on association of eye movements with cognitive processes, they suggested that mental or cognitive processes are performed during a gaze (series of fixations). Hence, considering eye movement time to, and gaze time in, respective areas gives us the opportunity to study cognitive action durations for that particular activity. According to activity theory, the sensory–perceptual process includes a decision-making stage at the sensory–perceptual level (Bedny and Meister, 1997). In the MTM-1 system, eye focus time (EF), which requires approximately 300 msec, includes a decision-making stage about what should be done at the following stage after receiving information (Karger and Bayha, 1977). Viviani (1990) suggested that three processes take place during a fixation (250 to 300 msec) before a saccade. These three processes, include the analysis of the visual stimulus in the fovea field, the sampling of the peripheral field and planning of the next saccade.

Based on this study, formalized rules can be introduced so that the eye movement data can be separated into movement and gaze pairs. These rules are as follows:

1. Since saccades are very quick, it is not possible to execute complex mental operations in such short durations.
2. A mental stage performed during a gaze comprises different operations associated with receiving information, interpretation, etc.
3. The final stage of a gaze also includes a program of performance for the corresponding saccades. This is the point of separation of the two corresponding actions. As a result, one complete eye movement, and one complete gaze duration, following this eye movement is estimated as one complete action.
4. In the case of gaze durations that are longer and include multiple fixations, consideration has to be given to three aspects: the type of eye cursor movement at that point, the actions preceding the gaze and the action following the gaze. Using these three aspects, a fair idea about the type of actions can be obtained.

Hence, summation of the movement time and the associated gaze time provides us with the total approximate time of action. However, if, for example, the subject performs successive perceptual actions involved in extracting information from unfamiliar stimuli requiring the creation of a perceptual image, a series of eye movement and gaze pairs can be integrated into one complete perceptual action. And, in this situation, their components are then regarded as operations for this complicated action. In this study, complex image feature and image formation are not encountered and hence an eye movement gaze pair, as one complete action, is used.

For example, the movement time to the tool area is 100 msec, and the gaze time in the tool area is 300 msec; in order to complete that action of activating the tool, the user has to locate the tool, then select it mentally, and execute the action, while gazing at the particular area. Hence, the total time of action is given by movement time, plus the gaze time, which in this case is 400 (100+300) msec. Table 6.3 presents symbols used for the representation of different elements of a task during its formalized description.

The positions of the interface elements and their symbolic representations are depicted in Figure 6.8.

TABLE 6.3
Symbols Used for Representation of Different Elements of Task

	Tools									
Description	Position			Color				Format		
Elements	Horizontal	Vertical	Diagonal	Red	Green	Yellow	Blue	Bold	Underline	Strike
Representation on the interface	⇄	⇅	⤧	■	□	□	□	**B**	<u>U</u>	~~S~~
Symbols in the table	T_{PH}	T_{PV}	T_{PD}	T_{CR}	T_{CG}	T_{CY}	T_{CB}	T_{FB}	T_{FU}	T_{FS}
	Object elements						Goal elements			
Elements	Q	W	S	D			Q	W	S	D
Representation on interface	Q	W	S	D			Q	W		**D**
Symbols in the table	O_Q	O_W	O_S	O_D			G_Q	G_W	G_S	G_D

This coding system will be used later for development of an action classification table.

The action classification table is based on the qualitative analysis of eye movement data from the video and the duration of gazes in different areas. Since the dwell time is associated with a particular area on the screen, the click that follows it gives us the opportunity to interpret this time in connection with the duration and content of the mental actions that can be performed during this dwell time. Therefore, the basic principles to "penetrate users mind during task performance" and uncover mental components of activity are as follows:

1. The first principle is associated with breaking down of complex unobservable cognitive processes into more elementary mental actions or operations
2. The second principle includes chronometric microanalysis to uncover cognitive actions and their duration
3. The third principle is associated with the unity of cognition and behavior and, derived from this is the method of combination of eye movement and motor movement registration and mutual comparison of obtained data

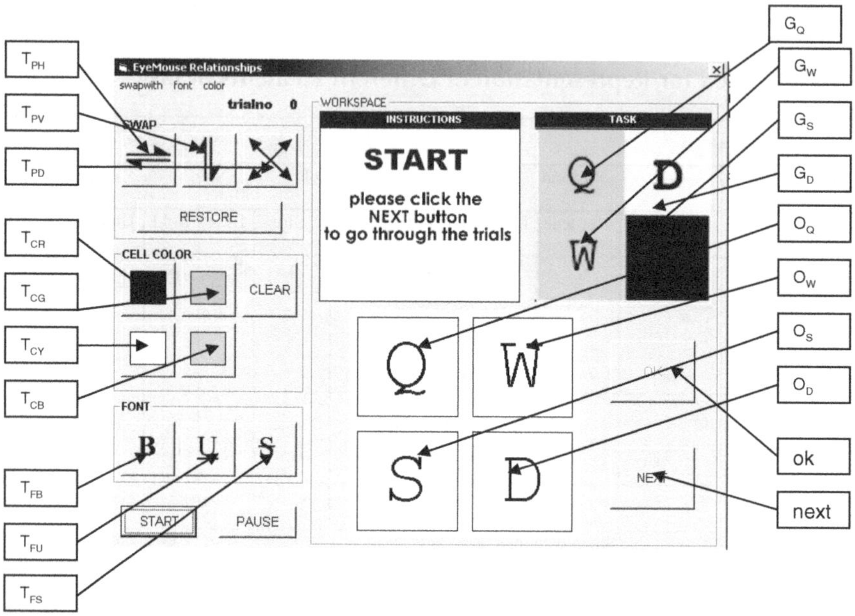

FIGURE 6.8 Coded interface objects for mouse event logging and eye POR qualification.

4. The fourth principle includes combination of eye movement registration and motor registration with qualitative methods of study. Several of these can be mentioned:
   (a) Perform concurrent or retrospective verbal protocol analysis during, or after, task performance
   (b) Cross-examine an expert as to how they typically perform a task or solve a problem; cross-examine a novice about task performance
   (c) Compare novice-expert differences in task performance (differences in strategies, difficulties, typical errors, etc.)
   (d) Change the conditions of task performance from those under which the task was previously learned, and measure task performance in new conditions, and then question the users about performance under different conditions
   (e) Introduce new elements into the task performance, or eliminate some of them, increase or decrease the speed of performance, change sequence of task elements, etc.

In our experiment, the more important methods of study are eye movement and motor movement registrations. The sequence of gazes and movements hence provide us with logical organization of mental actions. For example, consider the action performed in clicking the vertical positioning tool for changing the position of object D, which was previously selected. Here, in order to initiate action, the subject must perform a decision-making action owing to available choices with the position tool group.

Again, in addition to this decision making, he must perform a motor movement of moving the mouse to the tool and then activate the tool. Hence, the actions performed prior to the clicking of the vertical positioning tool can be attributed to thinking and decision making as well as to simultaneous perceptual actions.

Eye movement registration, or derived measures of eye movement data, is sometimes not sufficient in explaining the elements of activity. On the other hand, using scan path images would provide the path of eye fixations but not the dwell times registered at these points. But, in this case, a combination of the two had to be used and associated with activity elements. This is due to the fact that activity theory's basic premise of studying human performance is associated with task, under the principle of unity of cognition and behavior. The use of eye movements, along with events in the activity, is based on the unity of cognition and behavior principles. The rationale for using mouse events was the fact that every action in a computer task is based on previous mental processing. Hence comparison of eye movements with motor actions helps us to infer the purpose of cognitive actions.

Using every action or eye gaze and movement pair in the action classification table is not an economical method. In order to simplify this procedure the eye movement images were associated with the division of the task into segments, which included a logically completed set of actions. Other issues influencing this subdivision were easiness and precise interpretation of the events, as well as the notion of completion of logically related subtasks in an activity and, finally, incorporating the minimum number of subtasks, involving logically completed activity elements of the task in question. This method of associating eye movements and mouse actions with a complete set of cognitive and motor actions in task performance can be easily performed using high-end eye movement equipment available. Hence, the method requires only association of eye movements and mouse actions with the various elements of activity, using relationships of eye and mouse movement, and qualification of the corresponding duration. Combination of mouth clicks with scan path and eye movement dwell times, based on a combination of point of regard data and video recording, supplement each other to correspond to systemic principles of study of activity in HCI.

Eye movement registration and frame-by-frame analysis of video are both related to detail and microanalysis of cognitive and motor activity of a subject. They both can be related to the microstructural level analysis of activity.

Table 6.4 is the action classification table, with associated durations of gazes and movements, as well as the scan path of the point of regard data, showing the visited elements on the screen as presented. The columns of the table are explained as follows:

Column 1 represents the start and end position of the eye during one complete movement and dwell, which changes the focus of the eye. The association of the eyes with the position of the interface elements is based on an approximation of the position of the eye to the nearest element on the screen. For designation of the start and the end position symbols, the interface elements explained earlier (see Table 6.3, Figure 6.8) have been used. For example, the first transition represents the movement of the eye from the "start" position to the element G_Q (goal area — final state of Q). So the total time for movement of the eye from the start position to the position G_Q is 150 msec

TABLE 6.4
The Action Classification Table

1		2	3	4		5	6	7
Eye move and final position		Activity between successive mouse events (clicks) mental/ motor actions involved	Mouse events	Time (ms)		Total action time (a + b)	Classification of actions	Scan path generated/duration
From	To			a. Approx eye movement time to reqd. position (ms)	b. Approximate dwell time at position (ms)			
Start	G_Q	Goal acceptance and formation and creation of subjective model of situation; selection of object (O_D) for subsequent subtask execution. (Includes simultaneous perceptual actions, with explorative thinking. Comparison of object and goal in relation to the program of performance.)		150	180	330	Simultaneous perceptual actions	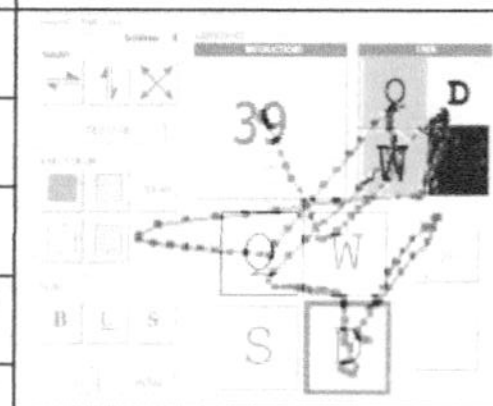 Image 1
G_Q	O_q			150	220	370	Simultaneous perceptual actions	
O_q	T_{CB}			180	150	330	Simultaneous perceptual actions	
T_{CB}	G_S			180	220	400	Thinking action based on visual information	
G_S	G_D			150	190	340	Thinking action based on visual information	
G_D	O_q			210	220	430	Thinking action based on visual information	
O_q	O_W			150	330	480	Thinking action based on visual information	
O_W	O_D			150	190	340	Thinking action based on visual information	

O_D	G_D	Decision on program of performance and motor action based on decision		210	190	400	Thinking action based on visual information	
G_D	O_D	Perceptual action with motor action and thinking based on program of performance	Click object element D with mouse	210	630	840	Decision-making action at sensory perceptual level. With simultaneous motor action	
O_D	T_{PV}	Motor action of eye and mouse along with selection from choice of tools		210	220	430	Simultaneous perceptual action with motor action	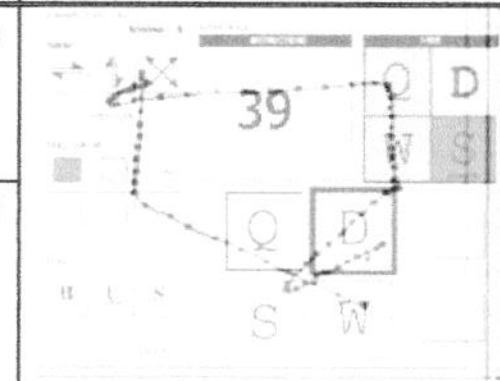 Image 2
T_{PV}	G_D	Eye move to goal area with mouse stationary at tool. Use of peripheral vision for mouse control while focus on goal area	Click vertical positio ning tool.	150	420	570	Simultaneous perceptual action decision-making action during visual assessment. Motor action	
G_D	O_D	Perceptual action on object for developing and		180	220	400	Decision-making action at sensory perceptual level	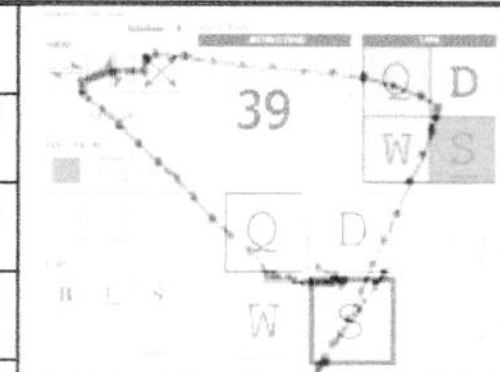 Image 3
O_D	O_Q			90	220	310	Thinking action based on visual information	
O_Q	O_D			150	250	400	Thinking action based on visual information	
O_D	O_S			150	250	400	Thinking action based on visual information	
O_S	G_W	Decision involving program of performance and tool choice for the next course of action		150	180	330	Thinking action based on visual information	
G_W	T_{PH}			150	330	480	Thinking action based on visual information	
T_{PH}	O_Q			150	180	330	Thinking action based on visual information	

(Continued)

TABLE 6.4
Continued

O_Q	O_S	Initiate motor action and start of next subtask based on accepted program of performance based on intermediate stage	Click of object S	60	220	280	Decision-making action at sensory perceptual level. With simultaneous motor action	
O_S	T_{PH}	Tool selection and thinking action for assessing the outcome of action		330	870	1200	Thinking action, Decision-making action at sensory perceptual level. With simultaneous motor action	
T_{PH}	G_S	Perceptual action for desired outcome	Click horizontal position tool	180	220	400	Simultaneous perceptual action with motor action	
G_S	O_Q	Comparing desired outcome to actual using simultaneous perceptual action	Click to select object Q	150	180	330	Decision-making action at sensory–perceptual level	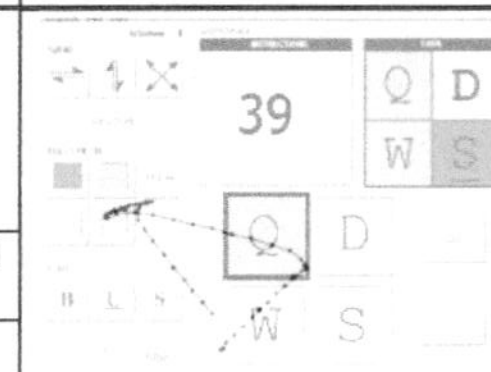 Image 4
O_Q	G_S	Visit goal for next stage of performance		150	220	370	Thinking action based on visual information	
G_S	T_{CB}	Tool selection based on accepted program of performance		180	240	420	Decision-making action at sensory–perceptual level	

T_{CB}	G_Q	Comparing desired outcome to actual using simultaneous perceptual action	Click blue coloring tool	180	220	400	Decision-making action at sensory–perceptual level. With simultaneous motor action	
G_Q	O_D	Perceptual action for identification of next set of actions to be performed for completion	Click object D	150	330	480	Simultaneous perceptual action with motor action	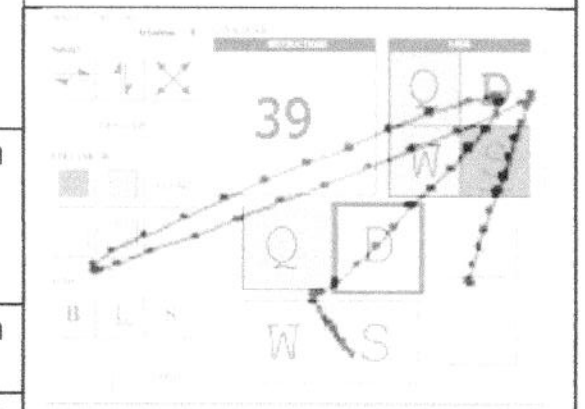
O_D	T_{CY}	Tool selection and initiate motor action		180	330	510	Simultaneous perceptual action with motor action	
T_{CY}	O_D	Perceptual action for getting feedback on action	Click yellow color tool	210	210	420	Simultaneous perceptual action	Image 5
O_D	O_W	Moving to the next object towards completion of task and subsequent motor action for selection	Click object W	150	390	540	Simultaneous perceptual action with motor action Decision-making action at sensory–perceptual level	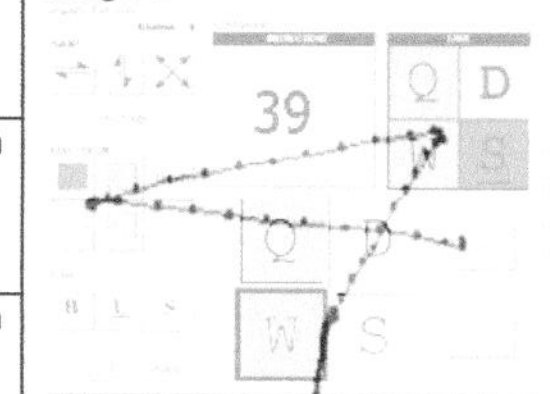
O_W	T_{CG}	Tool selection based on program of performance	Click green color tool	150	720	870	Simultaneous perceptual action with motor action Decision-making action	Image 6
T_{CG}	O_S	Less cognitive overload on feedback requiring less time and includes the corresponding action of moving on with the next stage of task	Click object S	150	330	480	Simultaneous perceptual action with motor action	
O_S	G_S	Compares goal		150	330	480	Thinking action based on visual information	

(Continued)

TABLE 6.4
Continued

G_S	T_{CR}	Tool selection based on desired program of performance		180	780	960	Decision-making action at sensory–perceptual level. Motor action Memory action	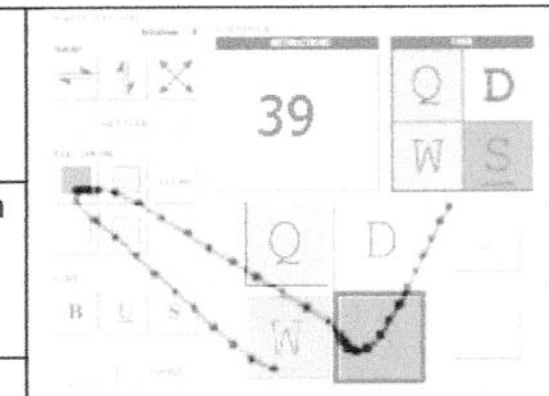 Image 7
T_{CR}	G_w	No feedback required from object directly goes to the goal for formulating next set of program	Click red color tool	180	190	370	Sensory–perceptual action with motor action	
G_w	G_S			60	190	250	Thinking action	
G_S	T_{FU}, T_{FS}	Misinterpretation of feature and improper tool selection	Click under line tool	180	900	1080	Thinking action Decision-making Motor action	
T_{FU}, T_{FS}	O_S	Corrective action based on feedback and proper tool selection	Remove under line	180	220	400	Sensory–perceptual action with motor action	
O_S	G_S		Click strike through	120	220	340	Thinking action	
G_S	G_Q			120	250	370	Thinking action	
G_Q	G_D			180	240	420	Thinking action	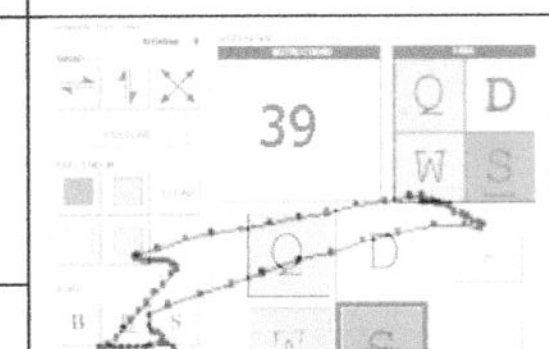 Image 8
G_D	O_D	Object selection based on program of performance	Click object D	240	250	490	Decision–making Simultaneous perceptual and motor action	

O_D	T_{FB}	Tool perception and decision on selection		150	250	400	Decision-making with simultaneous perceptual action and feedback	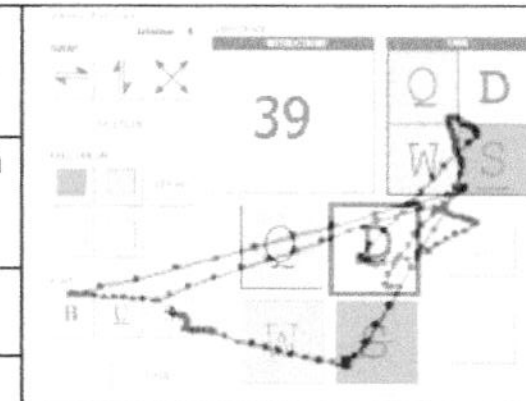 Image 9
T_{FB}	G_W	Tool action with focus on feedback	Click bold tool	120	240	360	Simultaneous perceptual action with motor action	
G_W	O_D	Assessing status of objects for finalizing		150	930	1080	Memory action Thinking action	
O_D	OK	Decision-making for end of task. Subjective assessment of completion of task	Finish	120	250	370	Decision-making action with simultaneous perceptual action motor action	
OK	G_S	Corresponding perceptual action on feedback and goal for assessing incorrectness and incompleteness in task		150	220	370	Simultaneous perceptual actions	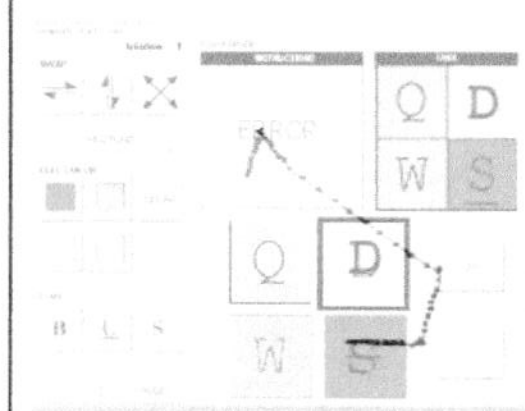 Image 10
G_S	Feed			150	210	360	Simultaneous perceptual actions	
Feed back	G_W			120	540	660	Thinking action with simultaneous perceptual actions	

(Continued)

TABLE 6.4
Continued

G_W	O_Q	Revisit objects for detection of error perceptual actions with memory action		150	600	750	Memory action, thinking action with simultaneous perceptual actions	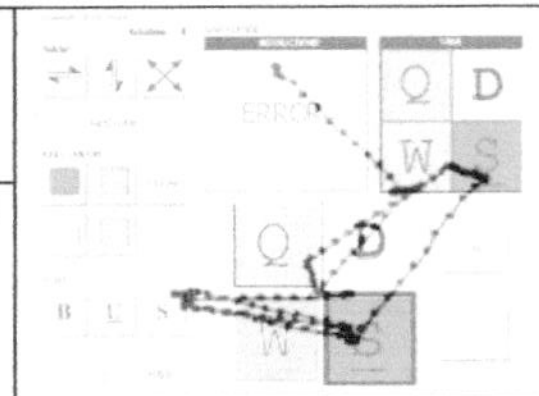
O_Q	G_S	Visit goal for finalizing program of performance based on error detection Decision on program of performance		180	220	400	Simultaneous perceptual action	
G_S	O_S	Motor action for executing program of performance	Select object S	150	220	370	Decision-making with motor action	Image 11
O_S	T_{FU}	Tool selection for correction of error	Remove strike	150	220	370	Simultaneous perceptual action with motor action	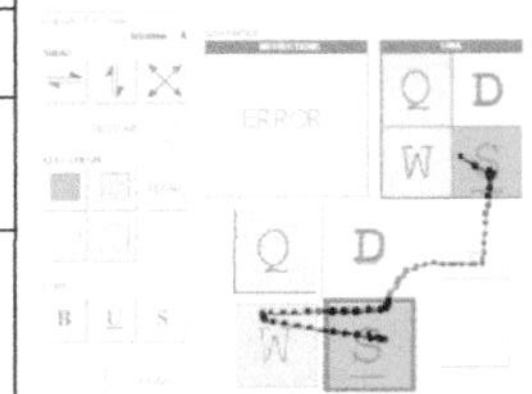
T_{FU}	O_D	Final comparison of goal and acceptance of completion of task	Click under line	150	280	430	Simultaneous perceptual action with motor action	
O_D	OK		Finish	120	390	510	Memory action thinking action and decision making at sensory perceptual level	Image 12
OK	Finish						Simultaneous perceptual action with motor action	

(given in column 4a). The dwell times at the end position (i.e., at G_Q), or the fixation time is given in column 4b. The summation of all the elements of times, represented in columns 4a and 4b, is given in column 5, which represents the total action time, since the total action time is given by the sum of the gaze time at the particular area or the element and the movement time to the particular element. Column 3 represents mouse actions in the same time line. Column 2 describes segment of activity between clicks. The scan path images in column 7 represent the total scan path generated between mouse actions and hence provide an understanding of the occurrence of the actions in the visual field. Description of the eye movements between motor actions and their time measures are given in Table 6.4. A total of 19 actions were done in order to generate 8 features, with 4 additional actions due to errors. The images in column 7 represent the eye movement data during task performance. Column 6 represents standardized description of actions according to developed principles of their classification described in this book.

Let us consider the first fragment of the task that is associated with image 1 in column 7 (Table 6.4). This image demonstrates that after performing a number of cognitive actions the subject selects element D in the object area and clicks on it. As a result this element is activated and its frame is highlighted. This means that the subject wants to shift this element into another position. In this case the subject spends most of his/her time on O_D, O_W, O_D and G_D and finally selects O_D. The sequence of an eye movement along with time of dwell suggests that the subject is more inclined to act on object D and might consider switching the position of the objects D and W.

The following is the sequence of actions: The eye moves from the start position to the position G_Q (goal area element Q). Dwell time of this movement is 180 msec and the total time for the action is 330msec. At this stage the subject just wants to identify the goal of task. Therefore this is perceptual action. According to the existing in SSAT action classification system it is a simultaneous perceptual action. The angle of the operative visual field is $\alpha \approx 10°$, therefore the subject can simultaneously perceive 4–6 elements. Hence the subject can perceive not only one letter but all four. The next eye movement begins from element G_Q to O_Q (see column 1 and column 7, where O_Q means object area of letter Q). This is the first shift of eyes into object area. Duration of the dwell time is 220 msec and the total time of this action is 370msec. The subject attempts to receive some general information. Therefore he moves his/her eyes to the tool (see column1, and image 1), Eyes move from element O_Q to element T_{cB} (tool element, color blue). This is also a perceptual action. At the next stage eyes shift from the tool area to the goal area again (from T_{cB} to G_s where the last symbol means goal area of letter S). The purpose of this movement is not only for receiving information (perception). For the first time the subject starts to pay attention to the relationship between the elements of the situation. The relationship between elements of the situation and the task goal is not a perceptual feature of the task but rather is the feature of the task that requires involvement of the thinking process that is performed based on visual information. The duration of the eye movement increases as a result. According to the existing classification system of actions it is a thinking action that is performed based on visual information (see Section 2.2.5). This is an example of the simplest thinking action. Eyes move from element G_s to element G_D. Similarly the purpose of this eye movement is to find out relationship between elements of the situation. Hence, it is also a thinking action performed based on visual information. Let us considered the last action in this fragment of the task.

The subject activates element D in the object area, and moves it up as demonstrated in image 2. This element has been previously selected and activated (highlighted by bold line, Table 6.4, image 1). Before activating element D in the object area the subject should perform a decision-making action due to choice the position in the tool group. In addition to this decision-making, the subject performs a motor action (move the mouse to the tool and execute the tool). Hence, mental action before the clicking of a vertical positioning tool can be classified as a decision-making action that has been based on visual information. According to the existed system of action classification (see Section 2.2.5) this is a simple decision-making action at a verbally thinking level. During this decision-making the subject moves the mouse to the tool area.

The scan path does not give the total picture about possible cognitive actions. The dwell times that follows the movement defines how much importance or attention each place requires. The longer the dwell time the higher the probability that the thinking process is involved in the task performance at that period of time. The information about the duration of eye movement, dwell time and cognitive action in general can be obtained from columns 4a, 4b, and 5 that are adjacent to image 1. The scan path in image 1 suggests the following sequence Start-G_Q-O_q-T_{CB}-G_S-G_D-O_q-O_W-O_D. This sequence suggests that the subject is trying to change the position from O_D to O_W. Evidently O_D is selected and the subject is switches the position of O_D and O_W. We can not consider details of all eye movements. The other fragments of the task will be discussed below in abbreviated manner.

The eye movement data in column 7 can be used for the functional analysis of activity during task performance. This stage of analysis is dedicated not so much to separate actions but rather to general strategies of performance and their relation to such functional mechanisms as a goal, subjectively relevant task conditions (dynamic mental model), formation of a program of task performance, etc. For example analysis of image 1 demonstrates that the subject first attempts to receive information about the goal, and then about the object area. The eye scan path suggests that the comparisons between the final required state (goal) and the initially given state in the object area is taking place. Thinking actions are required in order to evaluate the goal in a more specific manner through comparison of subjectively accepted task requirements (goal) with an initial state of the situation. Subjects attempt to develop a mental picture or model of initial situation of the task based on comparison of the goal and the object areas. Eye movement registration demonstrates how mental model of a situation is developed. There is considerable dwell time on the object and the goal area at the first stage of the task performance (image 1). In the next image (image 2) the focus shifts to the tool area in order to develop a plan of execution and to choose the tools that fit the corresponding actions. Therefore the general strategy of task performance includes interpretation and acceptance of the goal, development of a mental model of the situation based on comparison of a goal and object areas, evaluation of the tool area and development of the plan of actions accordingly.

By shifting eyes into a goal area a subject can simultaneously receive information about all four elements located in the goal area (perceptual action). However, while switching to the thinking actions the subject's attention is concentrated on the individual elements of the goal area and their functional relationship to the elements in the object area. Hence, image 1 in table 3 demonstrates that a subject at this stage does not simply receive information about a goal and an object areas but rather attempts to find out the functional relationship between different elements and comprehend what is the final goal of performance and the initial state of the object at this stage. Then a subject formulates a sub-goal of the task. So, the goal can not be considered simply as the end state to which behavior is directed. An objectively given goal is subjectively interpreted and accepted. During functional comparison of the final goal elements with the object area elements a subject formulates a more specific goal, evaluates initial state and develops a mental model of the situation. Based on these mental actions and operations the decision is made about what the intermittent goal should be. Such intermittent goals description and classification requires morphological analysis.

An analysis of the above eye movements in the action classification table (Table 6.4) demonstrates that during development and interpretation of this data some combination of functional and morphological analysis elements is used. A functional analysis involves paying attention to eye movements in different areas of interest and strategies of task performance. When a specialist studies action classification it is a morphological analysis. These methods are very often difficult to separate from each other. A functional analysis helps to discover preferable strategies of task performance. The mental models demonstrate that the subject disengages him/herself from such features of the object as color and symbol formation and decides to transform a situation according to the space criterion. This stage in task performance is shown in image 2 where the scan path reveals more transitions of the eyes to the tool area G_D-O_D-T_{PV}. Here the subject is more concerned with the completion of the task and decides on a course of actions based on the importance of the assessed tools. The third image reflects the subject's engagement in the completion of the arrangement of the letters' positions: G_D-O_D-O_Q-O_D-O_S-G_W-T_{PH}-O_Q-O_S-T_{PH} (see image 1 and 2 in Table 6.4). The next four images show the manipulation with the color features on the objects: G_S-O_Q-G_S-T_{CB} (image 4 in Table 6.4), G_D-O_D-T_{CY} (image 5 in Table 6.4), O_D-O_W-T_{CG} (image 6 in Table 6.4) , and O_S-G_S-T_{CR} (image 7 in Table 6.4). Although the strikethrough formation is wrong the subject overlooks the mistake and goes on through the next trial G_W-O_D-O_K (image 10). The action is completed by using the bold and the underline tools. Thus an error is reported in terms of using the underline tool where the subject mistakenly uses the strikethrough tool. When the subject issues the commit command that is the OK button (G_S-Feedback) the error is subsequently detected and the subject's focus immediately shifts to the object area. Here most of the dwell time is spent once again in the object and the goal areas in order to detect the difference in the letters positions (G_W-O_Q-G_S-O_S-T_{FU}-O_D). Finally once the difference is detected the subject easily furnishes the task using the complete feature. However, before going on to the next trial the user checks the arrangement in comparison to the given goal. It is an evaluative stage of the subtask performance. It can be observed that eye movements basically follow the natural pace of the task performance.

6.3.5 Algorithmic Description of HCI Task at Microstructural Level of Analysis

In contrast to mathematical and computer algorithms, a human algorithm describes the logical organization of actions and operations performed by a human subject. Human algorithms depend on how the subject chooses a strategy of task performance, which very often is optional for that user during the performance of HCI tasks. The selection of a possible strategy is frequently based on intuitive criteria and depends on both the significance and the difficulty of the task. For example, when the time of performance is a significant factor, the user can select a strategy that increases the speed of performance. In contrast, when precision is important, the user can prefer to use a strategy associated with a slower pace of performance. If the user decides to work for long periods of time, a strategy can be selected that helps to sustain the effort during the required period of time. Past experience is critically important in this process. In most cases, more representative strategies should be selected. For example, this can be either the slowest or the fastest possible strategies in particular situations (Bedny and Meister, 1997). More detailed analysis of possible strategies of activity can be performed based on analysis of the mechanisms of self-regulation (Bedny and Meister, 1997; Bedny and Karwowski, 2004).

Knowing possible strategies of activity is especially important for algorithmic analysis of computer based tasks. Algorithmic description of task performance consists of the subdivision of an activity into qualitatively distinct psychological units, with the following determination of their logical organization. Such elements are called members of the algorithm. They consist of one or several cognitive or motor actions, integrated through a higher order goal or the goal of this algorithm member. Owing to capacity limits of working memory, a member of an algorithm is usually restricted by 1 to 4 integrated actions. This is particularly relevant for situations when the worker coordinates a sequence of actions.

The member of an algorithm can be classified according to qualitative characteristics. First of all, operators and logical conditions need to be distinguished. Operators associated with receiving information are called afferent operators and are designated by symbol O^{α}. Operators associated with extraction of information from long-term memory, or keeping information in working memory, are designated as O^{μ}. Operators involved in executive components of the activity, for example, any motor actions, are called efferent operators and designated by O^{ε}. Any member of an algorithm which includes comparison between elements of a situation, discovering the functional purpose of symbols on the screen, their interrelationship, performance of logical actions, etc., is designated as O^{th}. This means that this member of the algorithm describes thinking activity. Thinking actions, based on visual information, are very often performed during human–computer interaction and are based on visual information. Hence, the member of an algorithm can include thinking actions that are performed based on visually presented data. In this situation, the symbol $O^{\alpha th}$ is used.

Logical conditions determine the logic of selection and the realization of different members of an algorithm and also include a decision-making process. They can be designated as l or L (based on a combination of several logical conditions). The symbols l or L have an associated arrow with a number on top, for example, $l_1 \uparrow^1$. Number one means that it is the first logical condition, and the arrow also has the number one at the top because it belongs to the first logical condition. This logical condition has only two outputs, respectively, 0 and 1. More complicated logical conditions have more then two outputs. Each output can occur with a different probability. Therefore, one can distinguish two basic kinds of algorithms. One is a deterministic algorithm. It includes only simple logical conditions with two outputs, respectively, 0 and 1. The second kind is a probabilistic algorithm, which includes logical conditions with more then two outputs, which can occur with different probabilities. It is a probabilistic algorithm. Computer-based tasks can usually be described by a probabilistic algorithm. For example, $l_1 \uparrow^{1(1-6)}$ indicates that this is the first complicated logical condition, one which will have six possible outputs as $\uparrow^{1(1)}, \uparrow^{1(2)}, \uparrow^{1(3)} \ldots \uparrow^{1(6)}$. The logical formula of an algorithm is presented in the left column of the algorithmic description table. The arrow after logical conditions $\uparrow^{(1)}$ indicates a transition from one member of an algorithm to another (e.g., $\uparrow^1 \downarrow^1$).

This means that a logical condition, according to output address, is indicated by a downward arrow, which is associated with another member of the algorithm. The tabular form of an algorithm is read from top to bottom. The left column with symbolic description is called the formula of the algorithm (see Table 6.5).

Algorithmic description of task can be performed with different levels of detail. If most actions are cognitive, and of very short duration, then it is appropriate to use algorithmic description at a microlevel of analysis. In the work of Bedny and Karwowski (2003), a macro level of algorithmic description is presented. Algorithmic description is a model of activity during task performance, which approaches real

TABLE 6.5
Symbols for Generating the Algorithm of Task Performance

Symbol	Denotes	Description
O	Operator	Consists of actions that transform objects energy and information
l, L	Logical operation	Determine the logic of selection and realization of different members of algorithm and include a decision-making process. Can be designated as l or L (based on a combination of several logical conditions)
α	Perceptual operator qualifier	Receiving information for example O^{α}
μ	Memory involvement qualifier	Involvement of memory actions O^{μ}
ε	Executive operator qualifier	Executive action in terms of motor performance O^{ε}
th	Thinking involvement qualifier	Involvement of thinking action O^{th}

performance of users. Following the development of the algorithm, an expert performs psychological analysis of the algorithm, returning to a qualitative stage of analysis. Each member of algorithm can be evaluated as a subsystem of activity.

During an experimental study, three more representative strategies were discovered. Table 6.6 shows the algorithm of performance for the task presented in Table 6.4. In this study, only one representative strategy of performance is described. Therefore there is no need to consider logical conditions and hence there is no error due to the existence of different output, that leads to the next member of the algorithm. The probability of this algorithmically described strategy is 0.4. From the action classification table, sets of actions can be distinguished, which can be attributed, respectively, to individual algorithms of performance. This is one way to describe the structure of holistic activity, in contrast to analysis of separate aspects of activity during task performance. Therefore, this is an example of the systemic description of activity during task performance. Presented algorithmic model of user activity utilizes different units of analysis. The first column in the left describes members of the algorithm in symbolic form. They are classified according to standardized psychological principles. The second column describes members of the algorithm as user actions by utilizing typical elements of task or technological units of analysis. This method is simply a description of what user did at different stages of task performance. The third column described members of the algorithm by utilizing psychological units such as actions or operations. These units are also classified according to standardized principles. The very right column presents the duration of different members of the algorithm or if required duration of the separate actions or operations. This allows to develop time structure of activity and evaluate task complexity and reliability of task performance at the further steps of analysis. The algorithmic description gives a fairly good idea of human performance for a particular situation. In cases where existing designs need to be evaluated for changes, this technique represents an ideal solution to comprehensively analyze for faults in an existing design.

6.3.6 Improvement of the Algorithm

The algorithm presented in Table 6.6 can be studied for improvement of the task sequence and, hence, the strategy used by users. In this case, for example, look at the members of algorithm O^{ε}_{18} and O^{ε}_{22}. Both these members are related to the selection of the element O_Q. But the respective member of the algorithm is performed at two different stages of the sequence. This results from the fact that the user performs the same member of the algorithm repeatedly while performing this task. All the algorithm members of following these two selections, could have been done with one single member of algorithm O^{ε}_{18}, followed by the sequence of all the other algorithm elements (O^{α}_{19}; l_5; O^{α}_{20}; and O^{α}_{21}). Then, for another example, consider algorithm O^{ε}_{28}. This is, once again, a repetition of the algorithm element O^{ε}_{5}. Hence, all the algorithm elements after O^{ε}_{28} can be done after O^{α}_{17}, once the position of the element O_D is changed to the desired location. This is appropriate sequence of actions for task performance. Considered strategy reduces the number of mental actions required for subjects to carry out the task. Examining the algorithm on this basis can help remove a significant number of elements in order to complete the task using fewer members

TABLE 6.6
Algorithm of Task Performance

Algorithm	Description	Actions obtained from action classification table	Time (ms)
O_1^{α}	Look at goal area and initial state of object area	Simultaneous perceptual actions (3 actions)	1030
$O_2^{\alpha th}$	Find out differences between goal area and object area	Thinking actions based on visual information (4 actions)	1650
$O_3^{\alpha th}$	Find out differences between goal area and object area and simultaneously perform O_4^{ε}	Thinking actions based on visual information (2 actions)	740
O_4^{ε}	Move cursor close to object area	Simple motor action	
l_1	Decide to click object (element O_D) and simultaneously perform O_5^{ε}	Decision-making action based on information from memory	840
O_5^{ε}	Simultaneously with l_1, click object element O_D	Simple motor action	
O_6^{α}	Look at tool area and simultaneously perform O_7^{ε}	Simultaneous perceptual action	430
O_7^{ε}	Simultaneously with O_6^{α} move cursor closer to tool area	Simple motor action	
l_2	Decide to click tool (element T_{PV}) and simultaneously perform O_8^{ε}	Simultaneous perceptual action Decision-making action during visual assessment	570
O_8^{ε}	Simultaneously with l_2 move cursor close to specific icon and click icon	Average precision motor action	
$O_9^{\alpha th}$	Evaluate how object area matched to goal area	Thinking action based on visual information	400
$O_{10}^{\alpha th}$	Evaluate intermittent state of object area	Thinking actions based on visual information (4 actions)	1740
O_{11}^{α}	Look at goal area and then look at tool area	Simultaneous perceptual actions (2 actions)	
l_3	Decide to click object (element O_S) and simultaneously perform O_{12}^{ε}	Decision-making action at sensory–perceptual level	280
O_{12}^{ε}	Simultaneously with l_3, click object element O_S by using mouse	Simple motor action	
O_{13}^{α}	Look at tool area and simultaneously perform O_{14}^{ε}	Simultaneous perceptual action with partly overlapping motor action (see below)	1200
O_{14}^{ε}	Simultaneously with O_{13}^{α} move cursor close to specific icon (horizontal position tool)	Average precision motor action	

(Continued)

TABLE 6.6
Continued

Algorithm	Description	Actions obtained from action classification table	Time (ms)
O^{α}_{15}	Look at object area to evaluate change of position of objects (O_S and O_W — horizontal shift) and perform O^{ε}_{16}	Simultaneous perceptual action with motor action (see below)	400
O^{ε}_{16}	Click tool to activate action simultaneously performed with O^{α}_{15}	Simple motor action	
O^{α}_{17}	Continue looking at object area	Simultaneous perceptual action	
l_4	Decide to click object (element O_Q) and simultaneously perform O^{ε}_{18}	Decision-making action at sensory–perceptual level	330
O^{ε}_{18}	Click object element O_Q	Simple motor action	
O^{α}_{19}	Look at goal area to evaluate color of elements	Simultaneous perceptual action	370
l_5	Decide to click blue icon tool	Decision-making action at sensory–perceptual level	420
O^{α}_{20}	Look at tool area and simultaneously perform O^{ε}_{21}	Simultaneous perceptual action with motor action	400
O^{ε}_{21}	Move cursor to tool area	Average precision motor action	
l_6	Decide to click object (element O_Q) and simultaneously perform O^{ε}_{22}	Simultaneous perceptual action with motor action Decision-making action at sensory–perceptual level	330
O^{ε}_{22}	Click object element O_Q	Simple motor action	
O^{α}_{23}	Look at tool area and simultaneously perform O^{ε}_{24}	Simultaneous perceptual action with motor action	420
O^{ε}_{24}	Move cursor to blue color tool	Average precision motor action	
O^{α}_{25}	Look at object area and simultaneously perform O^{ε}_{26}	Simultaneous perceptual action with motor action	400
O^{ε}_{26}	Click blue color tool performed simultaneously with O^{α}_{27}	Simple motor action	
O^{α}_{27}	Continue looking at object area	Simultaneous perceptual action	
l_7	Decide to click object O_D and simultaneously perform O^{ε}_{28}	Decision-making action at sensory–perceptual level	480
O^{ε}_{28}	Move cursor to O_D and click O_D	Average precision motor action	
O^{α}_{29}	Look at tool area and simultaneously perform O^{ε}_{30}	Simultaneous perceptual action with motor action	510
O^{ε}_{30}	Move cursor to yellow color tool	Average precision motor action	

(Continued)

TABLE 6.6
Continued

Algorithm	Description	Actions obtained from action classification table	Time (ms)
O^{α}_{31}	Look at object area and simultaneously perform O^{ε}_{32}	Simultaneous perceptual action with motor action	420
O^{ε}_{32}	Click yellow color tool	Simple motor action	
O^{α}_{33}	Continue looking at object area	Sensory perceptual action with motor action	540
O^{α}_{34}	Look at tool area (green color tool) and simultaneously perform O^{ε}_{35}	Sensory perceptual action with motor action	870
O^{ε}_{35}	Move to green color tool and click green tool	Average precision motor action	
O^{α}_{36}	Look at object area and simultaneously perform O^{ε}_{37}	Sensory–perceptual action with motor action	
O^{ε}_{37}	Click green color tool	Simple motor action	
O^{α}_{38}	Continue looking at object area and simultaneously perform O^{ε}_{39}	Simultaneous perceptual action with motor action	480
O^{ε}_{39}	Move cursor to object area	Simple motor Action	
O^{α}_{40}	Look at goal area and perceive color of goal element S; simultaneously perform O^{ε}_{43}	Simultaneous perceptual action with motor action	480
O^{ε}_{41}	Click object element O_S and simultaneously perform O^{α}_{42}	Simple motor action	
O^{α}_{42}	Look at tool area (red color tool) and simultaneously perform O^{ε}_{45}	Simultaneous perceptual actions	960
O^{ε}_{43}	Move cursor to red color tool	Average precision motor action	
O^{α}_{44}	Look at object area (O_S) and simultaneously perform O^{ε}_{45}	Simultaneous perceptual action with motor action	370
O^{ε}_{45}	Click red color tool	Simple motor action	
O^{α}_{46}	Look at goal area	Simultaneous perceptual action	250
O^{α}_{47}	Look at tool area and simultaneously perform O^{ε}_{48}	Simultaneous perceptual action with motor action	1080
O^{ε}_{48}	Move cursor to tool area	Simple motor action	
$O^{\alpha th}_{49}$	Evaluate function of tools	Thinking action under visual information	400
l_9	Decide to use strike tool	Decision-making action	340
O^{α}_{50}	Look at goal area and simultaneously perform O^{ε}_{51}	Simultaneous perceptual action with motor action	370
O^{ε}_{51}	Move cursor to object area	Simple motor action	
O^{α}_{52}	Continue looking at goal area	Simultaneous perceptual action	420
l_{10}	Decide to change object O_D	Decision-making action	245

(Continued)

TABLE 6.6
Continued

Algorithm	Description	Actions obtained from action classification table	Time (ms)
O^{α}_{53}	Look at object O_D and simultaneously perform O^{ε}_{55}	Simultaneous perceptual action with motor action	245
O^{ε}_{54}	Click object element O_D	Simple motor action	
O^{α}_{55}	Look at tool area and simultaneously perform O^{ε}_{57}	Simultaneous perceptual action with motor action	400
O^{ε}_{56}	Move to tool element bold	Average precision motor action	
O^{α}_{57}	Look at goal area and simultaneously perform O^{ε}_{59}	Simultaneous perceptual action with motor action	360
O^{ε}_{58}	Click bold tool	Simple motor action	
$O^{\alpha th}_{59}$	Look at goal and object area for finalizing status of objects as per the requirements of the goal	Thinking action	1080
$l_{10}\uparrow^{1}$	Look at object area and simultaneously decide to perform either O^{α}_{61} or O^{ε}_{62}	Decision-making for finalization of task completion	370
O^{α}_{60}	Look at object area and simultaneously perform O^{ε}_{62}	Simultaneous perceptual action with motor action	
O^{ε}_{61}	Move cursor to finish button	Simple motor action	
O^{α}_{62}	Continue looking at object area and simultaneously perform O^{ε}_{64}	Simultaneous perceptual action with motor action	370
O^{ε}_{63}	Click finish for feedback	Simple motor action	
O^{α}_{64}	Look at feedback area (area showing error)	Simultaneous perceptual action	360
$O^{\alpha th}_{65}$	Evaluate error information	Thinking action	660
$O^{\alpha th}_{67}$	Look at object area to detect source of error and evaluate error	Thinking action	750
O^{α}_{68}	Look at goal area	Simultaneous perceptual action	400
O^{α}_{69}	Look at object area and simultaneously perform O^{ε}_{70}	Simultaneous perceptual action with motor action	370
O^{ε}_{70}	Click object element O_S	Simple motor action	
O^{α}_{71}	Continue looking at tool area and simultaneously perform O^{ε}_{72}	Simultaneous perceptual action with motor action	370
O^{ε}_{72}	Click strike to remove effect	Simple motor action	
O^{α}_{73}	Look at object area and simultaneously perform O^{ε}_{74}	Simultaneous perceptual action with motor action	430
O^{ε}_{74}	Click tool underline	Simple motor action	
O^{α}_{75}	Look at object area and simultaneously performO^{ε}_{76}	Simultaneous perceptual action with motor action	510
O^{ε}_{76}	Click OK to complete trial	Simple motor action	

of algorithms or introducing more efficient methods of performance using different members of algorithm.

Now comes the turn of each member of algorithm, step by step. In this case, it can be observed that the users devote quite a lot of time to thinking actions. It is not the total cumulative duration of the thinking actions that is important but the average times required for each thinking action to be focused. In this case, the thinking actions can be first approached in order to understand any difficulty the users are facing. For example, the initial stage of the task shows a high increase in the number of thinking actions owing to the comparison of the object and the goal areas elements. However, at a later stage of the task, these thinking action durations are quite low. Hence, it can be suggested that using cues and instructions, at least at the initial stage of task performance, can reduce the thinking actions. However, it should be noted that the use of cues makes it additionally burdensome to understand the instructions. Consequently, the complexity involved in the total task may be reduced, even though the time may increase.

Perceptual actions are the most difficult to reduce in this case, as they are, first of all, the lowest complexity ones and are required for understanding the changes the user is imparting to the features of the different task elements. However, in the case of simultaneous perceptual actions, the mouse movement, or the motor action, is also included in the algorithm. This may prove to be a source of difficulty when the users face problems with the interface, in terms of accessing the elements (e.g., when the elements are very small) or when using unknown elements. However, as discussed before, when the context of the task is new the users utilize more decision making actions, rather than perceptual one.

Studying the algorithm on the basis of this approach can be used to develop the perfect algorithm that can be used to achieve the same task. This can be seen in Table 6.7.

It can be observed that the algorithm mentioned in this table is developed based on expert's analysis. However, the perfect algorithm in this case is not achieved by any of the users.

It was observed that most of the users did not approach the perfect algorithm. This can be explained based on functional analysis of activity. Activity is a self-regulative system. The final stage of self-regulation is the evaluative stage, which includes those function blocks or mechanisms as "subjective standard of successful result," "subjective standards of admissible deviation," "negative evaluation of result," and "positive evaluation of result." During independent learning processes the user, hopefully, shifts from less efficient to more efficient strategies. At this period of learning, he develops his own understanding of "good strategy" and permissible deviation from this strategy. Usually users never achieve a perfect strategy, because this strategy is unknown. Based on repetitive trials, the user gradually develops his own understanding of a "good" standard of task performance. Sometimes what is good is unclear to the user. Emotional–motivational factors, such as level of aspiration, can be important at this stage. Users can simply feel emotional satisfaction when they select a particular strategy of performance. As a result, the user stops to improve his performance based on his subjective criteria. If this strategy becomes habitual, then users may resist any changes in strategies of performance. Hence, subjective criteria of success during the

TABLE 6.7
Perfect Algorithm

Algorithm	Description	Actions obtained from action classification table	Time (ms)
O_1^{α}	Look at goal area and initial state of object area	Simultaneous perceptual actions (3 actions)	1030
$O_2^{\alpha th}$	Find out differences between goal area and object area	Thinking actions based on visual information (4 actions)	1650
$O_3^{\alpha th}$	Find out differences between goal area and object area and simultaneously perform O_4^{ε}	Thinking actions based on visual information (2 actions)	740
O_4^{ε}	Move cursor close to object area	Simple motor action	
l_1	Decide to click object (element O_Q) and simultaneously perform O_5^{ε}	Decision-making action at sensory–perceptual level	330
O_5^{ε}	Click object element O_Q	Simple motor action	
O_6^{α}	Look at goal area to evaluate color of elements	Simultaneous perceptual action	370
l_2	Decide to click blue icon tool	Decision-making action at sensory–perceptual level	420
O_7^{α}	Look at tool area and simultaneously perform O_8^{ε}	Simultaneous perceptual action with motor action	400
O_8^{ε}	Move cursor to tool area	Precise motor action	
l_3	Decide to click object (element O_Q) and simultaneously perform O_{22}^{ε}	Simultaneous perceptual action with motor action Decision-making action at sensory–perceptual level	330
O_9^{ε}	Click object element O_Q	Simple motor action	
O_{10}^{α}	Look at tool area and simultaneously perform O_{11}^{ε}	Simultaneous perceptual action with motor action	420
O_{11}^{ε}	Move cursor to blue color tool	Precise motor action	
O_{12}^{α}	Look at object area and simultaneously perform O_{13}^{ε}	Simultaneous perceptual action with motor action	400
O_{13}^{ε}	Click blue color tool	Simple motor action	
l_4	Decide to click object (element O_D) and simultaneously perform O_{14}^{ε}	Decision-making action based on information from memory	840
O_{14}^{ε}	Simultaneously with l_1, click object element O_D	Simple motor action	
O_{15}^{α}	Look at tool area and simultaneously perform O_{16}^{ε}	Simultaneous perceptual action	430
O_{16}^{ε}	Simultaneously with O_{15}^{α} move cursor closer to tool area	Simple motor action	

(Continued)

TABLE 6.7
Continued

Algorithm	Description	Actions obtained from action classification table	Time (ms)
l_5	Decide to click tool (element T_{PV}) and simultaneously perform O^{ε}_{17}	Simultaneous perceptual action Decision-making action during visual assessment	570
O^{ε}_{17}	Simultaneously with l_2 move cursor close to specific icon and click icon	Precise motor action	
$O^{\alpha th}_{18}$	Evaluate how object area matched to goal area	Thinking action based on visual information	400
$O^{\alpha th}_{19}$	Evaluate intermittent state of object area	Thinking actions based on visual information (4 actions)	1740
O^{α}_{20}	Continue looking at object area	Simultaneous perceptual action	
l_6	Decide to click object O_D and simultaneously perform O^{ε}_{21}	Decision-making action at sensory–perceptual level	480
O^{ε}_{21}	Move cursor to O_D and click O_D	Precise motor action	
O^{α}_{22}	Look at tool area and simultaneously perform O^{ε}_{23}	Simultaneous perceptual action with motor action	510
O^{ε}_{23}	Move cursor to yellow color tool	Precise motor action	
O^{α}_{24}	Look at object area and simultaneously perform O^{ε}_{25}	Simultaneous perceptual action with motor action	420
O^{ε}_{25}	Click yellow color tool	Simple motor action	
O^{α}_{26}	Continue looking at object area	Sensory–perceptual action with motor action	540
l_7	Decide to change object O_D	Decision-making action	245
O^{α}_{27}	Look at tool area and simultaneously perform O^{ε}_{28}	Simultaneous perceptual action with motor action	400
O^{ε}_{28}	Move to tool element bold	Precise motor action	
O^{α}_{29}	Look at goal area and simultaneously perform O^{ε}_{30}	Simultaneous perceptual action with motor action	360
O^{ε}_{30}	Click bold tool	Simple motor action	
O^{α}_{31}	Look at goal area and then look at tool area	Simultaneous perceptual actions (2 actions)	
l_8	Decide to click object (element O_s) and simultaneously perform O^{ε}_{31}	Decision-making action at sensory–perceptual level	280
O^{α}_{31}	Simultaneously with l_8, click object element O_s using mouse	Simple motor action	
O^{α}_{32}	Look at tool area and simultaneously perform O^{ε}_{33}	Simultaneous perceptual action with partly overlapping motor action (see below)	1200

(Continued)

TABLE 6.7
Continued

Algorithm	Description	Actions obtained from action classification table	Time (ms)
O^{ε}_{33}	Simultaneously with O^{α}_{32} move cursor close to specific icon (horizontal position tool)	Precise motor action	
O^{α}_{34}	Look at object area to evaluate change of position of objects (O_S and O_W — horizontal shift) and perform O^{ε}_{35}	Simultaneous perceptual action with motor action (see below)	400
O^{ε}_{35}	Click tool to activate action simultaneously performed with O^{α}_{34}	Simple motor action	
O^{α}_{36}	Continue looking at object area	Simultaneous perceptual action	
O^{α}_{37}	Look at goal area and perceive color of goal element S; simultaneously perform O^{ε}_{38}	Simultaneous perceptual action with motor action	480
O^{α}_{39}	Look at tool area (red color tool) and simultaneously perform O^{ε}_{40}	Simultaneous perceptual actions	960
O^{ε}_{40}	Move cursor to red color tool	Precise motor action	
O^{α}_{41}	Look at object area (O_S) and simultaneously perform O^{ε}_{42}	Simultaneous perceptual action with motor action	370
O^{ε}_{42}	Click red color tool	Simple motor action	
O^{α}_{43}	Look at goal area	Simultaneous perceptual action	250
O^{α}_{44}	Look at tool area and simultaneously perform O^{ε}_{45}	Simultaneous perceptual action with motor action	1080
O^{ε}_{45}	Move cursor to tool area	Simple motor action	
$O^{\alpha th}_{46}$	Evaluate function of tools	Thinking action under visual information	400
O^{α}_{47}	Look at object area and simultaneously perform O^{ε}_{48}	Simultaneous perceptual action with motor action	430
O^{ε}_{48}	Click tool underline	Simple motor action	
O^{α}_{49}	Look at tool area (green color tool) and simultaneously perform O^{ε}_{50}	Sensory–perceptual action with motor action	870
O^{ε}_{50}	Move to green color tool and click green tool	Precise motor action	
O^{α}_{51}	Look at object area and simultaneously perform O^{ε}_{50}	Sensory–perceptual action with motor action	
O^{ε}_{52}	Click green color tool	Simple motor action	
O^{α}_{53}	Continue looking at object area and simultaneously perform O^{ε}_{54}	Simultaneous perceptual action with motor action	480
O^{ε}_{54}	Move cursor to object area	Simple motor action	
O^{α}_{55}	Continue looking at object area and simultaneously perform O^{ε}_{56}	Simultaneous perceptual action with motor action	370
O^{ε}_{56}	Click finish for feedback	Simple motor action	

human–computer task guide the assessment of strategies of performance of the task by the user. During objective experimentation and formalized description of task performance it is important to understand what the more plausible strategy for the user is and what their subjective standard for success is. Sometimes individual differences of users should be taken into account at this step of the analysis. Strategies of performance also depend on the number of trials used by users. Therefore, the designer should to take into consideration how often this task can be performed by a user. Perfect strategy can be achieved usually only under supervisory training. This does not always take place during the performance of computer-based tasks. In addition, the perfect algorithm may not be achieved because subjectively it may be of more complexity to the user than the one he uses during the performance of the task. User interface design may be attributed to the design of computer interfaces for reducing the cognitive overload on the users by satisfactory interaction design. However, it is critically important to provide for the fact that users may be accustomed to different modes of task performance, which can have varying implications for the estimation of usability as well as for identification of appropriate design features. The solution is to study representative algorithms so as to obtain alternative design solutions. The best possible design addressing a variety of users, can be created thereby developing what is known as the real or "optimal algorithm" rather than the "perfect algorithm." The concept of perfect algorithm can be useful for the evaluation of the real algorithm of performance. Finally, attention is paid to the fact that human algorithms can be either deterministic or probabilistic. Probabilistic algorithms can describe sufficiently flexible strategies of performance. This is important for the study of HCI tasks.

6.3.7 Analytical Evaluation of HCI Task Complexity

Task complexity evaluation is based on the assumption that the more complex a task is, the higher the probability that it will be difficult for a performer and will increase errors. Complexity is an objective characteristic of task. The difficulty is with subjective evaluation of task complexity. The more complex a task, the greater the probability that it will be difficult for the user to perform it (Bedny and Meister, 1997).

The more important problem of task complexity evaluation is the correct selection of units of measure. Task complexity cannot be evaluated by calculating the number of controls, indicators, reactions, etc. For example, in one task a subject could use fewer controls that require very precise and cautious actions. In another task, a subject could use more controls and instruments that require using simple, automatic actions. Hence, the task that involves a smaller number of instruments and controls can be more complex. Utilizing measurable features of a task, such as task solving time, total number of states, or transitions as units of measure, as recommended by Rautenberg (1995), for evaluation of the complexity of computer-based tasks, is also not viable as such. For example, sometimes a more complex task can be performed in a shorter time because some components of it can be performed simultaneously. However, this can be achieved with higher levels of concentration of attention, or emotional tension. Therefore, tasks with shorter solving time can be more complex. Units of measure such as the number of actions performed by a subject cannot be used either. For example, in one task an operator can perform four simple decision-making actions based on "if-then" rules. In another task, he can perform only one complex and dangerous decision-making actions. This means that the task which includes fewer actions can be more complex. All the above examples demonstrate how scientists attempt to use noncommensurable units of measure. The quantitative method of evaluation of task complexity requires choosing units of measure that permit a comparison of different elements of activity. In other words, it is necessary to transfer different elements of activity into one plane of measurement. Activity is a multidimensional system and, hence, multiple measures should be used for evaluation of task complexity. Typical elements of activity (psychological units of analysis) rather than task elements (technological units of analysis) should be used as units of measure. Activity is a process and therefore an interval of time devoted to different components of activity should be used as units of measure. Any quantitative measure of complexity should reflect the possibility of simultaneous and sequential performance of elements of activity and their probability of occurrence (for more details, see Chapter 4).

Below, the authors demonstrate how by utilizing developed principles it becomes possible to evaluate the complexity of computer-based tasks. In this example, only one strategy is considered and all members of an algorithm occur with probability one. Therefore, there is no need to calculate probabilistic characteristics of task performance. The fact that some elements of task can be performed simultaneously can also be ignored in this analysis. This statement requires some explanation and will be discussed later. For developing measures of complexity, one should extract interval of time for typical elements of activity are devoted to motor or cognitive actions. The time intervals associated with cognitive activity are classified based on the dominant psychological process, for example, intervals of time for perceiving of information, memorizing, decision-making, or thinking in goal, object, and tool areas. The complexity of time intervals depends on the level of concentration of attention during this interval, the character of combination of elements of activity, and existing emotional stress. An ordered scale for evaluation of the complexity of the time interval, which depends on the level of concentration of attention, is thus developed (see Chapter 4). The greater the levels of concentration of attention required, the more complex the time interval for a particular fragment of activity. The simplest level of complexity for motor activity is denoted by level 1 and the more complex by 3 (low, average, and high level of concentration of attention). For cognitive activity, the simplest level of complexity has number 3 and the higher level has number 5 (see for details, Bedny and Meister, 1997; Bedny and Karwowski, 2001). Emotional stress and simultaneous performance of elements of activity also influence time interval complexity. As an example, two rules associated with evaluation of complexity, or cognitive actions, or motor motions can be presented.

1. Time intervals for motions requiring a lower (A), average (B), or higher (C) level of concentration of attention can be related to the first, second, and third categories of complexity.
2. If two actions are performed simultaneously and one requires a high level of concentration of attention (third category of complexity) and the second requires an average level of concentration of attention (the second category

of complexity), or the first category of complexity (low level of concentration of attention), the complexity of the time interval for these simultaneous elements of activity is determined by the complexity of the more difficult element (for more details see Chapter 4).

In this study, the factor associated with complexity evaluation of simultaneously performed actions is neglected. This can be explained by the following data. Most of the mouse movement distance is associated with low or average concentration of attention. Only the last phase of the distance (slowing phase of movement), when the pointer approaches the tool, requires the third category of complexity (high level of concentration of attention). This can be explained by the fact that the icons on the screen have a small size. In this task, the icons in the goal, tool, and object areas are fairly large (50 × 50 pixels on a 1024 × 768 resolution screen of 17 in.). Therefore, we consider the slowing stage (adjustment phase) of movement as the second category of complexity and the ballistic stage (acceleration stage) as the first category of complexity. This means that the ballistic stage of movement requires a low level of concentration of attention, and the slowing stage of movement requires an average level of concentration of attention. A combination of the elements of activity with one to two categories of complexity (motor activity), with the third category of complexity of cognitive actions encountered in this task, does not change the complexity of time interval. Based on this preliminary theoretical discussion, task complexity can be evaluated. Complexity is a multidimensional phenomenon. When developing measures of complexity, one should take into consideration that activity is a process and a structure that consists of qualitatively different elements. All these elements have a definite duration and are organized in time in a certain way. Hence, the measures of complexity should reflect the duration of qualitatively different elements and their interrelationship with other elements of activity. Qualitative content, duration of elements of activity, probability of occurrence of elements of activity, and the opportunity for their performance in sequence, or simultaneously, are the basis for developing these measures. Table 6.8 presents a list of complexity measures and their meanings.

Based on the foregoing developed principles and procedures, a specialist can develop other measures of complexity. Measures that can be used to evaluate stereotypical or repetitive components of activity components during task performance are presented in Chapter 4. Sometimes, it is necessary to evaluate the active waiting period. For example, the operator performs the first task and waits to begin the second task. These measures have been described in previous chapter. However, they are not important for computer-based tasks. Hence, in this work measures that are more important for computer-based tasks are described.

In Table 6.9 measures of complexity of the computer-based tasks are described. In Section 6.3.5 are presented time performance of different components of activity in sec (see Table 6.6).

As already noted, complexity is a multidimensional phenomenon. If a designer changes some features of the task, measures associated with these features can also be changed. Time of task performance is (T) 26.5 sec. From this, 14.7 sec are devoted to afferent operators (sensory–perceptual actions, T_{α}), 7.5 sec is related to thinking

TABLE 6.8
Measures of the Complexity of Task Performance and Their Psychological Interpretations

Name of measure 1	Formula for calculation 2	Variables 3	Psychological meaning 4
Time for algorithm execution (total time of task performance)	$T = \Sigma P_i t_i$	P_i — occurrence probability t_i — occurrence time of ith member of algorithm	Duration of activity during task performance
Time for performance of logical conditions (decision making)	$L_g = \Sigma P_i t_i$	P_i — occurrence probability, t_i — occurrence time of ith logical conditions	Duration of decision-making component of activity
Time for performance of afferent operators (sensory–perceptual actions)	$T_\alpha = \Sigma P^\alpha t^\alpha$	P^α — occurrence probability, t^α — occurrence time of rth afferent operators	Duration of perceptual component of activity
Time for performance of efferent operators (motor activity)	$T_{ex} = \Sigma P_j t_j$	P_j — occurrence probability, t_j — occurrence time of jth efferent operators	Duration of executive components of activity
Time for discrimination and recognition of distinctive features of task approaching threshold characteristics of sense receptors	$'T_\alpha = \Sigma P_{r'} t_{r'}$	$P_{r'}$ — occurrence probability, $t_{r'}$ — occurrence time of r'th afferent operators, characteristics of which approach threshold value	Duration of sensory–perceptual components of activity connected with processing of threshold data
Total time for performance in goal area (cognitive component)	$T_{gol} = \Sigma P_{gol} t_{gol}$	P_{gol} — occurrence probability, t_{gol} — occurrence time in goal area	Duration of cognitive components of activity in goal area
Total time for performance in object area (cognitive component)	$T_{obj} = \Sigma P_{objl} t_{obj}$	P_{obj} — occurrence probability, t_{obj} — occurrence time in object area	Duration of cognitive components of activity in object area
Total time for performance in tool area (cognitive component)	$T_{tool} = \Sigma P_{tool} t_{tool}$	P_{obj} — occurrence probability, t_{tool} — occurrence time in tool area	Duration of cognitive components of activity in tool area

Proportion of time for cognitive activity in goal area to total time of task performance	$N_{gol} = T_{gol}/T$	T_{gol} — total time for performance in goal area, T — total time of task performance	Relationship between cognitive activity in goal area to total time of task performance (complexity of goal-interpretation or goal-formation stage)
Proportion of time for cognitive activity in object area to total time of task performance	$N_{obj} = T_{obj}/T$	T_{obj} — total time for performance in object area, T — total time of task performance	Relationship between cognitive activity in object area to total time of task performance (complexity of comprehension initial stage of situation)
Proportion of time for cognitive activity in goal and object areas to total time of task performance	$N_{golobj} = T_{gol} + T_{obj}/T$	$T_{gol} + T_{obj}$ — cognitive activity in goal and object areas	Complexity of creation of mental model of situation
Proportion of time for cognitive activity in tool area to total time of task performance	$N_{tool} = T_{tool}/T$	T_{tool} — total time for performance in goal area, T — total time of task performance	Relationship between cognitive activity in tool area to total time of task performance (complexity of executive stage)
Proportion of time for logical conditions to total time for task performance	$N_l = L_g/T$	L_g — time for performance of logical conditions, T — total time for task performance	Relationship between decision-making process and total time for task performance (complexity of decision making stage)
Time for performance of operators associated with thinking process.	$T^{th} = \Sigma P^{th} t^{\alpha th}$	P^{th} — occurrence probability, t^{th} — occurrence time for thinking components of activity	Duration of thinking components of activity largely associated with manipulation of information presented through interface elements

(Continued)

TABLE 6.8
Continued

Name of measure 1	Formula for calculation 2	Variables 3	Psychological meaning 4
Time for performance of operators associated with thinking process based on external features presented through interface elements	$T^{\alpha th} = \Sigma P^{\alpha th} t^{\alpha th}$	$P^{\alpha th}$ — occurrence probability, $t^{\alpha th}$ — occurrence time for thinking components of activity whose operational nature is predominantly governed by information presented externally	Duration of thinking components of activity largely associated with manipulation of information presented through interface elements
Time for performance of operators associated with thinking process based on data extracted from memory	$T^{\mu th} = \Sigma P^{\mu th} t^{\mu th}$	$P^{\mu th}$ — occurrence probability, $t^{\mu th}$ — occurrence time for thinking components of activity whose operational nature is predominantly governed by information extracted from memory	Duration of thinking components of activity largely associated with manipulation of information in memory
Proportion of time for performance of operators associated with thinking process based on external features presented through interface elements	$N^{\alpha th} = T^{\alpha th}/T$	$T^{\alpha th}$ — time for performance of operators associated with thinking process based on external features presented through interface elements, T-total time of task performance	Relationship between thinking process depending largely on external features presented through interface elements and total time for task performance
Proportion of time for performance of operators associated with thinking process based on data extracted from memory	$N^{\mu th} = T^{\mu th}/T$	$T^{\mu th}$ — time for performance of operators associated with thinking process based on data extracted from memory, T — total time of task performance	Relationship between thinking process depending largely on data extracted from memory and total time for task performance
Proportion of time for thinking components of activity to total time of task performance	$\Delta T_{th} = T_{th}/T$	T^{th} — time for performance of operators associated with thinking process, T — total time of task performance	Relationship between thinking components of activity and total time for task performance

Proportion of time for logical components of work activity depending largely on information selected from long-term memory rather than external features presented through interface elements	$L_{\mathrm{ltm}} = l_{\mathrm{ltm}}/L_g$	l_{ltm} — time for logical components of activity whose operational nature is predominantly governed by information retrieved from the long-term memory	Level of memory workload and complexity of decision-making process
Proportion of time for retaining current information in working memory	$N_{\mathrm{wm}} = t_{\mathrm{wm}}/T$	t_{wm} — time for activity related to storage in working memory of current information concerning task performance	Level of workload of working memory
Proportion of time for discrimination and recognition of distinct features of task approaching threshold characteristics of sense receptors	$Q = T\alpha/T$	$T\alpha$ — time for discrimination and recognition of different features of task approaching threshold characteristics of sense receptors	Characteristics of complexity, sensory, and perceptual components of activity
Proportion of time for efferent operators (motor activity)	$N_{\mathrm{mot}} = T_{\mathrm{ex}}/T$	T_{ex} — time required for efferent operators (motor activity)	Relationship between motor components of activity and total time for task performance
Proportion of time for afferent operators (sensory–perceptual activity)	$N_\alpha = T_\alpha/T$	T_α — time required for afferent operators	Relationship between sensory–perceptual components of activity and total time for task performance
Scale of complexity (a) algorithm (b) member of algorithm	Xr — level of complexity (1,2,...5)	Level of concentration of attention during task performance (1 — minimum concentration, 5 — maximum)	Level of mental effort during task performance and performance of different elements. Unevenness of mental effort and critical points of task performance

Note: Proportion of time refers to the ratio of the time of the element to the total time required.

TABLE 6.9
Measures of Task Complexity for Computer-Based Task

Name of measure	Value of measure
Time for algorithm execution T (total time of task performance)	26.5
Time for performance of logical conditions L_g (decision making)	4.3
Time for performance of afferent operators T_α (sensory–perceptual actions)	14.7
Time for performance of efferent operators T_{ex} (motor activity)	10
Time for discrimination and recognition of distinctive features of task approaching threshold characteristics of sense receptors T_α	0
Total time for performance in goal area T_{gol} (cognitive component)	4.68
Total time for performance in object area T_{obj} (cognitive component)	6.8
Total time for performance in tool area T_{tool} (cognitive component)	5.01
Proportion of time for cognitive activity in goal area to total time of task performance N_{gol}	0.18
Proportion of time for cognitive activity in object area to total time of task performance N_{obj}	0.26
Proportion of time for cognitive activity in goal and object areas to total time of task performance N_{golobj}	0.43
Proportion of time for cognitive activity in tool area to total time of task performance N_{tool}	0.19
Proportion of time for logical conditions to total time for task performance N_l	0.16
Time for performance of operators associated with thinking process based on external features presented through interface elements $T^{\alpha th}$	7.5
Time for performance of operators associated with thinking T^{th}	0.28
Time for performance of operators associated with thinking process based on data extracted from memory $T^{\mu th}$	0
Proportion of time for performance of operators associated with thinking process based on external features presented through interface elements $N^{\alpha th}$	0.28
Proportion of time for performance of operators associated with thinking process based on data extracted from memory $N^{\mu th}$	0
Proportion of time for thinking components of activity to total time of task performance ΔT_{th}	0.28
Proportion of time for logical components of work activity depending largely on information selected from long-term memory rather than external features presented through interface elements L_{ltm}	0
Proportion of time for retaining current information in working memory N_{wm}	0
Proportion of time for discrimination and recognition of distinct features of task approaching threshold characteristics of sense receptors Q	0

(Continued)

TABLE 6.9
Continued

Name of measures	Value of measures
Proportion of time for efferent operators N_{mot} (motor activity)	0.4
Proportion of time for afferent operators N_{α} (sensory–perceptual activity)	0.55

Order scale of complexity (Xr)

a) Algorithm → 3
b) Member of algorithm → 20 efferent operators as O_4^ε, O_5^ε, O_7^ε, etc. belong to the first category of complexity; 9 operators as O_8^ε, O_{14}^ε, O_{21}^ε, etc. belong to the second category of complexity. All members of algorithm associated with cognitive components of activity have the third category of complexity.

or analysis of the problem (T^{th}), and 4.3 sec are associated with logical conditions L_g (decision making). Analysis of temporal data of task performance demonstrates that motor actions, in most cases, are performed simultaneously with perceptual actions. Time of performance of efferent operators (motor actions) T_{ex} requires 10 sec. Of the 10 sec for motor actions, 9.3 sec motor actions are performed simultaneously with perceptual actions. Motor actions, however, are seldom combined with thinking and decision-making actions. This can be explained by the logical structure of tasks. During the analysis of situation and decision making, the mouse is stationary or performs jitter (tremor) movements.

For the separation of involuntary tremor from goal-directed actions, motor actions, when the subject moves the mouse more than the width of the icon (in this case 50 pixels), are registered. In the study of the operator's task, when motor activity dominates, other kinds of strategies of performance are possible. If the user performs simple or average complexity motor actions, it is possible to carry out thinking or decision-making actions for future motor activity. Hence, in this strategy, the subject combines ongoing motor actions with cognitive actions, which are needed for the future motor components of activity. The width of the icon in our task is wider than the width of the regular icons on the screen. Hence, movements of the pointer to the particular icon require less precision. As a result, 20 of the efferent operators belong to the first category of complexity, and 9 efferent operators belong to the second category of complexity. Therefore, the third category of complexity of motor operators does not exist in this task. From 10.01 sec associated with motor activity, 9.3 sec motor actions are performed simultaneously with perceptual actions, which have the second category of complexity. When motor activity elements, with the one to two categories of complexity, are combined with perceptual actions of the third category of complexity, according to formalized rules developed according to Bedny (Bedny and Meister, 1997), this interval of time is related to the more difficult category. Hence, only a 0.71 sec (10.01–9.3) interval of time associated with motor activity can be related to the second category of complexity, and 9.3 sec are related to the first category. From this it follows that almost all periods of time during task performance can be related to the third category of complexity, based on evaluation of cognitive components. A combination of motor activity with cognitive activity does not change the complexity and, in general, this type of task can be assigned to the third category of complexity Xr. (see Table 6.9).

Let us consider some other measures. The proportion of time for logical conditions N_l (decision making) is 0.16, for operators associated with thinking components of activity N^{th} — 0.28, and for afferent operators N_α (sensory–perceptual actions) — 0.55. It is interesting to evaluate measures that require functioning of memory. For example, the proportion of time for performance of operators associated with thinking processes is based on data extracted from memory, which equals 0. This means that during thinking the user, first of all, operates with data presented externally on the screen and does not operate with data extracted from memory ($N^{\text{th}} = N^{\alpha\text{th}}$). This makes the task easier. The proportion of time for retaining current information in working memory N_{wm} also is 0. This means that the user does not keep intermediate data in memory during task performance. The same can be discovered during the performance of logical conditions (decision making) $N_l = 0.16$ ($L_{\text{ltm}} = 0$). This means that decision making is performed, first of all, based on external information. There is no need to operate with visual information that requires functioning sensory–perceptual processes in the threshold area. As a result, measure as a proportion of time for discrimination and recognition of distinct features of task approaching the threshold characteristics of sense receptors (Q) equals 0. All of these make tasks much easier and increase the usability of performance.

The proportion of time for cognitive activity in goal area N_{gol} is 0.18, in tool area N_{tool} is 0.19, and in object area N_{obj} is 0.26. In spite of the fact that the general goal was given in advance, and the specific goal of the task was presented in ready form externally, a user devotes approximately the same portion of time to goal area as to tool area. This means that goal interpretation, acceptance and formulation are important for task performance, and goal cannot be regarded merely as an externally given standard to which human performance is directed. Moreover a user constantly develops own intermediate goals of task performance. The proportion of time for cognitive activity in goal and object areas together, N_{golobj}, is 0.43. This proportion of time for orienting activity is associated with the creation of a mental model of a situation. Even a brief discussion of these measures demonstrates that they can very precisely describe the internal structure of cognitive activity during the performance of HCI tasks. If one changes the structure of the task, cognitive measures of complexity are also changed. In this chapter, it was demonstrated that systemic-structural activity theory suggests a totally new approach of usability evaluation and optimization of human performance. The suggested approach can be successfully applied in the study of HCI tasks.

We'll now briefly compare the GOMS (goals, operators, method, selection rules) method (Card et al., 1983) with SSAT. Various GOMS systems have emerged in literature in the last two decades (John and Kieras, 1996). According to Kieras (1994) GOMS method and its versions are based on constructing an explicit model of user's procedural knowledge. Description of human performance is considered as a model of the knowledge that user should have in order to perform the task. The description of models employs such terms as goals, operations, methods and selection rules. The difference in theoretical background of GOMS's models and models developed in SSAT is obvious because GOMS is based on cognitive psychology principles. GOMS method has explicitly defined mentalistic orientation. User's models describe procedural knowledge that resides in human memory. In contrast activity theory models utilize cognitive and behavior actions as basic units of analysis. The actions can be decomposed into operations and other units. This kind of units of analysis makes it possible to describe models of activity during interaction with objective reality. Object-orientedness is an important feature of activity models. GOMS' terminology sounds sometimes similarly to AT terminology. However this terminology has totally different meaning in activity theory. For example, in GOMS the goal is defined as "something that user tries to accomplish" (Kieras, 1994). We've discussed this definition of a goal before. Here we will just mention again that such understanding of a goal ignores origins of the goal in human activity. The goal does not exists outside of activity. It can not be simply considered as an end state of the situation toward which the human behavior is directed. There are objective requirements of the task. These requirements should be transferred by a subject into her/his goal of activity. This can be explained based on the interpretation of the goal, acceptance of the goal, its formulation, etc.

Kieras (1994) describes the difference between the goal and operator as being relative. The goal is something to be accomplished while the operator is executed. Operators are actions. In activity theory goal, actions and operators are different entities. In GOMS there are no definitions for actions and methods of their extraction from activity flow. This is just an intuitive terminology. GOMS' techniques are adopted for very narrow purposes. In SSAT, there are well developed unified and standardized methods of action extraction and classification along with other units of analysis. These units of analysis have a hierarchical organization and form a unified system, which facilitates a standardize description of holistic activity during task performance. SSAT is high level generality psychological theory for study of work behavior and learning. Let's consider some examples of GOMS analysis. Below we'll give a description of several "primitive mental operators" (Kieras, 1994).

Accomplish the goal of <goal description>

Report <goal accomplished>

Decide: If <operator…> Then <operator>

Else <operator >

Forget that < WM-object-description>

This example demonstrate that GOMS attempts to describe human behavior in terms of computer like logic and the algorithm of performance as per GOMS resembles a computer algorithm. The last operator can not be considered as a human voluntary action. Forgetting is involuntary process and can not be considered as voluntary, goal directed action. In activity theory "Report <goal accomplishment>" can be introduced only in a situation when a subject should inform somebody about goal accomplishment. “Else <operator>” also can not be considered as an elements of activity and can’t be used as a units of analysis for algorithmic description of human performance. In general GOMS and its versions do not possess required units of analysis and language of description to create adequate models of human behavior or activity. Human activity is not similar to computer functioning. In SSAT, one employs the concept of the human algorithm. The most important distinguishing feature of a human algorithm is the fact that the basic unites of analysis are cognitive and motor actions. In addition SSAT utilizes the functional analysis of activity when major units of analysis are function block. SSAT also distinguishes technological and psychological units of analysis. Psychological units of analysis are standardized.

SSAT considers activity as a multidimensional system. It requires different units of analysis and language of description. SSAT offers unified and standardized stages and levels of analysis. As a result it becomes possible to describe the same activity utilizing different interdependent models that consider activity from a different point of view. Models of activity are described not in term of production rules but in terms of human cognitive and behavioral actions. Activity theory puts emphasis on the consciousness of goal and its relation to the motives. Farther activity is described as a flexible self-regulative system that includes divers strategies of goal attainment. Activity is concerned with mediating role of artifacts and socio-historical development of human activity and mind. Subject-object and subject-subject relationships are critically important for the description and analysis of a task structure. All of this is ignored by GOMS method. GOMS describes activity and behavior only as a sequence of elements. However some elements of activity can be performed simultaneously or in parallel. Therefore the structure of activity can not be precisely defined by the GOMS system. SSAT considers the concept of complexity in a totally different light. SSAT sees complexity as an objective multidimensional characteristic of the task. In GOMS complexity is considered as a characteristic of internal knowledge. In general without understanding basic activity theory principles it is impossible to utilize correctly such concept as goal, action, operation, operator, etc.

7 Study of Personality in Activity Theory

7.1 GENERAL CHARACTERISTICS OF PERSONALITY IN ACTIVITY THEORY

7.1.1 SYSTEMIC-STRUCTURAL ANALYSIS OF PERSONALITY

There are two approaches to ensuring the effectiveness of the man–machine system. One of these approaches involves adapting the technical components of the system to the operator. This approach involves design decisions that correspond to the psychological, physiological, and anthropometrical characteristics of man. This approach of ensuring the effectiveness of the man–machine system cannot always be realized to the fullest extent. Consequently, the second approach is also important and involves adaptation of the operator to work conditions and the technical components of the man–machine system. This direction includes selection of personnel and training. These two approaches are interdependent and should often be used in unity. In those cases when the man–machine system involves extreme work conditions, the role of personnel selection is increased. However, issues of personnel selection are addressed in relationship to training within the framework of activity theory. In cases of extreme work conditions of special importance are individualized training methods, directed towards the formation of individual strategies of activity which provide the adaptation of the individual to the objective conditions of activity. Both in the training and selection of personnel, the study of individual characteristics of personality and their compatibility with the objective demands of work activity becomes critically important. It should be noted that the study of the individual characteristics of activity is also important for the design of the man–machine system. In this case the system is designed in such a way that the various function regimes of the man–machine system can be adapted to the individual personality characteristics of personnel. In this way, the study of personality and individual differences is of critical importance in the study of human work.

The psychology of personality is one of the most complicated and less well-understood areas of psychology. The complexity of the field of personality begins with the attempt to define personality. Allport (1937) counted more than fifty definitions of personality. Since then, the number of definitions have increased. Therefore, some psychologists prefer to infer a definition of personality through empirical studies (Hogan and Shelton, 1998). Former Soviet psychology also had myriad definitions of personality.

Personality may be treated as a system with a complex structure. This system is composed of discrete cognitive–affective units that include goal-oriented feed-forward and feedback components. Such a complicated dynamic requires a systemic method of study. One of the important principles of systemic approaches is to describe and analyze objects from multiple perspectives, thereby, uncovering different aspects of the object under consideration.

Currently, there is no consensual theory of personality or universal model that represents personality. The value of any theory of personality determines how adequately, from the theoretical and practical perspectives, a suggested theory or model captures what psychologists mean by personality. Five-Factor Models (FFM) of personality must similarly be evaluated inasmuch as there is widespread agreement that they do not exhaust what is meant by personality. For example, FFM do not entirely supersede the three-component, dynamic model offered by Freud (1916, 1917). Here, from a systemic approach, we see totally different descriptions of the same object.

Where the major purpose is the study of personality in terms of job performance, it is obvious that FFM is not sufficient. For example, the FFM model of personality does not specify the relationships among these factors and the notion of self-concept (Judge et al., 1998). Thus, Hogan et al. (1998) state that personality should include the perspectives of both the observer and the actor. Each theoretical concept of personality reveals distinct aspects of personality. It should be noted that the prevailing technique for the study personality is paper-and-pencil questionnaires. This method was also used in the former Soviet Union; however, questionnaires were regularly combined with experimental methods of study.

In activity theory special attention is given to the interrelationship of personality and activity. Rubinshtein wrote that the psychological characteristics of personality are expressed during activity and are simultaneously formed through it. From the activity theory point of view, a person is developed during his activity. Therefore human activity is a major determinant of the development of the personality. According to Rubinshtein, work activity is of special importance in this process. Development of one's personality through one's activity was viewed as one of the main tenets of activity theory formulated by Rubinshtein in 1940 (Rubinshtein, 1940). Another closely related principle suggested by Rubinshtein is the unity of consciousness (human cognition) and behavior. Both of these principles were later generalized into one principle, the unity of personality, consciousness, and behavior. This principle was formulated as follows: "personality as the carrier of consciousness, along with consciousness is discovered and formed during activity" (Platonov, 1982). Analyzing the interrelationship of external influences on personality, Rubinshtein (1940) wrote that everything in the psychology of the forming of personality, in one way or another is the result of the external environmental influences, but none of the development of personality results directly from external influences. External influences have one or another effect mediated by psychological features and states of the subject. Later, this idea was formulated as the principle of the personality within activity theory. The basic idea of this principle is that activity changes not only the external world but the person as well. Briefly, it can be formulated in the following way: personality can be understood as a system of internal conditions, which consists of its structure of properties, qualities and specificities, through which all external influences are mediated. Persons determine themselves through objects that they create. Rubinshtein emphasized the interaction of environmental and innate characteristics of personality in the formation of personality. In activity, the subject not only changed a situation or object, but also developed his own personality features. From this it follows that mental development cannot be understood as a simple internalization of ready-made social rules, norms, and standards. All external influences are mediated by internal conditions. Human activity has creative character; therefore external influences always interact with internal personal conditions. At the same time intersubjective aspects of personality development have not been sufficiently investigated in Rubinshtein's work. These aspects of human development in a more detailed way were described by Vygotsky (1960). These two aspects of personality development should be considered in unity. Therefore, personality is regarded as a dependent variable of human activity and social interaction.

The personality approach is related to the individual approach. The individual approach is narrower; it focuses on isolated aspects of personality and their interaction with particular external conditions. However, in all cases when the individual approach is applied one should not ignore the holistic aspects of personality, in other words, the structural interaction of different aspects of personality. Very often during the use and design of psychological tests, only isolated aspects of personality are taken into consideration, and their structural interrelationship is ignored. Activity theory resists reducing the study of personality to a list of personal traits and factors. Rather, it emphasizes the structure and organization of the various components of personality. This framework supplies the basis for the systemic-structural approach to the study of personality and human performance. An individual is considered an agent of activity who continuously interacts socially with others.

Individual personalities are formed through activity and social transactions. From this process, we can extract "subject-object" and "subject-subject" relationships that are intimately interconnected. Thus, the person is often treated as stable role-relationships that develop through activity and social interactions. Such stable role-relations determine particular strategies of behavior, which, in turn, are conditioned to adapt to changing environments as well as the character of the other person in a relationship. The inner relationship between social role and activity is a function of the fundamental strategies of self-regulation.

As a result, different situations frequently elicit distinct individual features. For example, in one study, observation and personality inventories reveal that the levels of conformity exhibited by subjects, as a feature of personality, vary as a function of the significance of the situation to the subject. In nonsignificant situations they exhibit conformity, while in significant situations they do not tend to conform. In other experiments, with the assistance of personal inventories, athletes were selected in order to study competitiveness. It was discovered that this tendency can be significantly changed, depending upon the significance of the particular games in which they were involved. In important games they were very collegial. However, during unimportant games they demonstrated their more harshly competitive tendencies (Petrovsky, 1986). This example demonstrates that individual features of personality can be changed depending on the significance of the situation and the social interactions.

Some aspects of the study of personality within activity theory overlap with the study of personality in the west. However the term "personality" within activity theory has a broader meaning. For example, it encompasses the individual features of cognitive processes. In studying cognitive features of personality psychologists rely on a variety of methods such as those employed by cognitive psychologists. However, in cognitive psychology these studies are conducted without considering personality variables. Nevertheless, when we seek to study cognitive styles as personality variables cognition does become an aspect of personality psychology.

Specificity of memory and thinking and other psychological processes and features in a particular individual may affect the strategy of an individual activity. Strategies of activity derived from idiosyncratic features of personality are called individual style of performance. Individual style of performance is a critically important notion that links an individual to job performance (Bedny and Seglin, 1999). In this case developing strategies of an activity permits a particular person to perform job, duties and tasks efficiently. Other important aspects of the study of personality and individual style of activity concern the development of individualized training. Knowledge of the specific cognitive processes of a particular person and his other features becomes important. As may be seen, the suggested individual psychological model does not reject the approaches to personality in the west. We believe these approaches can be constructively combined. It is, as noted, merely a different position for considering the same objective.

Thus, strategies of behavior derive from the operation of the functional mechanisms of self-regulation (Bedny and Karwowski, 2003). In this case the functional mechanism related to determining the significance of the situation is most relevant to personal social traits. This data must be taken into consideration in interpreting the behavioral manifestations of measured traits in specific situations. These experimental methods enhance the power of personality inventories. The wide range of personality features exhibited in our work empowers the expansion of the application of personality psychology to performance situations. In the former Soviet Union, the important role of employee selection was less pertinent than the development of rubrics for individual training, performance evaluation, and job design. Such applications enabled the deployment of a wide array of individuals more efficiently into the labor market. The central notion in these studies is the individual style of activity that connects features of personality with performance (Bedny and Seglin, 1999).

In the study of personality all psychological phenomena can be separated into three groups, psychological processes, psychological states, and psychological features. Psychological processes include sensation, perception, memory, and thinking. All these processes are tightly interconnected, for example, the specifics of memory processes affect the thinking process. Similarly, emotional and volitional processes affect intellectual ones and vice versa. Thus, the individual has a particular idiosyncratic structure of psychological processes.

Psychological states have a restricted duration that varies over time. The persistence of these states is affected both by external contextual and internal psychological factors. For example, visual sensitivity may be altered by ambient lighting. Similarly, the intrinsic interest of a lecture affects attention processes. Repeated contextual factors may continuously elicit specific psychological states. Sufficient repetition tends to inculcate stable psychological features (Kovalev, 1965). It is well known that vigilance, normally mobilized only in emergency situations, may become a personality feature as a result of some traumatic episode or experience. In short,

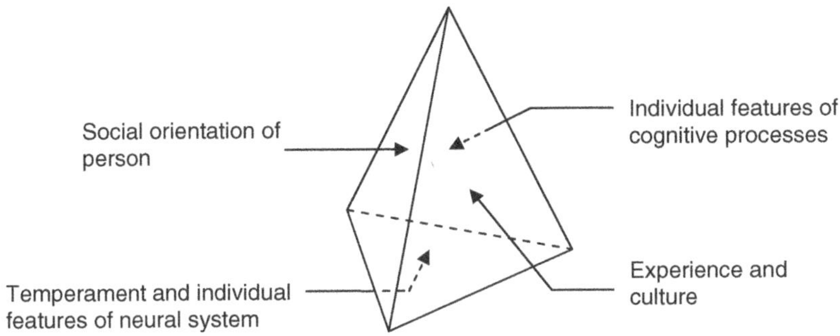

FIGURE 7.1 Structure of personality.

we "tenure" certain psychological features based on repeatedly elicited temporal states.

Personality features interact with one another in various ways. They evolve as a complex substructure of:

1. Abilities, developed from an integration of intellectual and emotional–volitional features
2. Character, which refers to a style or manner of relationship and behavior in diverse situations

Abilities and character are complex substructures of personality which depend on the specific components of personality that determine stable individual social orientation as well as the developmental phenomenology of personality. What we call specific components of personality determine the psychological processes — as a system of mental actions — in a concrete individual as well as features of the neural system that affect temperament. The individual psychological structure of personality, according to Platanov, is presented in Figure 7.1 (Platonov, 1982). The first component of personality is the stylistic qualities of cognitive processes. Those processes such as memory, thinking, emotion, perception, etc. emerge as various forms of reflection of reality. However, each form of psychological reflection is fixed in memory and takes on an individual character that becomes a property of personality. The specific nature of the interaction between the various psychological processes within a person comprises an individual personality feature in and of itself.

The second component of personality is called the social orientation of personality or directness. This component of personality includes relatively stable personal motivational tendencies, values, predispositions, interests, ideology, desires, and so on. Within the theory of activity, dispositions and attitudes are viewed through the prism of the social determinants of these characteristics. Directness can be regarded as a relatively stable, functionally autonomous system of motives that gives individual coloring to a person's orientation to a situation. Directness also includes the set, that determines the tendencies of interpretation of the situation, specificity of emotional evaluation of it and the readiness to act in a particular way. Taking together the set determine not precisely consciously control manner in which a person will act in relation to an object or situation.

The third component of personality is past experience or culture. It includes knowledge, skills and habits. However, this component includes only well-developed knowledge, learned abilities, skills, and habits which are specific for particular individuals. This component also includes preferable strategies of social interaction, worldview, morality, ethical position, general culture, accepted norms, and requirements. Such experience is determined not only by external training but also by internal characteristics of personality. This past experience has an individual specificity which distinguishes the past experience of one individual from another.

The fourth component refers to biological features and natural inclinations which include the individual features of the nervous system which is the basis for the formation of the temperament structure. This fourth component is sometimes referred to as the biopsychological component of personality.

The other component of personality is the self. This feature of personality is not listed in Platonov's model of personality. Self determines the relationship of the individual to himself/herself, subjective vision of himself/herself and ability to carry on an internal dialogue with one's self and based on this, develop a person's own relation to the other and situation. The self is a prerequisite and a consequence of activity and social interaction. It appears from the dynamic relation of human material practice, social interaction, and person's subjectivity. Therefore, it is not only an individual but also a social phenomenon. The self as a personality component is associated with those concepts as level of aspiration, self-esteem, self-efficacy, etc. The self is an important mediated component of motivation and performance (Petrovsky, 1982; Bundura, 1982). The self plays an important role in the organization and direction of human activity.

The above outlined substructures reflect both the biological and the social aspects of personality. For example, the temperament of the individual is primarily a biological aspect of personality, while the social orientation is a social aspect of personality. The specificity of activity is determined not only by the properties of the above discussed components but also by their interaction.

Zarakovsky (2004) describes the personality as a subsystem of activity. Activity includes different subsystems and they are designated as features of activity (Figure 7.2).

There are five basic features or subsystems. The feature or subsystem 1 is a result (outcome) of activity. Knowledge of the result is important for the analysis of activity. In the case of social interaction the result of activity is evaluated according to solidarity, empathy, subjective judgments, etc. Feature 1 or subsystem associated with the activity result focuses on stages of object transformation according to the goal of activity. Information about the initial state of object, intermittent states and final state is important for understanding procedural aspects of activity. The result of an activity is evaluated on the basis of objective and subjective criteria.

Operational (informational) subsystem of an activity (Feature 2) includes cognitive processes, cognitive and motor actions, and operations and their logical

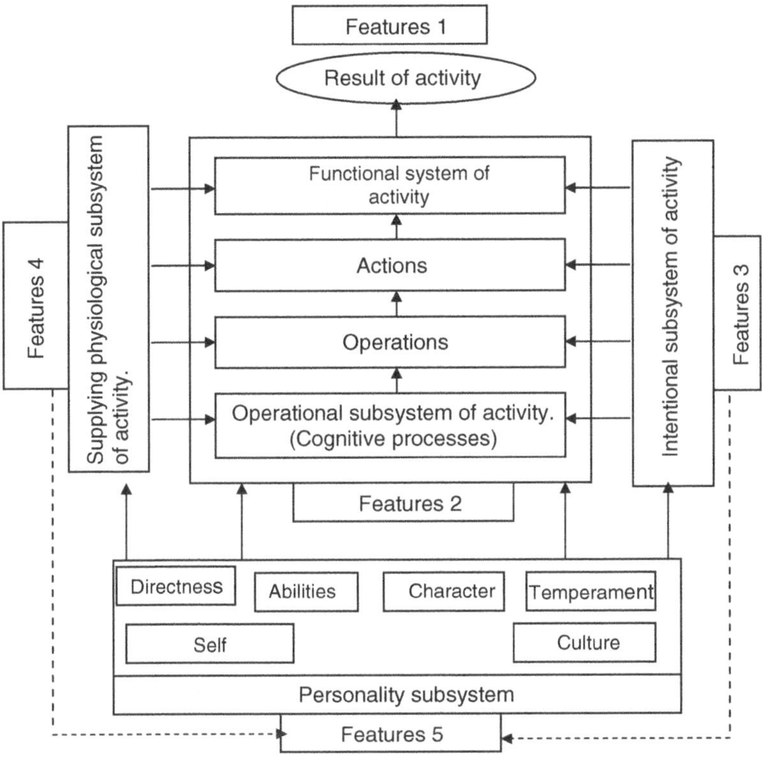

FIGURE 7.2 Activity subsystems.

organization. One should distinguish, in this regard, goal-directed and basic functional systems. Subsystem 2 is organized as a goal-directed self-regulation system. The goal-directed operational subsystem (Feature 2) is directly involved in activity performance. Operational subsystem includes analysis of a situation, goal-formation process, program formation or selection ready program, and algorithms and execution of actions. This subsystem also includes mechanisms of correction through proper feedback. A goal-directed or an operational functional system develops and exists only during the process of planning and realization of the conscious goal.

There are two energetic subsystems. One of these subsystems is psychophysiological (Feature 3) and the other physiological or supplying one (Feature 4). The psychophysiological subsystem (Feature 3) is responsible for intentional aspects of activity. Another energetic subsystem has pure physiological characteristics (Feature 4). This is the basic functional system that integrates physiological mechanisms and provides energetic resources of activity.

The personality subsystem (Feature 5) is the highest level of activity regulation. This subsystem provides a level of activity regulation, taking into consideration its subjective and social significance. All other subsystems are subordinated to this subsystem. This subsystem is the basis for the study of individual style of performance.

According to Merlin (1986) the system of individuality can be grouped into the following three hierarchically organized levels and their sublevels:

1. The level of individual features of organism
   (a) biochemical
   (b) general somatic
   (c) features of nervous system (neuro-dynamic)
2. The level of individual psychological features
   (a) psychodynamic (temperament)
   (b) psychological features of personality
3. The level of sociopsychological individual features
   (a) motives, relations of personality, personal status
   (b) social roles in groups
   (c) social roles in sociohistorical societies (nations, classes)

All levels interact with each other. Between them there is stochastic or many-to-many relationships. Inside of one level there is a causal relationship between different sublevels. The specificity of the relationship can be changed during personal development. Levels of individuality immerge during phylogenetic and ontogenetic development. All levels of described system are organized according to principles of self-regulation. Therefore, personality can be described as a self-regulative system. This system continuously develops and constructs different strategies of activity. These strategies can be regarded as an individual style of activity which we will consider in detail later.

Based on analysis of data presented by different authors we can list the following basic components of personality: (a) the self; (b) individual features of cognitive processes; (c) culture or past experience; (d) temperament and individual features of the nervous system; (e) directness or social orientation. These components can be combined into a more complicated structure. These combinations determine the abilities and character of personality. Therefore, each component of personality has its own internal organization which may itself be presented as a substructure appropriate to this aspect of personality. Personality may be presented as hierarchically organized substructures. It follows from this that the study of isolated features of personality will not suffice. It is also important to study how these features are integrated into separate substructures. These substructures are themselves organized into a holistic system that determines the specificity of personality. The content of a person's substructure may be altered through personal development. This is especially true for the high level substructures such as "social orientation" or "goal orientation" and complex substructures such as ability and character. The study of personality is not restricted to the study of separate psychological processes or features of temperament. Persons possess distinctive configurations that determine their individuality. Their individuality appears in diverse traits of temperament and character, habits and skills, in dominant interests, specific kinds of abilities, and in individual style of activity, etc.

We have discussed only the systemic-structural approach to personality. There are also other approaches to personality in the former Soviet Union (Kovalev, 1965; Mel'nik and Yampol'sky, 1985) and others. However, the systemic-structural approach to personality is best suited to the study of human work. The bases of the systemic-structural approach to the study of personality are its view of personality not as a series of independent characteristics but as a structural interrelationship of these characteristics.

Mel'nik and Yampol'sky (1985) proposed a structural and hierarchical model of personality on the basis of a factor analysis of the interconnections of the scales of MMPI and 16PF. Their concept of activity demonstrates the interconnection between the study of personality in the former Soviet Union and data collected in the United States. However, this conceptualization of personality is more oriented towards the application to clinical psychology rather than towards work activity and therefore will not be considered in detail here.

7.1.2 Structural Relationship between Temperament and Features of the Nervous System

The quest for traits has been tied to the research method of factor analysis. Factor analysis is a "data reduction" technique that reduces a manifold of correlations among variables — in this case personality descriptors — to a simple structure of underlying factors to which each observable variable has a linear relationship. Early factor analysis appeared to reveal an isomorphism between the factor structures of traits and temperament (Sheldon, 1942; Cattell, 1965). Accordingly, Anglo-American psychology has tended to merge the formulation of temperament with the formulation of traits. Two generations of empirical and statistical research have converged on "The Five-Factor" solution to the intercorrelations of personality descriptors — Extroversion, Neuroticism, Agreeableness, Openness to Experience, and Conscientiousness. (Brody, 1989). The achievement of such consensus is a major accomplishment. However, factor analysis is a fundamentally atheoretical approach to personality that conflates motivation, attitude, temperament, and values. Moreover, the mathematics of factor analysis precludes identification of qualitative differences in personality types, the effects of complex moderator relationships among the variables, nonlinear relationships among variables and factors or consideration of unusual or unique standings on the measured variables.

The nonlinear relationship between personality factors and subsystems was noted by Merlin (1986).

In the description of the systemic characteristics of personality, linear relationships are describable through linear correlations and regressions; however both linear relationships and normal distributions of data are infrequently encountered in the study of personality. Complex self-regulating systems with relatively enclosed, automatic subsystems of personality are possible only with nonlinear relationships. The mathematical descriptions of systemic aspects of personality employ not only two variable correlations and linear regressions but also multidimensional relationships and mathematical models with a large number of variables. Merlin noted that these statistical methods model additive relationships but in personality relationships of actual variables are often nonlinear and nonadditive.

Any given innate property of the nervous system correlates with several properties of temperament. For example, the property of the strength of the excitation process is correlated with emotional reactivity, strength of emotion, agitation, rigidity, fatigability, extraversion, and introversion. At the same time, one aspect of temperament correlates with several properties of the nervous system. For example, rigidity correlates with the strength of the excitability process, balance of strength (excitability and inhibition), and flexibility of the nervous process (Merlin, 1986). In general each property of temperament correlates with several properties of the nervous system. The above examples demonstrate the "many-to-many dependence of the properties of temperament on the nervous system." Between the properties of the nervous system and temperament there exists both a linear and a nonlinear relationship. Nonlinear relationships can be described through a curve relationship. For example, Paley (1976) demonstrated that neuroticism is not linearly related to any of the indicators of the strength of the excitable process. A nonlinear relationship was discovered between these variables.

Eysenck attempted to reconcile the temperamental typologies and dimensional trait theories of personality. (Eysenck and Eysenck, 1985) Beginning with the "Big Two" of the "Big Five," Introversion/Extroversion and Stable/Unstable, Eysenck proposes that the quadrants formed by these two dimensions constitute the temperamental types that have long been posited by observers. Kagan (1995) argues that the "simple structure" of factor analysis obscures the subtle but decisive qualitative differences in personality and temperament that emerge from developmental and neuropsychological research. He notes that scientific psychology originated in the 19th Century European quest for clarification of the substance and boundaries around psychological types. He attributes the eclipse of this project to an emphasis on continuous dimensions, caused by methodological and mathematical imperatives. Kagan (1995) further points out that current neurophysiological studies resonate more closely with the 19th Century European approach. Endorsing this approach, he urges that the study of temperament, as well as personality as a whole, return to the definition of qualitative distinctions and identifications of categorical types.

Soviet psychology maintained a closer tie to the 19th century paradigm, never abandoning the quest for the neuropsychological correlates of mental and behavioral functioning. In the Soviet research tradition temperamental features determine specific dynamics of psychic activity such as pace, speed, rhythm, and intensity of psychic processes and states. One individual is lethargic and apathetic; another is sensitive and reactive; while a third is frenetic, and yet another is very cautious and deliberate in his or her style. These dynamic features of personality are jointly determined by personality "structure" and transitory psychological states related to the events in the psychological or physical environment. For example, restraint is a feature of personality that depends upon temperamental tendencies in the absence of conscious intent. As is obvious to anyone who has been involved in a court case, irascible and unstable individuals regularly exhibit decorum and restraint in the courtroom situation. This tendency to adapt to prevailing norms and incentives is, of course, a general function behavior. Thus, overt behavior may be attributable to either "native" features or situational induced states.

There are three major, interconnected components of temperament that determine the dynamic features of activity — general psychological activation, motor activation, and emotionality (Merlin, 1964; Nebilitsin, 1976). All of these components are intimately related. The first component determines tendencies to exhibit energy and dynamic flow of mental processes. The second component is manifested in motor activity such as speed, strength, sharpness, amplitude, rhythm, etc. During the study of this component of temperament, muscular and verbal behavior are easier to observe.

The third component of temperament is emotionality. This is a broad nexus of qualities and features characterized by the style of emergence, flowing and halting of different feelings, mood or affects. Impressionability, sensitivity, impulsiveness, emotional irritability and lability are included here. Sensitivity informs the affective tone of social relationships as well as overall sentimentality. Impulsiveness designates the suddenness with which these processes emerge, their strength, and the speed with which this state is triggered. Emotional lability designates the speed with which a state can be interrupted or changed.

The studies of Merlin (1964), Klimov (1969), and Strelau (1982) demonstrate that temperament affects the individual style of performance; however, it does not predetermine mental abilities of the individual. Merlin provides the following definition of temperament: a stable psychological feature that determines the dynamics and intensity of individual psychological activity. It has a tendency to be relatively stable during the pursuit of various goals, governed by varying motives and exhibited through different content of activity in accordance with the specific structural syndromes defining various temperaments.

Here we present an attempt to develop a theory of temperament based on neurobiology. This is a theory that relates temperament to various features of the nervous system. The first to attend to the relationship of temperament to nervous features was Pavlov (1927). Pavlov's ideas were taken up by Eysenck (1967). He utilized Pavlov's idea regarding the relationship between excitation and inhibition processes. Later, Eysenck elaborated this approach with his theory of arousal.

Like the psychologists of the former Soviet Union, Eysenck focuses on the interrelationship between properties of the nervous system and temperament. However, these psychologists interpret the interrelationship between the nervous system and temperament in different ways. Eysenck views variation in temperament as occurring on a continuum. He viewed individual differences as a result of two independent variables: neuroticism versus stability (strong–weak emotions) and extraversion versus introversion. Temperament was described in terms of these two scales, in a two-dimensional schema.

In activity theory researchers focus on the multiplicity of temperament types that exist in a complex interrelationship with properties of the nervous system. In this case it becomes possible not only to consider continuous variations in temperament but also to pay attention to interindividual discrete variation in temperament. The type of temperament of any given person emerges from the structural combination of its properties.

On the basis of experimental studies psychologists have isolated such properties of temperament as anxiety, extraversion/introversion, and rigidity (Belous, 1967). According to Eysenck, anxiety is an independent factor in relationship to introversion and extraversion. In the work of Belous, anxiety is in the same group of properties as introversion and extraversion. This can be explained by the fact that Eysenck views anxiety more broadly and identifies it with emotionality. In contrast, Belous views anxiety more narrowly in the context of threat and tension. Emotionality, by contrast, is considered more broadly and can be expressed in any context. In this understanding emotionality is independent of (orthogonal to) extraversion and introversion. Therefore anxiety and emotionality are independent characteristics of personality.

Activity theory in different ways regarded those features as rigid. For example rigidity was viewed by Freud as a result of a contradiction of motives. In activity theory rigidity is viewed as depending on the characteristics of the nervous system that determine temperament. Rigidity has two subtypes; the first is the rigidity of automatic responses such as that found in habit changing, which can either be difficult or easy, the second subtype is the rigidity of motives. The last includes variables expressed in repeated attempts in the face of persistent failures. In all cases temperament is regarded as a structural organization of elements or systems. In brief these are some differences between the western and activity approaches to temperament.

Pavlov's idea that features of temperament depend on features of the nervous system was generally accepted by psychologists in the former Soviet Union. Meanwhile in the former Soviet Union, Pavlov's ideas regarding basic features of the nervous system were significantly reevaluated and revised. New features of the nervous system were discovered. Qualitative characteristics of the nervous system and their dependence upon particular nervous structures were also extensively reconsidered (Teplov, 1961; Nebylitsin, 1965). While a number of unresolved issues remain, the data point to some theoretical and practical conclusions. The more thorough studies revealed such features of the nervous system as strength, mobility, dynamics, and liability. We will, next, provide a short description of these features.

The strength of the nervous system refers to the robustness and endurance of the cortical neural cells, their structure, as well as their ability to perform activity in the face of overload. The opposite pole of "strength," "weakness" of the nervous system, correlates with limited robustness, poor endurance, and so on.

The biological meaning of the strength of the nervous system is straightforward. Under equivalent conditions a person with a strong nervous system exhibits a greater ability to function in stressful situations. Such functioning under stress is, of course, moderated by motivational states. Sometimes individuals with weak nervous systems can perform better in a stressful situation than persons with a strong neural system. However, such highly motivated behavior by a person with a weak nervous system will soon show the effects of exhaustion and fatigue. Further, it should be noted that the weakness of the nervous system cannot be simply treated as a negative factor, because the weakness of the nervous system correlates with the sensitivity of the nervous system (Teplov, 1985). Nervous systems gain from strength, but lose in sensitivity. Sensitivity has an adaptive meaning to a living being in its environment. Animals with a weak nervous system and high sensitivity have a greater attunement to their young, better chances of detecting danger early, more advantages in discovering food, etc. Gurevich (1970) discovered that under monotonous work conditions, individuals with weak nervous systems tend to develop conditioned responses more quickly. On the other hand, in accidents, or crises, people with stronger nervous systems exhibited better performance.

Such features of the nervous system as mobility are connected with "speed of reconditioning" when the meaning of conditioning stimuli is altered. The opposite of "mobility" is "inertness." "Inert" nervous systems have reduced mobility. Mobility and inertness are most evident in situations demanding frequent changes of actions or reactions contingent upon changes in the external environment.

Another important feature of nervous systems is dynamic quality. The speed of conditioning derives from this feature. It determines how quickly and easily excitatory or inhibitory reflexes may be shaped. Nervous systems that lead to repeated formations of positive associations are considered dynamic in relation to excitation. A nervous system that provides quick formation of an inhibitory reflex will be treated as dynamic in relation to inhibition. Dynamic qualities provide rapid formation not only of elementary reflexes but of more complicated connections. Accordingly, these features of the nervous system correlate with the ability to learn.

A full discussion of the features of the nervous system and the methods of their study is beyond the scope of this book. Therefore, we can only touch on some of the other features. Other features of the nervous system derive from these more basic features. The "Balance" of the nervous system may be formulated in terms of "Strength," "Mobility," and "Dynamics." These attributes may be further differentiated. The nervous system may be strong with respect to excitability of nervous processes, while being weak with respect to inhibition of expressive processes. This research further shows that speed in adapting to the reversal of the meaning of stimuli from excitatory to inhibitory is independent of the speed of adapting to the reversal of the meaning of stimuli from inhibitory to excitatory. In other words, the reversal of the conventional signals of green for "go" into "stop," may well show a different pace of adaptation than the reversal of red for "stop" into a signal to "go." By the same token "balance" may be formulated in terms of the "dynamics." This implies that an individual may quickly acquire connections between inhibitory and excitatory stimuli.

Of course, in the study of the features of the nervous system there remain many stubborn difficulties. Not all of the methods utilized to attack these problems in the former Soviet Union may be used in practical situations. However, there are many straightforward, practical, and theoretical implications for this work for Western psychologists.

A complex relationship exists among the features of the nervous system and temperament. Personality is viewed as a complex self-regulating system with different components and many-to-many[1] interrelationships between them. This interrelationship is such that a variable from set "a" is interconnected to several variables in set "b," while each variable in set "b" is interconnected with several variables in set "a." According to Merlin (1986) there exists a many-to-many relationship between properties of the nervous system and features of temperament. The features of temperament depend upon the structural relationship among diverse features of the nervous system. An individual may possess both strong excitatory and strong inhibitory processes. When strong excitatory processes dominate strong inhibitory processes an individual may demonstrate unrestrained passion. By the same token, both excitatory and inhibitory processes may be weak. When weak excitatory processes dominate

[1] There are other terms for description of this relationship. For example Ashbey (1959) described this connection as stochastically deterministic.

weak inhibiting processes, the person may exhibit a agitation and impulsiveness with a histrionic quality.

The psychological characteristics of temperament may be formulated in terms of the following major features that are similar to Western approaches (1) Sensitivity (alertness to various stimuli); (2) Reactivity (emotional intensity); (3) Flexibility (adaptation to changes in the environment); (4) Rigidity (the opposite of flexibility); and (5) Extroversion/Introversion and Neuroticism. As noted, these may be found in Eysenck and more recent "Five Factor" solutions to personality theory. Features of temperament that are basic and biosocial affect the development of other features of personality but do not exactly determine them. For example, temperament affects the development of character. Based on experimental findings Merlin (1986) generalized the following relationship between temperament and character: (1) each characteristic of temperament determines several different, sometimes contradictory properties of character; (2) a character feature that determines the approach to other people may depend on several different properties of temperament; (3) depending on the social conditions of upbringing connections between various properties of temperament and character may form and disappear. We will consider the relationship between temperament and other properties of character in the following section in a more detailed manner.

7.1.3 Character and Ability

Character and ability are treated as important aspects of personality distinct from temperament. They are treated as elements embedded in a structural foundation of the above-mentioned four components of personality. They are not treated as a distinct domain of personality. There exists a Russian saying that "You plant character and harvest destiny." Character represents the totality of individual features that are elicited in typical circumstances occupied by the person. Character further determines the style of behavior and attitudes in those circumstances. As was shown in the Authoritarian Personality Studies, individual timidity may be exhibited during interactions with superiors, while, at the same time, arrogance may be exhibited with subordinates (Adorno et al., 1950). Here timidity and its opposite, arrogance, are features of a single attribute of character rather than temperament.

Individual specificity of character determines the relation of people to situations in two ways. Emotional feelings and individual or specific styles of behavior in particular situations depend upon character. If one knows the character of a person one can predict how he will behave in various circumstances.

Every feature of character is at the same time a feature of personality. However, not every feature of personality is a feature of character. In order to be a feature of character, the peculiarity should be clearly expressed and systematically demonstrate itself in different circumstances. For instance, such polar features as diligence and laziness can be expressed so weakly that they cannot be regarded as character features. Only when such features are clearly expressed and demonstrated in different situations can they be regarded as features of character. Character also includes stable habits that can be demonstrated in different circumstances. The saying stays that "Habit is second nature."

The behavior of a person derives, in the first instance, from those goals he sets him self. This determines the social orientation of the individual. However, two persons with the same goals may pursue them through radically divergent behaviors and activity. Frequently, the style of behavior depends upon the character features of the person. Character is a trigger for programs of performance in typical situations. Character possesses inducing and motivational forces that become salient in overcoming obstacles or in dealing with stressful situations. The Achievement Motivation literature addresses the social forces required to inculcate a character trait (Heckhausen, 1991). As this literature makes clear, these character variables are a function of acquired interpretative canons and criteria as well as of acquired behavioral repertoires.

A prelogical function, personal "sense," may engender distinct meanings for different persons (Bedny and Karwowski, 2003). By acquiring the socially approved "sense" the person also acquires moral outlook, principles, norms and customs. Through their personal "sense" an individual's motivation is attached to the situation. It guides and colors the interpretation of the situation, and, a fortiori, the manner of behaving in a particular situation. In the process of social learning individuals acquire not only meanings of different situations but also methods of emotional evaluation of the situation. These in turn affect the acquisition of the socially prescribed "sense" of the situation.

Different character traits do not exist independently. They are organized into a particular structure. The structure of character is revealed through the interdependence of its aspects. For example, if a person is very anxious, he is unlikely to exhibit autonomy because of the fear of consequences. In view of this interdependence, character traits may be divided into several basic groups. The first group (Platonov, 1982) contains general traits of character, such as honesty, fortitude, conformity, and deliberateness. The second group consists of relationship traits such as friendliness, empathy, civility, or hostility. The third group is made up of work orientations such as attitudes to work requirements, initiative, persistence, ambition or lassitude, task anxiety, passivity, etc. The fourth group of character traits consists of how people relate to themselves, dignity, modesty, pride, narcissism, etc. Supplementary groups exist, but it must be emphasized that independent analyses of these aspects encounters the fact that the character traits interact to create a personality syndrome. From this it follows that character needs to be cultivated as a system.

Character may also be classified in terms of such criteria as relation to will, feelings and aptitudes. Thus we can distinguish will, intellectual, and emotional character traits. Different character traits may contradict one another. In this situation we speak of contradictions within character. For example, friendliness may contradict with being principled, feelings of humor with responsibility, etc.

Variations of character may be manifest not only in multiple qualities but also in the intensity of the expression of specific features. If character traits exhibit extreme expression we may speak of accentuation of character. Sometimes this may be connected with pathological features. Thus we speak in clinical terms of schizoid features of character, agitated characters, etc. However, for pedagogical situations psychological terminology is used more often — introverted or extroverted, balanced or unbalanced, neurasthenic, sensitive, and demonstrative.

While character is treated as the socially acquired features of personality, it also depends on features of the nervous system and temperament. Temperament, of course, renders more or less probable the development of certain features of character. Persons with phlegmatic temperament are less likely to develop initiative and determination than someone with choleric temperament. By the same token a melancholic is less likely to develop persistence and proactivity. However, sometimes because of acculturation, certain features of character may develop that contradict or inhibit temperament. Features of temperament more frequently determine the dynamic features of character.

Merlin (1986) compared the features of character with the attitude toward people, with such features of temperament as extraversion-introversion, anxiety, and extrapunitiveness. Introversion is the tendency of the subject to regulate his/her activity based on his/her internal feelings. Extraversion is expressed when the subject regulates his activity based on the external situation. Anxiety is demonstrated by the tendency of a person to avoid active involvement in situations under threat. It is often distinguished from fear in that an anxiety state is often objectless whereas fear assumes a specific feared object, person or event. The extrapunitiveness (aggression) is a tendency to react to frustration by showing anger toward and investing blame in others. These psychodynamic features of temperament can in some cases be similar to attitude to others as a feature of character. For example, such features of character as friendliness can be expressed in the same way as sociability of the extravert in the group. Anxiety as a feature of character when the person is dealing with people of much higher social status can be expressed the same way as anxiety caused by the psychological feature of temperament. Aggression toward others as a feature of character can be exhibited in the same way as extrapunitiveness which is a feature of temperament when the person is frustrated. However the extrapunitive person would feel sorry after the fact. In specific situations people with different temperaments may develop different types of character. Similar stimulus influences can lead to different psychological consequences. By the same token, similar interpersonal relations may result in distinct responsive reactions. We should also note that in "similar social environments"; different character features may develop in individuals with similar individual features. What constitutes similarity for one individual may not constitute similarity for another. For example, there may be a pair of twins; one may consider himself the leader and try to govern the behavior of his brother. The other tries to avoid this despotism. From this it follows that in objectively similar social environments different forms of character may evolve. This requires consideration of the meaning and sense of the same situation for each.

Temperament traits combine with character traits to shape some unity that is important for the study of individuality. However, this does not imply that they are equivalent. For example, the same trait of personality as restraint can be either a temperamental trait or a character trait. If an individual exhibits restraint in different social situations and this does not require a great deal of "will power," such traits are probably related more to temperament. Later, these temperamental features may become habitual and thereby become features of character. However, restraint as a personal feature may also be socialized into the individual through family influences. Persons with impulsive temperament may suppress their impulsiveness by using "will power"

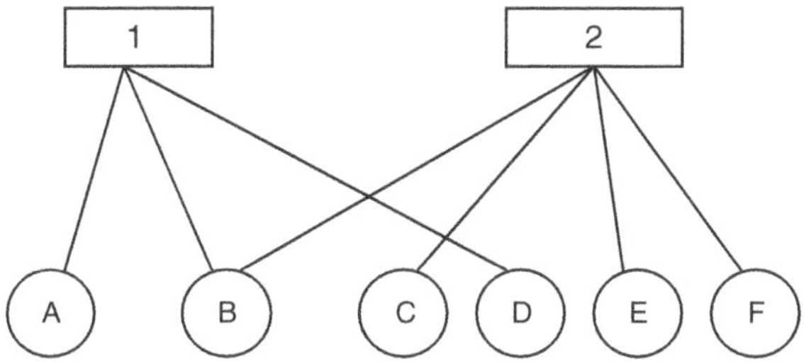

FIGURE 7.3 Relationship between features of temperament and relations of personality to different situations. (1) Features of temperament (boxes): 1 — extroversion–introversion; 2 — extrapunitiveness during frustration; (2) features of character (circles): A — social self-identity; B — aggressive relation to others; C and D — authoritarianism in relation to students in pedagogical work; E — control of own hostility in frustrated situation; F — predisposition to dominate on others.

and thereby exhibit restraint. However, this is a character trait feature. The same features of character possess different coloring contingent upon their temperamental features. For example, the determination of a person with choleric temperament may be accounted for by an absence of insight into the consequences of his acts. On the other hand, the determination of a person who has a balanced neural system or a sanguine temperament relies on a careful diagnosis of the associated features. Interesting empirical evidence for this comes from the study of identical twins who sometimes have similar temperament but different character (Petrovsky, 1982).

Many-to-many relationships between temperament and some other features of personality can be better understood if this relationship is described graphically (Syvorova, 2003). As an example, it will be considered a many-to-many relationship between character and temperament. Figure 7.3 presents a relationship between features of temperament and relations of personality to different situations.

Boxes 1–3 in Figure 7.3. demonstrate features of temperament. The circles below demonstrate those features of character that demonstrate personal relations to specific situations. One can see that on the same feature of temperament depend several sometimes opposite features of character. For example on extraversion-introversion (box 1) depend three different features of character: social self-identity (A), aggressive relation to others (B), authoritarianism in relation to students in pedagogical work (C and D).

Further, from the above presented figures one can see that the same feature of character depends on different features of temperament. For example, in Figure 7.3 it can be seen that aggressive relation to other people (circle B) is connected with extraversion-introversion (box 1) and with extrapunitiveness (box 2).

In Figure 7.4 eight boxes designate the features of temperament and eleven circles designate "style" of behavior in social situations. One can see that on the same feature of temperament (box 1) depend nine other features of character which define "style" of behavior in social situations (excluding features E and F).

Figure 7.4 demonstrates, for example, that such features of characters as frankness–openness (circle A) are connected with extraversion–introversion (box 1) and with those features of temperament as impulsiveness (box 4).

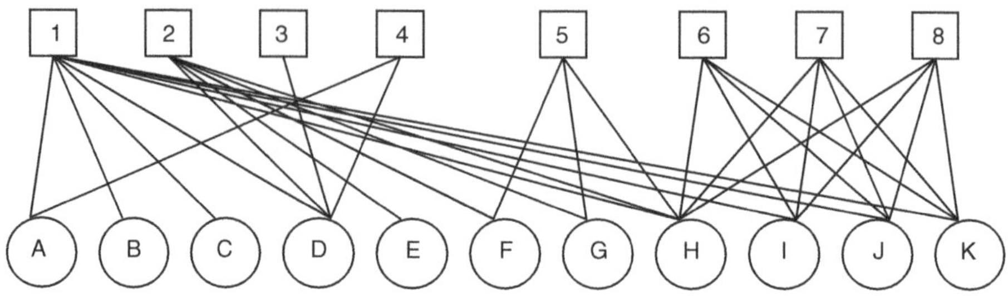

FIGURE 7.4 Relationship between features of temperament and "style" of behavior in social situations. (1) Features of temperament (boxes): 1 — extroversion–introversion; 2 — psychodynamic anxiety; 3 — reactivity; 4 — impulsiveness; 5 — energetic; 6 — emotional stability; 7 — emotional excitability; 8 — rigidity. (2) features of character (circles): A — frankness-openness; B — complaisance-hostility; C — talkativeness, trustfulness, taciturnity, suspiciousness; D — courtesy-hardness; E — loyalty-evasiveness; F — confidence in social interaction; G — sensitiveness in situation of social interaction; H — good breeding; I — organizational abilities; J — communicative features; K — emotionally volitional features.

Interconnections between temperament and character can be changed depending on life conditions, specificity of profession, age, etc. For example, if a person has a clearly defined feature of temperament as psychodynamic anxiety, this feature interacts with anxiety as a feature of his character when the subject has developed a feeling of anxiety in some particular situations. After the development of individual style of performance or psychotraining this undesirable connection can be eliminated. Therefore the feature of temperament can be the same but the feeling of anxiety in some particular situations can be eliminated.

Finally, from Figure 7.4 one can see that such features as courtesy-steadiness (circle D) depend on four features of temperament, and feature of character as "breeding" (circle H) depend on six features of temperament (boxes 1, 2, 5, 6, 7, 8). The wider the range of many-to-many interconnections between the features of temperament and the features of character, the greater is the possibility to compensate negative features of character. The negative influence of one feature of temperament on a person's character can be compensated by positive influences of other features of temperament. If these interconnections are narrow, the ability to compensate becomes more difficult. The features of temperament depend more on inborn natural features of the individual than on character. Hence the features of temperament are more difficult to change than features of character.

It is obvious that the temperament type can influence strategies of human performance. From functional analysis perspective a specialist can outline three facets of temperament that originate from individual differences: 1) relationship between stages of self-regulation and temperament. 2) subject-object or subject-subject oriented aspects of temperament; 3) psychic activation, psychic plasticity, psychic temp and emotional sensitivity.

During the analysis of the self-regulation process one can extract an orienting stage, a decision making and a programming stage, an execution and evaluation stages of performance. These stages have their own specifics features depending on

TABLE 7.1
Self-Regulation and Temperament

		Subject-object oriented aspects of temperament		
1	Oriented aspect (OA)	Specificity of programming of executive strategies of activity (SPES)	Executive aspect (EA)	Sensitivity to feedback (SF)
2	Psychic activation (PA)	Psychic plasticity (PP)	Psychic temp (ST)	Emotional sensitivity (ES)
		Subject-subject oriented aspects of temperament		
1	Socially-oriented psychic activation (SOPA)	Social plasticity (SP)	Socially-oriented psychic temp (SOST)	Social emotional sensitivity (SES)

the temperament type. At the first stage of self-regulation the speed of interpretation or formulation of goal and the speed of reformulation of goal if environmental conditions change depend on the temperament type. The temperament type can also influence the speed of formation of the dynamic model of situation and its reconstruction in the dynamically changing environment. At the second stage of self-regulation the speed of reconstructing the performance program can change depending on the subject's temperament. The third stage of self-regulation that also depends on temperament determines the speed and energetic quality of activity execution. Finally, the forth and last stage of self-regulation that depends on temperament reflects different sensitivity of subjects to the inconsistency between the activity result and the subjective criteria of success. These four stages of self-regulation are described in Table 7.1.

As has been suggested by Shevandrin (2003) we have introduced a number of different indexes of temperament:

General emotionality - ES + SES;
Emotional imbalance - ES − SES;
Readiness to object oriented activity - PA + PP;
Index of general pace - ST + SOST;
Readiness to socially oriented activity - SOPA + SP;
Index of object-oriented activation - PA + PP + ST;
Index of social activation - SOPA + SP + SOST;
Index of general activation - PA + PP + ST + SOPA + SP + SOST;
Activation imbalance - (PA + PP + ST) − (SOPA + SP + SOST);
Index of adaptability - (PA + PP + ST) − (ES + SES) + (SOPA + SP + SOST)

The relationship between the types of temperament an self-regulation and the proposed indexes of temperament is only preliminary and requires future empirical and theoretical validation.

Another important substructure of personality is ability. We will not discuss this problem in detail, as it intersects with the huge Western literature on abilities. It suffices to say that ability is treated as the substructure of personality, characterized by the dynamic of acquisition of knowledge and skills that are important for a particular kind of practical and theoretical performance. Ability is exhibited through knowledge, skills, competence and aptitudes as well as through predispositions. However, ability may not be reduced to this. Ability determines the dynamics of acquiring knowledge and skills as well as the overall levels of accomplishment. Any four substructures of personality can be regarded as abilities if they positively or negatively influence the acquisition of a particular kind of activity.

In contrast to the notion of ability there exists the notion of deficits. Deficits are particular features of personality that inhibit the successful acquisition of a particular performance or restrict the level of accomplishment (Platonov, 1982). Identifying the specific or particular kind of deficits is important to industrial personnel testing. Therefore test procedures must be tuned not only to abilities but also to specific deficits.

Deficits can also be regarded as negative abilities. We can also distinguish between actual and potential abilities. Actual abilities are those that have already been demonstrated in a particular kind of activity. Potential abilities are those that have not yet been displayed in any kind of activity. However, it is taken as given that a person has certain characteristics that can be important to acquire a particular kind of activity. Therefore abilities are studied not only in regard to particular features of personality but also in relation to certain types of activity. Abilities, as with other features of personality, are formed and demonstrated through activity. There are only potential abilities outside of activity. However not everybody has potential abilities to a particular kind of activity. Hence the dynamic of developing different types of abilities can differ from individual to individual. Moreover, there are deficits that preclude the development of certain abilities. Finally, it may be noted that, similar to Western psychology, activity theory divides abilities into specific and general.

Analyses of presented data demonstrate that there are not only similarities but also differences between the description of the structure of personality in the West and those concepts that are derived from activity theory. For example, during the study of personality psychologists who utilize activity theory pay attention to those important aspects of personality as the relationship between the features of the nervous system and those of temperament. A critically important component of personality is character. It is regarded as the socially formed aspect of personality. However, it also depends on more natural features of personality as temperament. Personality is considered in relation to activity of a person. The features and structure of personality are developed through practical human activity and social interaction. A person develops different strategies of activity, which derive from self-regulative mechanisms. From these follow the internal or mental operations as a result of active formation of personality. Object related activity, social interaction and individual features of personality are interdependent and influence each other.

7.1.4 Social–Psychological Features of Personality

The social–psychological and individual–psychological features of personality are intimately related. On the one hand, a person may be regarded as an individual agent of activity. On the other hand, a person may be treated as a concrescence of social traits that are developed through interaction among people. The first approach of studying personality is called the intra-individual approach. The second is called the inter-individual approach. In preliminary sections we pay the most attention to intra-individual features of personality. Here we emphasize the inter-individual aspects of activity.

People may be developed or shaped through their involvement in different kinds of activity. Activity theory provides a basis for comprehending personality structure and dynamics. People are engaged in mutual interactions during joint performance of activities. At such times specific kinds of activity may emerge, which in Russian are termed "*obshenya*," which may roughly be translated as "social interaction." This kind of activity has an array of specific features and may be regarded as an independent psychological category. Social interaction may emerge as part of the activity or may be a relatively independent type of activity performing specific functions. Social interaction may carry out communication functions as well as provide information among people. Here we can outline verbal and nonverbal communication. The goal of communication in this case is the transfer of information among people that is required for joint activity. Other types of social interaction include intimate interactions among people. The major goal here is shaping social behavior and socializing the personal dispositions of individuals. With respect to this process, norms and standards play a major role in promulgating a shared culture. Next, "social interaction" may be regarded as a social-perceptual program. In this case interpersonal perception and cognition are significantly implicated. As may be seen, not only outward activity in the material world but also processes of social interaction are very important for the shaping of personality.

However, not all types of social interaction are equally important for shaping personality. Those social interaction episodes that are more personally significant and therefore possess great personal "sense" exercise a more profound influence. During such "social interaction" individuals acquire moral standards, social norms and requirements, etc. Here we again encounter the fundamental notion of "meaning" and "sense" and the function of these concepts in explicating the shaping of personality according to activity theory (Bedny and Karwowski, 2003). For example, observations of other people involve mental activity and "social interaction." In regard to both this process may be represented as a system of mental actions that are directed to achieve conscious goals. The goal of observing others' behavior may serve the purpose of acquiring professional skills. Or observation may promote understanding of the relationships among physical phenomena or the context within which a person operates. At the same time, observation of others is very important for understanding others, their temporal state and desire, acquiring social norms, standards of behavior, etc. Through the use of different systems of mental actions, people can extract entirely different meaning from the same context. For example, depending on the observer's goals and motives, observation of the aggressive behavior of others may result in internalizing and expressing such aggressiveness, or in developing negative reactions to the aggressor, or different kinds of aggressive behavior. This depends upon what components of the context, or methods of interpreting the situation, are significant to the observer.

Developing different kinds of features of personality depends on both involvement in specific types of "activities" or "social interaction" and on the goals and significance of the different types of "activity" and "social interaction."

In addition to intra and inter-individual features of personality, the analysis of personality formation reveals supra-individual, charismatic features of personality. In this case it is important how a particular persons influences others. This constitutes a third aspect of personality. Individuals are treated as the subjects that active influence intellectual and emotional-willing sphere of people engaged in social interaction with them (Petrovsky, 1982). In some cases influence on other people is a major goal of personality. Power motives, among others, play an important role in supra-individual features of personality.

An important aspect of the study of personality is the social orientation and activation of the person. The greatest influence on this topic is psychoanalysis. According to Freud, the source of activation and social orientation of personality is unconscious impulses (Freud, 1916, 1917). In spite of its profound influence, psychoanalysis is subjected to extensive criticism. For a long time in the former Soviet Union, Freud's theory was prohibited for ideological reasons. According to activity theory, a person is shaped through joint activities and social interaction, both of which processes include conscious and unconscious components and both influence the development of personality. However, from the activity point of view conscious components are more emphasized. The more detailed specification of unconscious components was treated within the concept of "set" originally studied by the Georgian psychologist Uznadze (1966). Set is formulated as "Readiness" and "Predisposition" in a specific way to perceive and apprehend situations or actions with objects. Set is connected with past experience of the subject. For example, Set is related to uncritical stereotyped thinking. Set may be more or less conscious. Psychological studies describe three components of set: (1) cognitive components that determine readiness to comprehend and perceive others; (2) emotional–evaluative components consisting of the complex of a subject's sympathies or antipathies toward different objects; and (3) behavioral components considered as a readiness to act in a particular way to a set of objects. Taken together, these determine the manner in which a person will act in relation to an object or situation.

This aspect of personality is closely connected with the directedness or social orientation of personality. Under the notion of "social orientation" in activity theory we may understand a syndrome of relatively stable, functionally autonomous motives that orient the person to a situation. Social orientation includes such variables as "interest," "desirability," "values," etc.

From an applied perspective, it is important to know how people who possess stable individual features may adapt to different requirements of activity. It is important particularly to know how people with different features of neural system and temperament can adjust to different objective requirements. Russian psychology adduces several methods of adaptation to objective requirements for people with different individual features. One of the more efficient methods of adaptation to objective requirements is the individual style of performance (Klimov, 1969; Merlin, 1986). To some extent this represents the opposite application of personality to the "screening out" of individuals with specific attributes.

Social–psychological study of personality is closely connected with the sociocultural school of psychology founded by Vygotsky (1956). This approach in psychology explains individual differences in personalities through differences in culture, social groups, and specificity of practical domain in a particular society. People's activity becomes social when each person forms and coordinates his goals with the goals of other people involved in social practice. It is a collective activity that influences individual activity and formation of the individual. Collective activity sustains the development of the features of personality of the individual and his abilities.

7.2 INDIVIDUAL STYLE OF ACTIVITY AND PERFORMANCE

7.2.1 Individual Style of Activity and Functional Mobility of Psychological Processes

In this chapter we attempt to examine how an individual adapts to objective requirements of activity. The major principle governing adaptation of individuals with different features of personality to objective requirements is "individual style of activity." In activity theory a lot of research has been devoted to the study of individual style of activity. The concept of individual style of activity was first introduced by Merlin (1964) and Klimov (1969). Different individuals can perform the same work with equal efficiency through the use of their own strategies of performance which are more suitable for their personal features. People attempt to compensate for individual weaknesses with their personal strengths in a given task situation. In this way they diminish the impact of the negative features of personality. Individual style of performance may be shaped both consciously and unconsciously. Very often through inadequate training which ignores individual features of personality, the subject acquires methods of performance that contradict individual features of personality. It is an important goal of psychology to identify the individual style of activity that is helpful for a particular individual who interacts with the task situation. In our work we associate individual style of performance with mechanisms of self-regulation. Individual style of activity may be regarded as strategies derived from individual features of the performer (Bedny and Voskoboynikov, 1975; Bedny and Seglin, 1999). Through individual style, which corresponds to individual features of personality, the subject can more efficiently adapt to a situation. Individual style of performance is important not only in different production situations but also in learning. Through individual style of performance students comprehend situations better and acquire new knowledge and skills. Cognition and activity in general are not only specific to the situation but also adequate to our individual features.

We need to distinguish individual style of activity from individual method of job performance. The latter is not dependent upon individual features of personality. Individual method of job performance depends on organizational factors, supervisory procedures that are imposed, etc. Sometimes these methods of performance that derive from organizational factors can run counter to features of individual personality, and this is not desirable.

Individual style of activity facilitates adaptation to objective requirements of activity, enabling individuals to use their strengths and compensate for their weaknesses. Individual style evolves from extended self-training. However, in some cases such adaptation entails trade-offs among requirements. For example, a worker may neglect safety considerations to ease the execution of tasks. The role of a psychologist in these settings is to identify the individual features of personality that are relevant to task performance and those that contradict job requirements. Based on this they help develop individual styles of performance that enhance productivity and quality of performance. This issue is fundamental for connecting personality theory with performance psychology, thereby enabling psychologists to deploy assessment procedures not only for selection purposes, but also to support job design, instructions, training, and supervisory procedures.

It is helpful to outline the following methods of formation of individual style of activity:

1. A person involuntarily and unconsciously utilizes methods preferable for his individual strategies of performance. For example, a person with an inertial nervous system develops a predisposition to organize and plan work in advance and attempts to utilize a stereotyped method of performance. This way, the person compensates inertial features of her/his nervous system. One should pay attention to the fact that in some situations inertial nervous system can be considered a positive factor. For example, during the performance of a monotonous job, a person can in an easier way utilize a stereotyped method of performance.
2. A person through blind trial and error and feedback corrections (unconscious level of self-regulation) attempts to develop strategies of performance that help him to overcome individual weaknesses. For example, a person with sluggishness as an individual feature, after several failures to react to the required signal, can unconsciously develop strategies of attention that help him to be ready for reaction.
3. A person understands that he possesses some negative feature for a particular situation, and attempts to consciously overcome this feature. For example, a student who has not sufficiently developed memory starts his preparation for an exam much earlier.
4. A person consciously or unconsciously attempts to utilize his positive features for particular situations to compensate his negative features. For example, the tallest boxer with a sufficiently strong punch attempts to use distance fating.

There are different features of personality which can be critically important in different situations for formation of individual style of performance. One important aspect of the study of individual style of activity is associated with mobility of nervous system and cognitive processes. As an example, we consider individual style of activity first of all in relation to this important feature of personality. Mobility of the nervous system is associated with "speed of reconditioning" when the meaning of conditioning stimuli are altered. The opposite of "mobility" is "inertness." "Inert" neural systems have reduced mobility. Mobility and inertness are most evident in situations demanding frequent changes of actions or reactions contingent upon changes in the external environment.

Mobility is a bipolar dimension; high levels of mobility the termed "flexible nervous system"; low level of mobility is referred to as "inertial nervous" system. These features of nervous systems must be distinguished from other temporal aspects of the operation of the nervous system. For example, "dynamicness," which is another "temporal" feature, is inferred from the speed of conditioning. It is, however, unrelated to the speed with which reactions to stimuli can be altered when the signal character of a stimulus is altered. In other words, dynamicness refers to the speed of acquisition of conditioned responses, whereas mobility demonstrates how a subject can change his actions or strategies in a changeable situation. The study of mobility helps to forecast how individuals can adapt to a changeable environment on the basis of altering of nervous processes and cognitive mechanisms associated with them.

There exist a number of methods for studying this feature of the nervous system. One practical experimental technique was developed by Khilchenko (1960), which is illustrative of the ways in which Soviet psychologists drew on their theories for applied practice. In Pavlovian terms this technique consists of differentially processing stimuli either by the "first," or by the "second" nervous systems with the presentation on a screen of three distinct stimuli. Stimuli, such as words, address the second nervous system, while graphic symbols such as different kinds of geometric shapes — for example, squares, triangles, and circles — address the first nervous system. The procedure consists of subjects who hold two handles with buttons in the right and left hands. They put their right and left index fingers onto these two buttons. If, for example, a square appears, the subject is required to press the left button. If a triangle appears, the subject is required to press the right button. If a circle is presented, the subject is required not to press any button. The pace of the presentation of stimuli may vary from slow to fast. The pace is typically increased until the rate of errors exceeds 5%. The duration of each level of pacing persists for one minute. After two minutes of work the subject rests for about five minutes. In this way the ability of subjects to shift strategies from one signal to another either within or across nervous systems is assessed by the mobility of nervous system.

Analysis of this method from the self-regulation of activity point of view allows one to conclude that this method is more complicated than was suggested by Khilchenko and his followers. In the experiment that uses this method scientists deal not so much with the nervous system as with the subject. In this situation, such aspects of activity as goal, motivation, criteria of success, and significance of task for subject become very important. Of course, mobility of the nervous processes is important to perform the task at hand. However, this method is not physiological but psychological. This method models real situations that requires mobility in processing of information. Thus in using this method of study instead of the term "mobility of nervous system" we use the term "functional mobility of psychological processes." This psychological feature determines complicated integral characteristics of human activity associated with speed of information processing in changing environmental conditions. It is obvious that the Khilchenko method is important for application and we will consider a number of examples that demonstrate its application.

The Khilchenko method has been improved to allow the use of different so-called programs of stimuli presentation to evaluate the mobility of cognitive processes. For each program participants received instructions according to which on each trial they reacted in a specific manner to the presented stimuli. Here we briefly describe each program of stimuli presentation and then the procedure of the experiment which applies to all of the programs:

1. Program one — black depiction of geometrical figures (circle, square, and triangle; these are nonverbal stimuli, that is, part of the program which addresses stimuli to the first signaling system according to Pavlov) on a white background.
2. Program two — simple words composed of several syllables — including three categories: names of animals, plants, and objects. These are considered verbal stimuli, belonging to the second signaling system according to Pavlov.
3. Program three — depictions of various animals, plants, and objects, stimuli mainly addressed to the first signaling system according to Pavlov. These should be classified into three categories: animals, plants, and objects through the second signaling system. Therefore this program is addressed to the first and second signal systems.
4. Program four — three words, names of animals, plants and objects. These words are composed of the same number of syllables and begin with the same letter. This is a more complicated program addressed to the second signal system.
5. Program five — each presentation contains three geometrical figures in a row. Two of them are identical, for example, one or two squares and one circle or two circles and one triangle. Subjects, depending on instructions should react to stimuli in different way.

For all of the above programs participants were given instructions which included three potential responses: pressing the left button with the thumb of the left hand, pressing the right button with the thumb of the right hand or withholding a response. The proportion of responses was 64% left or right button presses and in 36% of the trials, a response was to be withheld. Approximately 50% of the responses were left button presses and approximately 50% right. The order of the responses was pseudo-random such that there were no more then 4 button presses in a row (left and right), and no more than 4 withheld responses in a row. At the beginning of the experiment the stimuli were presented at a slow rate (children 20 stimuli/min; adults 40 to 50 stimuli/min). Error rates were registered automatically and the task went on as long as the subjects' rate of error was below 5%; the difficulty was increased, by increasing the frequency of the presentation with each consecutive block of 50 trials. The maximal rate of presentation that the participant could reach, keeping error rates below 5% was a measure of the functional mobility of that participant's cognitive processes.

Using the above methodology it is possible to study the influence of changing the meaning of stimulus on performance (e.g., making a stimulus that was previously designated as warranting a response now signaling the withholding of a response, or stimuli which require responses by the left hand in new trials require responses by the right hand). After the maximal speed of that participant was preliminarily determined, the experiment in which the meaning of the stimuli was changed began with presenting stimuli at a rate 10% below the maximal speed for that participant. If the participant was able to perform at this speed with an error rate below 5% he was considered to have successfully adapted to the change. If not, the speed was reduced every 50 trials until the participant reached 5% an error rate of less than. One variant was to maintain the same rate of presentation for three blocks to measure how quickly the participant adapted to the change in the meaning of the stimulus. If the participant was unable to reach below 5% error rate within 3 blocks of trials he/she was considered to have failed at adapting to the meaning change.

Using the same methodology one can measure the resistance to of the nervous system fatigue. Resistance to fatigue is regarded as a strength of the nervous system. The above methodology does not account for motivational factors that also can influence resistance to fatigue. Therefore we consider it to measure psychological and physiological factors. However, in this case the strength of the nervous system is most important. To measure the fatigability and strength of the nervous system the subjects are presented with 200 to 800 stimuli. The experiment begins by presenting stimuli at a rate 10% below the maximal speed for that participant. The level of fatigability is measured through error rates every 50 stimuli and over the whole set of trials. It is also possible to graph the growth of the error rate throughout the course of the experiment. Another method for measuring the fatigability of the nervous system is to present stimuli at a maximal rate for that subject for 3 to 4 min. The above method can also be used to measure context specific fatigability in the process of actual labor. Contamination that is determined by motivational factors leads us to the conclusion that this method of study can be designated simply as a method of measuring resistance to fatigue. Therefore the above methodology is a psychophysiological one. In addition to measuring all error rates, the experimenters can classify error types (1) pressing a button when on a withhold trial (failure of inhibition and prevailing excitatory error or + error); (2) failure to press a button in the presence of a go trial (inhibitory error or a – error); (3) pressing of the wrong button on a go trial (error ++). Error type 3 is described as a failure in the flexibility of cognitive proccsses.

Let us consider a practical example from military aviation, the study of a radio operator. Usually the selection of personnel was done by criteria such as short-term memory, both auditory and visual, for words, numbers, and sounds, sense of rhythm, ability to distinguish tones, and motor flexibility of the hand and figures. The functional mobility of psychological processes was not studied. To remedy this, the methodology of Khilchenko was integrated into the study process. The age of the participants was between 18 and 22 years. Visual stimuli were used, either verbal stimuli or geometrical figures. Through experimental studies the participants were divided into several groups according to how quickly they could perform tasks without going above an error rate of 5%. During the presentation of verbal stimuli 110 stimuli/min was considered a high rate of performance, 109 to 90 stimuli/min was considered average, and anything below that a low rate of performance. During studies with geometrical figures, a high rate was 140 stimuli/min and above; an average rate was between 139 and 120 stimuli/min; 119 and below was considered low. Based on these studies it was later demonstrated through prospective analysis that those participants who performed at the highest rates on these tasks were most successful during the training process. It was further concluded that the use of acoustical stimuli in experiments proved a better indicator of further performance in training of aviation radio operators. Analogous experiments were performed with professional drivers. It was demonstrated that individuals with low indicators of psychophysical mobility (50 to 60 stimuli/min) in 70% of the cases turned out to be unable to perform their jobs as drivers. Research demonstrated that those with an extremely low mobility of cognitive process are a small proportion of the individuals who are interested in becoming professional drivers, 8% of the studied population. This research showed that the above methodology is suited to use in personnel selection for those professions for which psychological mobility is important.

The following are the stages of formation of individual style of performance:

1. Description critical for task performance conditions on which depends success of task performance
2. Description of major possible strategies of task performance
3. Discovering the more important (positive) individual features of personality that are required for a particular task or situation
4. Discovering negative features of personality for a particular task or situation
5. Developing strategies of performance that help to utilize by person their positive features and overcome their negative features during performance

The following three are the more important ways of adaptation to objective requirements:

1. Deficiency in required abilities can be partly compensated through development of adequate knowledge and skills
2. Insufficiently developed abilities can be compensated based on the development of adequate individual style of performance
3. Selection of people with required individual features for jobs

7.2.2 Adaptation of Individual to Objective Requirements during Skill Acquisition

One of the most important aspects of studying the individual style of activity is the analysis of how well the individual adapts to the requirements of activity during the training period. It is important to observe how people with different individual characteristics acquire knowledge and skills. One difficulty in the study of this phenomenon in an experimental setting is ensuring that the participants are not aware of what the experiment is trying to measure. Due to this, in one of our experiments that will be

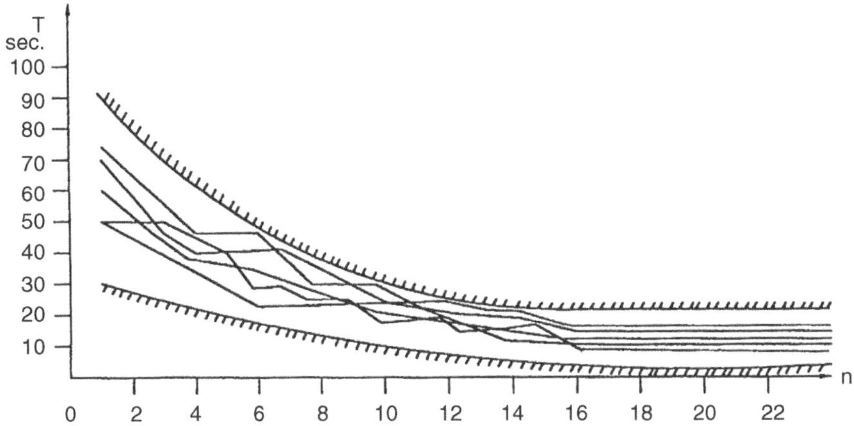

FIGURE 7.5 Time reduction of task performance by pilots during training.

described later our participants were elementary school students who we felt would be less attuned to the goal of the experiment. The obtained data was used to make projections about how individuals adapt to the objective task requirements in real work situations.

Adaptation of persons to the objective requirements of an activity may be seen with special clarity during training and skill acquisition. The time of performance for the same task is gradually reduced during training. This may be represented graphically in terms of curves that illustrate the skill acquisition process. If a number of different subjects take part in an experiment, we can obtain a number of distinct curves. Let us consider as an example Figure 7.5.

This figure demonstrates how task performance of military pilots changes over the training process. The task involves sustaining a flight pattern during the failure of a generator (Lomov et al., 1970). Each curve in Figure 7.5 represents the time task performance is reduced over training. The relative position of the individual curves is a function of the individual features of the performers. Those pilots whose personalities are more suited to the requirements of the activity adapt more rapidly to the task. Since all pilots were screened based on their personal features, their curve of skill acquisitions falls into an area of acceptability, designated by the dashed line in Figure 7.5. If someone is outside of the range of acceptability, indexed by the upper dashed line in Figure 7.5, then this means that some features of personality fail to meat objective requirements of activity. If someone's curves fall beneath the lower boundary represented in Figure 7.5, this means that this subject has individual features of personality that help him to adapted to the most complicate activity requirements. The very substantial convergence of the skill acquisition curves is noteworthy.

In the current example the divergence among the curves is reduced by a factor of three during the period of training. Similar results were obtained by Bedny and Zelenin (1989), in a study of skill acquisition of blue-collar workers, where the divergences were reduced by a factor of six. The magnitude of this reduction is attributed to the fact that blue-collar workers are not preselected for their personal characteristics. Unfortunately, the training literature rarely analyzes the interaction between the curves of skill acquisition in conjunction with the individual features of the learners and the individual style of performance.

It is important to go beyond merely studying how people adapt to standard requirements of activity. Studies are also needed to identify how different subjects may be subsumed into distinct groups with respect to their patterns of skill acquisition as their capacity to adjust to the requirements of the activity is reduced as a function of increasing task complexity.

Below, we present one experiment demonstrating this principle in school students (Bedny and Voskoboynikov, 1975). The study examined how students fall into distinct groups depending upon the complexity of the task and the individual features of their personality. Frequently, in simple situations individuals exhibit similar levels of achievement. However, when the task becomes complicated, individuals begin to vary more in their performance. For example, high-level gymnasts may perform comparably with average gymnasts on a relatively easy task. However, when these two groups perform complex tasks, the average gymnasts cannot maintain tasks at the same level of efficiency as high-level gymnasts.

We selected ten students who had completed the first grade. They were required to perform simple mathematical tasks. Teacher observations provided the basis for three subgroups: superior, average, and poor in their mathematics ability. Based on these expert evaluations, three were exceptionally gifted, four were average, and three were depressed performers, who were being considered for special educational placement. The question at issue was how quickly the students performed in repetitive mathematical calculations.

For these purposes we use the simplest mathematical problems — additions or subtractions of single digit numbers. Altogether, students should solve twelve of these arithmetic problems. The operand symbols were not presented in the problems. Rather, the subjects were instructed to alternate their problems from additions to subtractions. If subjects made errors the experimenter signaled that an error had been made, and the subjects were to correct their calculations. When they completed all twelve problems, their task was finished. They were timed in minutes. Based on the obtained data we generated learning curves for each individual (Figure 7.6).

The experimenter worked with each subject individually in order to make careful behavioral observations. The students performed each task three times per day, over a period of three days. In general they performed 30 tasks. This experiment was selected as an illustration because learning and training are crucial to the study of human performance. Further, primary school students are naive subjects who perceive the experiment as an extension of their schoolwork. Thus, they are less prone to be affected by the demand characteristics of the experimental situation. Finally, the tasks they performed and their results were accessible to measurement.

During the experiments, we expected stabilization of calculating skill for each group at their characteristic level. In the first several days of experiments, the results appeared to vindicate our expectation of distinctive levels of performance for each group. Weaker subjects spent up to 40 min to complete the first task. On the other hand, superior subjects spent less than four minutes. However, the overall experimental results were contrary to our expectations. Day-by-day, weaker subjects

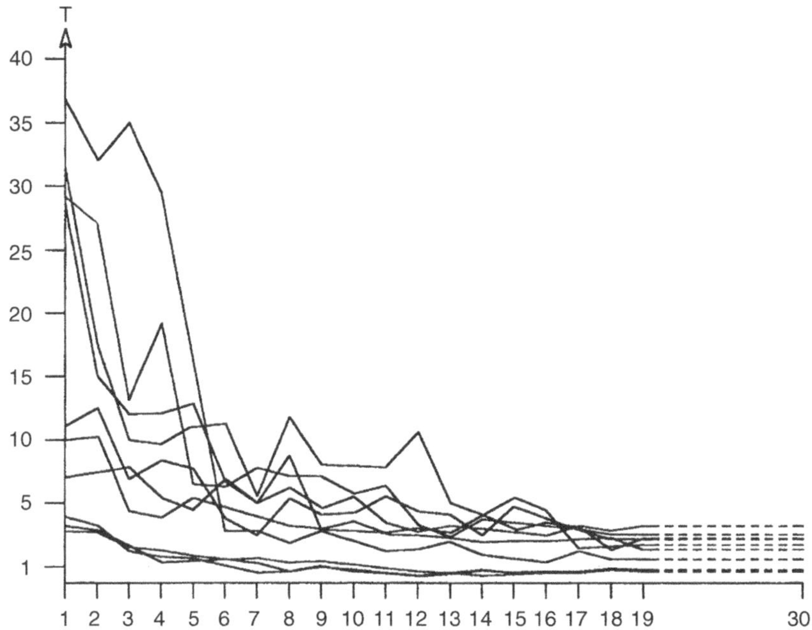

FIGURE 7.6 Time reduction of task performance of students with different individual features.

sharply increased their speed of performance. Subjects with superior ability exhibited a much flatter learning curve showing only slight improvement, suggesting that they were near a ceiling at the outset. Average and weak subjects continued their approach to the task performance times of the superior students. By the 18th and 19th trials (over a six-day experiment), both stabilized the times of their performance. The point of stability nearly converged with those of the high-ability students, who had achieved stabilization at thirteen trials. From 19 to 30 trials no improvement in time performance was observed among any of the students. This point at which there is no further improvement is designated by the dashed lines in Figure 7.6. At stability, the range of individual differences across the ten students was from 3 min to 40 sec. This range was substantially reduced from initial conditions of 37 min, 10 sec to 3 min, 30 sec (see Figure 7.6). In other words, within reasonable parameters it can be said that all of the subjects reached approximately similar times. Thus, we can infer that each of the students had individual features that facilitated their adaptation to the requirements of performance.

This suggests the following question. How, through training, do subjects with such substantial differences in their mental development improve their performance? For these purposes, we compared our observational data, discussion with students, and analysis of their results. Our observations demonstrate that in the first stages, subjects with weaker abilities used their fingers as well as external speech. This means that their mental operations depended on external practical actions. Only after multiple executions of the task with the help of the experimenter did they start to rely

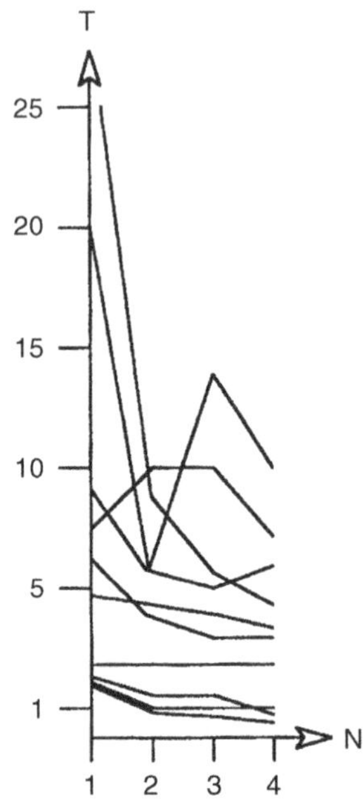

FIGURE 7.7 Time performance of the second task for the same students.

exclusively on internal mental operations. Subjects with average ability also had an initial tendency to rely on external practical actions to guide solutions. They performed calculations accompanied by whispers that were barely audible and produced a higher level of quality. Thereby, they facilitated slow mental operations with external actions. The students with superior abilities immediately started their calculations without any relation to external practical actions.

Our observations revealed the interesting fact that some subjects from the weaker group relinquished external practical actions even before the average students demonstrated such internalized performance. Our observations suggest that this occurred through the utilization of rote memorization by the weaker students. In order to examine the significance of rote memorization for the diverse groups of students, we conducted the following brief experiment. Subjects were instructed in a similar task of addition and subtraction with new numbers. Each subject performed four trials in a single day. The results are presented in Figure 7.7.

From this figure we may see that the performance of the weaker students deteriorates more precipitously. The average students also deteriorated in their performance, but less so. Only the superior students failed to show substantial deterioration. Moreover, the superior students demonstrated the same result in the last trials of the first experiment. The results of this series of experiments help us understand the internal, adaptive mechanisms of the styles of performance. Adaptive effects in the academically weaker subjects occurred through more efficient use of rote memorization. Among average students adaptations follow the route of refinement of external and internal calculative mental actions. The superior students immediately focus on consolidating their internal mental operations.

It is noteworthy that memorization occurred in all the subjects during the first series of experiments. However, better memorization was demonstrated in the weaker subjects. Average subjects succeeded in memorization strategies later than the weaker ones because they were attempting to mobilize internal mental operations. Superior students almost immediately rely on internal mental operations. They do not rely on rote memory. After the subjects found their individual style of performance, they quickly improved and stabilize their level of performance. Consequently, through such identification of characteristic styles they were able to arrive at surprisingly comparable results. We noted that only superior subjects achieved meaningful adaptation to the objective requirements of activity; they reached a high level of automatization of mental operations with single-digit numbers enabling them to transfer this skill to other examples with unfamiliar digits. Weaker subjects demonstrated adaptation to requirements in a more circumscribed way because they could operate efficiently only by using particular numbers that they memorized.

Taking the results into consideration, we decided to test the strategies of the subjects by reducing the viability of memorization by increasing the complexity of the task. Accordingly, we conducted a third series of experiments with the same subjects. The task complexity was increased in the following way: the subjects were asked to perform addition and subtraction of two pairs of numbers. If the preliminary result of a first operation was greater than the results of a second operation, then the subjects were to subtract the second from the first. Conversely, if the second was larger than the first, then they were to add them. In order to frustrate rote memory, the same numbers were used in different positions across several versions of the task. The number of trials was 30. From the 19th to the 30th trials, no further improvement was observed (see Figure 7.8).

The results of this experiment also demonstrate adaptation to requirements. However, these adaptations of abilities are more restricted in comparison with the first experiment (see Figure 7.6). We note that weaker and average subjects at the outset could not perform the task without the assistance of the experimenter. Only after that did they start to perform unassisted. Only the superior students were immediately able to perform the task independently. They demonstrated high levels of performance in initial trials and rapidly stabilized the times of their performance. They stabilized their results within 15 trials and completed the entire task in 1 to 1.5 min. In the average group, stabilization occurred after 20 trials and was achieved in four to six min. Among the weaker students, stabilization was in the range of eight to ten minutes after 20 trials. After 20 trials, time performance stabilized without further improvement. We designated performance for trial 24 to trial 30 with a dashed line, since no one demonstrated improvement in their performance (Figure 7.8). In contrast to the first series of experiments, where all groups are converged to similar level of performance, these experiments resulted in the diverse groups converging around different levels of performance specific to each group. Thus, with increasing complexity of task requirements, subjects who demonstrated similar results in more simple tasks now separate into distinct groups with distinct levels of performance for each group. From this, it follows that the selection of individuals for a particular work should be

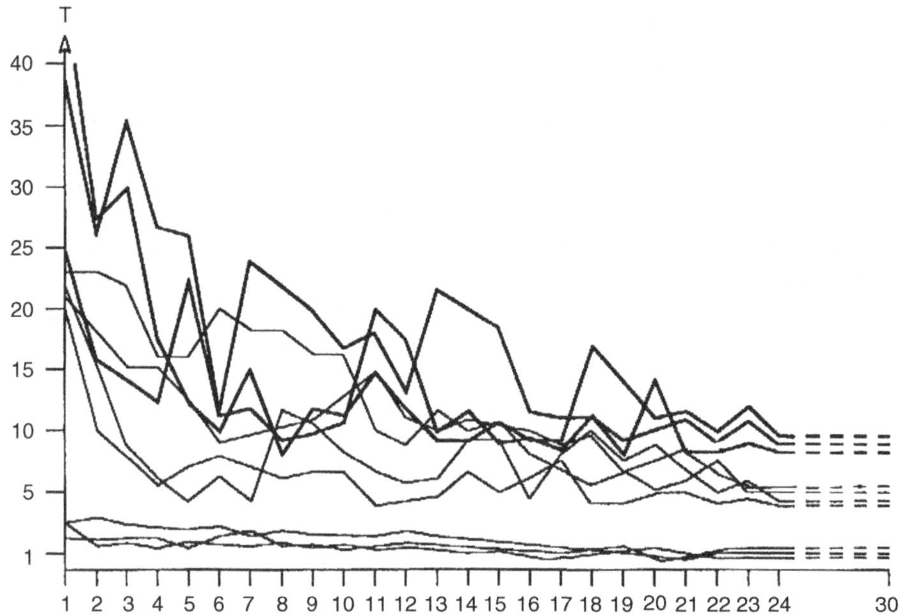

FIGURE 7.8 Time reduction of task performance of the more complicated task by the same students.

performed on the basis of the complexity of the tasks to be performed and their ability to utilize compensatory strategies.

The process of improvement among the weaker and average students demonstrates that these students do not exhibit a smooth learning curve. There are dramatic improvements followed by significant regressions, all of which is constrained by a very palpable limit. Observation of the subjects reveals that they sample different strategies until they acquire the most suitable strategy for their individual style. Individual style of performance depends on their personal characteristics and past experience. Each individual style has a level of adaptation that cannot be exceeded by using other strategies potentially available to them. Weaker subjects can acquire a method of rapid calculation on their fingers, but this method has restrictions in its speed of performance of mathematical tasks. For this purpose a process of internalization is required, when physical and verbal external operations should be transformed into internal mental operations, which is critically difficult for weaker students.

7.2.3 Systemic Approach to Study of Individual Style of Activity

The individual style of activity is a result of different aspects of personality; however, in any given situation some aspects of personality are more significant than others. The individual style of activity can have a large impact on cognitive activity such as strategies of attention, and specificity of approach to a particular solution. It further affects the emotional–motivational and motor aspects of activity. For example, during the work of an operator in bringing the lathe tool to the turning part, jerky and fast movements are characteristic of an individual with a mobile nervous system. However, for a worker with an inertial nervous system, more characteristic is a smooth and slow trajectory of the lathe tool to the surface of the turning part (Klimov, 1969). In gymnastics, those individuals with a more inertial nervous system are more successful at vault performance with a longer running start (Merlinkin, 1967). Individual style was also studied in relation to the properties of temperament. Those properties of temperament that are not in agreement with the objective requirements of activity are suppressed through the internalization of an individual style of activity. The properties of temperament that do not contradict or are in agreement with the objective requirements of activity are used in the individual style of activity. For example, impulsiveness as a property of temperament leads to reactions and actions that occur before the appropriate time. As a consequence, often a sprinter with such a temperament, without awareness will lean the body back at the start of a race to prevent a false-start.

The individual style of activity cements adequate methods and strategies of activity that are formed through self-regulation. An individual style of activity is composed of a specific relationship of orienting, executive, and evaluative components of activity. An adequate individual style of activity is most quickly formed by individuals with a positive outlook on an activity. It is important to keep in mind that the individual style is largely formed unconsciously.

The individual style is manifest not only in object-oriented activity but also in situations of social interactions. During the process of social interaction more often strategies can be formed that will facilitate the achievement of intended results. Noting these aspects of social behavior, Petrovsky (1982) described it as a subject-subject-object relationship. For example, the demanding nature of a manager's behavior can be the result of his/her temperament, a stable characteristic of behavior. However the same manner of behavior can be temporary and caused by his/her frustration.

The analysis of research on the individual style of activity demonstrates that most of these studies considered relations between isolated features of personality and individual style of performance, such as how the strength of the neural system affects the method of performance, or how the mobility of the nervous system is connected with adaptation to a changing environment. This approach is valuable in activities where special isolated features of personality become particularly germane to concrete characteristics of tasks. However, this approach ignores the structure of personality where very important interactions of different features of personality are implicated in the development of individual styles of performance. This notion of structure of personality is crucial for dealing with compensatory trade-offs among diverse strengths and weaknesses of personality features and individual styles of performance. When studying individual style of activity, one can undertake the following:

1. A microstructural or analytic approach involving the study of isolated features of personality and how these features influence individual aspects of performance, if they are critically important for a particular task

2. A macrostructural or holistic approach in which one discovers more salient relationships among different features of personality and individual style of activity

In the latter case, it is noteworthy that the same individual style of performance may be based on different features of personality or their combination. Only consideration of all aspects of adaptation of personality to objective requirements can guide individualization of training and education. We suggest that analyzing critical situations during students' performance, and following comparison with individual features of personality, enable more efficient individualization of training and teaching. In these studies we regard the individual style of activity as a series of activity strategies that are formed based on mechanisms of self-regulation and depend on the individual characteristics of personality.

In the following experimental study, we proposed that the same adaptive behavioral act might be elicited by different features of personality. "Mobility of the neural system" has a fundamental meaning in the study of personality in activity theory. Mobility is a bipolar dimension; high levels of mobility are characterized by the term "flexible nervous system"; low levels of mobility are referred to as "inertial nervous" system. These features of nervous systems must be distinguished from other temporal aspects of the operation of the nervous system. For example, "dynamicness," which is another "temporal" feature, is inferred from the speed of conditioning. It is, however, unrelated to the speed with which reactions to stimuli can be altered when the signal character of a stimulus is altered. In other words, dynamicness refers to the speed of acquisition of conditioned responses, while mobility tends to refer to speed reconditioning and decision-making under changing environmental contingencies. According to Nebylitzin (1965), mobility essentially influences the decision-making process. We have already considered this feature of the nervous system.

Mobility is typically identified and indexed in two steps: an initial screening entails natural observations and questionnaires, and the second step entails experimental procedures to identify and index mobility (Merlin, 1964; Klimov, 1969). The Hilchenko (1966) method, which is discussed below, is one suitable and widely used such experimental procedure that we recommend to American practitioners. As noted earlier in this section, rather than the notion of "nervous system mobility," we prefer the term "functional mobility of cognitive processes." However, because this section is a comparison of our data with that of other authors we also use the terminology proposed by those authors.

Flexibility is an important neurocognitive characteristic that may be used in personnel selection, training, and other disciplines including clinical psychology. Toward elaborating this concept we replicated experiments by Merlinkin (1977), that selected two groups of students. One group had an "inertial nervous system," while the other had a "flexible nervous system." The first group had difficulty in adapting to changing environments. The second adapted more quickly to changing environments. Both groups were required to perform three consecutive forward rolls, finishing at upright attention. Merlinkin filmed their performances. He discovered that both groups performed correctly, but each group utilized a distinct individual style of activity. Students with "flexible" nervous systems completed all three forward rolls at the same speed, and immediately stopped. Those with "inertial" systems performed each forward roll at a different speed. Initially, they performed the first one quickly. By the second, they began to slow down. On the third, they were definitely slowing down in order to facilitate a crisp finish at attention. Thus they achieved the same result with the same quality by utilizing a different style of performance. Difference in style of performance was outside the consciousness of the students.

The same individual style of performance can be caused by different nervous features of personality and by their idiosyncratic combinations (Bedny, 1976). In order to evaluate this hypothesis we conducted the following experiment. First we selected highly experienced gymnasts and tumblers (henceforth to be known as gymnasts), so that only students with inertial nervous systems were included. Second, we selected sprinters (henceforth known as nongymnasts). Only nongymnasts with flexible nervous systems were selected. All of them had some experience with forward rolls from general physical education classes.

We defined "inertial" or "flexible" nervous systems through a two-step procedure. The first step included observations during their physical education classes. Second, we used Klimov's (1969) special questionnaire. Based on this preliminary study we selected 16 "inertial" gymnasts and 18 "flexible" nongymnasts. All of them were assigned three tasks. In the first task the subjects executed three forward rolls at their own pace. In the second task they performed the three forward rolls with a precise stop. In the third task they executed the rolls and precise stop blindfolded. Each subject performed each task three times, and then his or her results were averaged. We also recorded their performance with a 16 frame/second movie camera. The results are presented in Table 7.2.

We compared differences in time to perform the forward rolls. We performed a one-way analysis of variance corrected for interdependencies among data taken over time. Once the null hypothesis was rejected, we used the multiple comparison procedure on mean differences. As may be seen from this table, there are only small differences among the average speeds for all three rolls for the gymnasts. This difference was not statistically significant even in blindfolded conditions ($p > .05$). On the

TABLE 7.2
Strategies of Performance of Forward Roll by Subjects with Inertial and Flexible Neural Systems Evaluated by Observation and Questionnaire

Task	Without precise stop			With precise stop			With precise stop, blindfolded		
Number of forward roll	First roll	Second roll	Third roll	First roll	Second roll	Third roll	First roll	Second roll	Third roll
Inertial gymnast	14.3	14.3	14.2	14.4	13.5	15.0	14.6	13.6	15.2
Flexible nongymnast	15.9	15.8	16.8	16.0	16.1	18.8	16.7	16.4	20.7

other hand, for the nongymnasts the second and third forward rolls were significantly slower when the precise stop was introduced. This data was statistically significant ($p < .05$). This effect was even more pronounced under the blindfolded conditions.

Differences between the two groups' individual styles of performance were especially pronounced under blindfolded conditions. This may be explained as follows. The regulation of movements depends on feedback. When a skill is not fully automaticized, exteroreceptive feedback or "external contours of regulation" dominate the performance. Visual control is especially important at this step. When a skill becomes automated, vestibular and proprioceptive feedback, or "internal contours of regulation" assume greater importance (Bernshtein, 1966). The gymnasts execute forward rolls in the same way whether blindfolded or not. Nongymnasts have not automized the skill. Accordingly, in order to perform under blindfolded conditions, it becomes more complicated. Their second and third rolls — especially their third — demonstrate greater slowing under blindfolded conditions. In Merlinken's study, the slow down in the final roll forward was attributed to inertial features of the nervous system. In our study the same adaptive behavior is caused by the lack of past experience.

The next step was to evaluate the quality of "flexibility" and "inertial" nervous system using experimental procedures. For this purpose, we selected Hilchenko's method (1966), a simple, practical method, that calls for the presentation in a random sequence, at increasing pace, of two or three stimuli each of which requires from the subject two or three different responses. Mobility refers to the maximum pace achieved in this alternation of responses to the stimuli with minimum errors. Experimenters present the subjects three different kinds of stimuli on a special screen or monitor. In our experiment we used stimuli that address the first signal system, like a triangle, a circle and a square. These signals are presented to the subjects at a varying pace, and each signal is presented on a separate slide. The pace varies from slow to fast and is defined by the subjects reacting with an error rate of no more than 5%. This is treated as the maximum individual speed of the subject. The instruction might call for the subject to press a right-hand button when presented with a square, a left-hand button when presented with a triangle, and to inhibit a response when presented with a circle.

Gymnasts identified as "inertial" by nonexperimental studies have a speed of 102 to 134 frames per minute before reaching the "maximum" speed. Nongymnasts identified as "flexible" attain a speed of 116 to 144 frames/min before reaching "maximum." From these two groups we selected the five most "inertial" gymnasts and the five most "flexible" nongymnasts. Three conditions were studied (without precise stop, with precise stop and with precise stop blindfolded). Three trials of three forward rolls under the three conditions were used. In Table 7.3 we present average results from three trials.

We applied the same statistical technique. The experimental data show that for the inertial gymnasts the average differences in the performance speed between their first and third forward roll was not statistically significant even in blindfolded conditions ($p > .05$). The differences between the first and third forward roll for flexible nongymnasts was statistically significant when they were required to make a precise stop, both with their eyes open and blindfolded ($p < .05$). The more significant differences were found in the blindfolded conditions. Here we again see that performance

TABLE 7.3
Strategies of Performance of Forward Roll by Subjects with Inertial and Flexible Neural Systems Evaluated by Hilchenko Method

Task subject	Without precise stop			With precise stop			With precise stop, blindfolded			
	Roll 1	Roll 2	Roll 3	Roll 1	Roll 2	Roll 3	Roll 1	Roll 2	Roll 3	Flexibility
Inertial gymnast										
1	14	14	15	14	14	16	14.5	14.5	15.5	116
2	14	13	14	14	13	15	13.5	13	14.5	102
3	14	13	14	15	14.5	16	14	14.5	16	116
4	14	13	14	14.5	13.5	15.5	15	13.5	15.5	102
5	14.5	14	15	15	14	15.5	15.5	14.5	16	105
Flexible nongymnast										
1	15.5	15	16	15.5	15	17.5	14.5	14.5	17.5	144
2	15	15	16	16	16	19	16	16	20	144
3	16	16.5	18	15	15	19.5	15	15.6	20.5	134
4	15	15	17	15.5	15	18	15.5	15.5	20	134
5	16	16	17	16.5	16.5	19	17	18.5	22	134

differences emerge with the increased complexity of the task. From Table 7.3 we also see that, in the group of "inertial" gymnasts, the maximum differences in the speed between the second and third roll in the second series is equal to two frames. At the same time, for flexible nongymnasts, two frames are the minimum differences between the second and third row. In the third series (blindfolded, inertial gymnasts) the difference between the second and third role does not increase. For the flexible nongymnast these differences increase from 3.1 to 4.0 frames on average. When we compare our results with Merlinken's study, we find that the flexible subjects in our study use the same individual style as the inertial subjects in his study. In other words, experienced inertial gymnasts did not slow down on their final forward roll. They used a strategy that is usually acceptable for flexible nonexperienced subjects who perform the same tasks. At the same time flexible nongymnasts use strategies that are usually suitable for inertial subjects without experience. This was especially evident when they performed under the blindfolded conditions. This implies that the same adaptive strategy, or individual style of activity, may be invoked by distinct features of personality. (In the current case this would mean past experience and flexibility of nervous system). Our study shows that past experience with the assigned tasks eclipses in importance the trait of inertia or flexibility of the nervous system. Further, these data imply that the same adaptive action, or individual style of performance, may be explained through psychological or physiological features of personality. In Merlinken's study the slow roll forward may be explained by the inertial features of nervous system. In our study the same result may be explained through the amount of past experience interacting with the complexity of task requirements. These results underscore that analyzing problems of individual style of activity requires going

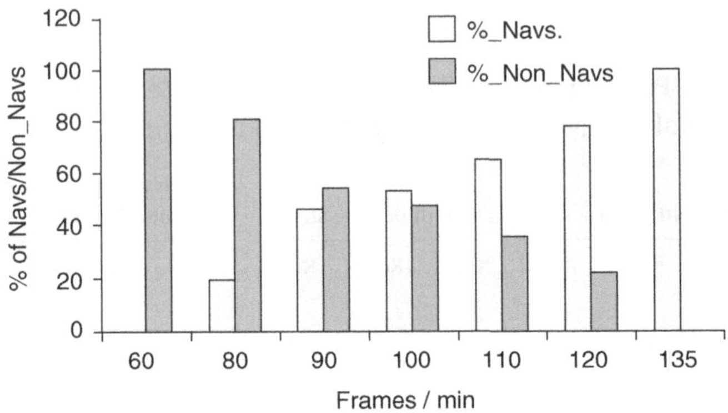

FIGURE 7.9 Proportion of navigators and nonnavigators with different levels of flexibility according to method of Hilchenko.

beyond isolated features of personality to study the structural relations among them. This structural relationship also determines the style of activity.

In the Black Sea laboratory of psychophysiology an assessment of flexible–inertial nervous systems was conducted in distinct categories of merchant marine officers serving in the Black Sea Fleet (Bedny and Zelenin, 1977). Two of these groups — navigators and nonnavigators — were selected. The navigator had to react to situations in a flexible and dynamic manner. The others did not require rapid adaptation to situations. They were evaluated on the quality of the "flexibility" and "inertia" of their nervous system through the Hilchinko methods described above. All of the signals were presented and varied from slow to maximum speed in accordance with the Hilchenko method. The slowest speed was 60 frames/min. The fastest was 135 frames/min. Those who successfully passed at the slowest speed graduated to the next speed. For example, those who successfully passed a speed of 60 frames/min proceed to 80 frames/min, then to 90, up to 135 frames/min. Different numbers of subjects successfully passed the tests at each speed. Accordingly, the results of test evaluation of subjects were analyzed in the following way: the overall number of subjects who fall into a particular level according to the Hilchenko method was defined as 100%. Then we calculated the proportion of navigator and nonnavigators at each level. The results of this study are presented in Figure 7.9.

The shaded column corresponds to the navigators. The nonshaded columns refer to nonnavigators. From this figure we may see that those reaching their limit at the speed of 60 frames/min are only nonnavigators. On the other hand, only navigators attained the maximum speed of 135 frames/min. Intermediate levels showed a mixture of both, with the proportion of nonnavigators diminishing as the speed increased. Therefore, a flexible neural system is highly associated with being a navigator. It is interesting that no psychological tests were used to select for these positions. This suggests that people intuitively self-select in some professions requiring special individual features.

Material presented in this chapter demonstrates that activity theory introduced some distinctive features in the study of personality. According to this approach the subject, through diverse actions, changes his or her environment, as well as modifies the structure of his or her own personality in a dialectical dynamic. Further, these diverse actions engender a social psychology that shapes idiosyncratic configuration of individual psychological features. As can be seen, material activity and social "community" are fundamental in the determination of personality. Through involvement in different types of activity and social interaction individuals change themselves and influence others. This socialization process orients individuals toward tasks that are significant for them as well as society.

The notions "individual features of personality" are distinct from the traditional understanding of traits. The former includes not only those features included in the Five-Factor model of personality but emergent and additional factors as well. Moreover, most traits that are described by the Five-Factor Model from the perspectives of the Russian Theory of Activity relate principally to temperament. According to activity theory, personality has a much broader meaning than measuring global traits and abilities. Individuals are not treated merely as manifestations of temperament, character, ability, and other idiosyncratic characteristics. Rather they are a complicated structure wherein diverse aspects of personality are treated holistically in terms of the interactions of their above-mentioned components. The specificity of temperament affects intellectual ability and vice versa. Specific character affects acquisition of skill. Individuals adapt to various situations and requirements of activity in diverse ways. Individual style of activity constitutes one of the more dominant methods of adjustment to and adaptation of the environment. This is especially pertinent to work and educational settings. Instruction based on individualized approach is regarded as the most powerful approach of guiding individuals in their interactions with the environment.

7.3 CONCLUDING COMMENTS

There is a substantial gulf between theory and theoretical concepts in psychology and applied studies in ergonomics and industrial/organizational psychology. This can be explained by the fact that cognitive psychology, with its mentalistic orientation, concentrates on studying separate cognitive processes. The proof of the pudding is that ergonomics and industrial/organizational psychology studies holistic human work and activity theory creates a framework for practitioners to assess human work within the context in which it occurs. Currently, the field of human factors and applied psychology offers a plethora of disconnected and not always sufficiently systemized data. This complicates both the process of theoretical interpretation of obtained data and their use in practice, which is particularly important in the study of human work.

Material presented in this book demonstrates that systemic-structural activity theory (SSAT), which is derived from general activity theory, has not only theoretical but also practical meaning for both ergonomics (human factors), including industrial/organizational psychology, and psychology in general. SSAT helps professionals to eliminate significant gaps in psychological theory and practice as can be observed today in study of human work. SSAT has been used to examine problems in task analysis and design, learning and training, as well as work motivation and the study of personality. General activity theory is a theoretical framework which emerged about 70 years ago in opposition to the behavioral approach then dominant in the former Soviet Union. In the West, general activity theory has been utilized during the last two decades as an alternative to cognitive psychology. In contrast to behaviorism, which considers the subject as a reactive system, or cognitive psychology, which concentrates on the study of separate cognitive processes, activity theory promotes fundamental ideas, such as unity of cognition and behavior, mediation, history, and culture. Activity theory studies human practice in a manner that requires us to study human actions as basic units of analysis. The concept of action corresponds well to work analysis requirements. The personality principle in activity theory states that through activity a subject changes the world, and as a result changes himself. General activity theory, which was founded by Rubinshtein and Leont' ev, does not offer methods and procedures which can be directly utilized in practice. Therefore, in this book we advocated the adoption of SSAT.

From the systemic perspective, activity is a goal-directed system which integrates cognition, behavior, and motivation. Basic concepts such as a goal, action, and self-regulation are described more precisely in SSAT than in other approaches. It was also demonstrated how basic concepts can be utilized in practice. Different units of analysis were suggested. Cognitive and behavior actions and operations, functional mechanisms or blocks, and members of algorithms can be related to these units of analysis. Classification of actions and procedures for extraction of cognitive and behavior actions were described. Activity is regarded as a multidimensional system and, therefore, systemic principles of activity study are presented in this book. SSAT does not reject data from the field of cognitive psychology. However, the cognitive approach is regarded as a stage of analysis. In contrast to general activity theory, SSAT suggests very precise principles, including general methods and techniques of work analysis. SSAT stresses the importance of their logical organization during the study of human work. Described principles and methods of study are not only qualitative but also formalized and quantitative. As a result, SSAT can be turned into a fundamental scientific approach in psychology, which can be utilized for different purposes and, particularly, for ergonomic design and human performance.

Glossary

Activity — a goal-directed system, where cognition, behavior and motivation are integrated and organized by the mechanism of self-regulation toward achieving a conscious goal (activity has a recursive, loop structure organization).

Systemic-structural approach — studies structural systems. In the study of these systems, particular attention should be paid to units of analysis, relationship between elements of the system, stages and levels of analysis, their relationship and transition between them. Genesis of these systems is also an important aspect of systemic-structural analysis.

Organizational system — consists of different elements that have no relation to each other. Any changes in one element of the system can change the system but does not change other elements of the system.

Structural system — consists of different elements that are interrelated. Any changes of one element of the system can change the system, their elements and relation between them. The latest can be dynamic and develops over time.

Situated system of activity — activity are constructed or adopted to situations according to mechanisms of self-regulation. It includes flexible reconstructive strategies (situated components) and preplanned and preprogrammed (prespecified) components.

Functional systems — dynamic self-regulative entities that are mobilized, formed, and disappear upon consummatory activities. It can be described as the process of self-regulation at physiological and psychophysiological levels.

Functional analysis — analysis of the structure of activity is performed based on different functional models of self-regulation of activity. The major units of analysis are functional mechanisms or function blocks.

Self-regulation — an influence on a system that derives from the system in order to correct its behavior or activity. It includes cognitive (informational) and motivational (energetic) components and has a loop structure organization. Self-regulation includes orienting, programming, executive, and evaluative components and can be performed at conscious or unconscious levels which interact during the process of self-regulation.

Strategy — a plan or program of performance that are responsive to external contingencies, as well as to the internal state of the system. Strategy has a dynamic and adaptive character enabling changes in the goal attainment as a function of external and internal conditions of self-regulative system.

External and internal contours of self-regulation — External contours feed-forward and external feedback from external receptors during activity performance. External feedback provides meaningful interpretation of events. An internal contour includes feed-forward and feedback in proprioceptive systems of motor activity that are typically unconscious. The interrelation among these two contours has a dynamic character. Some internal components of regulation can be transferred into external contours, enabling more exact conscious control of behavior.

Plan and program of performance — the content and sequence of the deferent components of an activity or separate actions (mental or behavior) by means of which an activity or separate actions should be performed. We use the term plan when the subject deliberately and consciously determines the sequence of the elements of an activity in a particular situation. The term program is used in situations when planning is unconscious and has a very short duration.

Physiological type of self-regulation — is based on homeostasis. The purpose of this type of self-regulation is to reduce the discrepancy between the optimal state of the physiological system and real state of the system, in order to reduce disturbances on the system and restore balance. Many physiological imbalances are corrected automatically. The structure of physiological self-regulation processes is completely predetermined.

The psychological type of self-regulation — is a goal-directed process. This system can change its own structure based on its experience. Such a system can form its own goals and sub-goals and its own criteria for an activity evaluation. Psychological type of self-regulation provides the integration of cognitive, executive, evaluative, and emotional aspects of activity. The major problem facing the self-regulation system is the process of continuing reconsideration of activity strategies or even changing the goal of activity when internal and external conditions or situation have changed.

Conscious and unconscious levels of self-regulation — these levels of self-regulations are interdependent. In the conscious level of self-regulation, leading role plays goal and verbally logical components of activity. In unconscious level of self-regulation, leading role plays set, imagination, intuition, and nonverbalized meaning.

Functional mechanisms — integration of cognitive processes or actions for particular purpose during process of self-regulation. Functional mechanisms can be considered as particular subsystem with specific regulatory functions within a structure of activity.

Function block — represents a coordinate system of subfunctions with specific purposes in the structure of activity. A function block is considered as a functional mechanism that has specific relation with other functional mechanisms. Each function block mediates a particular function in the regulation of activity. The interrelationship among the functional blocks is critical to the understanding of activity regulation. The content of the function block can change, but the purpose of each function block in the self-regulative model is constant.

Functional microblock — decomposition of activity at the microlevel. Chronometrical studies are important at this stage of analysis. This method is applied for description of functional structure of cognitive and motor actions. Actions are described as self-regulative systems which are comprised from different functional microblocks. Each microblock describes psychological microprocesses such as iconic memory, mechanism of scanning information, etc.

Functional macroblocks — decomposition of activity at the macrolevel of analysis and description of activity as a whole. The function block in this analysis has a much more complex architecture. At this stage of analysis we do not apply chronometrical studies of very short duration cognitive processes. Each function block requires much more time for its realization. This method of analysis of activity is particularly important for description of prescribed and real strategies of performance and their acquisition during training process. The same components of activity depending on their role in the process of self-regulation can be included in different function blocks.

Self-regulative model of formation and acceptance of goal — describes the process of goal formation and acceptance of the goal from the standpoint of self-regulation.

General model of self-regulation — describes the process of the self-regulation of activity during task performance. This model describes all stages of self-regulation including executive stage of activity associated with transformation of situation (object) according to required goal of activity. Behavior actions and their interrelationship with cognitive actions are important at this stage of analysis.

Model of self-regulation of orienting activity — orienting activity is explorative or gnostic, in nature and, therefore, is very flexible. The main characteristic of orienting activity is its dynamic reflection of the situation. Dynamic reflection of the situation, developing dynamic mental model, and interpretation of the situation is the major purpose of orienting activity. One should distinguish between orientation as a stage of activity and orienting activity when reflection of reality is major purpose of the activity. Model of self-regulation of orienting activity describes the type of activity where executive components of activity are significantly reduced.

Function block "goal" — an integrative mechanism of self-regulative process which interacts with motivation and creates vector motive → goal. The relationship "motive–goal" provides direction to the self-regulative process. Studies show that different individuals may have an entirely different understanding of a goal, even if objectively identical situation or instruction is given. As a result of that we distinguish "subjective" and "objective" understanding of the goal. Goal cannot be considered as something externally given, in ready form, to the subject. The goal is always associated with some stage of activity. It includes stages such as goal recognition, goal interpretation, goal reformulation, goal formation, etc.

Function block "meaning"— involved in interpretation of input information (can be extracted not only from external data but from memory as well). It provides a relationship not only between the sign and its referent but also a relationship of sign and activity. Function block "meaning" in its study of the relationship between the sign and meaning considers not only the individual but also the world of culture created by human activity.

Conscious and unconscious meaning — the verbal and conscious aspects of the meaning. When one considers these aspects of self-regulation function block "meaning," associated with conscious goal. When block "meaning" works together with function block "set" it is involved in extraction of "nonverbalized situational meaning."

Function block "sense" — those aspects of sense that are involved in emotionally evaluative aspects of activity and produce personal significance of different components of activity. Personal significance within the goal-directed activity leads a person to interpret the meaning of the presented information and transfer it into the subjective sense.

Function block "assessment of task difficulty" — is cognitive component of activity that involves consciousness of the objective complexity of the task, as well as some intuitive assessment of complexity. The more complex the task is, the greater the probability will be that the task will be difficult for a subject. The subject can evaluate the same task as more or less difficult depending on his/her past experience or individual features. Therefore, the cognitive effort and inducing motivational components of activity depend on the task difficulty. The individual may under or overestimate the objective complexity of the task and this has influence on strategies of task performance. Incorrect assessment of difficulty can result in inadequate personal sense or motivation to sustain the efforts for completing the task.

Function block "orienting reflex" — creates conditions for a heightened receptivity of the organism to sudden changes in the situation. This is accomplished through the development of a complex, short-lived and transitory physiological processes, the change of the activation level in the neural system with a general inhibition of conquering ongoing activity.

Function block "afferent synthesis" — provides analysis, comparison, and synthesis of all data that the organism needs in order to perform adoptive response in the given circumstances. Major stimulus, which causes a reaction, never exists in isolation. It interacts with supplementary environmental stimuli that influence response, information extracted from memory that is relevant to this response, and current motivational state.

The function block "set" — the set is characterized by the role it plays in the formation of the purposeful behavior. The set is responsible for creation of internal state of the organism that determines purposefulness of human behavior, but this state is not conscious. The set creates a predisposition to processing incoming information in a particular way, or predisposition to perform particular actions. The set can be transferred into conscious goal and vice versa. Therefore the set performs similar functions at the unconscious level of self-regulation as the goal performs at the conscious level of self-regulation.

The function block "subjectively relative task conditions" — is responsible for the creation of "dynamic model of situation." This block is involved in the creation of the holistic mental model of reality. This block includes two subblocks. Subblock "operative image" to a large extent provides unconscious dynamic reflection of the situation. Subblock "situation awareness" includes a logical and conceptual subsystem of the dynamic reflection of situation in which the operator is very conscious of the information processing. These two subsystems of dynamical reflections overlap. The subject is also very conscious of the information processing in the overlapping part of the imaginative subsystem. Conscious and unconscious components of dynamic reflection can to some degree transform into each other.

The function block "conceptual model" — is responsible for developing a broad and relatively stable mental model, which serves as a general framework for understanding different situations relevant to particular professional duties. It includes different scenarios of possible work situations. Although this model is general, and is stored in long-term memory, it is more specific than past experience. Imaginative components are one of the distinguishing characteristics of this model, and play an important role in its functioning.

The function block "past experience" — is the general background of the subject that also influences strategies of performance, and therefore can be considered as functional mechanism. It includes general and professional knowledge of the subject, knowledge of culturally accepted norms for behavior, and customs that described how the community functions. Past experience is acquired through activity that evolve over time within a culture. The interaction of past experience and new input information results in the assessment of the meaning of the immediate input information. Past experience includes not only cognitive but also emotionally motivational components and evaluation of task difficulty.

The function block "formation of a program of task performance" — involves the development of the program of execution of actions directed to achieve the accepted goal. This mechanism represents information regarding the method to be used in achieving the task goal and may or may not be conscious. This program is developed prior to the task or actions performance, and can be modified during that performance. Program of performance can be comprised from hierarchically organized subprograms. Some of them can be conscious, others, unconscious.

The function block "motivation" — is responsible for the development of inducing components of motivation. While the block "sense" refers to emotionally evaluative components of the activity, motivational block determine the directness and energy for attaining a specific goal. Sense block and motivational block are intimately connected, but some times emotionally evaluative components of motivation can be in conflict with inducing components.

The program performance block — is involved in the execution of required activity according to the plan. Realization of the developed program of performance does not necessarily exactly match the developed program.

Evaluative stage of self-regulation includes a number of different blocks:

1. *Subjective standard of successful result* — is responsible for development of subjective criteria for success which can deviate from objective requirements
2. *Subjective standard of admissible deviation* — subjects define which errors are significant and which are not. If these deviations do not exceed subjective tolerance, subject do not correct their actions
3. *Positive or negative evaluation of result* — final evaluation of result based on subjective and objective criteria
4. *Information about interim and final result* — is the subjective interpretation of obtained data at the different stages of self-regulation process

Function block making decision about correction is involved in correction of self-regulative process and modification of the goal.

Note: Some functional blocks have abbreviated definition.

Bibliography

Ackoff, R. (1980). Towards a system of systems concepts. In Smith, H.R., Carroll, A.B., Kefalas, A.G., and Watson, H.J. (Eds.), *Management: Making Organizations Perform* (pp. 43–87), New York: Macmillan.

Adorno, T.W. Brunswik, E.F., Levinson D.J., and Sanford, R.N. (1950). *The Authoritarian Personality*, New York: Norton Simon.

Akhutin, V.M. (1977). Stage modeling and synthesis of adaptive biotechnical and ergotical systems. In Lomov, B.F., Rubakhin, V.F., and Venda, V.F. (Eds.), *Engineering Psychology: Theory, Methodology, Practical Applications* (pp. 149–181), Moscow: Manufacturing Publishers.

Aladjanova, N.A., Slotintseva, T.V., and Khomskaya, E.D. (1979). Relationship between voluntary attention and evoked potentials of brain. In Khomskaya, E.D. (Ed.), *Neuropsychological Mechanisms of Attention* (pp. 168–173), Moscow: Science.

Allen, G., Buxton, R.B., Wong E.C., and Courchesne, E. (1997). Attentional activation of the cerebellum independent of motor involvement. *Science*, 275, 1940–1943.

Allport, G.W. (1937). *Personality: A Psychological Interpretation*, New York: Holt.

Anan'ev, B.G., Vekker, L.M., Lomov, B.F., and Yarmolenko, A.V. (1959). *Sensing by Touch in Cognition and Work*, Moscow: Academy of Pedagogical Science RSFSR.

Anokhin, P.K. (1935). *The Problem of Center and Periphery in the Physiology of Higher Nervous Activity*, Gorky: Gorky Publishers.

Anokhin, P.K. (1955). Features of the afferent apparatus of the conditioned reflex and their importance in psychology. *Probl. Psychol*, 6, 16–38.

Anokhin, P.K. (1962). *The Theory of Functional Systems as a Prerequisite for the Construction of Physiological Cybernetics*, Moscow: Academy Science of the USSR.

Anokhin, P.K. (1969). Cybernetic and the integrative activity of the brain. In Cole, M. and Maltzman, I. (Eds.), *A Handbook of Contemporary Soviet Psychology* (pp. 830–857).

Ausubel, D.P. (1968). *Educational Psychology: A Cognitive View*, New York: Holt Rinehart and Winston.

Bakhtin, M.M. (1982). *The Dialogic Imagination*, Austin: University of Texas Press.

Bakhtin, M.M. and Voloshinov, V.N. (1973). *Marxism and Philosophy of Language*, Cambridge, MA: Harvard University Press.

Bandura, A. (1977). *Social Learning Theory*, Englewood Cliffs, NJ: Prentice-Hall.

Bandura, A. (1982). Self-efficacy mechanism in human agency. *Am. Psychol.*, 37, 122–147.

Bardin, K.V. (1982). The observer's performance in a threshold area. *Psychol. J.*, 1, 52–59.

Bardin, K.V. and Zabrodin, T.A. (1988). Changing sensory sensitivity when solving problems. *Probl. Psychol.*, 1, 149–154.

Barnes, R.M. (1980). *Motion and Time Study Design and Measurement of Work*, New York: John Wiley & Sons.

Basov, M.Y. (1931). *General Foundation of Pedagogy*, Moscow-Leningrad: Gosizdat.

Bassin, F.V. (1973). The problem of meaning and sense. *Probl. Psychol.*, 6, 14–23.

Bateson, G. (1972). *Steps to an Ecology of Mind*. New York: Ballantine.

Batishev, S.Y. (1977). Teaching workers in industry. In Batishev, S.Y. and Shaporinsky, S.A. (Eds.), *The Fundamentals of Vocational Pedagogy* (pp. 412–503), Moscow: Higher Education Publishers.

Bedny, G.Z. (1979). *Psychophysiological Aspects of a Time Study*, Moscow: Economika Publishers.

Bedny, G.Z. (1981). *The Psychological Aspects of a Timed Study During Vocational Training*, Moscow: Higher Education Publishers.

Bedny, G.Z. (1985). Some mechanisms of self-regulation in learning professional pace of activity. *Questions of Psychology*, 74–81.

Bedny, G.Z. (1987). *The Psychological Foundations of Analyzing and Designing Work Processes*, Kiev: Higher Education Publishers.

Bedny, G. (2000). Activity theory. In Karwowski, W. (Ed.), *International Encyclopedia of Ergonomics and Human Factors* (Vol. 1, pp. 358–362), London, UK, Taylor & Francis Ltd.

Bedny, G.Z. and Harris, S. (2005). The systemic-structural theory of activity: Application to the study of human work.

Bedny, G. and Karwowski, W. (2000). Theoretical and experimental approaches in ergonomics design: towards a unified theory of ergonomics design. *Proceedings of the 14th Triennial Congress of the International Ergonomics Association and Human Factors and Ergonomics Society* (pp. 197–200), San Diego, California.

Bedny, G. and Karwowski, W. (2001). Activity theory as a framework of study of human–computer interaction. *HCI International Conference* (pp. 342–346).

Bedny, G.Z. and Karwowski, W. (2003). A systemic-structural activity approach to the design of human-computer interaction tasks. *Int. J. Hum. Comput. Interac.* 2, 235–260.

Bedny, G.Z. and Karwowski, W. (2004a). A functional model of human orienting activity. In Bedny, G.Z. (invited Ed.). *Theor. Issues Ergonomics Sci.*, 5 (Special Issue), 255–274.

Bedny, G.Z. and Karwowski, W. (2004b). Activity theory as a basis for the study of work. *Ergonomics*, 47, 134–153.

Bedny, G.Z. and Karwowski, W. (2004c). Meaning and sense in activity theory and their role in study of human performance. *Ergonomia*, 26, 121–140.

Bedny, G.Z. and Karwowski, W. (2004d). Functional analysis of orienting activity and study of human performance. *Proceedings of the 1Vth Triennial Congress of the International Ergonomics Association and Human Factors and Ergonomics Society* (Vol. 6, pp. 443–445), Seoul, Korea.

Bedny, G.Z.and Karwowski, W. (2004e). Concept of orienting activity and situation awareness. *Proceedings of the 1Vth Triennial Congress of the International Ergonomics Association and Human Factors and Ergonomics Society* (Vol. 6, pp. 447–450), Seoul, Korea.

Bedny, G. and Meister, D. (1994). The Russian approach to task analysis and ergonomics design. *Human Factors and Ergonomics Society 38th Annual Meeting* (pp. 993–997), Nashville, Tennessee. Human factors and ergonomics society. Human Factors Psychology.

Bedny, G. and Meister, D. (1995). Temporal characteristics of work activity. *Human Factors and Ergonomics Society 39th Annual Meeting* (pp. 1152–1154), San Diego, CA. Human factors and ergonomics society.

Bedny, G. and Meister, D. (1996). Comparative analysis Russian and western psychological theories. Abstracts of the XXVI international congress of psychology. *International Journal of Psychology*. 16–21, Montreal, Canada.

Bedny, G. and Meister, D. (1997a). *The Russian Theory of Activity: Current application to Design and Learning*, Mahwah, NJ: Lawrence Erlbaum Associates, Publishers.

Bedny G. and Meister, D.(1997b). The factor of significance in the operator activity. *Human Factors and Ergonomics 41st Annual Meeting*. Albuquerque, New Mexico. Human factors and ergonomics society (pp. 1075–1078), Human Factors Psychology and work motivation.

Bedny G. and Meister D. (1999). Theory of activity and situation awareness. *Int. J. Cogn. Ergon.* 3, 63–72.

Bedny G. and Seglin M. (1997). The use of pulse rate to evaluate physical work load in Russian Ergonomics. *Am. Ind. Hyg. Assoc. J.*, 58, 335–379.

Bedny G. and Seglin M. (1999a). Individual style of activity and adaptation to standard performance requirement. *Hum. Perform.*, 12, 59–78.

Bedny G. and Seglin M. (1999b). Individual features of personality in former soviet union. *J. Res. Pers.* 546–563.

Bedny, G.Z. and Voskoboynicov, F. (1975). Problems of how a person adapts to the objective requirements of an activity. In Aseev, V.G. (Ed.), *Psychological Problems of Personality* (Vol. 2, pp. 18–30), Irkutsk, Russia: Irkutsk University Press.

Bedny, G.Z. and Zelenin, M.P. (1977). *Personality, Leadership and Groups*, Moscow: Marchent Marin Publishers.

Bedny, G.Z. and Zelenin, M.P. (1988). *Ergonomic Analysis of Work Activity and the Problem of Safety in Merchant Marine Transportation*, Moscow: Merchant Marine Publishers.

Bedny, G., Stolberg V., and Polyakov L. (1996). Russian psychological concept of work activity. *National Social Science Journal. Cognit. Exp. Psychol.* 8, 145–158.

Bedny, G., Seglin, M., and Meister, D. (2000). Activity theory, history, research and application. *Theor. Issues Ergonomics Sci.*, 1–39.

Bedny, G., Karwowski, W., and Bedny, M. (2001a). The principle of unity of cognition and behavior: Implications of activity theory for the study of human work. *Int. J. Cogn. Ergon.* 5, 401–420.

Bedny, G., Karwowski, W., and Young-Guk Known. (2001b) A methodology for systemic-structural analysis and design of manual-based manufacturind operations. *Hum. Factors Ergonomics Manuf.*, 11, 233–253.

Bedny, G, Karwowski, W., and Seglin, M. (2001c). A hart evaluation approach to determine cost-effectiveness an ergonomics intervention, *Int. J. Occup. Saf.*, 7, 121–133.

Bedny, G., Karwowski, W., and Seglin, M. (2001d). Activity theory as a basis for the study and redesign of computer based task. *HCI International Conference* (pp. 342–346).

Bedny, G., Karwowski, W., and One-Jang Jeng. (2001e) Activity theory as a basis to study human-computer interaction. *National Conference of HCI*, Hawaii.

Bedny, G.Z., Karwowski, W., and Jeng, O.-J. (2004). The situation reflection of reality in activity theory and the concept of situation awareness in cognitive psychology. In Bedny, G.Z. (invited Ed). *Theor. Issues Ergonomics Sci.* 5 (Special Issue), 275–296.

Bedny, G., Chebykin, O.Ya., and Karwowski, W. (2005). The basic characteristic of task from systemic-structural activity theory perspectives. 2005 *HCI International Conference*, Las Vegas, Nevada, USA, e-book.

Bedny, G.Z., Zelenin, M.P. (1989). *Vocational training in industry, Text of lecture*, Odessa: Merchant Marine Publishers.

Bellamy, R.K.E. (1997). Design of educational technology: Computer-mediated change. In Nardi, B.A. (Ed.), *Context and Consciousness: Activity Theory and Human-Computer Interaction* (pp. 123–146), Cambridge, MA, London: The MIT Press.

Belous, V.V. (1967). Experimental psychophysiological characteristics of some features of temperament. Merlin, V.S. (Ed.), *Typological Studies of Personality* (Vol. 4, pp. 104–135), Perm'.

Beregovoy, G.T., Zavalova, N.D., Lomov, B.F., and Ponomarenko, V.A. (1978). *Experimental-Psychology in Aviation and Aeronautics*. Moscow: Science Publishers.

Bernshtein, N.A. (1935). The problem of the relationship between coordination and localization. *Arch. Biol. Sci.*, 38, 1–34.

Bernshtein, N.A. (1947). *On the Structure of Movement*, Moscow: Medical Publishers.

Bernshtein, N.A. (1966). *The Physiology of Movement and Activity*. Moscow: Medical Publishers.

Bernshtein, N.A.(1996). On dexterity and its development. In Latash, M.L. and Turvey, M.T. (Eds.), *Dexterity and its Development* (pp. 1–244), Lawrence Erlbaum Associates, Publishers. Mahwah, New Jersey.

Bloch, V. (1966). Level of wakefulness and attention. In Fraisse, P. and Piaget, J. (Eds.), *Experimental Psychology* (Vol. 3, pp. 97–146). Paris: University Press of France.

Bohlen, G.A., and Barany, J.W. (1966). A learning curve prediction model for operators performing industrial bench assembly operations. *Int. J. Product. Res.*, 14, 295–302.

Braha, D. and Maimon, O. (1998). The measurement of a design structural and functional complexity. *IEEE Trans. Systems, Man, Cybernet-Part A: Systems Humans*, 28, 527–535.

Broadbent, D.E. (1958). *Perception and Communication*, London: Pergamon.

Brodesky, A.Ya. (1998). Foundation of visual communication. *Appl. psychol.* Russia.

Brody, N. (1989). *Personality: Research and Theory*, New York: Academic Press.

Brushlinsky, A.V. (1979). *Thinking and Forecasting*, Moscow: Thinking Press.

Brushlinsky, A.V. (1987). Activity, action, and mind as a process. *Soviet Psychol.*, 4, 59–81.

Cattell, R.B. (1965). *The Scientific Analysis of Personality*, Chicago, IL: Aldine.

Chebisheva, V.V. (1969). *The Psychology of Vocational Training*, Moscow: Education Publishers.

Chmiel, N.R.J. and Wall, T. (1994). Fault prevention, job design, and adoptive control of advanced manufacturing technology. *Appl. Psychol. Int. Rev.* 43, 455–473.

Christensen, E. H. (1953). Psychophysiological Valuation of Work in the Nykroppa Iron Works. In Floyd, W.F. and Welford, A.T. (Eds.), *Ergonomics Society Symposium on Fatigue*, London: Lewis.

Cooper, C.S., and Shepard, R.N. (1973). Chronometric studies of the rotation of mental images, In Chase, W.G. (Ed.), *Visual Information Processing*, New York: Academy Press.

Daniels, H. (2001). *Vygotsky and Pedagogy*, RoutlegeFalmer, Taylor & Francis Group.

Dobrolensky, U.P., Zavalova, N.D., Ponomarenko, V.A., and Tuvaev, V.A. (1975). In Dobrolensky, U.P. (Ed.), *Methods of Engineering Psychological Study in Aviation*. Moscow: Manufacturing Publishers.

Donders, F.S. (1862). Die schnelligkeit psychischer processe. *Arch. Anat. Physiol.* 657–681.

Endsley, M.R. (1995). Toward a theory of situation awareness in dynamic systems. *Hum. Factors*, 37, 32–64.

Endsley, M.R. (2000). Theoretical underpinnings of situation awareness: P A critical review. In Endsley, M.R. and mGarland, D.J. (Eds.), *Situation Awareness Analysis and Measurement* (pp. 3–32).

Engestrom, Y. (1987). *Learning by Expending*. Helsinki, Finland: Orienta-Konsultit.

Engestrom, Y. (1999). Changing practice through research: Changing research through practice. Keynote Address, *7th Annual International Conference on Post Compulsory Training and Education*. Griffith University. Australia.

Engestrom, Y. (2000). Activity theory as a framework for analyzing and redesigning work. *Ergonomics*, 7, 960–974.

Engestrom, Y. and Escalante, V. (1997). Mundane tool or object of affection? The rise and fall of the postal buddy. In Nardi, B.A. (Ed.), *Context and Consciousness: Activity Theory and Human–Computer Interaction* (pp. 325–374), Cambridge, MA London: The MIT Press.

Engestrom Y, Miettinen R., and Punamaki R.-L. (Eds.) (1998). *Perspectives on Activity Theory*. Cambridge: Cambridge University Press.

Eysenck, H.J. (1967). *The Biological Basis of Personality*, Springfield, IL: C.C. Thomas.

Eysenck, H.J. and Eysenck, M.W. (1985). *Personality and Individual Differences A Natural Sciences Approach*, New York: Plenum.

Fitts, P.M. (1954). The information capacity of the human motor system in controlling the amplitude of movement. *J. Exp. Psychol.*, 47, 381–391.

Fitts, P.M. and Patterson, J.R. (1964). Information capacity of discrete motor responses. *J. Exp. Psychol.*, 67, 103–112.

Frege, G. 1948 (1892). Sense and reference. *Philos. Rev.*, 57, 207–230.

Freud, S. (1916, 1917). *A General Introduction to Psychoanalysis*. New York: Pocket Books.

Frese, M. and Zapf, D. (1994). Action as a core of work psychology: a German approach. In Triadis, H.C., Dunnette, M.D., and Hough, L.M. (Eds.), *Handbook of Industrial and Organizational Psychology* (pp. 271–340), Polo Alto, CA: Consulting Psychologists Press.

Frolove, N.I. (1976). *Physiology of Movement*, Leningrad: Science Publishers.

Gellershtein, S.G. (1966). Anticipation and the problem of unconsciousness. In Bassin, F.V. (Ed.), *The Problem of Unconsciousness* (pp. 305–360), Moscow: Science Publishers.

Gallwey, T.J. and Drury, G.G. (1986). Task complexity in visual inspection. *Hum. Factor*, 28, 596–606.

Gal'perin, P.Y. (1966). The psychology of thinking and study of the stages in the formation of cognitive actions. In Lent'ev, A.N. (Ed.), *Research on Thinking in Soviet Psychology* (pp. 236–237), Moscow: Academy of Pedagogical Science RSFSR.

Gal'perin, P.Y. (1969). Stages in the development of mental acts. In Cole, M. and Maltzman, I. (Eds.), *A Handbook of Contemporary Soviet Psychology* (pp. 249–273), New York: Basic Books.

Gal'perin, P.Y. and Sachko, N.N. (1968). Formation of motor skills. In Gal'perin, P.Y. (Ed.), *Stage by Stage Formation of Knowledge and Skills* (pp. 3–54), Moscow University.

Gal'sev, A.D. (1973). *Time Study and Scientific Management of Work in Manufacturing*. Moscow: Manufacturing Publishers.

Genisaretsky, O.I. (1975). Methodology of organization of activity system. In Yudin, E.G. (Ed.), *Development of Automatic Systems in Design*, Moscow: Science Publisher.

Gibson, J.T., and Chandler, L.A. (1988). *Educational Psychology*, Boston: Allyn and Bacon, Inc.

Gibson, J.J. (1966). *The Senses Considered as Perceptual Systems*, Boston: Houghton Mifflin.

Gibson, J.J. (1979). *The Ecological Approach to Visual Perception*, Boston: Houghton Mifflin.

Gilbreth, F.V. (1911). *Motion Study*, Princeton, NJ: D. van Nostrand Company.

Gilbreth, F.V. and Gilbreth, L.M. (1920). *Motion Study for Handicapped*, London, George Routlege and Sons.

Glezer, V.D. and Nevskaya, A.A. (1964). On the simultaneous and sequential processing of information in a visual system. In Glezer, V.D. (Ed.), *Collected Papers of the Academy of Science of the USSR* (Vol. 155, pp. 97–105), Moscow: The Academy of Science of the USSR.

Gielo-Perszak, Karvovski, W.

Goldberg, J., Kotval, P. (1999). Computer interface evaluation using eye movements: methods and contrasts. *International Journal of Industrial Ergonomics*, 24, 631–645.

Goldstein, I.I. (1974). *Training Program Development and Evaluation*, Monterey, CA: Brook/Cole.

Gollwitzer, P.M. (1966). The volitional benefits of planning. In Gallwitzer, P. and Bargph, J.P. (Eds.), *The Psychology of Action* (pp. 287–312), New York: The Gilford Press.

Golubev, G.G. (1953). *Issues in the Methods of Flight Instruction*, Moscow: Defense Publishers.

Gorbynova, I.M. and Rokotova, N.A. (1970). On rhythmic organization of sequential motor actions. In Alekseev, M.A., Voronzov, F.S., and Gorbunova, I.M. (Eds.), *Control of Motor Actions* (pp. 7–17), Leningrad, Science Publisher.

Gordeeva, N.D. and Zinchenko, V.P. (1982). *Functional Structure of Action*, Moscow: Moscow University Publishers.

Gordeeva, N.D., Devishvily, V.M., and Zinchenko, V.P. (1975). In Zinchenko, V.P. (Ed.), *The Microstructural Analysis of Executive Activity*, Moscow: All Union Research Institute of Technical Esthetics.

Green, D.M. and Swets, J.A. (1966). *Signal Detection Theory and Psychophysics*, New York: Wiley.

Gurevich, K.M. (1970). *Professional Fitness and Basic Features of Neural System*, Moscow: Pedagogical Academy of Sciences.

Hacker, W. (1985a). Activity: a fruitful concept in industrial psychology. In Frese, M. and Sabini, J. (Eds.), *Goal Directed Behavior. The Concept of Action in Psychology* (pp. 262–284), Hillsdale, NJ: Erlbaum.

Hacker, W. (1985b). *Work Psychology and Engineering Psychology. Psychological Structure and Regulation of Work Activity*. Moscow: Manufacturing Publishers. (Translation from German).

Hacker, W., Ivanova, A., and Richter, P. (1983). *Activity System*. Berlin: Diagnostic Center.

Hancock, P.A. and Caird, J.K. (1993). Experimental evaluation of a model of mental workload. Human Factors, 35(3), 413–429.

Hassan, M. and Block, S.M. (1968). Study of simultaneous motions. *L'Etude du Travail*, 196, 29–36.

Heckhausen, H. (1991). *Motivation and Action*. Berlin, New York: Spring-Verlag.

Heckhausen, H. and Gollwitzer, P.M. (1987). Thought contents and cognitive functioning in motivational versus volitional states of mind. *Motivat. Emotion*, 11, 101–120.

Hick, W.E. (1952). On the role of gain of information. *Q. J. Exp. Psychol.*, 4, 11–26.

Hilchenko, A.E. (1966). Study of the mobility of neural processes in man. *Physiol. J.*, 6, 67–81.

Hilgard, R., Atkinson, R.L., and Atkinson, R.C. (1979). *Introduction to Psychology*, New York: Harcourt, Brace, Jovanovich.

Hogan, R. and Shelton, D. (1998). A socioanalytic perspective on job performance. In Hogan, J. (Ed.), *Human Performance* (Vol. 11, pp. 129–144).

Hogan, J., Rybicki, S., Motoviedlo, S.J., and Borman, W. (1998). Relations between contextual performance, personlaity, and occupational advancement. In Hogan, J. (Ed.), *Human Performance* (Vol. 11, pp. 129–144).

Hornof, A.J. (2001). Visual search and mouse pointing in labeled versus unlabeled two-dimensional visual heirarchies. *ACM Transactions on Computer-Human Interaction*, 8(3), 171–197.

Hrunove, L.S., Hkachaturanz, L.S., and Popov, V.A. (1974). *Human operator in cosmos flight*. Moscow: Manufacturing Publishers.

Hyman, R. (1953). Stimulus information as determined of reaction time. *J. Exp. Psychol.*, 45, 188–196.

Jacko, J.A. (1997). An empirical assessment of task complexity for computerized menu systems. *Int. J. Cogn. Ergon.*, 1, 137–147.

Janet, P. (1928). Le Volution de la memoire et de la notion du temps. Paris, A. Chahine.

Judge, J.A., Erez, A., and Bono, J.E. (1998). Power of being positive: the relationship between positive self-concept. In Hogan, J. (Ed.), *Human Performance* (Vol. 11, pp. 167–188).

Kagan, J. (1995). *Galen's Prophecy: Temperament in Human Nature*, New York: Basic Books.

Kahneman, D. (1973). *Attention and Effort*, Englewood Cliffs, NJ: Prentice-Hall.

Kamishov, I.A. (1968). The methodology of eye movement registration and determination of the operator's eye movement. *Questions Psychol.*, 4, 62–81.

Kanfer, R. (1996). Self-regulation and other non-ability determinants of skill acquisition. In Gollwitzer, P.M. and Bargph, J.P. (Eds.), *The Psychology of Action* (pp. 404–423), New York: The Gilford Press.

Kantorer, S. (1977). Calculation of the economic efficiency in the construction industry. In Kantorer, S. (Ed.), *Using Machine in Building Construction* (pp. 273–311), Moscow: Building construction Publishers.

Karger, D.W. and Bayha, F.H. (1977). *Engineering Work Measurement* (3rd ed.), New York: Industrial Press.

Karsten, A.Q. (1928). Psychology saturation. *Psychol. Res.*, 10, 612–628 (German).

Karwowski, W. (1991). Complexity, fuzziness and ergonomic incompatibility issues in the control of dynamic work environments. *Ergonomics*, 34, 671–686.

Karwowski, W. (1992). The human world of fuzziness, human entropy, and the need for general fuzzy system theory. *J. Japan Soc. Fuzzy Theory Systems*, 4, 591–609.

Karwowski, W. (2000). Cognitive ergonomics, requisite compatibility, fuzziness and nonlinear dynamics. *Proceedings of the IEA 2000/HFFS 2000 Congress*. Santa Monica: Human Factors and Ergonomics Society. 1, 580–583.

Khilchenko, A.E. (1960). Study of flexibility of nervous system. *Physiological Journal*. 6, #1, 21–28.

Khodikina, S.L. and Portnoy, M.P. (1985). Applying rhythmical stimulation while training on a keyboard. *The Problems of Engineering Psychology* (pp. 142–155), Moscow: Higher Educational Publishers.

Khrunov, E.V., Khachatur'yanz, L.S., Popov, V.A., and Ivanov, E.A. (1974). *Man-Operator in Astronautics Flights*, Moscow: Manufacturing Publishers.

Kieras, D. and Polson, P.G. (1985). An approach to the formal analysis of user complexity. *Int. J. Man. Mach. Stud.* 22, 365–394.

Kirwin, B. and Ainsworth, L.K. (Eds.) *A Guide to Task Analysis*, London: Taylors & Francis.

Kleinback, U. and Schmidt, K.H. (1990). The translation of work motivation into performance. In Kleinback, V., Quast, H.-H., Thierry, H., and Hacker, H. (Eds.), *Work Motivation* (pp. 27–40), Hillsdale, NJ: Lawrence Erlbaum Associates, Inc.

Klimov, E.A. (1969). *Individual Style of Activity*. Kazan: Kazahnsky State University Press.

Kochurova, E.I., Visyagina, A.I., Gordeeva, N.D., and Zinchenko, V.P. (1981). Criteria for evaluating executive activity. In Wertsch, V. (Ed.), *The Concept of Activity in Soviet Psychology*, (383–433), M.E. Sharpe, Inc., Armonk, New York.

Konopkin, O.A. (1980). *Psychological Mechanisms of Regulation of Activity*, Moscow: Science.

Konopkin, O.A., Engels, I.L., and Stephansky, V.T. (1983). On shaping of subjective standards of success. *Probl. Psychol.* 109–114.

Kosslyn, S.M. (1973). Scanning visual images: some structural implications. *Percep. Psychophys.*, 14, 90–94.

Kotik, M.A. (1974). *Self-Regulation and Reliability of Operator*, Tallin: Valgus.

Kotik, M.A. (1978). *Textbook of Engineering Psychology*, Tallin: Valgus.

Kotter, J.P. and Heskett, J.L. (1992). *Cooperative Culture and Performance*, New York: The Free Press.

Kovalev, A.G. (1965). *Psychology of Personality*, Moscow: Education Publisher.

Kozeleski, J. (1979). *Psychological Theory of Decision Making*, Moscow: Progress. (Translation from Poland).

Kozemchenko, A.F. (1962). *Funny Stories*. Pedagogical Publisher.

Kozulin, A. (1986). *The concept of activity in Soviet Psychology: Vygotsky, his disciples and critics*. American Psychologist, V41, #,3, 264–274.

Krukov, A.A. and Kremen, M. (1983). Method of supporting points. *Aviat. Cosmonaut*, 6,7 26-27, 7, 27-28.

Kuhl, J. (1992). A theory of self-regulation: action versus state orientation, self-discrimination and some applications. *Appl. Psychol. Int. Rev.*, 41, 97–129.

Kuutti, K. (1997). Activity theory as a potential framework for human–computer interaction research. In Nardi, B.A. (Ed.), *Context and Consciousness: Activity Theory and Human–Computer Interaction* (pp. 17–44), Cambridge, MA, London: The MIT Press.

Kuzemchenko, A.F. (1982). *Funny Stories*, Moscow: Pedagogical Publisher.

Landa, L.M. (1976). *Instructional Regulation and Control: Cybernetics, Algorithmization and Heuristic in Education*, Englewood Cliffs, NJ: Educational Technology Publication. (English translation).

Lazareva, V.V., Svederskaya, N.E., and Khomskaya, E.D. (1979). Electrical activity of brain during mental workload. In Khomskaya, E.D. (Ed.), *Neuropsychological Mechanisms of Attention* (pp. 151–168), Moscow: Science Publisher.

Lee, T.W., Locke, E.A., and Lathman, G.P. (1989a). Goal setting, theory and job performance. In Pervin, A. (Ed.), *Goal Concepts in Personality and Social Psychology* (pp. 291–326), Hillsdale, NJ: Lawrence Erlbaum Associates.

Lee, T.W., Locke, E.A., and Lathman, G.P. (1989b). Goal setting theory and job performance. In Pervin, A.L.A. (Ed.), *Goal Concept in Personality and Social Psychology* (pp. 291–326).

Lehmann, G. (1962). *Practical Work Physiology*, Stuttgart: George Theme Verlag.

Leont'ev, A.N. (1947). *Essay of Development of Psychic*, Moscow University Publishers.

Leont'ev, A.N. (1978). *Activity, Consciousness and Personality*, Englewood Clifts: Prentice Hall

Leontév, A.N. The problem of activity in psychology. *Questions Psychol.* 5–19.

Lewin, K. (1951). *Field Theory in Social Science*, Chicago, IL: University of Chicago Press.

Lindsay, P., Norman, D.A. (1992). *Human information processing*. (2 ed.). San Diego: Harcourt Brace.

Linhart, J. (1970). *Process and Structure of Human Learning*, Moscow: Progress. (Translation from Chechoslovakia).

Locke, E.A. and Lathman, G.P. (1984). *Goal Setting: A Motivational Techniques that Works*, Englewood Cliffs, NJ: Prentice Hall.

Locke, E.A. and Lathman, G.P. (1990). Work motivation: the high performance cycle. In Kleinbeck, V. et al. (Ed.), *Work Motivation* (pp. 3–26), Hillsdale, NJ: Lawrence Erlbaum Associates.

Lomov, B.F. (Ed.) (1982). *Handbook of Engineering Psychology*. Moscow: Manufacturing Publishers.

Lomov, B.F. (Ed.) (1986). *Engineering Psychology Textbook*, Moscow: Higher Education.

Lomov, B.F. and Surkov, E.N. (1980). *Anticipation in Structure of Activity*, Moscow: Science Publisher.

Lomov, B.F., Vasiliev, A.A. , Oficerov, V.V., and Rubakhin, V.F. (Eds.). (1970). *Military Engineering Psychology*, Moscow, USSR: Moscow Defense Publishers.

Lomov, B.F., Zavalova, N.D., and Ponomarenko, V.A. (1977). Principle of the active operator. In Lomov, V.F., Rubakhin, and Venda, V.F. (Eds.), *Engineering Psychology* (pp. 119–134), Moscow: Science.

Luria, A.R. (1975). Cognitive development its cultural and social foundations. In Cole, M. (Ed.), Cambridge: Harvard University.

Luria, A.R. (1977). *The Making of the Mind*, Harvard: Harvard University Press.

Marchenko, E.N., Kandror, I.S., and Ryazanov, S. (1972). Principle of classification of physical work according to physiological criteria. *Work Hygiene Professional Sickness*, 3, 30–36.

Marx, K. (1969). Kapital, V. 1, Political Publisher, Moscow, p. 358.

Maslow, A.H. (1954). *Motivation and Personality*, New York: Harper and Row.

Maynard, H.B., Stegemerten, G.J. (1948). *Method-Time Measurement*. McGraw-Hill, New York.

Mead, G.H. (1944). *Mind, Self and Society*: Chicago, University of Chicago Press.

Mead, G.H. (1945). *The Philosophy of the Act*, Chicago: University of Chicago Press.

Medvedev G. (1991). *The Truth about Chernobyl*, NJ: Harper Collins. (Translated from Russian).

Meister, D. (1999). *The History of the Human Factors and Ergonomics*. Mahwah, NJ: Lawrence Erlbaum Associates, Publishers.

Mel'nik, V.M. and Yampol'sky, L.T. (1985). *Introduction into Experimental Psychology of Personality*, Moscow: Educational Publisher.

Merlin, V.S. (1964). *Outlines of a Theory of Temperament*, Perm, Russia: Perm Pedagogical Institute.

Merlin, V.S. (1986). *Outlines of integral study of individuality*, Moscow: Pedagogy.

Merlinkin, V.P. (1977). Some neuropsychological differences in development novice tumblers' skills. In Klimov, E.A. (Ed.), *Neuropsychological Study of Personality*, Perm, Russia: Perm University Publishers, pp. 72–87.

Miller, G.A., Galanter, E., and Pribram, K.H. (1960). *Plans and the Structure of Behavior*. London: Holt.

Miller, L., Staney, K., Guckenberger, D., and Guckenberger, E. (1997). Above real-time training. *Ergonomics in design*, pp. 21–24.

Mirabella, A. and Wheaton, G.R. (1974). *Effects of Task Index Variations on Transfer of Training Criteria*. Report NAVTRAEQUIPEN 72-C-0126-1, Naval Training Equipment Center, Orlando, Fla. January (AD 773-047/7GA).

Mittkin, A.A. (1976). *Electrooculography in Engineering Psychology*, Moscow: Science.

Morosanova, V.I., Stepansky, V.I., and Sablin, V.N. (1980). Calculation of objective indices in quality of gun shooting proficiency. *Collection of Scientific Work of VNIIFK* (pp. 52–62), Moscow: Higher Education.

Morosanova, V.I. and Stepansky, V.I. (1982). Relationship between subjective and objective criteria of success and its influence on result of performance. Collective Works, *New Study in Psychology* (pp. 14–17), Moscow: Pedagogy.

Morris, C.W. (1946). *Signs, Language, and Behavior*, New York: Braziller.

Murrell, K.F.H. (1965). *Human Performance in Industry*, New York: Reinhold.

Nardi, A. (1997). Same reflection on the application of activity theory. In Nardi, B.A. (Ed.), *Context and Consciousness: Activity Theory and Human–Computer Interaction* (pp. 17–44), Cambridge, MA, London: The MIT Press.

Nayanko, N.I. (1976). *Psychic Tension*, Moscow: Moscow University.

Nebilitsin, V.D. (1976). *Psychological study of individual differences*. Moscow: Pedagogy.

Nebylitsin, V.D. (1965). *Basic Features of Man's Neural System*, Moscow: Education Publishers.

Neisser, U. (1967). *Cognitive Psychology*, New York: Appleton.

Newtson, D. (1976). Foundation of attribution, the perception of ongoing behavior. In Harvey, J.H., Ickes, W.V., and Kidd, R.F. (Eds.), *New Direction in Attribution Research* (pp 172–187). Hillsdale, NJ: Lawrence Erlbaum Associates.

Nielsen, J. (1994). *Usability Engineering*. San Francisco: Morgan Koufman.

Nojivin, U.S. (1974). On psychological self-regulation of sensory motor actions. In Shadrikov, V.D. (Ed.), *Engineering and Psychology* (Vol. 1, pp. 206–210), Yaroslav: Yaroslav University.

Norman, D.A. (1976). *Memory and Attention: An Introduction to Human Information Processing* (2nd ed.), NY: Wiley.

Norman, D.A. (1986). Cognitive engineering. In Norman, D. and Draper, S. (Eds.), *User Centered System Design* (pp. 31–61), Erlbaum: Hillsdale, NJ.

Novikov, A.M. (1986). *Process and Method of Formation of Vocational Skills*, Moscow: Higher Education.

Ormrod, J.E. (1990). Human learning. *Theories, Principles and Educational Applications*, New York: Macmillan.

Oshanin, D.A. (1976). Dynamic operative system. *The Problems and Results in Psychology*, 59, 37–48.

Oshanin, D.A. (1977). Concept of operative image in engineering and general psychology. In Lomov, B.F., Rubakhin, V.F., and Venda, V.F. (Eds.), *Engineering Psychology*, Moscow: Science.

Paley, I.M. (1976). Psychological characteristic of personality and prerequisite of its development. In Bodalev, A.A. (Ed.), *Social Psychology*: Leningrad, Leningrad University publishers. pp. 90–105.

Parsons, J.W. (1937). *The Structure the Social Action*. New York: Grune and Stratton.

Pavlov, I.P. (1927). *Conditioned Reflex*, London: Oxford University Press.

Payne, W. (1997). Task complexity and contingent processing in decision-making: An information search and protocol analysis. *Organizational Behavior Human Performance*, 16, 366–387.

Pervin, L.A. (1989). Goal concepts, themes, issues and questions. In Pervin, L.A. (Ed.), *Goal Concepts in Personality and Social Psychology* (pp. 173–180), Hillsdale, NJ: Lawrence-Elbaum Associates.

Petrovsky, A.V. (1982). *Personality, Activity, Groups*, Moscow: Education.

Petrovsky, A.V. (Ed.) (1986). *General Psychology*, Moscow: Education Publishers.

Piaget, J. (1952). *The Origins of Intelligence in Children*, New York: International University Press.

Platonov, K.K. (1982). *System of Psychology and Theory of Reflection*, Moscow: Science Publishers.

Platanov, K.K. and Adaskin, B.M. (1977). *Study of Personality of Students in Vocational School*, Moscow, USSR: Higher Education.

Ponomarenko, V.A. (1998). Image of spirit in subjective life of person. *Appl. Psychol.*, 1–19.

Ponomarenko, V.A. and Zavalova, N.D. (1970). Readiness to danger. *Aviat. Aeronaut.*, 3, 11–13.

Ponomarenko, V.A. and Zavalova, N.D. (1981). Study of psychic image as regulator of operator actions. In Lomov, B.F. and Venda, V.F. (Eds.), *Methodology of Engineering Psychology and Psychology of Work of Management* (pp. 30–41), Moscow: Science.

Ponomarenko, V.A., Vorona, A.A., and Zatsarny, N.N. (1993). Psychological substantiation of principle directions of creation and use of technical training devises in aviation. *Ergonomics*. All-Russia Scientific-Research Institute of Information. Issue 1–2, 36–43.

Ponomarev, Y.A. (1976). *Psychology of Creativity and Education*, Moscow: Pedagogy.
Preece, J., Rogers, Y., Sharp, H. Benyon, D. Holland, S. and Carey, T. (1994). *Human-computer interaction*. Addison-Wesley.
Pushkin, V.N. (1965). *Operative Thinking in Large Systems*, Moscow: Science.
Pushkin, V.N. (1978). Construction of situational concepts in activity structure. In Smirnov, A.A. (Ed.), *Problem of General and Educational Psychology* (pp. 106–120), Moscow: Pedagogy.
Pushkin, V.N. and Nersesyan, L.S. (1972). *Psychology of the Railroad*, Moscow: Transportation.
Rasmussen, J. (1986). *Information Processing and Human–Machine Interaction: An Approach to Cognitive Engineering*, Amsterdam: Elsevier.
Rasmussen, J. (1990). Mental models and the control of action in complex environments. In Ackermanj, D. and Tauber, M.J. (Eds.), *Mental Models of Human–Computer Interaction* (Vol. 1, pp. 41–69), Amsterdam: North-Holland.
Rasmussen, J. Goodstein, L.P. (1988). Information technology and work. In Helander (Ed.), *Handbook of human-computer interaction* (pp. 175–201). Amsterdam: Elsevier.
Rautenberg, M. (1996). How to measure of cognitive complexity in human-interaction tasks. In Trappll, R. (Ed.), *Cybernetic and Systems*, V2, Austrian Society for Cybernetic Studies, 815–820.
Rayner, K. (1992). Eye movement and visual cognition: Scene perception and reading. In Rayener, K. (Ed.), *Eye movements and visual cognition: Introduction*, (1-7). Springer-Verlag.
Reason, J.T. (1990). *Human Error*, New York: Cambridge University Press.
Reykovski, J. (1979). *Experimental Psychology of Emotion*, Moscow: Progress. (Translation from Poland).
Rouse, W.B. and Morris, N.M. (1986). On looking into the black box. Prospects and limits in the search for mental models. *Psychol. Bull.*, 100, 349–363.
Roze, N.A. (1963). Experimental analysis of finger micro-movements in manipulating control apparatus. In Ananév, B.G., and Lomov, B.F. (Eds.), *Problems of General and Industrial Psychology* (pp. 35–43), Leningrad: Leningrad University.
Rozenblat, V.V. (1961). *The Problem of Fatigue*, Moscow: Medicine.
Rozenblat, V.V. (1975). Principle of physiological evaluation of hard labor based on pulse measurement procedures. In Rozenblat, V.V. (Ed.), *Function of Organism in Work Process* (pp. 112–126), Moscow: Medicine.
Rubinshtein, S.L. (1935). *Foundation of Psychology*. Moscow: Pedogogy Publishers.
Rubinshtein, S.L. (1940). *Problems of General Psychology*, Moscow: Academic Science.
Rubinshtein, S.L. (1946). *Foundations of General Psychology*, Moscow: Academic Pedogogical Science.
Rubinshtein, S.L. (1957). *Existence and Consciousness*, Moscow: Academy of Science.
Rubinshtein, S.L. (1958). *About Thinking and Methods of its Development*, Moscow: Academic Science.
Rubinshtein, S.L. (1959). *Principles and Directions of Developing Psychology*, Moscow: Academic Science.
Rubinshtein, S.L. (1973). *Problems of General Psychology*, Moscow: Academic Science.
Rumelhart, D.E. (1975). Notes on a schema for stories. In Bobrov, D.G. and Collins, A. (Eds.), *Representation and Understanding* (pp. 211–236), New York: Academic Press.
Schmidt, R.A. (1975). A schema theory of discrete motor skill learning. *Psychol. Rev.*, 82, 225–260.
Schneider, W. and Shiffrin, R.M., (1977). Controlled and automatic human information proccessing I: detection, search, and attention. *Psychol. Rev.*, 84, 1–66.

Seashore, H. and Bavelas, A. (1941). The functioning of knowledge of results in Thorndike's experiment. *Psychol. Rev.*, 84, 155–164.
Sechenov, I.M. (1968). *The Reflex of the Brain*, Amstredam: Bonset.
Sengupta, T. and Jeng, O.-J. (2003). Activity based analysis for drawing task. *The Proceedings of the XVth Trienial Congress of the International Ergonomics Association and the 7th Joint Conference of Ergonomic Society of Korea/Japan Ergonomic Society* (pp. 455–458), The Ergonomic Society of Korea.
Shchedrovitsky, G.P. (1995). *Selective Works*, Moscow: Cultural Publisher.
Sheldon, W.H. (1942). *The Varieties of Temperament*, New York: Harper.
Shevandrin, N.I. (2003). *Foundation of psychodiagnostics*. Moscow: Humanitarian Publishing Centre.
Shmelev, A.G. (1983). Concept of meaning in experimental psychosemantic. *Questions Psychol.*, 4, 16–29.
Siegal, A.I. and Wolf, J.J. (1969). *Man–Machine Simulation Models*, New York: Wiley.
Skinner, B.F. (1974). *About Behaviorism*, New York: Knopf.
Smidtke, H. and Stier, F. (1961). An experimental evaluation of validity of predetermined elemental time system. *J. Ind. Eng.*, 3, 182–204.
Smirnov, S.D. (1985). *Psychology of Image: The Problem of Active Psychic Reflections*, Moscow: Moscow University.
Smith, B.A., Ho, J., Ark, W., and Zhai, S. (2000). Hand eye coordination patterns in target selection, Proceedings of the symposium of *Eye tracking research and applications*. p. 117–122, Novermber 06–08, 2000, Palm Beach Gardens, Florida, United States.
Sodhi, M., Reimer, J.L., Vastenburg, E., Kaaras, R., Kirshenbaum, S. (2002). On-road eye movement tracking using head-mounted devices. *Proceedings, ETRA, Eye Tracking Research and Applications Symposium.*
Sokolov, E.N. (1960). Neural models and the orienting reflex. In Braziere, A.B. (Ed.), *The Central Nervous System and Behavior* (pp. 187–276), New York: Josiah Macy, Jr. Foundation.
Sokolov, E.N. (1963). *Perception and Conditioned Reflex*, New York: Macmillan.
Sokolov, E.V. (1974). *Culture and Personality*, Moscow: Science Publisher.
Sperling, G. (1960). The information available in brief visual presentations. *Psychol. Monogr.*, 74 (Whole No. 11).
Sternberg, S. (1969a). The discovery of processing stages: extension of Donder's method. *Acta Psychol.*, 30, 276–315.
Sternberg, S. (1969b). Memory-Scanning, mental processes revealed by reaction-time experiments. *Am. Sci.*, 57, 421–457.
Sternberg, S. (1975). Memory scaning: new findings and current controversies. *Q. J. Exp. Psychol.*, 27, 1–32.
Strelau, J. (1982). *Role of Temperament in Psychic Development*, Moscow: Progress Publisher.
Strumilin, S.G. (1983). About of work classification. In Zinchenko, V.P. and Noskova, O.G. (Eds.), *History of Soviet Work Psychology* (pp. 104–118), Moscow: Moscow University Press.
Suchman, L.A. (1987). *Plans and Situated Actions: The Problem of Human–Machine Interaction*, Cambridge: Cambridge University Press.
Suh, N.P. (1990). *The principles of Design*, New York, Oxford: Oxford University Press.
Swets, J.A. (1964). *Signal Detection and Recognition by Human Observers*, New York: Wiley.
Syvorova, G.A. (2003). *Psychology of Activity*, Moscow: PERSE Publisher.
Talizina, N.F. (1975). *The Management of Processing Skills and Knowledge*, Moscow: Moscow University.
Taylor, F.W. (1911). *The Principles of Scientific Management*, New York: Harper and Brothers.

Telegina, E.D. (1925). The interrelation of conscious and unconscious actions in the thinking process. *Questions Psychol.*, 2, 15–32.
Teplov, B.M. (1961). The theory of types of higher nervous activity and psychology. *Probl. Psychol.* 1, 3–16.
Teplov, B.M. (1985). Examination of the features of the neural system as a method for approaching individual psychological differences. In Teplov, B.M. (Ed.), *Individual Differences. Collected Works* (Vol. 2, pp. 137–168), Moscow, USSR: Pedagogy.
Tikhomirov, O.K. (1984). *Psychology of Thinking*, Moscow: Moscow University.
Tolman, E.C. (1932). *Purposive Behavior in Animals and Men*, New York: Century.
Tomazevski, T. (1979). *Activity and Consciousness*, Weinheim, Germany: Beltz.
Troshihkina, B.A., Kol'chenko, S.I., and Moldavsky, S.I. (1970). New methods for studying typological specificity of man's higher neural system. *Pavlov's J. Higher Neural Systems*, 4, 37–42.
Turvey, M.T. (1996). Dynamic touch. *Am. Psychol.* 51, 1134–1152.
Uznadze, D.N. (1961). *The Psychology of Set*, New York: Consultants Bureau.
Vasil'ev, I.A., Popluzhny, V.L., and Tikhomirov, O.K. (1980). *Emotions and thinking*. Moscow: Moscow University Publishers.
Vekker, L.M. and Paley, I.M. (1971). Information and energy in psychological reflection. In Ana'ev, B.G. (Ed.), *Experimental Psychology* (Vol. 3, pp. 61–63), Leningrad: Leningrad University.
Venda, V.F. and Ribalchenko, M.V. (1983). Transformational theory of learning and design of engineering — psychological experiment. In Venda, V.F. and Vavilov, V.A. (Eds.), *Theory and Experiment in Analysis of Operator Works* (pp. 147–157), Moscow: Science.
Venda, V.F. (1975). *Engineering psychology and synthesis of informational sources*. Moscow: Manufacturing Publishers.
Vicente, K.J. (1999). *Cognitive Work Analysis: Toward Safe, Productive, and Healthy Computer-Based Work*, Mahwah, NJ: Lawrence Erlbaum Associates, Publishers.
Viviani, P. (1990). Chapter 8 in *Eye Movements and Their Role in Visual and Cognitive Processes*, Kowler, K. (Ed.), Elsevier Science Publishers.
Vygotsky, L.S. (1926). *Pedagogical Psychology*, Moscow: Educational Worker Publisher.
Vygotsky, L.S. (1930). The problem of higher intellectual functions in system of psychotechnical studies. *Psychotechnic and Psychophysiology of Work*, 5, 373–384.
Vygotsky, L.S. (1956). *Selected Psychological Research*, Moscow: Academy of Pedagogical Science. RSFSL.
Vygotsky, L.S. (1960). *Developing Higher Order Psychic Functions*, Moscow: Academia of Pedagogical Science. RSFSR.
Vygotsky, L.S. (1962). *Thought and Language*, Cambridge: MIT Press.
Vygotsky, L.S. (1971). *The Psychology of Arts*. Cambridge: MIT Press.
Vygotsky, L.S. (1978). Mind in society. *The Development of Higher Psychological Processes*, Cambridge, MA: Harvard University Press.
Wade, C. and Tavris, C. (1990). Psychology. *Task Bank to Accompany* (2nd ed.), New York: Harper Collins.
Weber, M. (1947). *Theories of Social and Economic Organizations*, New York: Wiley.
Wehner, T., Clases, C., and Backmann, R. (2000). Co-operation at work: a process-oriented perspective on joint activity in inter-organizational relations. *Ergonomics*, 43, 983–998.
Weiner, N. and Rosenblueth, A. (1950). *Purposeful and Non-Purposeful Behavior*, Philosophy of Science, 17, 20–36.
Welford, A.T. (1960). On the human demands of automation, mental work, conceptual model, satisfaction and training. *Industrial and Business Psychology* (Vol. 3, pp. 182–194), Compehagen: Munksgaard.

Wertsch, V. (Ed.) (1981). *The Concept of Activity in Soviet Psychology*. New York: M. E. Sharpe, Inc., Armonk.

Wiener, N. (1958). *Cybernetics*, Moscow: Soviet Radio. (Russian translation).

Yarbus, A.L. (1965). *The Role of Eye Movements in the Visual Process*, Moscow: Science Publishers.

Yaroshevsky, M.G. (1985). *History of Psychology* (3rd ed.), Moscow: Thinking.

Yaroshevsky, M.G. (1992). Vygotsky, L.S. and Marxism in Soviet psychology. Response to Brushlinsky. *Psychol. J.*, 5, 97–98.

Yastrebov, S.Y. (Ed.) (1977). *Underwater Works*, Leningrad: Ship Construction.

Yur'nev, A.P. (1975). *Unmanned underwater apparatus*. Moscow: Science Publishers.

Zaporozhets, A.V. (1960). *Development of Voluntary Motions*, Moscow: Academy of Pedagogical science publishers.

Zaporozhets, A.V. (1969). Some of the psychological problems of sensory training in early childhood and the preschool period. In Cole, M. and Maltzman, I. (Eds.), *Handbook of Contemporary Soviet Psychology* (pp. 86–120), Basic Books Publishers.

Zaporozhets, A.V. and El'konin, D. (1971). *The Psychology of Preschool Children*, Cambridge, MA: MIT Press.

Zaporozhets, A.V. and Zinchenko, V.P. (1982). *Perception, Motions and Actions*, Moscow: Pedagogy.

Zarakovsky, G.M. (1966). *Psychophysiological Analysis of Work Activity; Logical-Probability Approach*, Moscow: Science.

Zarakovsky, G.M. (2004). The concept of theoretical evaluation of operators' performance reliability derived from activity theory. In Bedny, G. (Ed.), Special Issue. *Theoretical Issues in Ergonomics Science* (pp. 313–337).

Zarakovsky, G.M. and Magazannik, V.D. (1981). Psychological criteria of complex decision making processes. In Lomov, B.F. and Venda, V.F. (Eds.), *Methodology of Engineering Psychology and Psychology of Work Management* (pp. 63–79), Moscow: Science.

Zarakovsky, G.M. and Medvedev, V.I. (1971). *Psychological Evaluation of Efficiency of Man–Machine System. 3rd Conference of Operator Reliability*, Leningrad: Psychology Society.

Zarakovsky, G.M. and Pavlov, V.V. (1987). *Laws of Functioning Man–Machine Systems*, Moscow: Soviet Radio.

Zarakovsky, G.M., Korolev, B.A., Medvedev, V.I., and Shlaen, P.Y. (1974). *Introduction to Ergonomics*, Moscow: Science.

Zavalova, N.D. and Ponomarenko, V.A. (1980). Structure and content of psychic image as mechanisms of regulation of actions. *Psychol. J.*, 1, 5–18.

Zavalova, N.D., Lomov, B.F., and Ponomarenko, V.A. (1971). Principle of active operator and function allocation. *Probl. Psychol.*, 3, 3–12.

Zavalova, N.D., Lomov, B.F., and Ponomarenko, V.A. (1986). *Image in Regulation of Activity*, Moscow: Science.

Zeigarnik, B. (1927). Remembering completed and uncompleted actions. *Psychol. Res.*, 9, 1–85 (German).

Zinchenko, P.I. (1961). *Involuntary Memorization*, Moscow: Pedagogy.

Zinchenko, V.P. (1978). Functional structure of executive perceptual-motor actions. In Zinchenko, V.P. and Munipov, V.M. (Eds.), *Study of Cognitive and Executive Actions* (Vol. 16, pp. 3–40), VNITE.

Zinchenko, V.P. (1995). Vygotsky's idea about units for the analysis of mind. In Wertsch, J. (Ed.), *Culture, Communication and Cognition: Vygotskian Perspectives* (pp. 94–118).

Zinchenko, V.P. and Vergiles, N.Y. (1969). *Creation of Visual Image*, Moscow: Moscow University.

Zinchenko, V.P., Gusheva, T.M., and Gordon, V.M. (1971). Investigation of the temporal and functional structure response dependencies in the recognition of graphic representations. In Kotik, M.A. (Ed.), *Problems of the Experimental Investigation of Response Time* (pp. 41–53), Tallin, Estonia: Valgus.

Zinchenko, V.P., Munipov, V.M., and Smolyan, G.L. (1974). *Ergonomic Foundation of Work Organization*, Moscow: Economic Publishers.

Author Index

A

B

C

D

E

F

G

H

I

J

K

L

T

U

V

W

Y

Z

Subject Index

A

B

C

D

E

F

G

H

I

Q

R

S

T

U

For Product Safety Concerns and Information please contact our EU representative GPSR@taylorandfrancis.com Taylor & Francis Verlag GmbH, Kaufingerstraße 24, 80331 München, Germany